Multiphase Homogeneous Catalysis

Edited by
B. Cornils, W. A. Herrmann,
I. T. Horváth, W. Leitner, S. Mecking,
H. Olivier-Bourbigou, D. Vogt

Further Titles of Interest

B. Cornils, W. A. Herrmann (Eds.)

Aqueous-Phase Organometallic Catalysis
2nd Completely Revised and Enlarged Edition

2004
ISBN 3-527-30712-5

J. A. Gladysz, D. P. Curran, I. T. Horváth (Eds.)

Handbook of Fluorous Chemistry

2004
ISBN 3-527-30617-X

W. Ehrfeld, V. Hessel, H. Löwe

Microreactors

2000
ISBN 3-527-29590-9

J. G. Sanchez Marcano, T. T. Tsotsis

Catalytic Membranes and Membrane Reactors

2002
ISBN 3-527-30277-8

Multiphase Homogeneous Catalysis

Volume 1

Edited by
B. Cornils, W. A. Herrmann, I. T. Horváth, W. Leitner,
S. Mecking, H. Olivier-Bourbigou, D. Vogt

WILEY-VCH

WILEY-VCH Verlag GmbH & Co. KGaA

Editors

Professor Dr. Boy Cornils
Kirschgartenstrasse 6
65719 Hofheim
Germany

Professor Dr. Dr. h. c. mult.
Wolfgang A. Herrmann
Anorganisch-Chemisches Institut
der Technischen Universität München
Lichtenbergstrasse 4
85747 Garching
Germany

Professor Dr. István T. Horváth
Eötvös University
Pázmány Péter sétány 1/A
1117 Budapest
Hungary

Professor Dr. Walter Leitner
Technische und Molekulare Chemie
RWTH Aachen
Worringer Weg 1
52074 Aachen
Germany

Professor Dr. Stefan Mecking
Universität Konstanz
Universitätsstrasse 10
78484 Konstanz
Germany

Professor Dr. Hélène Olivier-Bourbigou
Institut Français du Pétrole
PO Box 3
69390 Vernaison
France

Professor Dr. Dieter Vogt
Eindhoven University of Technology
PO Box 513
5600 MB Eindhoven
The Netherlands

Library of Congress Card No.: applied for
A catalogue record for this book is available
from the British Library.

Bibliographic information published by
Die Deutsche Bibliothek
Die Deutsche Bibliothek lists this publication
in the Deutsche Nationalbibliografie;
detailed bibliographic data is available in the
internet at http://dnb.ddb.de.

© 2005 Wiley-VCH Verlag GmbH & Co.
KGaA, Weinheim

Printed in the Federal Republic of Germany
Printed on acid-free paper

Cover Design   SCHULZ Grafik-Design,
Fußgönheim
Composition   Manuela Treindl, Laaber
Printing   betz-druck GmbH, Darmstadt
Bookbinding   J. Schäffer GmbH, Grünstadt

ISBN-13 978-3-527-30721-0
ISBN-10 3-527-30721-4

Preface

This book describes for the first time all homogeneously catalyzed reactions in *multiphase* operation. Thus it summarizes the progress which became possible by the introduction of separate phases in the context of "homogeneous" catalysis: an apparent contradiction with far-reaching consequences. The book reviews all the realistic possibilities described so far using multiphase operation of homogeneous catalysis: processes with organic/organic, organic/aqueous, or "fluorous" solvent pairs (solvent combinations), nonaqueous ionic solvents, supercritical fluids, and systems with soluble polymers. The accounts concentrate on the modification and the handling of homogeneous catalysts under multiphase conditions in general, and the removal and subsequent recycling of the catalyst in particular.

Why multiphase systems? This goes back to the 1980s and the enormous impetus which was given to the homogeneous catalysis community by the first realization of Ruhrchemie/Rhône-Poulenc's aqueous-phase oxo process at the Oberhausen plant site. Astonishingly, this fact (and *not* the earlier SHOP process of Shell) sensitized the scene to the possibilities of multiphase action to imitate the decisive advantage of heterogeneous catalysis: the immediate separation of catalyst and substrates/products just after reaction which makes it possible to avoid additional separation steps post-reaction, such as distillations and other thermally stressing procedures.

All proposals have the same target: to enable the homogeneous catalyst to be bound to a suitable "support", i.e., another phase, without losing its superior homogeneous catalytic activity and selectivity. Within the scope of this book the editors define "phase" not only thermodynamically (as uniform states of matter of *one* substance which are separated (and separable) from each other by unequivocal phase boundaries; for example, water–ice or normal–supercritical states) but also as *different* states of aggregation of different compounds, such as systems consisting of water–organic liquids. Thus this book deals with the various possibilities of homogeneous catalysts on *liquid* supports. Additionally, the processes described imply two- or three-phase reactions.

The status of the different variants of multiphase homogeneous catalysis is described with the state-of-the-art as at the end of 2004.

The editors, as well-known players in their respective fields of homogeneous and homogeneously multiphase catalysis, have tried to portray the scene from the basic idea through the development stage up to commercial applications and the

Multiphase Homogeneous Catalysis
Edited by Boy Cornils and Wolfgang A. Herrmann et al.
Copyright © 2005 Wiley-VCH Verlag GmbH & Co. KGaA, Weinheim
ISBN: 3-527-30721-4

processes which are on the track to economical realization. Within the definitions as given in the introductory Chapter 1, the contents of the individual sections are the responsibility of the respective editor. Some contradictory statements within the various chapters of the book may arise from the fact that the authors started from very differing experiences and used different focal points to emphasize the importance of "their" multiphasic approach. All sections give outlooks about the developments to come.

Once more, we have to express our thanks not only to the authors and co-authors but also to the team at Wiley-VCH at Weinheim, especially Claudia Grössl and Melanie Rohn for the production and Elke Maase, the publishing editor. Diana Boatman from Redhill, Surrey (UK), served as freelance copy-editor and was an invaluable help for all of us who write in English, but not as our first language, during the difficult process of completion.

August 2005

Boy Cornils
Wolfgang A. Herrmann
István T. Horváth
Walter Leitner
Stefan Mecking
Hélène Olivier-Bourbigou
Dieter Vogt

Contents

Multiphase Homogeneous Catalysis
Edited by Boy Cornils and Wolfgang A. Herrmann et al.
Copyright © 2005 Wiley-VCH Verlag GmbH & Co. KGaA, Weinheim
ISBN: 3-527-30721-4

Volume 2

Contributors

Dr. Juan Almena
Degussa AG-Projekthaus Katalyse
65926 Frankfurt/Germany
e-mail: juan.almena@degussa.com
(Section 7.5)

Prof. Dr. Jacques Augé
Université de Cergy-Pontoise
5, mail Gay-Lussac
95031 Cergy-Pontoise/France
e-mail: jacques.auge@chim.u-cergy.fr
(Section 2.2.1)

Dr. Helmut Bahrmann
Celanese GmbH/Werk Ruhrchemie
PO Box 130 160
46128 Oberhausen/Germany
(Section 2.4.1.1.3)

Prof. Dr. Willi Bannwarth
Universität Freiburg
Albertstrasse 21
79104 Freiburg/Germany
e-mail:
willi.bannwart@organik.chemie.uni-
freiburg.de
(Section 4.2.1)

Prof. Dr. rer. nat. Arno Behr
University of Dortmund
Department of Biochemical &
Chemical Engineering
Emil-Figge-Strasse 70
44227 Dortmund/Germany
e-mail: arno.behr@bci.uni-
dortmund.de
(Sections 2.3.2, 3.2.2, 3.2.3)

Prof. Dr. Matthias Beller
Universität Rostock
Buchbinderstrasse 5–6
18059 Rostock/Germany
e-mail: matthias.beller@ifok.uni-
rostock.de
(Section 2.4.1.2)

Dr. Claudio Bianchini
Institute of Chemistry of
Organometallic Compounds
(ICCOM-CNR)
Via Madonna del Piano 10
50019 Sesto Fiorentino (FI)/Italy
e-mail:
claudio.bianchini@iccom.cnr.it
(Section 2.4.2.2)

Multiphase Homogeneous Catalysis
Edited by Boy Cornils and Wolfgang A. Herrmann et al.
Copyright © 2005 Wiley-VCH Verlag GmbH & Co. KGaA, Weinheim
ISBN: 3-527-30721-4

Prof. Dr. Armin Börner
Universität Rostock
Albert-Einstein-Strasse 29a
18059 Rostock/Germany
e-mail: armin.boerner@chemie.uni-rostock.de
(Section 2.2.3.2.3)

Dr. Sandra Bogdanovic
Booz Allen Hamilton
Lenbachplatz 3
80333 München/Germany
(Section 2.4.1.1.3)

Dr. Philippe Bonnet
ARKEMA (formerly ATOFINA)
Centre de Recherches Rhône-Alpes
Rue Henri Moissan – BP 63
69493 Pierre-Benite Cédex/France
e-mail:
philippe.bonnet@arkemagroup.com
(Section 5.2.2.6)

Dr. Jesper Brask
Novozymes A/S
Krogshoejvej 36
2880 Bagsvaerd/Denmark
e-mail: jebk@novozymes.com
(Section 5.2.2.5)

Prof. Mario Bressan
Università G. d'Annunzio
Viale Pindaro 42
65100 Pescara/Italy
e-mail: bressan@sci.unich.it
(Section 2.4.8.2)

Dr. Johannes A.M. van Broekhoven
Shell Research and Technology
Centre Amsterdam
Postbus 38000
1030 BN Amsterdam/
The Netherlands
(Section 2.4.4.6)

Dr. Henry E. Bryndza
DuPont Nylon
PO Box 80328
Wilmington, Del 19880-0302/USA
e-mail:
bryndza@esvax.enet.dupont.com
(Section 2.4.4.1)

Dr. Peter H.M. Budzelaar
Shell Research and Technology Centre
Amsterdam
Postbus 38000
1030 BN Amsterdam/
The Netherlands
(Section 2.4.4.6)

Dipl.-Chem. Katja Burgemeister
Technische und Makromolekulare
Chemie, RWTH Aachen
Worringer Weg 1
52074 Aachen/Germany
(Section 6.4.4)

Prof. Dr. Yves Castanet
Université de Lille
ENSCL
PO Box 108
59652 Villeneuve d'Ascq Cedex/France
e-mail: yves.castanet@ensc-lille.fr
(Section 4.2.3)

Dr. Raghunath V. Chaudhari
Homogeneous Catalysis Division
National Chemical Laboratory
Pune 411 008/India
e-mail: rv.chaudhari@ncl.res.in
(Section 2.4.1.1.2)

Prof. Dr. Yves Chauvin
CPE/COMS Consultant
(formerly IFP)
10, Place François Sicard
37000 Tours/France
e-mail: chauviny@aol.com
(Section 5.1)

Prof. David J. Cole-Hamilton
School of Chemistry
Purdie Building
The University of St. Andrews
Fife, KY16 9ST/Scotland
e-mail: djc@st-and.ac.uk
(Section 6.4.3)

Dr. María Contel
Instituto de Ciencia de Matariales
de Aragón
Universidad de Zaragoza – CSIC
50009 Zaragoza/Spain
e-mail: mcontel@posta.unizar.es
(Section 4.2.4.2)

Prof. Dr. Boy Cornils
65719 Hofheim/Germany
e-mail: boy.cornils@t-online.de
(Chapter 1, Sections 2.1, 2.3.6,
2.4.1.1.1, 2.5.1, 2.5.2, 2.7, Chapter 8)

Prof. Dr. Henri Delmas
Department of Chemical Engineering
Ecole Nationale Superieure des
Ingenieurs en Art Chimiques
(ENSIACET)
118 Route de Narbonne
31077 Toulouse Cédex 04/France
e-mail: henri.delmas@ensiacet.fr
(Section 2.6)

Dr. Uwe Dingerdissen
Institut für Angewandte Chemie
Berlin-Adlershof e. V.
Postfach 96 11 56
12474 Berlin/Germany
e-mail:
uwe.dingerdissen@degussa.com
(Section 7.5)

Dr. Eite Drent
Shell Research and Technology Centre
Amsterdam
Postbus 38000
1030 BN Amsterdam/
The Netherlands
e-mail: eite.e.drent@opc.shell.com
(Section 2.4.4.6)

Prof. Dr. Birgit Drießen-Hölscher (†)
Formerly: Universität Paderborn
33098 Paderborn/Germany
(Section 2.4.4.5)

Prof. Paul J. Dyson
Institut des Sciences et Ingénerie
Chimiques
Ecole Polytechnique Fédérale de
Lausanne
1015 Lausanne/Switzerland
e-mail: paul.dyson@epfl.ch
(Section 5.2.2.3)

Dr. Frédéric Favre
Institut Français du Pétrole
IFP – Lyon
PO Box 3
69390 Vernaison/France
e-mail: frederic.favre@ifp.fr
(Section 5.3.1)

Prof. Dr. Richard H. Fish
University of California
Lawrence Berkeley National
Laboratory
1 Cyclotron Road Mail Stop 70-108 B
Berkeley, CA 94 720/USA
e-mail: rhfish@lbl.gov
(Section 4.2.4.2)

Dr. Alain Forestière
Institute Français du Pétrole
IFP – Lyon
PO Box 3
69390 Vernaison/France
e-mail: alain.forestiere@ifp.fr
(Section 5.3.1)

Dr. Carl-Dieter Frohning
Regnitstrasse 50
46485 Wesel/Germany
e-mail: Frohning@cityweb.de
(Section 2.4.1.1.3)

Dipl.-Chem. Mandy-Nicole Gensow
Universität Rostock
Albert-Einstein-Strasse 3a
18059 Rostock/Germany
e-mail: mandy-
nicole.gensow@chemie.uni-
rostock.de
(Section 2.2.3.2.3)

Dr. Charles M. Gordon
Pfizer Global Research
Ramsgate South
Sandwich
Kent CT13 9NJ/UK
(Sections 6.3, 6.4.1)

Dr. Lasse Greiner
Technische und Makromolekulare
Chemie, RWTH Aachen
Worringer Weg 1
52074 Aachen/Germany
e-mail: greiner@itmc.rwth-aachen.de
(Section 7.4.4)

PD Dr. Peter Härter
Technische Universität München
Lichtenbergstrasse 4
85747 Garching/Germany
e-mail: peter.haerter@ch.tum.de
(Section 2.4.4.4)

Dr. John A. Harrelson Jr.
DuPont Nylon
PO Box 80328
Wilmington, DE 19880-0302/USA
(Section 2.4.4.1)

Prof. Dr. Dr.h.c.mult.
Wolfgang A. Herrmann
President of the Technische
Universität München
80333 München/Germany
e-mail: sekretariat.ac@ch.tum.de
(Chapter 1, Sections 2.2.2, 2.4.4.3,
2.4.4.4, 2.4.8.3, 2.7, Chapter 8)

Dr. Dietmar Hoff
Rheinchemie
Paul-Ehrlich-Strasse 10
67132 Altrip/Germany
e-mail:
dietmar.hoff@rhenchemie.com
(Section 2.2.3.2.1)

Dr. Jens Hoffmann
Universität Jena
Lessingstrasse 12
07743 Jena/Germany
e-mail: jens.hoffmann@rz.uni-jena.de
(Section 5.4)

Prof. Dr. Eric G. Hope
University of Leicester
Department of Chemistry
University Road
Leicester LE1 7RH/UK
e-mail: egh1@leicester.ac.uk
(Section 4.2.2)

Prof. Dr. István T. Horváth
Eötvös University
Pázmány Péter sétány 1/A,1117
Budapest/Hungary
e-mail: istvan.t.horvath@hit-team.net
(Chapter 1, Sections 4.1, 4.3)

Dr. Nikolai V. Ignatev
Merck AG
New Ventures – Materials
Frankfurter Strasse 250
64293 Darmstadt/Germany
e-mail: nikolai.ignatiev@merck.de
(Section 5.3.3.2)

Prof. Kazuaki Ishihara
Nagoya University
Graduate School of Engineering
Furo-cho, Chikusa
Nagoya 464-8603/Japan
e-mail: ishihara@cc.nagoya-u.ac.jp
(Section 4.2.6)

Dr. Bernd Jastorff
Universität Bremen
Leobener Strasse
28359 Bremen/Germany
e-mail: jastorff@uni-bremen.de
(Section 5.4)

Dr. Ulises Jáuregui-Haza
Centro de Quimica Farmaceutica
Apdo. 16082
La Habana, Cuba
e-mail: ulises@cqf.co.cu
(Section 2.6)

Prof. Philip G. Jessop
Queen's University
Department of Chemistry
90 Bader Lane
Kingston, Ontario, K7L 3N6/Canada
e-mail: jessop@chem.queensu.ca
(Section 6.4.2)

Prof. Dr. Zilin Jin
Dalian University of Technology
116012 Dalian/P.R. of China
e-mail: hpcuo@mail.dlptt.ln.cn
(Section 2.3.5)

Prof. Ferenc Joó
Institute of Physical Chemistry
University of Debrecen
PO Box 7
4010 Debrecen/Hungary
e-mail: fjoo@delfin.unideb.hu
(Section 2.4.2.1)

Dr. Christel Thea Jørgensen
Novozymes A/S
Krogshoejvej 36
2880 Bagsvaerd/Denmark
e-mail: chjn@novozymes.com
(Section 5.2.2.5)

Prof. Dr. Philippe Kalck
Ecole Nationale Supérieure des
Ingénieurs en Arts Chimiques et
Technologiques
118 Route de Narbonne
31077 Toulouse Cédex 04/France
e-mail: philippe.kalck@ensiacet.fr
(Section 2.2.3.2.2)

Dr. Paul C.J. Kamer
University of Amsterdam
Nieuwe Achtergracht 166
1018 WV Amsterdam/
The Netherlands
(Section 2.2.3.2.1)

Dr. Ágnes Kathó
Institute of Physical Chemistry
University of Debrecen
PO Box 7
4010 Debrecen/Hungary
e-mail: katho@tigris.klte.hu
(Section 2.4.2.1)

Dr. Ole Kirk
Novozymes A/S
Krogshoejvej 36
2880 Bagsvaerd/Denmark
e-mail: oki@novozymes.com
(Section 5.2.2.5)

Prof. Dr. Shu Kobayashi
Graduate School of Pharmaceutical
Sciences
The University of Tokyo
Hongo, Bubkyo-ku
Tokyo 113-0033/Japan
e-mail: skobayas@mol.f.u-tokyo.ac.jp
(Section 2.4.8.1)

Dr. Christian W. Kohlpaintner
Chemische Fabrik Budenheim
Rheinstrasse 27
55257 Budenheim/Germany
e-mail: ckohlpaintner@budenheim-
cfb.com
(Section 2.4.1.1.3)

Dr. Jürgen G.E. Krauter
Degussa AG
Rodenbacher Chaussee 4
63457 Hanau-Wolfgang/Germany
e-mail: juergen.krauter@degussa.com
(Section 2.4.1.2)

Priv.-Doz. Dr. Fritz E. Kühn
Technische Universität München
Lichtenbergstrasse 4
85747 Garching/Germany
e-mail: fritz.kuehn@ch.tum.de
*(Sections 2.2.2, 2.2.3.1, 2.4.3.3,
2.4.8.3, 7.4.3)*

Dr. Thulani E. Kunene
Sasol Technology
PO Box 1
Sasolburg
1947/South Africa
e-mail:thulani.kunene@sasol.com
(Section 6.4.3)

Dr. Emile G. Kuntz
CPE-Lyon
43 Bd. du 11 novembre 1918
69616 Lyon/France
(Section 2.4.1.1.1)

Dr. Mariano Laguna
Instituto de Ciencia de Matariales
de Aragón
Universidad de Zaragoza – CSIC
50009 Zaragoza/Spain
e-mail: mlaguna@posta.unizar.es
(Section 4.2.4.2)

Dr. Dominique Lastécouères
Université de Bordeaux 1
UMR-CNRS 5802
351 Cours de la Libération
33405 Talence Cedex/France
e-mail:
d.lastécouères@lcoo.u-bordeaux1.fr
(Section 4.2.4.2)

Prof. Dr. Piet W.N.M. van Leeuwen
University of Amsterdam
Nieuwe Achtergracht 166
1018 WV Amsterdam/
The Netherlands
e-mail: pwnm@anorg.chem.uva.nl
(Sections 2.2.3.2.1, 2.4.1.1.3)

Prof. Dr. Walter Leitner
Technische und Makromolekulare
Chemie, RWTH Aachen
Worringer Weg 1
52074 Aachen/Germany
e-mail: leitner@itmc.rwth-aachen.de
(Chapter 1, Sections 6.3, 6.4.1, 6.4.4, 6.6)

Dr. Peter Licence
The University of Nottingham
Nottingham, NG7 2RD/UK
e-mail:
peter.licence@nottingham.ac.uk
(Section 6.5)

Prof. Dr. Andreas Liese
Universität Münster
Wilhelm-Klemm-Strasse 2
48149 Münster/Germany
e-mail: aliese@uni-muenster.de
(Section 7.4.4)

Prof. Dr. André Lubineau
Université de Paris-Sud
Bâtiment 420
91405 Orsay/France
e-mail: lubin@icmo.u-psud.fr
(Section 2.2.1)

Dr. Matthias Maase
BASF AG
67056 Ludwishafen/Germany
e-mail: matthias.maase@basf-ag.de
(Section 5.3.2)

Dr. Lionel Magna
Institut Français du Pétrole
PO Box 3
69390 Vernaison/France
e-mail: lionel.magna@ifp.fr
(Section 5.2.2.2)

Dr. Zai-Sha Mao
Institute of Process Engineering
Chinese Academy of Sciences
PO Box 353
Beijing 100080/China
e-mail: zsmao@home.ipe.ac.cn
(Section 2.3.1)

Prof. James McNulty
McMaster University
Department of Chemistry
1280 Main Street West
Hamilton, Ontario, L8S 4M1/Canada
e-mail: jmcnult@mcmaster.ca
(Section 5.2.2.7)

Prof. Dr. Stefan Mecking
Universität Konstanz
Universitätsstrasse 10
78457 Konstanz/Germany
e-mail:
stefan.mecking@uni-konstanz.de
(Chapter 1, Sections 7.1, 7.2, 7.3,
7.4.2, 7.6)

Christian P. Mehnert
ExxonMobil Chemical Company
Baytown Technology and Engineering
Complex West
5200 Bayway Drive
Baytown, TX 77520/USA
e-mail:
christian.p.mehnert@exxonmobil.com
(Section 5.2.1.3)

Dr. Andrea Meli
Institute of Chemistry of
Organometallic Compounds
(ICCOM-CNR)
Polo Scientifico Area CNR
Via Madonna del Piano 10
50019 Sesto Fiorentino (FI)/Italy
(Section 2.4.2.2)

Dr. Kerstin Mölter
Universität Bremen
Leobener Strasse
28359 Bremen/Germany
e-mail: kmoelter@uni-bremen.de
(Section 5.4)

Prof. Dr. Eric Monflier
Fac. des Sciences J. Perrin
Université d'Artois
Rue Jean Souvraz
Sac postal 18
62307 Lens Cédex/France
e-mail: monflier@univ-artois.fr
(Sections 2.2.3.3, 2.4.1.1.3, 2.4.3.2,
4.2.3)

Prof. Dr. André Mortreux
Université de Lille
PO Box 108
59652 Villeneuve d'Ascq Cédex/France
e-mail: andre.mortreux@enac-lille.fr
(Section 4.2.3)

Dr. Christian Müller
Eindhoven University of Technology
PO Box 513
5600 MB Eindhoven/The Netherlands
e-mail: c.mueller@tue.nl
(Section 7.4.1)

Prof. Dr. Yutaka Nakamura
Niigata University of Pharmacy and
Applied Life Sciences
Faculty of Applied Life Sciences
Department of Biomolecular Sciences
265-1 Higashijima
Niigata 956-8603/Japan
(Section 4.2.5.2)

Prof. Dr. Günther Oehme
Universität Rostock
Buchbinderstrasse 5–6
18055 Rostock/Germany
e-mail: guenther.oehme@ifok.uni-
rostock.de
(Section 2.3.4)

Prof. Dr. Hélène Olivier-Bourbigou
Institut Français du Pétrole
PO Box 3
63390 Vernaison/France
e-mail: helene.olivier@ifp.fr
*(Chapter 1, Sections 5.2.1.1, 5.2.2.1,
5.2.2.4, 5.3.4, 5.5)*

Prof. Dr. Bernd Ondruschka
Universität Jena
Lessingstrasse 12
07743 Jena/Germany
e-mail:
bernd.ondruschka@rz.uni-jena.de
(Section 5.4)

Prof. Dr. Martyn Poliakoff
University of Nottingham
Nottingham NG7 2RD/UK
e-mail:
martyn.poliakoff@nottingham.ac.uk
(Section 6.5)

Dr. Gianluca Pozzi
CNR – Instituto di Scienze e
Tecnologie Molecolari
Via Golgi 19
20133 Milano/Italy
e-mail: gianluca.pozzi@istm.cnr.it
(Section 4.2.4.1)

Dr. Silvio Quici
Istituto ISTM–CNR
Via Golgi 19
20133 Milano/Italy
e-mail: silvio.quici@unimi.it
(Section 4.2.4.1)

Prof. Dr. Peter J. Quinn
King's College London
Department of Life Science
150 Stamford Street
London SE1 9NH/UK
e-mail: p.quinn@kcl.ac.uk
(Section 2.4.7)

Dr. Johannes Ranke
Universität Bremen
Leobener Strasse 1
28359 Bremen/Germany
e-mail: jranke.@uni-bremen.de
(Section 5.4)

Dr. Joost N.H. Reek
University of Amsterdam
Nieuwe Achtergracht 166
1018 WV Amsterdam/
The Netherlands
e-mail: reek@science.uva.nl
(Section 2.2.3.2.1)

Dr. Claus-Peter Reisinger
Exatec LLC
31220 Oak Creek Drive
Wixom
Michigan 48393/USA
e-mail: www.exatec.biz
(Section 2.4.4.4)

Al Robertson
Cytec Canada Inc.
PO Box 240
Niagara Falls
Ontario/Canada L2E 6T4
e-mail: al.robertson@cytec.com
(Section 5.3.3.3)

Dr. Stefan Rossenbach
Bergische Universität
Gauss-Strasse 20
42047 Wuppertal/Germany
(Section 2.2.3.2.1)

Dr. José Sanchez Marcano
Institut Européen des Membranes
2 Place Eugène Bataillon
34095 Montpellier Cédex 5/France
e-mail: jose.sanchez@iemm.univ-
montp2.fr
(Section 2.3.3)

Dr. Ana M. Santos
Technische Universität München
Lehrstuhl für Anorganische Chemie
Lichtenbergstrasse 4
85747 Garching/Germany
e-mail: ana.kuehn@chem.tum.de
(Sections 2.2.3.1, 2.4.8.3, 7.4.3)

Dr. Marie-Christine Scherrmann
Université de Paris-Sud
Bâtiment 420
91405 Orsay/France
e-mail: mcscherr@icmo.u-psud.fr
(Section 2.2.1)

Dr. Ulf Schlotterbeck
BASF Coatings
Glasuritstrasse 1
48165 Münster-Hiltrup/Germany
(Section 7.4.2)

Dr. Siegfried Schneider
ALTANA Pharma AG
Byk-Gulden-Strasse 2
78467 Konstanz/Germany
e-mail:
siegfried.schneider@altanapharma.com
(Section 4.2.1)

Dr. Marcel Schreuder-Goedheijt
Dinosynth
PO Box 20
5340 H Oss/The Netherlands
e-mail: marcel.schreudergoedheijt
@dinosynth.com
(Section 2.2.3.2.1)

Prof. Dr. Aaron M. Scurto
University of Kansas
Department of Chemical and
Petroleum Engineering
NSF-ERC Center for Environmentally
Beneficial Catalysis
1530 W. 15th Street
4132 Learned Hall
Lawrence, KS 66045/USA
e-mail: ascurto@ku.edu
(Section 6.1)

Dipl.-Ing. Joachim Seuster
University of Dortmund
Department of Biochemical &
Chemical Engineering
Emil-Figge-Strasse 70
44227 Dortmund/Germany
e-mail: seuster@bci.uni-dortmund.de
(Section 2.3.2)

Prof. Dr. Roger A. Sheldon
Delft University of Technology
Julialaan 136
2628 BL Delft/The Netherlands
e-mail: R.A.Sheldon@tnw.tudelft.nl
(Section 2.4.3.1)

Prof. Dr. Denis Sinou
Université Claude-Bernard Lyon I
Bâtiment CPE
43 Boulevard du 11 novembre 1918
69622 Lyon/France
e-mail: sinou@univ-lyon.fr
(Sections 2.4.5, 2.4.6, 4.2.5.1)

Prof. Dr. Othmar Stelzer (†)
Formerly: Bergische Universität
Wuppertal/Germany
(Section 2.2.3.2.1)

Dr. Frauke Stock
Universität Bremen
Leobener Strasse
28359 Bremen/Germany
e-mail: fstock.@uni-bremen.de
(Section 5.4)

Dr. Reinhold Störmann
Universität Bremen
Leobener Strasse
28359 Bremen/Germany
e-mail: stoertebecker@uni-bremen.de
(Section 5.4)

Dr. Alison M. Stuart
University of Leicester
Department of Chemistry
University Road
Leicester LE1 7RH/UK
e-mail: amc17@leicester.ac.uk
(Section 4.2.2)

Prof. Seiji Takeuchi
Niigata University of Pharmacy and
Applied Life Sciences
Faculty of Applied Life Sciences
Department of Biomolecular Sciences
265-1 Higashijima
Niigata 956-8603/Japan
e-mail: takeuchi@niigata-pharm.ac.jp
(Section 4.2.5.2)

Dr. Nils Theyssen
Max-Planck-Institut für
Kohlenforschung
PO Box 10 13 53
45466 Mülheim/Germany
e-mail: theyssen@kofo.mpg.de
(Section 6.2)

Dr. Carl Christoph Tzschucke
Universität Freiburg
Albertstrasse 21
79104 Freiburg/Germany
(Section 4.2.1)

Dr. Marc Uerdingen
Solvent Innovation GmbH
Nattermannallee 1
50829 Köln/Germany
e-mail: marc.uerdingen@solvent-
innovation.com
(Section 5.3.3.1)

Dr. Martine Urrutigoïty
Ecole Nationale Supérieure des
Ingénieurs en Arts Chimiques et
Technologiques
118 Route de Narbonne
31077 Toulouse Cédex 04/France
(Section 2.2.3.2.2)

Dr. Christophe Vallée
Institut Français du Pétrole
PO Box 3
69390 Vernaison/France
e-mail: christophe.vallee@ifp.fr
(Sections 5.2.1.1, 5.3.4)

Dr. Jean-Marc Vincent
Université de Bordeaux 1
UMR-CNRS 5802
351 cours de la Libération
33405 Talence Cedex/France
e-mail: jm.vincent@lcoo.u-bordeaux.fr
(Section 4.2.4.2)

Prof. Dr. Dieter Vogt
Eindhoven University of Technology
Schuit Institute of Catalysis
Laboratory of Homogeneous Catalysis
PO Box 513
5600 MB Eindhoven/The Netherlands
e-mail: d.vogt@tue.nl
(Chapter 1, Sections 3.1, 3.2.1, 3.3, 3.4,
7.4.1)

Dr. Markus Wagner
Solvent Innovation GmbH
Nattermannallee 1
50829 Köln/Germany
e-mail: markus.wagner@solvent-
innovation.com
(Section 5.3.3.1)

Dr. Yanhua Wang
Dalian University of Technology
State Key Laboratory of Fine
Chemicals
Dalian 116012/P.R. of China
e-mail: yhuawang@online.ln.cn
(Section 2.3.5)

Dr. Paul B. Webb
Sasol Technology (UK)
Purdie Building
St. Andrews
Fife KY16 8PP/UK
e-mail: paul.webb@uk.sasol.com
(Section 6.4.3)

Prof. Tom Welton
Imperial College
Department of Chemistry
London SW7 2AZ/UK
e-mail: t.welton@imperial.ac.uk
(Section 5.2.1.2)

Dr. Urs Welz-Biermann
Merck KGaA
New Ventures – Materials
Frankfurter Strasse 250
64293 Darmstadt/Germany
e-mail: urs.welz-biermann@merck.de
(Section 5.3.3.2)

Ernst Wiebus
Celanese GmbH/Werk Ruhrchemie
PO Box 130 160
46128 Oberhausen/Germany
e-mail: ernst@wiebus.de
(Section 2.3.6)

Dr. Anne-Marie Wilhelm
Ecole Nationale Superieure
d'Ingenieurs de Genie Chimique
(ENSIGC)
31078 Toulouse Cédex 04/France
(Section 2.6)

Prof. Hisashi Yamamoto
University of Chicago
5735 S. Ellis Ave.
Chicago, IL 60637/USA
e-mail: yamamoto@uchicago.edu
(Section 4.2.6)

Dr. Chao Yang
Institute of Process Engineering
Chinese Academy of Sciences
PO Box 353
Beijing 100080/China
e-mail: chaoyang@home.ipe.ac.cn
(Section 2.3.1)

Dr. Noriaki Yoshimura
Kuraray Co. Ltd.
2045-1 Sakazu
Kurashiki
710 Okayama/Japan
(Section 2.4.4.2)

Dr. Dongbin Zhao
Ecole Polytechnique Fédérale
de Lausanne
1015 Lausanne/Switzerland
e-mail: dongbin.zhao@epfl.ch
(Section 5.2.2.3)

1
Introduction

Boy Cornils, Wolfgang A. Herrmann, István T. Horváth, Walter Leitner, Stefan Mecking, Hélène Olivier-Bourbigou, and Dieter Vogt (Eds.)

Multiphase Homogeneous Catalysis
Edited by Boy Cornils and Wolfgang A. Herrmann et al.
Copyright © 2005 Wiley-VCH Verlag GmbH & Co. KGaA, Weinheim
ISBN: 3-527-30721-4

1
Introduction

Boy Cornils, Wolfgang A. Herrmann, István T. Horváth, Walter Leitner,
Stefan Mecking, Hélène Olivier-Bourbigou, and Dieter Vogt

> "Sipping a cup of decaffeinated coffee the reader may wonder on the somewhat unusual classification of solvents as 'alternative': alternatives to what? And why would we need alternative media for doing chemistry or for any other purpose? These may be the first questions for those who are just starting to discover the existing new developments on using solvents other than volatile and often toxic organics for synthesis and especially for catalytic reactions. Yes, indeed, ..." [1].

Yes, indeed, similarly to an opening cornucopia, the arsenal of methods and techniques in homogeneous catalysis has offered remarkable progress in recent years. Above all, these improvements concentrate on the modification and the handling of homogeneous catalysts in general and the removal and subsequent recycling of catalysts in particular. The progress to be dealt with in this book has been rendered essentially possible by the introduction of separate phases in the context of "homogeneous" catalysis: an apparent contradiction with far-reaching consequences. For the first time, this book reviews all the realistic possibilities described so far using multiphase operation of homogeneous catalysis: processes with organic/organic, organic/aqueous, or "fluorous" solvent pairs (solvent combinations), nonaqueous ionic solvents, supercritical fluids, and systems with soluble polymers. In Figure 1, the family tree of homogeneous catalysis proves that this recent research extends considerably the scope of the work.

Following the logic of this tree, the multiphase processes on the left-hand side belong among the operations with "immobilized catalysts" but on "liquid supports". The topics of this book are the processes with the liquid supports water, supercritical fluids, ionic liquids, organic liquids, soluble polymers, and fluorous liquids; among these, only two processes (Ruhrchemie/Rhône-Poulenc and Shell SHOP) are operative industrially so far. The more important "leaves" of the family tree are shaded in gray.

In Figure 2, demonstrating the genesis of homogeneous and heterogeneous catalysis, the topical processes are on the borderline between heterogeneous and homogeneous catalysis (and catalysts).

Multiphase Homogeneous Catalysis
Edited by Boy Cornils and Wolfgang A. Herrmann et al.
Copyright © 2005 Wiley-VCH Verlag GmbH & Co. KGaA, Weinheim
ISBN: 3-527-30721-4

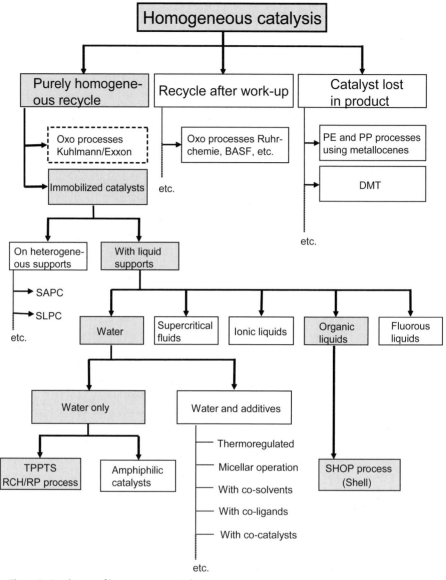

Figure 1 Family tree of homogeneous catalysis.

Why multiphase systems? This goes back to the 1980s and the enormous impetus which was given to the homogeneous catalysis community by the first realization of Ruhrchemie/Rhône-Poulenc's oxo process at the Oberhausen plant site. Astonishingly, it was this development (and *not* the earlier SHOP process of Shell) that sensitized the scene to the possibilities of multiphase action: only on the basis of the "aqueous" activities that so much widespread and multi-faceted research work, with effects on the newer areas mentioned has been accomplished success-

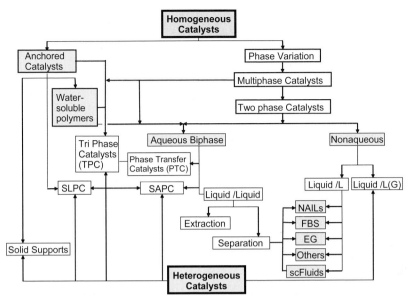

Figure 2 Genesis of homogeneous and heterogeneous catalysis.

fully. There was earlier work and proposals [2] to imitate the decisive advantage of heterogeneous catalysis: the immediate separation of catalyst and substrates/products just after reaction which makes it possible to avoid additional separation steps post-reaction, such as distillations and other thermally stressing procedures. All proposals have the same target: to enable the homogeneous catalyst to be bound to a suitable "support", i.e., another phase, without losing its superior homogeneous catalytic activity and selectivity. Within the scope of this book the editors define "phases" not only thermodynamically (as uniform states of matter of *one* substance which are separated (and separable) from each other by unequivocal phase boundaries; for example, water–ice or normal–supercritical states) but also as *different* states of aggregation of different compounds, such as systems consisting of water–organic liquids. Thus this book deals with homogeneous catalysts on *liquid* supports. Additionally, the processes described imply two- or three-phase reactions (the latter is the case if gaseous reactants complete the reaction scheme, e.g., hydrogen in hydrogenations or syngas in hydroformylations).

The use of liquids in homogeneous catalysis thus means not only a liquid support and from there a basic intervention in the handling and the operation of the catalyst, but also a modern separation technique for efficient work-up in organic synthesis [3]. Figure 3 illustrates the enormous importance of the biphasic technique for homogeneous catalysis: the catalyst solution is charged into the reactor together with the reactants A and B, which react to form the solvent-dissolved reaction products C and D. The products C and D have different polarities than the catalyst solution and are therefore simple to separate from the catalyst phase (which may be recycled in a suitable manner into the reactor) in the downstream phase separation unit.

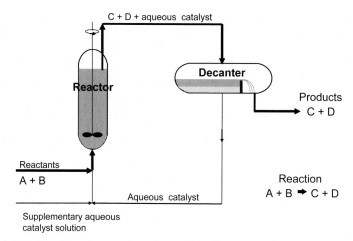

Figure 3 Principle of two-phase catalysis illustrated for the aqueous-phase reaction A + B → C + D (cf. also Chapter 2).

This double meaning of the multiphasic approach is specially visible in the case of fluorous liquids, where organic chemists are at least as interested as the catalytic community in the use of these fluids.

The ability of different combinations of solvent pairs to enable biphasic operation can be estimated on the one hand according to the principle of "like dissolves like" (*"similia similibis solvuntur"*, as the old alchymists used to say) in respect to the solvent for the catalyst and, on the other hand, with the help of diagrams as shown in Figure 4 [4] and of fundamental investigations.

The fundamentals of miscibility (solvation power, E_T^N) of various solvents from nonpolar, aprotic tetramethylsilane (TMS; with $E_T^N = 0$ as defined) to polar water ($E_T^N = 1$) are given by the solvent polarity scale in Figure 5 [5].

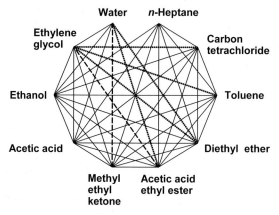

Figure 4 Miscibility of organic solvents: —— miscible in all proportions; – – – limited miscibility; ⋯⋯ little miscibility; no line: immiscible.

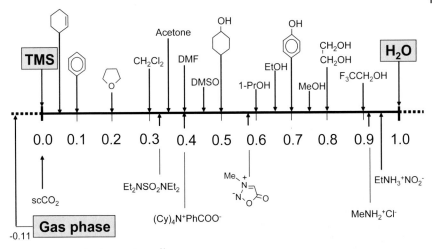

Figure 5 Empirical polarity values E_T^N.

This graph gives a selection of 14 (out of approximately 360) of the usual solvents above the baseline and seven more exotic solvents (supercritical CO_2 and ionic liquids included) below. The 14 compounds, from left to right with increasing solvent polarity, include apolar, aprotic (such as TMS, cyclohexene, or benzene), bipolar (such as acetone or DMF), and eventually bipolar, protic solvents (cyclohexanol, ethanol, phenol). Using the E_T^N values, numerous solvent-dependent processes may be correlated with each other. Other measures that can be used for the estimation of miscibility/solvent power are the cohesive pressures, solubility parameters, dispersive forces, Kamlet–Taft parameters, etc. [6a,b]. Solvent combinations of exotic members and systems with more than two members are known and have been recommended, but their application has been concentrated in the lab because of economic disadvantages with their handling and recyclability/separability [6b–e].

A recent proposal concerns mixed organic–aqueous tunable solvents (OATS) such as dimethyl ether–water, the solubility of which for substrates can be influenced by a third component such as carbon dioxide. CO_2 acts as a "antisolvent" and as a switch to cause a phase separation and to decant the phases from each other (preferably under pressure). This behavior makes the operation of bi- or multiphase homogeneous catalytic processes easier and more economic: the preferential dissolution at modest pressure of carbon dioxide causes phase separation which results in large distribution coefficients of target molecules in biphasic organic–aqueous systems. This extraordinary behavior lead to a sophisticated flow scheme (Figure 6) [7].

This operation, which needs at least two internal recycles, may be economic for special purposes (e.g., highly prized applications) such as enzyme–catalyst conversions. Indeed, it has been tested for the ADH-catalyzed reduction of hydrophobic ketones coupled with regeneration of the cofactor NADH. Another possibility discussed recently is the use of surface polarity-modified (heterogeneous) catalysts and their distribution between two immiscible solvents which occurs against gravity [8].

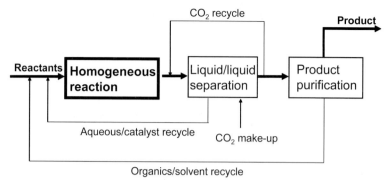

Figure 6 Schematic of the OATS procedure with a CO_2 switch for catalyst recycle.

It might be added that the multiphase operation offers more than the decisive separation between desired products and catalyst, although there are differences between the various multiphase liquids [9]. It cannot be emphasized enough that the use of polar multiphase liquids also separate the byproduct "heavy ends" from the catalyst in the system, thus avoiding a build-up in the catalyst recycle. In other processes (and probably also if very apolar fluorous liquids are used) an additional purge is needed to remove the high boilers from the catalyst, which then requires a further (and costly) separation or purification [10].

It is also worth mentioning that the multiphase approach has been used as a strategy to avoid undesired consecutive reactions [11a–c] or even to segregate two different and incompatible catalysts in one-pot or in tandem syntheses. In a typical example, Chaudhari et al. described the combination of a hydroformylation step – catalyzed by phosphine-modified rhodium complexes – with the Mannich reaction of the oxo aldehydes formed, catalyzed by tertiary amines. Thus the manufacture of methacrolein according to Figure 7 proceeds best in two different phases: the organic phase for hydroformylation and the aqueous phase in which the Mannich reaction is achieved (Figure 8) [11d].

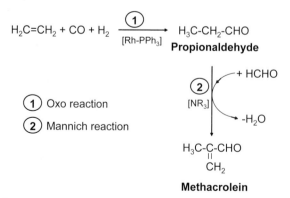

Figure 7 Overall reaction in two phases.

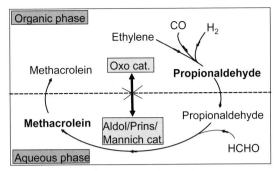

Figure 8 Tandem two-phase catalysis with incompatible catalysts.

Our definitions of phases or multiphases, respectively, include interfacial events, but it must be emphasized that the engineer's definition of "interphases" or "interfacial engineering" concentrates more on solid–solid phase boundaries such as in the manufacture of chips, photographic films, polymer composites, advanced ceramics, etc. [12], or on processes with heterogeneous catalysts and other phases [13]. Last, but not least, the possibility of placing the emphasis on "greener" techniques and thus so-called "sustainable" operations [14] may be mentioned in passing.

Although they also are "biphasic", this book does not include immobilized, modified homogeneous catalysts on *solid* supports ("anchored catalysts" [15]). There is little mileage in looking for ostensibly more and more effective ligands and better and better optimization of the solid support and/or catalyst precursor to ensure, on the one hand, adequate immobilization on the support (sufficient stability of the covalent bond between the support matrix and the central atom) and, on the other, adequate mobility for the catalytically active catalyst constituents (sufficient lability of the modified ligand sphere). All the results so far allow only the conclusion that with the heterogenizing techniques used significant problems remain to be solved. The reason for this is that the various catalyst species undergo changes in spatial configuration as they pass through the catalytic cycle typical of a homogeneous process.

The constant "mechanical" stress on the catalyst's central atom↔ligand bonds and the constant change in the bond angles and lengths ultimately lead also to a weakening of the central atom↔support bond. In hundreds of publications this is conveniently demonstrated using the hydroformylation reaction catalyzed by heterogenized cobalt or rhodium carbonyls as an example. The catalyst passes through the two forms of a trigonal-bipyramidal and a tetrahedral metal carbonyl, which overstresses and weakens the heterogenizing bond between the metal and the support – ending up with a considerable "leaching" of the catalysts's active components. This leaching always affects not only the usually costly central atoms but also the (frequently more costly) ligands of homogeneous metal complex catalysts. Therefore, it might be no coincidence that – apart from lab or pilot plant work [18] and apart from enzymatic techniques – no homogeneously catalyzed reaction attained large-scale, commercial status. Processes including the reaction in biphasic operation but on heterogeneous catalysts (as, for example, in [16]) are also not included.

According to our definition, homogeneous catalysis includes catalysts which, inter alia,

- are molecularly dispersed "in the same phase" or between two suitable phases (see above),
- are unequivocally characterized chemically and spectroscopically and can be synthesized and manufactured in a simple and reproducible manner,
- can be tailor-made for special purposes according to known and acknowledged principles and based upon a rational design, and
- permit unequivocal reaction kinetics related to each reactant catalyst molecule, in general each metal ion [17].

These definitions allow the well-known compilation of the advantages/disadvantages of both catalytic regimes (Table 1).

Entry 7 of Table 1 is the reason for the application of the idea to use multiphase techniques, probably first articulated and systematized by Manassen [19] and Joó [20]. In 1972, Manassen suggested

> "... the use of two immiscible liquid phases, one containing the catalyst and the other containing the substrate ..."

and hence the general form of biphase catalysis, which constitutes a logical development of the work in "molten salt media" (further developed and known today as "ionic liquids"; this term used to refer to high-melting, inorganic or organic/inorganic salts or salt mixtures [21]; cf. Chapter 5). The inventors of Shell's SHOP

Table 1 Homogeneous versus heterogeneous catalysis.

		Homogeneous catalysis	*Heterogeneous catalysis*
1	Activity	high	variable
2	Selectivity	high	Variable
3	Reaction conditions	mild	harsh
4	Service life of catalysts	variable	extended
5	Sensitivity against catalyst poisons	low	high
6	Diffusion/mass transfer problems	low	important
7	Catalyst recycling[a]	expensive	not necessary
8	Variability of electronic and steric properties of catalysts	possible	not possible
9	Mechanistic understanding	plausible under random conditions	extremely complex

a) *Catalyst recycling* has to be distinguished from *catalyst (or catalytic) cycle.* While the first-mentioned describes the way in which a catalyst is formed, employed, separated or deposited, made up, and regenerated or recovered, the "catalytic cycle" is the visual interpretation of a complex reaction mechanism by subdividing the overall reaction into a series of ad- and desorption steps (with heterogeneous catalysts) or arranging the intermediates of a homogeneously catalyzed reaction in a logical sequence to form a closed cycle [18].

process, who had already worked on soluble, homogeneous complex catalysts in a biphase system some years earlier [22], cited the special method without particular emphasis. Some years later, Joó concentrated on hydrogenations and Kuntz [23] published his work on aqueous-phase hydroformylation, i.e., reactions of commercial interest. Other historical roots may be found in [24].

Looking back, it must be stated that Manassen and Beck/Joó's ideas were developed independently of each other. Remarkably, the fundamental papers of Joó and Kuntz created little interest and only found a wider echo in academic research once Shell and Ruhrchemie had managed to achieve industrial scale-up of their biphase catalyses in organic/organic or in aqueous systems. In a drastic departure from the normal pattern, here basic research lagged considerably behind industrial research and application [25]. This has changed with the introduction of other liquid phases such as ionic liquids (as defined today), supercritical liquids, polymeric fluids, and fluorous liquids.

As a last definition the book concentrates on *organometallic* catalysts as one of the bases of homogeneous catalysis, although the introduction of strange additives like water and other immiscible ingredients such as other organic liquids, carbon dioxide, nonaqueous ionic liquids, solvent-miscible polymers, or even perfluorinated compounds was originally regarded as disturbing if not poisoning. So Cintas wrote in respect of the "additive" water [26]:

> "At first, the idea of performing organometallic reactions in water might seem ridiculous, since it goes against the traditional belief that most organometallics are extremely sensitive to traces of air and moisture and rapidly decompose in water".

Other statements voice people's doubts about the same fact [27]. Acid–base catalysts, the other well developed category of homogeneous catalysts, are mentioned when their action is typical or decisive for the demonstration of the respective multiphase action.

Besides one book summarizing *Chemistry in Alternative Reaction Media* [6a], which concentrates more on physical aspects rather than applications, the above-mentioned multiphase media have so far been dealt with only in monographs such as

- *Aqueous-Phase Organometallic Catalysis* [24, 26],
- *Applied Homogeneous Catalysis with Organometallic Compounds* [17],
- *Handbook of Fluorous Chemistry* [28],
- *Ionic Liquids in Synthesis* [29],
- *Chemical Syntheses Using Supercritical Fluids* [30],
- *Emulsion Polymerization and Emulsion Polymers* [31].

Except for transitions from heterogeneous to homogeneous catalysis, there is also common ground for the various methodologies described here: the application of $scCO_2$ is described in the presence of ionic liquids [32a] or of fluorous solvents [32b,c] as well in aqueous operation [33a–d]; and aerobic epoxidations have been attained in fluorous biphasic systems using ionic liquids [33e]. On the other hand, ionic liquids [34] or fluorous solvents [35a,c] have been used together with aqueous operations; and water-soluble polymers are the focal point of the application of

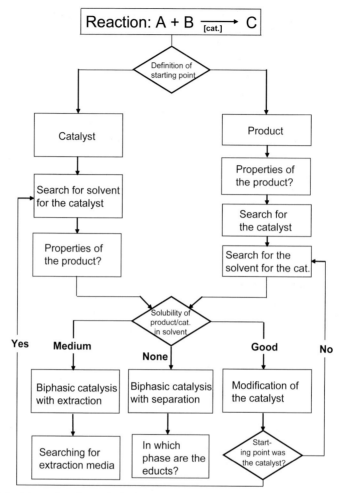

Figure 9 Decision-making for multiphase reactions in respect to the adjustment of their properties as far as catalyst, phase, and solvent are concerned.

polymers in homogeneous catalysis (see Chapter 7). Some of these hybrid systems are described in more detail, for instance in Sections 5.2.1.1 or 5.2.1.2. Membrane techniques, as described for aqueous applications in Chapter 2, may be applied to all other techniques where solvent-resistant membranes can be developed [35b,c] (Section 2.3.3). This is of decisive importance for the use of soluble, polymer-supported catalysts (cf. Chapter 7). Keeping in mind the intentions which underlie the use of nonmiscible solvents, as a paradox the development of supported ionic liquids may be mentioned [36].

Considering the transferability of conventional chemical processes to a multiphase technology, Behr [37] made proposals for a formal decision-making process on whether, and how, this transition could be achieved (Figure 9).

There are some examples of industrial applications of aqueous-phase homogeneous catalysis which will be described in Section 2.5.

The other process based on Manassen/Joò's ideas of liquid supports [19, 20] goes back to Keim's developments of *organic–organic biphase systems* for the Ni-catalyzed conversion of olefins to medium- and long-chain alpha- and internal olefins [38] by a combination of various process steps, among them oligomerization, metathesis, and isomerization. This is described in Chapter 3.

Based on Ziegler's work at the MPI für Kohlenforschung in Mülheim/Germany, Wilke and co-workers learned how to control the selectivity of nickel-catalyzed reaction by use of ligands. Keim, then at Shell at Emeryville, introduced P–O chelate ligands and carried out the basic research for the oligomerization process. The whole process was developed in a collaboration between Shell Development in the US and the Royal Shell Labs in Amsterdam/The Netherlands. Today there are two locations producing nearly 1 MMt of α-olefins per year.

Within the scope of this book, the first step of the organic–organic biphasic oligomerization is of interest. This process step is carried out in a polar solvent in which the nickel catalyst is dissolved but the nonpolar products – the α-alkenes – are almost insoluble. Preferred solvents are alkanediols, especially 1,4-butanediol (BDO). The nickel catalyst is prepared in situ from a nickel salt by reduction with $NaBH_4$ in BDO in presence of an alkali hydroxide, ethylene, and a chelating P–O ligand such as *ortho*-diphenylphosphinobenzoic acid.

The oligomerization is carried out in a series of reactors at 80–140 °C and pressures of 7–14 MPa. The α-alkenes produced have a distribution of up to 99% linearity. In a high-pressure separator the insoluble products and the catalyst solution as well as unreacted ethylene are separated; the catalyst is recycled into the oligomerization reactor.

Although a genuinely homogeneous reaction takes place without mass transfer limitation at the liquid–liquid phase boundary (as far as is known to date), it is noteworthy that the Shell SHOP process is up to now the only example of the biphasic organic/organic technology. This curious status may be the result of the fact that aqueous-biphasic processes received more attention or – at least – it is a consequence of the fact that the search for other applicable couples of organic/organic liquids was ineffective. It is only very recently that academic groups became interested again in this concept. One reason might be the fact that much more empirical data and knowledge are required for a thorough description of such a process (i.e., phase diagrams considering all the components of the system envisaged). Because there are only occasional publications (and no precise or informative ones) by Shell, the economics of organic/organic processes remain doubtful, at least if the first steps of the reaction sequence are not in the hands of the producers of petrochemicals (with their possibilities of cost distribution).

But once more, and similarly to aqueous-biphase operation, basic academic research also lagged behind the industrial research and application.

Even the use of perfluorinated alkenes, dialkyl ethers, etc., forming *fluorous biphase systems* because of their low miscibility with common organic solvents such as acetone, toluene, THF, and alcohols, is a subdivision of biphasic catalysis and,

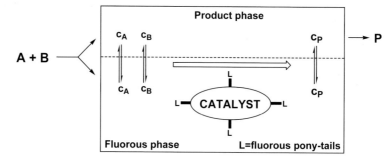

Figure 10 The fluorous-biphase concept for the catalytic reaction of substrates **A** + **B** → **P**. The attachment of appropriate fluorous ponytails **L** to a homogeneous catalyst ensures that the fluorous catalyst remains in the fluorous phase.

eventually, the application of Manassen/Joó's principle (see Chapter 4). The term *fluorous* was introduced as the analogue to the term *aqueous* to emphasize the fact that one of the phases of a biphase system is richer in fluorocarbon than the other. Such systems can be used in catalytic chemical transformations by immobilizing catalysts in the liquid, i.e., fluorous, phase. A fluorous catalyst system consists of a fluorous phase containing a preferentially fluorous-soluble catalyst and a second product phase which may be any organic or inorganic solvent with limited solubility in the fluorous phase.

The characteristic of fluorous phase operation is (like using supercritical carbon dioxide; see Chapter 6) that the catalysts have to be made fluorous-soluble by incorporating fluorocarbon moieties in their structure in appropripate size and number (Figure 10). The most effective fluorocarbon moieties are linear or branched perfluoralkyl chains with high carbon numbers that may contain other heteroatoms (the "fluorous ponytails"). The best fluorous solvents are perfluorinated alkanes, perfluorinated dialkyl ethers, and perfluorinated trialkylamines [28, 39].

A fluorous-biphase reaction can proceed either in the fluorous phase or at the interphase of two phases, depending on the solubilities of the substrates in the fluorous phase. It should be emphasized that a fluorous-biphase system might be converted to a one-phase system by increasing the temperature. Thus fluorous catalysis could combine the advantages of *monophasic* catalysis with *biphasic* product separation by running the reaction at higher temperatures and separating the products at lower temperatures (Figure 11).

In this respect, fluorous-phase operation is similar to temperature-regulated phase transfer catalysis (see Section 2.3.5) and to special versions of soluble polymer-bound catalysis (see Chapter 7). Alternatively, the temperature-dependent solubilities of *solid* fluorous catalysts in liquid substrates or in conventional solvents containing the substrates could eliminate the need for fluorous solvents.

Up to a certain degree the properties of fluorous solvents can be calculated and thus "tuned", especially when observing the strong electron withdrawing effects of "insulating" groups between the fluorous ponytails and the basic molecule [40] –

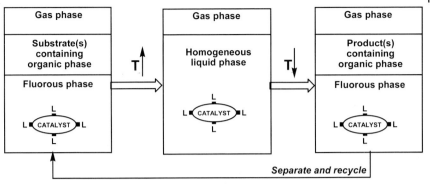

L = Fluorous solubilizing groups

Figure 11 The temperature-dependent fluorous-liquid/liquid-biphase concept.

a great advantage for well-aimed tuning to a special application and a disadvantage for the production of larger-scale tonnages of bulk chemicals.

Fluorous catalysis is now regarded by some specialists as a well-established area providing complementary approaches to aqueous (see Chapter 2) or ionic-biphase (see Chapter 5) catalysis. So far, no economical application of fluorous-biphase catalysis is known (except for highly specialized lab uses); but fluorous-biphasic separation techniques without catalytic reactions involved may be the trailblazers of later catalytic uses.

As far as *nonaqueous ionic liquids* (more shortly but erroneously also called "ionic liquids", or ILs) are concerned, they were just considered as potential alternatives for multiphase reactions and were mentioned in forward-looking chapters of books dealing with this area, while water was already in key transition metal-catalyzed processes; see, for example, [6a, 17, 24, 29] (cf. also Chapter 2). Ionic liquids also have recently attracted much interest: it is now possible to buy them, which probably promotes their use, and there are more and more physical data available for these solvents. The range of reactions that have been described in IL media is probably wider than in scCO$_2$ (and surely wider than in fluorous liquids). But it would not be realistic to say that all catalyzed reactions can be transferred to ionic liquids with benefits. The advantages of using ILs have been well described for some reactions which will be reported later in this book and just a few of them have been run on micro-pilot or pilot plant scale.

The main interest in ionic liquids was first to offer "green" alternatives to volatile organic solvents. But not only this: because of their unique set of physico-chemical properties they are very different from conventional organic solvents. They may give the opportunity to promote reactions that are not possible in other solvents. For example, they offer a nonaqueous environment to substrates and can be poorly miscible with organic compounds (cf. Section 5.2.1). But one would expect more than just a physical solvent: "new chemistry" may be foreseen. Indeed, it has been proven that the nature of the ionic liquid may influence the outcome of chemical reactions [41]. In many cases, ILs contribute to improving reaction rates and selectivities. This is the case for biphasic systems in which the organic reactants are more soluble in the catalyst ionic phase than are the products of reaction intermediates. Consecutive side reactions of products can be minimized (cf., for example, Sections 5.2 or 5.3).

Product separation can be made easier and less energy consuming. ILs may also stabilize catalysts and prevent their deactivation, leading to less catalyst consumption and waste. The overall process may be simplified if the ILs act simultaneously as solvent *and* as catalyst.

From an engineering point of view, ILs offer a huge potential (similar to the aqueous variants as described in Chapter 2) for separating reaction products and recycling the catalyst. New reactors may be smaller than in conventional homogeneous catalysis (e.g., the Difasol or Basil processes), which contributes to making processes "environmentally friendlier".

Their physical properties, such as viscosity, density, thermal stability, or surface tension, are important to consider during the design phase of new processes. The demonstration of the thermal and chemical stability of ILs and the recyclability can only be proven through continuous pilot plant runs. On the other hand, the sensitivity to feed impurities that can accumulate in the ionic phase requires feed pretreatments or guard beds for some scheduled applications.

Examples of commercial applications are scarce up to now (cf. Section 5.3; use of ILs has been considered for a series of specific questions). The scaling-up of IL synthesis procedures is normally without problems; however, the commercialization and/or transport of the ionic liquids raise the question of their registration (EINECS for Europe or equivalents; see Section 5.4). Disposal and recycle of ILs are important concerns and have to be considered on a case-by-case basis. And: "Ionic liquids are not always green" – as has been stated by Rogers et al. [42]. From the standpoint of "life cycle assessment" and "hazard analysis" ILs are clearly not recommendable for industrial use, especially if those with PF_6 or BF_4 as anions are concerned. And it is obviously no wonder that recent new developments such as BMIM octylsulfate have been emphasized as "even greener" ionic liquids [43].

Supercritical (sc) fluids have received considerable attention recently as new reaction media for chemical syntheses [30]. In particular, carbon dioxide in its supercritical state ($scCO_2$; $T_c = 31\,°C$, $p_c = 7.37$ MPa) appears to be a very attractive "green" solvent, owing to its lack of toxicity and ecological hazards, the mild critical data, and the unique combination of gas-like and liquid-like properties. The possibility of tuning its solubility properties by variation of pressure and temperature gives access to a wide range of fascinating applications. In particular, it opens the way to new methodologies for the immobilization of transition metal catalysts either solely with CO_2 or in multiphase combinations [32, 33]. In multiphase approaches, $scCO_2$ can be envisaged either as the substrate/product phase (mobile phase) or as the catalyst phase (stationary phase) [44]. The existent technology platform for use of CO_2 in natural product extraction and materials processing may serve as a promising basis for the development of integrated production schemes combining reaction and separation stages. So far, only one application of (heterogeneous) catalysis in $scCO_2$ phase operation has been described (see Section 6.5).

As a possible alternative to immobilized or heterogenized homogeneous catalysts, the binding of metal complexes to *soluble polymers* and other colloidally dispersed systems is interesting. The concept of binding catalytically active metal complexes to soluble metal complexes was brought forward by Manassen as early as the 1970s [45]. First results of homogeneously catalyzed reactions with rhodium complexes bound to phosphine-modified soluble polystyrenes were communicated by Bayer and Schurig soon afterward [2a]. It was not until ten years later that Bergbreiter initiated a broader investigation of the topic [2f], and since the mid-1990s the field has begun to attract a wider interest in academia as well as industry.

The separation of polymer-bound catalysts from a (single-phase, homogeneous) reaction solution after catalysis by means of properties specific to macromolecules is an alternative to homogeneous two- or multiphase catalysis. Such specific properties, allowing for differentiation between the polymer-bound catalysts and low molecular weight compounds in the reaction mixture (products, substrates, and solvent), can be the dependence of polymer solubility on solvent composition or temperature. Another option is the separation based on the large difference in "size" of the dissolved species by means of ultrafiltration. These approaches, the subject of Chapter 7, can be differentiated from the use of polymers as mere solubility-impeding groups in ligands for two- or multiphase catalysis. An example of the latter is the replacement of the sulfonate groups in the archetypical TPPTS by nonionic poly(ethylene glycol) chains.

In linear polymers, catalyst binding can occur by coordination of the metal centers with two functional end groups or with functional moieties (S) pendant to the polymer backbone chain (Figure 12). The latter approach obviously offers the advantage of higher possible loadings. A somewhat less common case is the

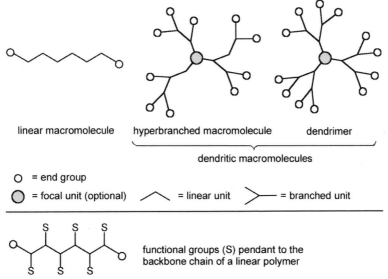

linear macromolecule   hyperbranched macromolecule   dendrimer

dendritic macromolecules

○ = end group
⬤ = focal unit (optional)   ⌃ = linear unit   ⪢— = branched unit

functional groups (S) pendant to the
backbone chain of a linear polymer

Figure 12 Schematic structures of linear and branched polymers.

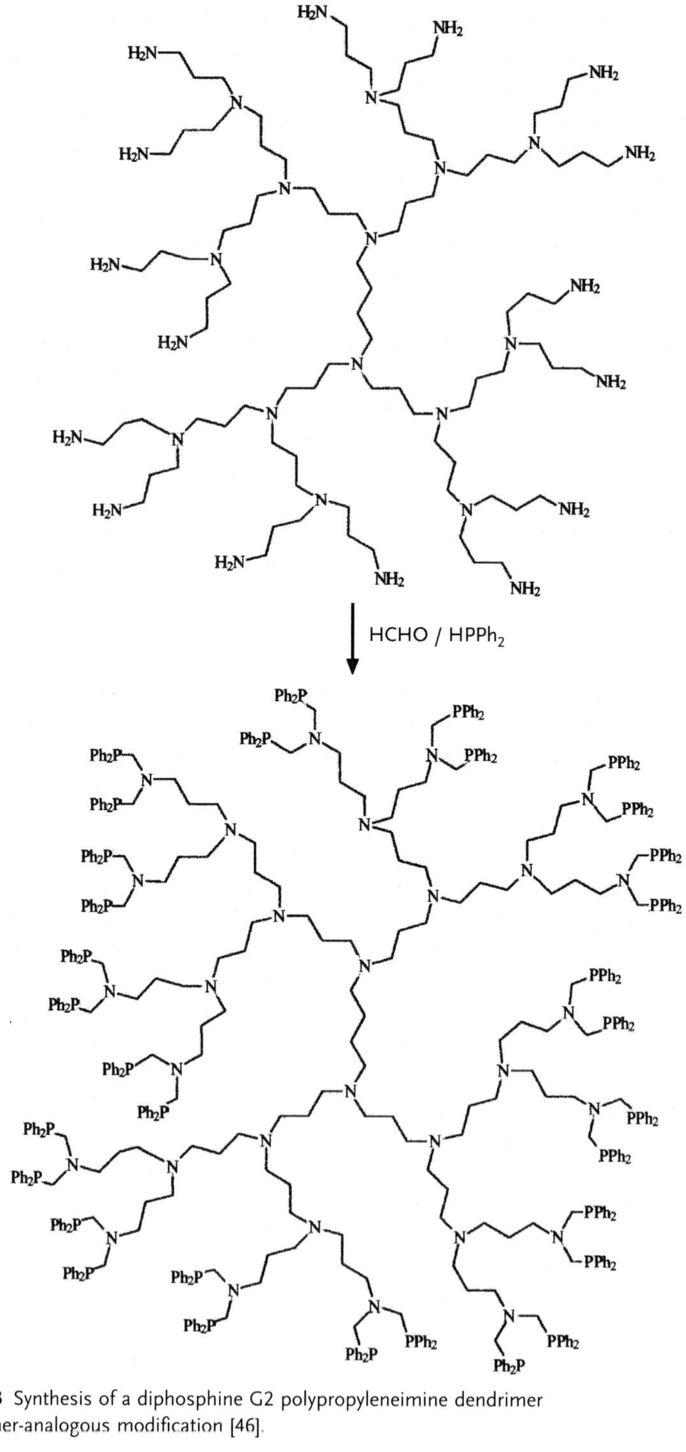

Figure 13 Synthesis of a diphosphine G2 polypropyleneimine dendrimer by polymer-analogous modification [46].

coordination of the metal centers by heteroatoms which are directly part of the polymer backbone. The most prominent example of the latter are (linear or branched) polyethylene imines with $-CH_2CH_2NR-$ repeat units.

A variety of linear polymers have been employed as catalyst supports with functional groups serving as ligands covalently attached as end groups or to the polymer backbone, e.g., polystyrene, low molecular weight polyethylenes, and also polymers which can be water-soluble such as poly(ethylene glycol)s or poly(n-alkyl-acrylamides). In addition to these organic polymers, polymers with an inorganic backbone have also been studied, most notably polysiloxanes.

Concerning polymer synthesis in general, the functional groups which will represent the ligands for the metal center at a later stage in catalysis can be introduced by homo- or copolymerization of corresponding substituted monomers, or by post-polymerization functionalization reactions of preformed polymers (also termed polymer-analogous reactions). Whereas linear polymers have been employed mostly, recently highly branched dendrimers (Figure 13) have also aroused considerable interest as soluble supports for metal complex catalysts.

By contast to dendrimers, hyperbranched polymers can be obtained in one-step procedures. As a drawback, molecular weight distributions are often extremely broad. However, certain hyperbranched polymers can be prepared with reasonably narrow molecular distributions if suitable reaction conditions are employed. Hyperbranched polyethyleneimine is produced on a large scale industrially (e.g., Lupasol from BASF). Hyperbranched polyesteramides with, e.g., terminal OH groups (tradename Hybrane® by DSM) and hyperbranched polyglycerol, a polyether-polyol with terminal OH groups, are available as a specialty product on a kilogramm scale currently. For molecular weights around M_n 5000 g mol^{-1}, polyethyleneimine and polyglycerol are available routinely with polydispersities of $M_w/M_n \leq 1.3$; higher molecular weight samples are more broadly distributed.

The attachment of metal complexes to the polymer most often occurs via coordinating functional groups (ligands) bound to the polymer in a covalent fashion as outlined in the earlier chapters of this book, but various types of noncovalent attachment are also well documented. The latter can be achieved, for instance, by means of electrostatic interactions, physisorption by amphiphilic polymer micelles (either as common association micelles or as unimolecular micelles), by hydrogen bonding, or by specific interactions of proteins with a molecule (Figure 14).

Noncovalent attachment can offer the advantage of lower synthetic effort in the catalyst preparation. On the other hand, it can be assumed that the resulting catalysts will often be restricted to a comparatively narrow range of organic solvents for reasons of solubility (electrostatic interactions) or leaching (physisorption and hydrogen bonding), and an enhanced sensitivity to temperature and changes in the solvent composition of the latter two types of binding compared with covalent attachment or electrostatic interactions must be considered.

Metal colloids, i.e., colloidally stable dispersions of metal particles in the size range 1–10 nm, are often considered to be an intermediate between classical homogeneous and heterogeneous catalysts. On the one hand, like heterogeneous catalysts such colloids contain more than one phase, i.e., the solid nanoparticle

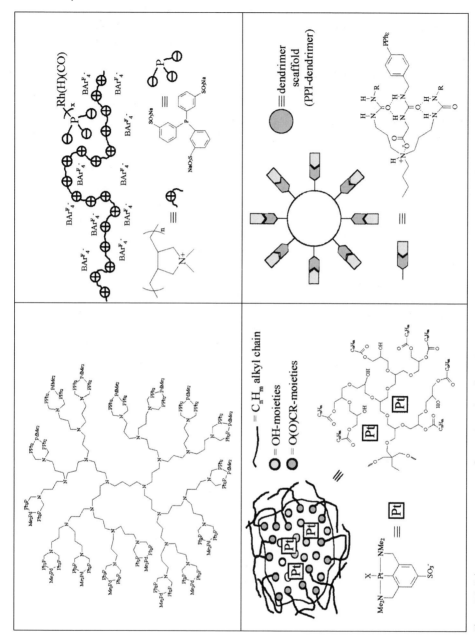

Figure 14 Schematic representation of various modes of attachment of catalytically active complexes to soluble polymers.

dispersed in a liquid medium. On the other hand, in terms of practical handling they are more similar to homogeneous catalysts in resembling single-phase "solutions". In terms of catalytic reactivity, metal colloids are known to catalyze a variety of reactions. Most interest to date has certainly been devoted to catalytic hydrogenation with metal colloids. Generally speaking, the reactivities and activities observed are roughly similar to hydrogenations with standard heterogeneous catalysts, albeit in detail colloids can offer advantages in some cases. In terms of reaction mechanisms, hydrogenation by soluble metal colloids does not appear to differ from the well-known mechanism of hydrogenation on metal surfaces of heterogeneous catalysts.

By comparison to homogeneous catalysts based on metal complexes, the properties of which can be tuned via the ligands coordinating to the metal center, the scope of reactions and viable substrates is certainly less broad for metal colloids.

Given the similarities to homogeneous complexes in some aspects, involving catalyst recovery, and the major importance of soluble polymers for the stabilization of metal colloids, the latter will also be considered throughout Chapter 7 where appropriate.

It is likely that the spread of biphase (multiphase) processes as described here will increase the proportion of homogeneously catalyzed reactions and hence the importance of homogeneous catalysis in general. It will then also be possible to demonstrate in full the great advantages of homogeneous catalysis over the rather empirical methods of heterogeneous catalysis and answer Heinemann's 1971 question, "Homogeneous and heterogeneous catalysis – common frontier or common territory?" [47], clearly in favor of the homogeneous version.

References

1 Ferenc Joó's preface to [6a].

2 For example, there are the publications by E. Bayer and N. Schurig (a) *CHEMTECH* **1976**, March, 212; (b) *Angew. Chem. Int. Ed. Engl.* **1975**, *14*, 493) or (c) Mathur et al. (N. K. Mathur, C. K. Narang, C. K. Williams, R. E. Williams, *Polymers as Aids in Organic Chemistry*, Academic Press, New York **1980**), by (d) Whitehurst (*CHEMTECH* **1980**, p. 44), or by (e) Bailey and Langer (*Chem. Rev.* **1981**, *81*, 109) with proposals to introduce polymer ligands or polymer-supported ligands as a way to make homogeneous catalysts separable without changing their activity [cf. the surveys Bergbreiter {(f) D. E. Bergbreiter, *J. Org. Chem.* **1986**, *51*, 4754; *CHEMTECH* **1987**, Nov., 686 and *Catal. Today* **1998**, *42*, 389}].

3 C. C. Tzschuke, C. Markert, W. Bannwarth, S. Roller, A. Hebel, R. Haag, *Angew. Chem. Int. Ed.* **2002**, *41*, 3964.

4 G. Duve, O. Fuchs, H. Overbeck, *Lösemittel Hoechst*, 6th ed., Hoechst AG, Frankfurt/M. **1976**, p. 49.

5 C. Reichardt, *Solvents and Solvent Effects in Organic Chemistry*, 3rd ed., Wiley-VCH, Weinheim **2003**.

6 (a) D. J. Adams, P. J. Dyson, S. J. Tavener, *Chemistry in Alternative Reaction Media*, Wiley, Chichester **2004**; (b) A. R. Katritzky, D. C. Fara, H. Yang, K. Tämm, *Chem. Rev.* **2004**, *104*, 175; (c) C. A. Eckert, C. L. Liotta, J. S. Brown, *Chem. & Ind.* **2000**, 7 Feb., 94; (d) F. Kong, J. Jiang, Z. Jin, *Catal. Lett.* **2004**, *96*, 63; (e) D.-H. Chang, D.-Y. Lee, B.-S. Hong, J.-H. Choi, C.-H. Jun, *J. Am. Chem. Soc.* **2004**, *126*, 424.

7 J. Lu, M. J. Lazzaroni, J. P. Hellett, A. S. Bommarius, C. L. Liotta, C. A. Eckert, *Ind. Eng. Chem. Res.* **2004**, *43*, 1586.

8 (a) S. S. Pröckl, W. Kleist, M. A. Gruber, K. Köhler, *Angew. Chem. Int. Ed.* **2004**, *43*, 1881; *Catal. Lett.* **2004**, *94*, 177; (b) M. Gruber, M. Wagner, R. Heidenreich, J. G. E. Krauter, N. Coskun, K. Köhler, *Catal. Lett.* **2004**, *94*, 177.

9 E. Lindner, T. Schneller, F. Auer, H. A. Mayer, *Angew. Chem. Int. Ed.* **1999**, *38*, 2154.

10 P. W. N. M. van Leeuwen, P. C. J. Kamer, J. N. H. Reek, *Cattech* **2000**, *3*(2), 64.

11 (a) T. Prinz, W. Keim, B. Driessen-Hölscher, *Angew. Chem.* **1996**, *108*, 1835; (b) Y. Zhang, Z.-S. Mao, J. Chen, *Catal. Today* **2002**, *74*, 23; (c) C. Yang, X. Bi, Z.-S. Mao, *J. Mol. Catal. A:* **2002**, *187*, 35; (d) R. M. Deshpande, M. M. Divekar, A. N. Mahajan, R. V. Chaudhari, *J. Mol. Catal. A:* **2004**, *211*, 49.

12 (a) R. J. Stokes, D. F. Evans, *Fundamentals of Interfacial Engineering*, Wiley-VCH, Weinheim **1997**; (b) A. W. Neumann, J. K. Spelt (Eds.), *Applied Surface Thermodynamics*, Surfactant Surface Sciences, Vol. 63, Dekker, New York **1996**; (c) H.-J. Butt, K. H. Graf, M. Kappl, *Physics and Chemistry of Interphases*, Wiley-VCH, Weinheim **2003**; (d) H.-M. Yang, H.-S. Wu, *Catal. Rev.* **2004**, *54*(3+4), 463; (e) A. G. Volkov, *Interfacial Catalysis*, Dekker, New York **2003**; (f) E. Lindner, T. Schneller, F. Auer, H. A. Mayer, *Angew. Chem. Int. Ed.* **1999**, *38*, 2155.

13 (a) M. P. Dudukovic, F. Larachi, P. L. Mills, *Catal. Rev.* **2002**, *44*, 123; (b) A. G. Volkov (Ed.), *Interfacial Catalysis*, Dekker, New York **2003**.

14 (a) R. T. Baker, W. Tumas, *Science* **1999**, *284*, 1477; (b) C. A. Eckert, C. L. Liotta, J. S. Brown, *Chem. & Ind.* **2000**, 7 Feb., 94.

15 R. R. Hartley, *Supported Metal Complexes*, Reidel, Dordrecht **1985**.

16 (a) Asahi Kasei KKK (K. Yamashita, H. Obana, I. Katsuta), EP 0.552.809 (**1992**); (b) *Stud. Surf. Sci. Catal.* **1994**, *92*, 375; (c) *Catal. Today* **2002**, *74*, 5; (c) R. R. Davda, J. W. Shabaker, G. W. Huber, R. D. Cortright, J. A. Dumesic, *Appl. Catal. B:* **2003**, *43*, 13 and *J. Catal.* **2004**, *222*, 180; (d) M. Marchetti, C. Botteghi, S. Paganelli, M. Taddei, *Adv. Synth. Catal.* **2003**, *345*, 1229; (e) H. Nur, S. Ikeda, B. Ohtani, *J. Catal.* **2001**, *204*, 402; (f) J. R. Anderson; E. M. Campi, W. R. Jackson, Z. P. Yang, *J. Mol. Catal. A:* **1997**, *116*, 109; (g) D. Sloboda-Rozner, P. Witte, P. L. Alsters, R. Neumann, *Adv. Synth. Catal.* **2004**, *346*, 339; (h) M. Salaices, B. Serrano, H. I. de Lasa, *Chem. Eng. Sci.* **2004**, *59*, 3.

17 B. Cornils, W. A. Herrmann (Eds.), *Applied Homogeneous Catalysis with Organometallic Compounds*, 2nd ed., Wiley-VCH, Weinheim **2002**, p. XIII.

18 (a) B. Cornils, W. A. Herrmann, R. Schlögl, C.-H. Wong (Eds.), *Catalysis from A to Z*, 2nd ed., Wiley-VCH, Weinheim **2003**, "catalytic cycle", "regeneration" entries; (b) B. Tooze, D. J. Cole-Hamilton (Eds.), *Recovery and Recycling in Homogeneous Catalysis*, Kluwer, Dordrecht **2005**.

19 J. Manassen in *Catalysis: Progress in Research* (Eds.: F. Bassolo, R. L. Burwell), Plenum Press, London **1973**, pp. 177, 183.

20 (a) F. Joó, M. T. Beck, *Magy. Kém. Folyóirat* **1973**, *79*, 189 as cited in [9b]; (b) F. Joó, *Aqueous Organometallic Catalysis*, Kluwer, Dordrecht **2001**.

21 G. W. Parshall, *J. Am. Chem. Soc.* **1972**, *94*, 8716.

22 (a) W. Keim, *Chem. Ing. Tech.* **1984**, *56*, 850; (b) W. Keim, *Stud. Surf. Sci. Catal.* **1989**, *44*, 321; (c) Shell (W. Keim, T. M. Shryne, R. S. Bauer, H. Chung, P. W. Glockner, H. van Zwet), DE 2.054.009 (**1969**).

23 (a) Rhône-Poulenc (E. G. Kuntz), FR 2.314.910 (**1975**); FR 2.349.562 (**1976**); FR 2.338.253 (**1976**); FR 2.366.237 (**1976**); (b) B. Cornils, E. G. Kuntz, *J. Organomet. Chem.* (**1995**), *502*, 177.

24 B. Cornils, W. A. Herrmann, *Aqueous-Phase Organometallic Catalysis*, 2nd ed., Wiley-VCH, Weinheim **2004**.

25 B. Cornils, *Org. Prod. Res. & Dev.* **1998**, *2*, 121.

26 P. Cintas, *CEN* **1995**, March 20, p. 4.

27 E.g., H. W. Roesky, M. G. Walawalkar, R. Murugavel, *Acc. Chem. Res.* **2001**, *34*, 201.

28 J. A. Gladysz, D. P. Curran, I. T. Horváth (Eds.), *Handbook of Fluorous Chemistry*, Wiley-VCH, Weinheim **2004**.

29 P. Wasserscheid, T. Welton (Eds.), *Ionic Liquids in Synthesis*, Wiley-VCH, Weinheim **2003**.

30 (a) P. G. Jessop, W. Leitner (Eds.), *Chemical Syntheses Using Supercritical Fluids*, Wiley-VCH, Weinheim **1999**; (b) W. Leitner, *Acc. Chem. Res.* **2002**, *35*, 746.

31 (a) P. A. Lovell, M. S. El-Asser (Eds.), *Emulsion Polymerization and Emulsion Polymers*, Wiley, Chichester **1997**; (b) D. Distler (Ed.), *Wäßrige Polymerdispersionen*, Wiley-VCH, Weinheim **1999**; (c) S. Mecking, A. Held, F. M. Bauers, *Angew. Chem. Int. Ed.* **2002**, *41*, 545; (d) D. E. Bergbreiter, *Catal. Today* **1998**, *42*, 389.

32 (a) P. B. Webb, D. J. Cole-Hamilton, *Chem. Commun.* **2004**, 612; (b) J. Zhu, A. Robertson, S. C. Tsang, *Chem. Commun.* **2002**, 2044; (c) Y. Hu, D. J. Birdsall, A. M. Stuart, E. G. Hope, J. Xiao, *J. Mol. Catal. A:* **2004**, *219*, 57.

33 (a) University of California, US 6.479.708 (**1999**); (b) M. McCarthy, H. Stemmer, W. Leitner, *Green Chem.* **2002**, *4*, 501; (c) Y. Kayaki, T. Suzuki, T. Ikariya, *Chem. Lett.* **2001**, 1016; (d) A. A. Gakin, B. G. Kostyuk, V. V. Lunin, M. Poliakoff, *Angew. Chem. Int. Ed.* **2000**, *39*, 2738; (e) G. Ragagnin, P. Knochel, *Synlett* **2004**, 951.

34 (a) P. J. Dyson, D. J. Ellis, W. Henderson, G. Laurenczy, *Adv. Synth. Catal.* **2003**, *345*, 216; (b) B. Wang, Y.-R. Kang, L.-M. Yang, J.-S. Suo, *J. Mol. Catal. A:* **2003**, *203*, 29; (c) G.-T. Wei, Z. Yang, C.-Y. Lee, H.-Y. Yang, C. R. C. Wang, *J. Am. Chem. Soc.* **2004**, *126*, 5036; (d) P. M. P. Gois, C. A. M. Afonso, *Eur. J. Org.* **2003**, 3798 and *Tetrahedron Lett.* **2003**, *44*, 6571.

35 (a) T. Mathivet, E. Monflier, Y. Castanet, A. Mortreux, *Tetrahedron* **2002**, *58*, 3877; (b) J. S. Kim, R. Datta, *AIChE J.* **1991**, *37*, 1675; (c) E. L. V. Goetheer, A. W. Verkerk, L. J. P. van den Broeke, E. de Wolf, B. J. Deelman, G. van Koten, J. T. F. Keurentjes, *J. Catal.* **2003**, *219*, 126.

36 J. Bodis, T. E. Müller, J. A. Lercher, *Green Chem.* **2003**, *5*, 227; V. Neff, T. E. Müller, J. A. Lercher, *J. Chem. Soc.,Chem. Commun.* **2002**, *8*, 906.

37 A. Behr, R. Ott, B. Turkowski, Poster presented to the XXXVI Jahrestreffen Deutscher Katalytiker, 19–21.3.2003 at Weimar, Preprints.

38 Complete references are given in [24], pp. 644 ff.

39 (a) I. T. Horváth, J. Rábai, *Science* **1994**, *266*, 72; (b) Exxon (I. T. Horváth, J. Rábai), US 5.463.082 (**1995**).

40 (a) I. T. Horváth, G. Kiss, R. A. Cook, J. E. Bond, P. A. Stevens, J. Rábai, E. J. Mozeleski, *J. Am. Chem. Soc.* **1998**, *120*, 3133; (b) L. J. Alvey, R. Meier, T. Soós, P. Bernatis, J. A. Gladysz, *Eur. J. Inorg. Chem.* **2000**, 1975; (c) H. Jiao, S. Le Stang, T. Soós, R. Meier, K. Kowski, P. Rademacher, L. Jafarpour, J.-B. Hamard, S. P. Nolan, J. A. Gladysz, *J. Am. Chem. Soc.* **2002**, *124*, 1516.

41 (a) M. J. Earle, S. P. Katdare, K. R. Seddon, *Org. Lett.* **2004**, *6*, 707; (b) D. Sémeril, H. Olivier-Bourbigou, C. Bruneau, P. H. Dixneuf, *Chem. Commun.* **2002**, 146.

42 R. M. Swatloski, J. D. Holbrey, R. D. Rogers, *Green Chem.* **2003**, *5*, 361.

43 P. Wasserscheid, R. van Hal, A. Bösmann, *Green Chem.* **2002**, *4*, 400.

44 P. G. Jessop, T. Ikariya, R. Noyori, *Chem. Rev.* **1999**, *99*, 475.

45 J. Manassen, *Plat. Met. Rev.* **1971**, *15*, 142.

46 M. T. Reetz, G. Lohmer, R. Schwickardi, *Angew. Chem. Int. Ed. Engl.* **1997**, *36*, 1526.

47 H. Heinemann, *CHEMTECH* **1971**, *5*, 286.

2
Aqueous-Phase Catalysis

Boy Cornils and Wolfgang A. Herrmann (Eds.)

Multiphase Homogeneous Catalysis
Edited by Boy Cornils and Wolfgang A. Herrmann et al.
Copyright © 2005 Wiley-VCH Verlag GmbH & Co. KGaA, Weinheim
ISBN: 3-527-30721-4

2.1
Introduction

Boy Cornils

Figures 1 and 2 of the Introduction to this book (Chapter 1) have given the orientation of aqueous-phase catalysis within the overall picture of homogeneous catalysis. Together with the other possibilities of multiphase operation as defined there, the aqueous option is one step toward the release of homogeneous catalysis from its system-immanent restriction, especially the necessity of costly catalyst recycling. It is no wonder that these limits were particularly disturbing in the upcoming and rapidly growing hydroformylation reaction [1] (current yearly output approx. 9 MM tons [2]). Thus it is also no wonder that the impressive advantages and the tremendous progress of the biphasic processes were first used commercially with aqueous- and aqueous/organic-phase operation [3]. Table 1 demonstrates that when using the aqueous-phase technique the expenditure for catalyst removal and recycling of homogeneous and of heterogeneous operations are for the first time at the same level.

The advantages of anchoring homogeneous catalysts on "liquid" supports (and thus the biphasic techniques) are obvious and render possible – besides the organic/organic operation; cf. Chapter 3 – the utilization of the principle which was expressed by Manassen ([5]:

"... the use of two immiscible liquid phases, one containing the catalyst and the other containing the substrate ...",

Table 1 Catalyst removal in homogeneous and heterogeneous catalysis (according to [4]).

	Homogeneous catalysis		Heterogeneous catalysis	
	In general	*Aqueous-phase*	*Suspension*	*Fixed bed*
Separation of catalyst/product	Expensive[a]	Decantation	Filtration	Simple
Additional equipment required?	Yes	No[b]	No[b]	No
Catalyst recycling	Expensive	Easy	Easy	Not necessary
Cost of catalyst losses	High	Minimal	Minimal	Minimal
Service life of catalyst	Just one cycle	——— Depending on conditions ———		

a) The normal recycling requires filtration or chemical/thermal deposition, distillation, evaporation, extraction, or similar means.
b) Except for decanter or filtrator.

Multiphase Homogeneous Catalysis
Edited by Boy Cornils and Wolfgang A. Herrmann et al.
Copyright © 2005 Wiley-VCH Verlag GmbH & Co. KGaA, Weinheim
ISBN: 3-527-30721-4

and

> "The two phases can be separated by conventional means and high degrees of dispersion can be obtained through emulsification. The ease of separation may be particularly advantageous in situations where frequent catalyst regeneration is required."

Figure 1 illustrates the enormous importance of the biphase technique for homogeneous catalysis: the aqueous catalyst solution is charged in the reactor together with the reactants A and B, which react to form the reaction products C and D. C and D are less polar than the aqueous catalyst solution and are therefore simple to separate from the aqueous phase (which is recycled directly into the reactor) in the downstram phase separator (the decanter).

The advantage of the "liquid support" water and of its high affinity toward homogeneous catalysts is evident. The catalyst is "heterogenized" with respect to the organic reaction products C and D and therefore can not only be separated from the products "in the other phase" (possibly including unconverted reactants A and B), but also immediately thereafter can start a new cycle of the catalytic cycle process. Aqueous biphase catalysis is therefore – intentionally – located between heterogeneous and homogeneous catalysis, as demonstrated in Figure 2 of Chapter 1. Special attention may be drawn to the somewhat confusing and ambiguous use of the terms "homogeneous" and "heterogeneous" in the context of homogeneous/heterogeneous *catalysis* and homoge-neous/heterogeneous *phase variation*: the catalytic system works mechanistically homogeneously although it is in heterogeneous phase.

The immobilization of the homogeneous catalyst with the aid of the liquid support water leads to appreciable technical – and consequently commercial [6] – simplifications, as illustrated for the example of large-scale oxo processes in Figure 2.

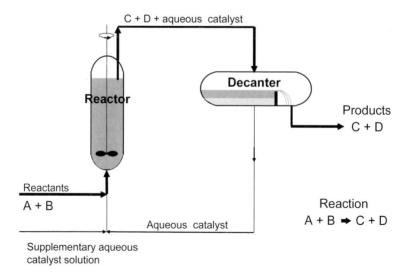

Figure 1 Principle of aqueous-phase catalysis illustrated for the reaction A + B → C + D.

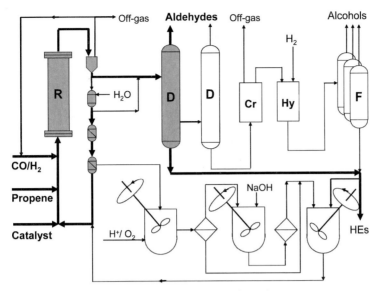

Figure 2 Schematic flowsheet of a "classical" oxo process, the earlier Ruhrchemie oxo process [7]. Those parts of the process and equipment which are dispensed with in the new RCH/RP process (with an aqueous catalyst system) are unshaded [6].

Table 1 predicts that, of the biphase processes, the aqueous version will attain particular importance because of the many advantages of water as the medium of the "liquid support". As a solvent, water has many anomalies (e.g., the density anomaly, being the only nontoxic and liquid "hydride" of the nonmetals, the pressure dependence of the melting point, the dielectric constant, etc.), and its two-/three-dimensional structure is still not well understood. The most important properties are listed below [8a]; recent results given in [8b–h].

1. Polar and easy to separate from apolar products; additionally, the polarity may influence (improve) reactivity (cf. Sections 2.2.1 and 2.2.2).
2. Nonflammable, not combustible, and nontoxic: a decisive advantage in terms of safety and occupational health (see Section 2.3.6).
3. Ubiquitous and widely available of suitable quality.
4. Odorless and colorless, making contaminations easy to recognize.
5. Forms a hexagonal two-dimensional surface structure, a tetrahedral three-dimensional molecular network, or clusters which influence the mutual (in)solubility significantly; chaotropic compounds lower the order by H-bond breaking.
6. High Hildebrand parameter as the unit of solubility of nonelectrolytes in organic solvents.
7. The density of 1.00 g cm^{-3} provides a sufficient difference from most organic substrates/products.

8. Very high dielectric constant.
9. High thermal conductivity, high specific heat capacity, and high evaporation enthalpy of water predestine it as a heat removal fluid; see Section 2.4.
10. Low refractive index n_D.
11. High solubility for many gases, specially CO_2.
12. Formation of hydrates and solvates with sometimes surprising properties (e.g., gas hydrates).
13. Highly dispersible and high tendency toward micelle and/or microemulsion formation, which can be stabilized by additives such as surfactants.
14. Amphoteric behavior in the Brønsted sense.
15. Advantageously influences chemical reactions.

For homogeneously catalyzed reactions of chemical compounds in a biphasic, aqueous manner, most of these properties are used advantageously. Except for the positive influence of water on the chemical reaction (no. 15; see also Sections 2.2.1 and 2.2.2) the peculiarities of water with direct significance for the aqueous-biphasic processes are the physiological (2, 4), economic (1, 3, 7), ecological (2, 3, 4), technical/engineering (1, 6, 7, 9, 11, 12, 13, 14), and special physical properties (1, 5, 6, 8, 10, 12, 14). The various properties have multiple effects and are mutually reinforcing. Astonishingly, the quality and significance of the properties listed above is debatable: some count the low solubility of organic substrates (and thus the basis for the aqueous-biphasic approach for homogeneously catalyzed processes) among the disadvantages. The different pros and cons of water as a reaction fluid are compiled in Table 2, which includes the alternative views of others on the whole picture.

Table 2 Pros and cons of water as liquid support [9].

Advantages	Disadvantages
Nonflammable	Large heat of evaporation[a]
Nontoxic	Difficult to detect in the case of leakage[b]
No smell[c]	Low solubility of many nonpolar substrates
Good separation with many organics	Hard to collect in the case of spills[d]
Cheap	No incineration of bleed streams[e]
Stabilization of certain organometallic complexes	Decomposition of water-sensitive compounds[f]

a) Which is an advantage within the economics of the heat compound.
b) Which – in contrast – is an advantage since contaminated water smells strongly in certain syntheses (for example, the manufacture of butyraldehydes).
c) See: the difference between "no smell" and the smell of "contaminated bleed streams is important and advantageous".
d) Which is true for all "solvents" and all organic liquids and is not a specific disadvantage of water.
e) Which in contrast is advantageous when containing the water-soluble catalysts: the system is self-extinguishing.
f) This is true but a matter of evaluation: nobody will recommend water in the case of decomposition of water-sensitive substrates or products.

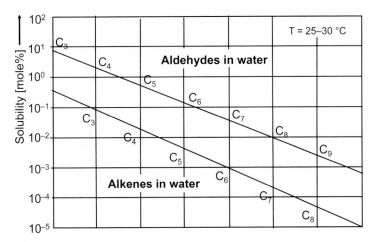

Figure 3 Solubility of alkenes and of the aldehydes obtained therefrom by hydroformylation in water [10].

Obviously, "misunderstandings continue to prevail", as van Leeuwen cautiously put it into words [9] although water does not ignite, does not burn, is odorless and colorless, and is ubiquitous: important prerequisites for the solvent of choice in homogeneous catalytic processes. The favorable thermal properties make water doubly exploited as a mobile support and as a heat-transfer fluid.

There is an important restriction regarding the general use of water in homogeneously catalyzed processes. This is that to attain adequate chemical reaction, there has to be a minimum solubility of the organic substrates in the aqueous catalyst phase. Normally, this solubility decreases with increasing carbon number of the substrates. This is illustrated in Figure 3 for the dependence of the solubilities of alkenes and aldehydes in water on their C number – thus an example is the homogeneously catalyzed conversion of alkenes to aldehydes (via oxo synthesis; cf. Section 2.4.1).

The differences in reactivity of the hydroformylation of the alkenes C_3 up to C_9 are readily explained by the solubility differences between the various olefins. This is also the case for other examples of aqueous-biphasic operation. All proposals to enhance the reactivity of higher alkenes by addition of solvents, co-ligands, co-solvents, tensides, counter ions, micelle-forming agents, surface-active ligands, etc. [11] (as a single measure or in combination: the record holder used three! [12]) are based on the improvement of the solubility of the feed alkene in the bulk of the aqueous catalyst solution. This is also important for the idea of reactivity, as shown in Figure 4.

On the one hand, the reaction can occur in the bulk of the liquid phase: the gaseous reactants have already become uniformly distributed in the liquid phase owing to the rapid mass transfer somewhat before the actual and slower chemical reaction commences. This model A contrasts with the idea of a relatively rapid chemical reaction as opposed to mass transport.

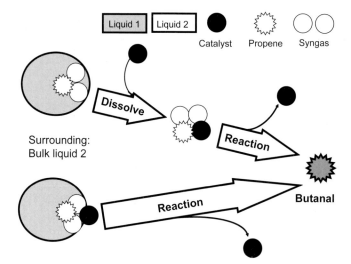

Figure 4 Comparison of the two models A (above: reaction in the bulk of the liquid) and B (below: reaction at the interfacial boundary) [13c].

The fact that the gaseous substrates react very quickly means that, in practice and according to model B, the reaction takes place at the phase boundary or in an interfacial layer with a relatively small thickness [13]. The latter has been proven and – via process modeling on the basis of appropriate kinetic models – renders possible a more optimal reactor and mixing design (cf. also Section 2.3.1). The dependence of the reactivity of aqueous systems on the solubility of the reactants in the aqueous catalyst solution is of appreciable importance for the problem of universal applicability.

Using water as the solvent not only has the advantage of employing a "mobile support" and hence of a de facto "immobilization" of the catalyst while retaining a homogeneous form of reaction, but also has positive repercussions on the environmental aspects of the different reactions to which it is applied, e.g., hydroformylation (see Section 2.4.1.1).

Aqueous processes will become more important in the future because of the immense advantages of this version of homogeneous catalysis in the multiphase version. In addition to the syntheses already utilized now, an enormous laboratory study has been done to convert conventional syntheses to the aqueous-multiphasic mode (see Section 2.5.2). A selection of recent work covering a multitude of reactions and and catalysts based on various transition metals is presented in Table 3.

Various protocols describe aqueous-phase processes with heterogeneous catalysts such as the reforming of oxygenated hydrocarbons over Sn-modified Ni catalysts [22a], the epoxidation of alkenes over zeolites [22b], etc. [22c–i,l,m].

Some authors reported commercial applications [22n] but this is rather uncertain, and in any case this would not be a "green application" as claimed.

Obviously, there are beneficial effects but the reason for this combination is not that heterogeneous catalysis is made much easier by bi- or multiphase operation.

Table 3 Research work in aqueous–phase reactions.

Type of reaction/Metal involved	References
Additions to alkynes/Pd,Rh,organo[a]	[19c], [20i], [21c,l]
Aldolization/Sc,In,Cu,Ln	[14a–d]
Alkylation/K,Pd,organo[a]	[14e,f], [18l]
Alkyl radical addition/In	[21o]
Alkynylation of acid chlorides/Pd	[23s]
Allylation/Ru,Pd,Zn	[14d,y,z], [17b], [20x]
Allylic amination	[23k]
Allylic benzoylation/Cu	[21m]
Amination/Pd	[14g], [21a]
Carbonylation/Pd,Rh	[14h–l]
Condensation reactions/heteropolyacids, Mo,W	[20s], [21d]
Conjugate addition/Rh	[20w]
Copolymerization/Pd	[20f], [23y]
Cross-couplings/Pd	[23v]
Cyanation/Pd,Ln	[14m,n]
Cycloaddition, cyclotrimerization/Co,Rh	[14o,p], [17v], [20m,n]
Dihydroxylation/Os	[17m], [20v]
Electrocatalysis/Mo	[21i]
Ene reactions/Pd,Rh,Ni	[19b]
Epoxidation/W,Re,Mn,Cr,Al,ionic liquids	[14q–u], [20o,p], [23n,z]
Hydration/Au,Pt,Rh	[14v], [17y], [18b]
Hydroaminomethylation/Rh,Ir	[14w]
Hydroamination/Ni	[20r]
Hydrocarboxylation/Pd	[14h], [15a–e], [18d]
Hydrodesulfurization/Ru,Rh	[15f,g]
Hydroformylation/Rh, Rh-α-CD	[23p,q,w]
Hydrogenation/Ru,Rh,Pd,Ir, Pt–Pd	[13a,b], [15h–t], [18m], [19a,d], [20j]
Hydrogenation of CO_2/Ru,Rh	[20a,u], [21j]
Hydrosilylation/Pt	[15u]
Hydroxycarbonylation/Pd,Ni	[20t]
Hydroxymethylation/organo[a]	[23l]
Isomerization/Ni,Ru,Rh,Pd,Ir	[15v,w], [21b,k,n,q]

Table 3 (continued)

Type of reaction/Metal involved	References
Metallo–ene reactions	[19b]
Metathesis/Ru,In	[15x–z], [16a]
Nucleotides/Fe	[23u]
Oligomerization/Pd,Al	[15c], [16b–f]
Oxidation/Ru,Pd,Os,W,Mn,Cu	[16g–l], [18a,e,h], [20b,d,q]
Oxidative dimerization/Pd	[20l]
Phenol synthesis/Fe	[16m]
Polymerization/Cu,Ti,Ni,Pd,Rh,Co,Ru	[16n–r], [21p]
Syntheses of various heterocycles/Rh,Ni	[16s], [18c,k]
Telomerization/Pd,Ni	[16t–v], [20e,y]
Thiolysis/In	[16w]
Transfer hydrogenation/Rh	[20k]
Use of carbene complexes/Ru	[23x]
Wacker reaction/Pd	[16x]
Water–gas shift reaction/Ru	[17t], [23r]
Examples of name reactions	
• Barbier-type allylation/In,Sn	[16y,z]
• Claisen rearrangement/Al	[17a]
• Diels–Alder/Si,Cu	[14d], [17c], [18j]
• Friedel–Crafts/Sc,Ga	[17d], [20h]
• Grignard/Sn, Sn–Rh	[17e]
• Heck/Pd,Rh	[17f–i], [18g], [23o]
• Kharasch/Pd	[17j]
• Knoevenagel/La	[23m]
• Mannich/Zn,Brønsted acid,NaOTf	[14d], [20c,z]
• Michael/Sc,Ag	[14d], [17k]
• Pauson–Khand/Rh	[14x], [17z]
• Reformatsky/Zn	[17l]
• Sharpless/Os	[17m–o], [18i]
• Sonogashira/Pd	[21f]
• Suzuki/Pd, organo[a]	[17p,q,x], [21g,r]
• Tsuji–Trost/Pd	[23t]
• Ullmann/Pd,Rh,Ni,Cu	[17u,w], [20g], [21e]
• Wittig/organic[a]	[17r]

a) Organocatalysts, cf. [17s].

The same is true for combinations of two or more measures to ensure biphasic homogeneous catalysis, such as aqueous-biphase + supercritical solvents [23a–d], aqueous-biphase + fluorous solvents [23e–h], supercritical solvents + ionic liquids [23i], fluorous + supercritical solvents [23j], or other combinations. By the way, some papers argue a new concept for the separation of heterogeneous catalysts by usage of two different and immiscible solvents. Against gravity, surface polarity modified catalysts are located only in one of the two phases. The catalyst phase can be separated by simple decantation [22j,k].

The application of aqueous-phase processes requires the rule over some new techniques. Systematic studies on the examples hydrogenation and hydroformylation have shown the vast improvements which can be achieved through variation of the ligands (cf. Section 2.2.3.2). Sections 2.2.3.1 and 2.4.8 prove the great variety of the central atoms that have been applied. Especially in the case of the ligands, it is true that a multitude of water-soluble compounds have been developed (and applied) but it is only TPPTS which is both active and inexpensive. With more exotic ligands, activity, productivity, and selectivity may be enhanced further but at the expense of a more costly and uneconomic operation. This is not true for enantioselective reactions, which are just to be applied. Last but not least, aqueous-phase operation requires some more techniques from the chemical engineering side. Except for the proven oxo process (see Sections 2.4.1.1.1 and 2.4.1.1.3 [24]) the developments under way will be described in Sections 2.3.1 through 2.3.5.

References

1 B. Cornils, in J. Falbe (Ed.), *New Syntheses with Carbon Monoxide*, Springer, Berlin **1980**.

2 H.-W. Bohnen, B. Cornils, *Adv. Catal.* **2002**, *47*, 1.

3 B. Cornils, W. A. Herrmann (Eds.), *Aqueous-Phase Organometallic Catalysis*, 2nd ed., Wiley-VCH, Weinheim **2004**.

4 J. Falbe, H. Bahrmann, *Chemie Unserer Zeit* **1981**, *15*, 87.

5 J. Manassen, in *Catalysis: Progress in Research* (Eds.: F. Bassolo, R. L. Burwell), Plenum Press, London **1973**, pp. 177, 183.

6 See [1], p. 162.

7 See [3], p. 10.

8 (a) D. J. Adams, P. J. Dyson, S. J. Tavener, *Chemistry in Alternative Reaction Media*, Wiley, Chichester **2004**; (b) B.-Q. Ma, H.-L. Sun, S. Gao, *Angew. Chem. Int. Ed.* **2004**, *43*, 1374; (c) T. Köddermann, F. Schulte, M. Huelsekopf, R. Ludwig, *Angew. Chem. Int. Ed.* **2003**, *42*, 4904; (d) L. A. Estroff, A. D. Hamilton, *Chem. Rev.* **2004**, *104*, 1201; (e) S. M. Biros, E. C. Ullrich. F. Hof, L. Trembleau, J. Rebek, *J. Am. Chem. Soc.* **2004**, *126*, 2870; (f) K. R. Asmis, *Science* **2003**, *299*, 1375; (g) W. H. Robertson, *Science* **2003**, *299*, 1367; (h) H. Fuhrmann, T. Dwars, G. Oehme, *Chem. Unserer Zeit* **2003**, *37*, 40.

9 P. W. N. M. van Leeuwen, P. C. J. Kamer, J. N. H. Reek, *Cattech* **2000**, *3*(2), 64.

10 (a) S. Partzsch, former Hoechst AG, private communication; (b) C. McAuliffe, *J. Phys. Chem.* **1966**, *70*, 1267; (c) R. M. Stephenson, *J. Chem. Eng. Data* **1993**, *38*, 630; (d) P. L. Davies, *J. Gas Chromatogr.* **1968**, *6*(10), 518.

11 (a) E. Lindner, T. Salesch, S. Brugger, F. Hoehn, P. Wegner, H. A. Mayer, *J. Organomet. Chem.* **2002**, *641*, 165; (b) Y. Huang, L. Min, Y. Li, R. Li, P. Cheng, X. Li, *Catal. Commun.* **2002**, *3*, 71; (c) M. Haumann, H. Yildiz, H. Koch, R. Schomäcker, *Appl. Catal. A:* **2002**, *236*, 173; (d) M. Li, Y. Li, H. Chen, Y. He, X. Li, *J. Mol. Catal. A:* **2003**, *194*, 13;

(e) J. T. Sullivan, J. Sadula, B. E. Hanson, R. J. Rosso, *J. Mol. Catal. A:* 2002, *214*, 213;
(f) P. Kalck, M. Dessoudeix, S. Schwarz, *J. Mol. Catal. A:* 1999, *143*, 41; (g) H. Chen,
Y. Li, J. Chen, X. Li, *Catal. Today* 2002, *74*, 131; (h) A. Riisager, B. E. Hanson, *J. Mol.
Catal. A:* 2002, *189*, 195; (i) L. Wang, H. Chen, Y. He, Y. Li, M. Li, X. Li, *Appl. Catal. A:*
2003, *242*, 85; (j) M. Gaumann, H. Koch, R. Schomäcker, *Catal. Today* 2003, *79–80*, 43;
(k) M. GHaumann, H. Koch, P. Hugo, R. Schomäcker, *Appl. Catal. A:* 2002, *225*, 239;
(l) A. Roucoux, J. Schulz, H. Patin, *Adv. Synth. Catal.* 2003, *345*, 222; (m) Y. Wang,
J. Jiang, Q. Miao, X. Wu, Z. Jin, *Catal. Today* 2002, *74*, 85; (n) F. Joó, E. Papp, A. Kathó,
Top. Catal. 1998, *5*, 113; (o) D. Sinou, C. Rabeyrin, C. Nguefak, *Adv. Synth. Catal.* 2003,
345, 357; (p) E. Paetzold, G. Oehme, *J. Mol. Catal. A:* 2000, *152*, 69 and 2002, *214*, 241;
(q) Y. Huang, L. Min, Y. Li, H. Chen, P. Cheng, X. Li, *J. Mol. Catal. A:* 2002, *185*, 41;
(r) M. Gimenez-Pedros, A. Aghmiz, C. Claver, A. M. Masdeu-Bulto, D. Sinou, *J. Mol.
Catal. A:* 2003, *200*, 157; (s) T. Ooi, Y. Uematsu, K. Maruoka, *Adv. Synth. Catal.* 2002,
344, 288.

12 W. J. Tic, *Przemyst Chem.* 2004, *83*(2), 87.

13 (a) S. Trinkhaus, R. Kadyrov, R. Selke, J. Holz, L. Götze, A. Börner, *J. Mol. Catal. A:*
1999, *144*, 15; (b) I. Grassert, J. Kovács, H. Fuhrmann, G. Oehme, *Adv. Synth. Catal.*
2002, *344*, 312; (c) O. Wachsen, K. Himmler, B. Cornils, *Catal. Today* 1998, *42*, 373.

14 (a) S. Nagayama, S. Kobayashi, *Angew. Chem. Int. Ed.* 2002, *39*, 567; (b) Y. Mori,
K. Manabe, S. Kobayashi, *Angew. Chem.* 2001, *113*, 2898; (c) K. Inoue, T. Ishida,
I. Shibata, A. Baba, *Adv. Synth. Catal.* 2002, *344*, 283; (d) K. Manabe, S. Kobayashi, *Chem.
Eur. J.* 2002, *8*, 4095; (e) T. Ooi, Y. Uematsu, K. Maruoka, *Adv. Synth. Catal.* 2002, *344*, 288;
(f) S. Shimizu, T. Suzuki, S. Shirikawa, Y. Sasaki, C. Hirai, *Adv. Synth. Catal.* 2002, *344*,
370; (g) G. Wüllner, H. Jänsch, S. Kannenberg, F. Schubert, G. Boche, *Chem.
Commun.* 1998, 1509; (h) G. Verspui, G. Papadogianakis, R. A. Sheldon, *Catal. Today*
1998, *42*, 449 and *J. Chem. Technol. Biotechnol.* 1997, *70*, 83; (i) J. Kiji, T. Okano,
H. Kimura, K. Saiki, *J. Mol. Catal. A:* 1998, *130*, 95; (j) C. W. Kohlpaintner, M. Beller,
J. Mol. Catal. A: 1997, *116*, 259; (k) R. Gomes da Rosa, J. D. Ribeiro de Campos,
R. Buffon, *J. Mol. Catal. A:* 2000, *153*, 19; (l) G. Papadogianakis, L. Maat, R. A. Sheldon,
J. Mol. Catal. A: 1997, *116*, 179; (m) J. Tian, N. Yamagiwa, S. Matsunaga, M. Shibasaki,
Angew. Chem. Int. Ed. 2002, *41*, 3636; (n) T. Okano, J. Kiji, Y. Toyooka, *Chem. Lett.* 1998, *5*,
425; (o) B. Heller, *Nach. Chem. Tech. Lab.* 1999, *47*, 9; (p) B. R. Eaton, M. S. Sigmann,
A. W. Fatland, *J. Am. Chem. Soc.* 1998, *120*, 5130; (q) Nagoya University, *Bull. Chem. Soc.
Jpn.* 1997, *70*, 905 [ref. in *CHEMTECH* 1997, (Nov.), 5]; (r) I. V. Kozhevnikov,
G. P. Mulder, M. C. Steverink-de Zoete, M. G. Oostwal, *J. Mol. Catal. A:* 1998, *134*, 223;
(s) H. Rudler, J. Ribeiro Grigorio, B. Denise, J.-M. Bregeault, A. Deloffre, *J. Mol.
Catal. A:* 1998, *133*, 255; (t) G. Pozzi, I. Colombani, M. Miglioli, F. Montanari, S. Quici,
Tetrahedron 1997, *53*, 6145; (u) B. Boyer, A. Hambardzoumian, G. Lamaty, A. Leydet,
J.-P. Roque, P. Bouchet, *New J. Chem.* 1996, *20*, 985; (v) E. Mizushima, K. Sato,
T. Hayashi, M. Tanaka, *Angew. Chem. Int. Ed.* 2002, *41*, 4563; (w) B. Zimmermann,
J. Herwig, M. Beller, *Angew. Chem. Int. Ed.* 1999, *38*, 2372; (x) K. Fuji, T. Morimoto,
K. Tsutsumi, K. Kakiuchi, *Angew. Chem. Int. Ed.* 2002, *42*, 2409; (y) X.-H. Tan, Y.-Q. Hou,
B. Shen, L. Liu, Q.-X. Guo, *Tetrahedron Lett.* 2004, *45*, 5525; (z) C. Zhou, Y. Zhou, J. Jiang,
Z. Xie, Z. Wang, J. Zhang, J. Wu, H. Yin, *Tetrahedron Lett.* 2004, *45*, 5537.

15 (a) F. Bertoux, S. Tilloy, E. Monflier, Y, Castanet, A. Mortreux, *J. Mol. Catal. A:* 1999,
138, 53; (b) E. Monflier, A. Mortreux, *Catal. Lett.* 1998, *50*, 115; (c) G. Verspui, J. Feiken,
G. Papadogianakis, R. A. Sheldon, *J. Mol. Catal. A:* 1999, *146*, 299; (d) F. Bertoux,
E. Monflier, Y. Castanet, A. Mortreux, *J. Mol. Catal. A:* 1999, *143*, 11; (e) F. Bertoux,
E. Monflier,. Castanet, A. Mortreux, *J. Mol. Catal. A:* 1999, *143*, 23; (f) I. Rojas, F. Lopez
Linares, N. Valencia, C. Bianchini, *J. Mol. Catal. A:* 1999, *144*, 1; (g) C. Bianchini,
A. Meli, V. Patinec, V. Sernau, F. Vizza, *J. Am. Chem. Soc.* 1997, *119*, 4945; (h) Z. Yang,
M. Ebihara, T. Kawamura, *J. Mol. Catal. A:* 2000, *158*, 509; (i) P. J. Baricelli, J. Lopez,
E. Lujano, F. Lopez-Linares, *J. Mol. Catal. A:* 2002, *186*, 57; (j) T. Malmström,
C. Andersson, *J. Mol. Catal. A:* 2000, *157*, 79; (k) A. Andriollo, J. Carrasquel, J. Marino,

F. A. López, D. E. Páez, I. Rojas, N. Valencia, *J. Mol. Catal. A:* **1997**, *116*, 157;
(l) G. Süss-Fink, M. Faure, T. R. Ward, *Angew. Chem. Int. Ed.* **2002**, *41*, 99 and *J. Mol. Catal. A:* **1998**, *132*, 5; (m) G. Papp, J. Elek, L. Nádasdi, G. Laurenczy, F. Joó, *Adv. Synth. Catal.* **2003**, *345*, 172; (n) G. Laurenczy et al., *Adv. Synth. Catal.* **2003**, *345*, 211, 216; (o) J. A. Loch, C. Borgmann, R. H. Crabtree, *J. Mol. Catal. A:* **2001**, *170*, 75; (p) H. Jiang, Y. Xu, S. Liao, D. Yu, H. Chen, X. Li, *J. Mol. Catal. A:* **1999**, *142*, 147; (q) D. Sinou, *Adv. Synth. Catal.* **2002**, *344*, 221; (r) F. López-Linares, M. G. Gonzalez, D. E. Páez, *J. Mol. Catal. A:* **1999**, *145*, 61; (s) K. Nomura, *J. Mol. Catal. A:* **1998**, *130*, 1; (t) A. W. Heinen, G. Papadogianakis, R. A. Sheldon, J. A. Peters, H. van Bekkum *J. Mol. Catal. A:* **1999**, *142*, 17; (u) A. Behr, N. Toslu, *Chem. Ing. Tech.* **1999**, *71*, 490; (v) H. Bricout, E. Monflier, J.-F. Carpentier, A. Mortreux, *Eur. J. Inorg. Chem.* **1998**, 1739 and *J. Organomet. Chem.* **1998**, *553*, 469; (w) C. de Bellefon, N. Tanchoux, S. Caraveilhes, P. Grenouillet, V. Hessel, *Angew. Chem.* **2000**, *112*, 3584; (x) J. Méndez-Andino, L. A. Paquette, *Adv. Synth. Catal.* **2002**, *344*, 303; (y) A. Romerosa, M. Peruzzini, *Organometallics* **2000**, *19*, 4005; (z) D. M. Lynn, R. H. Grubbs, *J. Am. Chem. Soc.* **1998**, *120*, 1627 and **2001**, *123*, 3187.

16 (a) M. C. Schuster, K. H. Mortell, A. D. Hegeman, L. L. Kiessling, *J. Mol. Catal. A:* **1997**, *116*, 209; (b) G. Verspui, F. Schanssema, R. A. Sheldon, *Angew. Chem. Int. Ed.* **2000**, *39*, 804; (c) G. Verspui, G. Papadogianakis, R. A. Sheldon, *Chem. Commun.* **1998**, *3*, 401; (d) F. Janin, E. G. Kuntz, J. B. Tommasino, O. M. Vittori, *J. Mol. Catal. A:* **2002**, *188*, 71; (e) Z. Jiang, A. Sen, *Macromolecules* **1994**, *27*, 7215; (f) IFP (Y. Chauvin, R. De Souza, H. Olivier, EP 0.753.346 **1997**; (g) C. Venturello, M. Gambaro, *J. Org. Chem.* **1991**, *56*, 5924; (h) P. L. Mills, R. V. Chaudhari, *Catal. Today* **1999**, *48*, 17; (i) G.-J. ten Brink, I. W. C. E. Arends, R. A. Sheldon, *Adv. Synth. Catal.* **2002**, *344*, 355 and *Science* **2000**, *287*, 1636; (j) Y. Kita, *Angew. Chem. Int. Ed.* **2000**, *39*, 1306; (k) J. Cornely, L. M. Su Ham, D. E. Meade, V. Dragojlovic, *Green Chem.* **2003**, *5*, 34; (l) L. Cammarota, S. Campestrini, M. Carrieri, F. Di Furia, P. Ghiotti, *J. Mol. Catal. A:* **1999**, *137*, 155; (m) Enichem, US 6.071.848 **2000**; (n) S. Mecking, A. Held, F. M. Bauers, *Angew. Chem. Int. Ed.* **2002**, *41*, 545; (o) B. Manders, L. Sciandrone, G. Hauck, M. O. Kristen, *Angew. Chem. Int. Ed.* **2001**, *40*, 4006; (p) D. M. Jones, W. T. S. Huck, *Adv. Mater.* **2001**, *13*, 1256; (q) M. A. Zuideveld, F. M. Bauers, R. Thomann, S. Mecking, XXXVI Jahrestreffen Deutscher Katalytiker, Weimar 19–21 März **2003**, *Preprints*, p. 385 and *Chem. Commun.* **2000**, 301; (r) A. Held, S. Mecking, *Chem. Eur. J.* **2000**, *6*, 4623; (s) D. Albanese, D. Landini, V. Lupi, M. Penso, *Adv. Synth. Catal.* **2002**, *344*, 299; (t) Elf–Atochem, US 5.345.007 **1995**; (u) T. Prinz, W. Keim, B. Driessen-Hölscher, *Angew. Chem. Int. Ed. Engl.* **1996**, *35*, 1708; (v) Bayer AG (B. Driessen-Hölscher, W. Keim, T. Prinz, H.-J. Traenckner, J.-D. Jentsch), EP 0.773.211 **1997**; (w) F. Fringuelli, F. Pizzo, S. Tortoioli, L. Vaccaro, *Adv. Synth. Catal.* **2002**, *344*, 379; (x) E. Monflier et al., *J. Mol. Catal. A:* **1996**, *109*, 27; (y) A. Lubineau, Y. Canac, N. Le Goff, *Adv. Synth. Catal.* **2002**, *344*, 319; (z) T. Okano, J. Kiji, T. Doi, *Chem. Lett.* **1998**, (1), 5.

17 (a) P. Wipf, S. Rodriguez, *Adv. Synth. Catal.* **2002**, *344*, 434; (b) S. Sigismondi, D. Sinou, *J. Mol. Catal. A:* **1997**, *116*, 289; (c) K. Itami, T. Nokami, J. Yoshida, *Adv. Synth. Catal.* **2002**, *344*, 441; (d) K. Manabe, N. Aoyama, S. Kobayashi, *Adv. Synth. Catal.* **2001**, *343*, 174; (e) C.-J. Li, Y. Meng, *J. Am. Chem. Soc.* **2000**, *122*, 9538; (f) S. Mukhopadhyay, G. Rothenberg, A. Joshi, M. Baidossi, Y. Sasson, *Adv. Synth. Catal.* **2002**, *344*, 348; (g) R. Amengual, E. Genin, V. Michelet, M. Savignac, J.-P. Genêt, *Adv. Synth. Catal.* **2002**, *344*, 393; (h) B. M. Bhanage, M. Shirai, M. Arai, *J. Mol. Catal. A:* **1999**, *145*, 69; (i) M. Beller, J. G. E. Krauter, A. Zapf, S. Bogdanovic, *Catal. Today* **1999**, *48*, 279; (j) D. Motoda, H. Kinoshita, H. Shinokubo, K. Oshima, *Adv. Synth. Catal.* **2002**, *344*, 261; (k) S. Kobayashi et al., *Tetrahedron Lett.* **2000**, *41*, 3107; (l) A. Chattopadhyay, A. Salaskar, *Synthesis* **2000**, 561; (m) P. Dupau, R. Epple, A. A. Thomas, V. V. Fokin, K. B. Sharpless, *Adv. Synth. Catal.* **2002**, *344*, 421; (n) G. M. Mehltretter, C. Döbler, U. Sundermeier, M. Beller, *Tetrahedron Lett.* **2000**, *41*, 8083, *J. Am. Chem. Soc.* **2000**, *122*, 10289 and *Angew. Chem. Int. Ed.* **1999**, *38*, 3026; (o) T. Wirth, *Angew. Chem. Int. Ed.* **2000**, *39*, 334; (p) M. Beller, J. G. E. Krauter, A. Zapf, *Angew. Chem. Int. Ed. Engl.* **1997**, *36*, 772;

(q) E. Paetzold, G. Oehme, *J. Mol. Catal. A:* **2000**, *152*, 69; (r) M. G. Russell, S. Warren, *J. Chem. Soc.,Perkin Trans.* **2000**, 505; (s) B. Cornils, W. A. Herrmann, R. Schlögl, C.-H. Wong, *Catalysis from A to Z*, Wiley-VCH, 2nd ed., Weinheim **2003**; (t) K. Nomura, *J. Mol. Catal. A:* **1998**, *130*, 1; (u) S. Venkatraman, T. Huang, C.-J. Li, *Adv. Synth. Catal.* **2002**, *344*, 399; (v) Y. Uozumi, M. Nakazono, *Adv. Synth. Catal.* **2002**, *344*, 274; (w) F. Raynal, R. Barhdadi, J. Pèrichon, A. Savall, M. Troupel, *Adv. Synth. Catal.* **2002**, *344*, 45; (x) C. Dupuis, K. Adiey, L. Charruault, V. Michelet, M. Savignac, J.-P. Gênet, *Tetrahedron Lett.* **2001**, *42*, 6523; (y) M. C. K. B. Djoman, A. N. Ajjou, *Tetrahedron Lett.* **2000**, *41*, 4845; (z) L. V. R. Bonaga, J. A. Wright, M. E. Krafft, *Chem. Commun.* **2004**, 1746.

18 (a) C. Dobler, G. M. Mehltretter, U. Sundermeier, M. Eckert, H.-C. Militzer, M. Beller, *Tetrahedron Lett.* **2001**, *42*, 8447; (b) L. W. Francisco, D. A. Moreno, J. D. Atwood, *Organometallics* **2001**, *20*, 4237; (c) N. Rosas, P. Sharma, C. Alvarez, A. Cabrera, R. Ramirez, A. Delgado, H. Arzoumanian, *J. Chem. Soc.,Perkin Trans.* **2001**, 2341; (d) I. del Rio, C. Claver, P. van Leeuwen, *Eur. J. Inorg. Chem.* **2001**, 2719; (e) R. A. Sheldon, I. W. C. E. Arends, G.-J. ten Brink, A. Dijksman, *Acc. Chem. Res.* **2002**, *35*, 774 and *Adv. Synth. Catal.* **2003**, *345*, 497; (f) A. Kruse, E. Dinjus, *Angew. Chem. Int. Ed.* **2003**, *42*, 909; (g) T. Koike, X. Du, T. Sanada, Y. Danda, A. Mori, *Angew. Chem. Int. Ed.* **2003**, *42*, 89; (h) Y. Uozumi, R. Nakao, *Angew. Chem. Int. Ed.* **2003**, *42*, 194; (i) T. Ishida, R. Akiyama, S. Kobayashi, *Adv. Synth. Catal.* **2003**, *345*, 576; (j) C. Loncaric, K. Manabe, S. Kobayashi, *Adv. Synth. Catal.* **2003**, *345*, 475; (k) C. C. Chapman, C. G. Frost, *Adv. Synth. Catal.* **2003**, *345*, 353; (l) D. Sinou, C. Rabeyrin, C. Nguefack, *Adv. Synth. Catal.* **2003**, *345*, 357; (m) F. Joó, *Acc. Chem. Res.* **2002**, *35*, 738.

19 (a) P. J. Baricelli, G. Rodríguez, M. Rodríguez, E. Lujano, F. López-Linares, *Appl. Catal. A:* **2003**, *239*, 25; (b) V. Michelet, J.-C. Galland, L. Charruault, M. Savignac, J.-P. Gênet, *Org. Lett.* **2001**, *3*, 2065; (c) M. Lautens, M. Yoshida, *J. Org. Chem.* **2003**, *68*, 762; (d) D. U. Parmar, S. D. Bhatt, H. C. Bajaj, R. V. Jasra, *J. Mol. Catal. A:* **2003** *202*, 9.

20 (a) A. Kathó, Z. Opre, G. Laurenczy, F. Joó, *J. Mol. Catal. A:* **2003**, *204–205*, 143; (b) V. O. Sippola, A. O. I. Krause, *J. Mol. Catal. A:* **2003**, *194*, 89; (c) T. Akiyama, J. Takaya, H. Kagoshima, *Adv. Synth. Catal.* **2002**, *344*, 338; (d) *Angew. Chem.* **2003**, *115*, 1072; (e) A. Behr, M. Urschey, *J. Mol. Catal. A:* **2003**, *197*, 101; (f) G. Verspui, F. Schanssema, R. A. Sheldon, *Angew. Chem. Int. Ed.* **2000**, *39*, 804; (g) M. Saphier, A. Masarwa, H. Cohen, D. Meyerstein, *Eur. J. Inorg. Chem.* **2002**, *6*, 1226; (h) G. K. S. Prakash, P. Yan, B. Torok, I. Bucsi, M. Tanaka, G. A. Oláh, *Catal. Lett.* **2003**, *85*, 1; (i) J. Zhang, C. Wei, C.-J. Li, *Tetrahedron Lett.* **2002**, *43*, 5731; (j) D. U. Parmar, S. D. Bhatt, H. C. Bajaj, R. V. Jasra, *J. Mol. Catal. A:* **2003**, *202*, 9; (k) A. N. Ajjou; J.-L. Pinet, *J. Mol. Catal. A:* **2002**, *214*, 203; (l) J. P. Parrish, Y. C. Young, R. J. Floyd, K. Woon Jung, *Tetrahedron Lett.* **2002**, *43*, 7899; (m) H. Kinoshita, H. Shinokubo, K. Oshima, *J. Am. Chem. Soc.* **2003**, *125*, 7784; (n) B. E. Ali, *J. Mol. Catal. A:* **2003**, *203*, 53; [o] B. Wang, Y.-R. Kang, L.-M. Yang, J.-S. Suo, *J. Mol. Catal. A:* **2003**, *203*, 29; (p) V. R. Chouhary, N. S. Patil, N. K. Chaudhari, S. K. Bhargava, *Catal. Commun.* **2004**, *5*, 205; (q) W. Sun, H. Wang, Ch. Xia, J. Li, P. Zhao, *Angew. Chem. Int. Ed.* **2003**, *42*, 1042; (r) V. Neff, T. E. Müller, J. A. Lercher, *Chem. Commun.* **2002**, 906; (s) Z. Hou, T. Okuhara, *J. Mol. Catal. A:* **2003**, *206*, 121; (t) F.-W. Li, L.-W. Xu, C.-G. Xia, *Appl. Catal. A:* **2003**, *253*, 509; (u) J. Elek, L. Nádasdi, G. Papp, G. Laurenczy, F. Joó, *Appl. Catal. A:* **2003**, *255*, 59; (v) T. Ishida, R. Akiyama, S. Kobayashi, *Adv. Synth. Catal.* **2003**, *345*, 576; (w) C. J. Chapman, C. G. Frost, *Adv. Synth. Catal.* **2003**, *345*, 353; (x) T. Hamada, K. Manabe, S. Kobayashi, *Angew. Chem. Int. Ed.* **2003**, *42*, 3927; (y) A. Behr, M. Urschey, *Adv. Synth. Catal.* **2003**, *345*, 1242; (z) C. Loncaric, K. Manabe, S. Kobayashi, *Adv. Synth. Catal.* **2003**, *345*, 1187.

21 (a) T. Gross, S. Seayad, M. Ahmed, M. Beller, *Chem. Ing. Tech.* **2002**, *74*, 554; (b) V. Cadierno, S. E. Garcia-Garrido, J. Gimeno, *Chem. Commun.* **2004**, 232; (c) M. Pal, V. Subramanian, K. Parasurama, K. R. Yeleswarapu, *Tetrahedron* **2003**, *59*, 9563; (d) M. M. Salter, M. Sardo-Inffiri, *Synlett* **2002**, 2068; (e) S. Venkatraman, T. Huang, C.-J. Li, *Adv. Synth. Catal.* **2002**, *344*, 399; (f) A. Mori, M. S. M. Ahmed, A. Sekiguchi, K. Masui, T. Koike, *Chem. Lett.* **2002**, 756; (g) E. Paetzold, I. Jovel, G. Oehme, *J. Mol.*

Catal. A: **2002**, *214*, 241; (h) L. R. Moore, K. H. Shaugnessy, *Org. Lett.* **2004**, *6*, 225; (i) R. Poli, *Chem. Eur. J.* **2004**, *10*, 332; (j) Y. Kayaki, T. Suzuki, T. Ikariya, *Chem. Lett.* **2001**, 1016; (k) F. Joó, E. Papp, A. Kathó, *Top. Catal.* **1998**, *5*, 113; (l) M. Lautens, M. Yoshida, *J. Org. Chem.* **2003**, *68*, 762; (m) J. LeBras, J. Muzart, *Tetrahedron Lett.* **2002**, *43*, 431; (n) D. A. Knight, T. L. Schull, *Synth. Commun.* **2003**, *33*, 827; (o) H. Miyabe, M. Ueda, A. Nishimura, T. Naito, *Tetrahedron* **2004**, *60*, 4227; (p) S. Ogo, K. Uehara, T. Abura; Y. Watanabe, S. Fukuzumi, *Organometallics* **2004**, *23*, 3047; (q) R. Abdallah, T. Ireland, C. de Bellefon, *Chem. Ing. Tech.* **2004**, *76*, 633; (r) N. E. Leadbeater, M. Marco, *Org. Lett.* **2002**, 2973 and **2003**, *888*, 1445.

22 (a) J. W. Shabaker, G. W. Huber, J. A. Dumesic, *J. Catal.* **2004**, *222*, 180; (b) H. Nur, S. Ikeda, B. Ohtani, *J. Catal.* **2001**, *204*, 402; (c) R. Abu-Reziq, J. Blum, D. Avnir, *Angew. Chem. Int. Ed.* **2002**, *41*, 4132; (d) M. Marchetti, C. Botteghi, S. Paganelli, M. Taddei, *Adv. Synth. Catal.* **2003**, *345*, 1229; (e) Y. Yuan, H. Zhang, Y. Yang, Y. Zhang, K. Tsai, *Catal. Today* **2002**, *74*, 5; (f) Asahi Kasei KKK (K. Yamashita, H. Obana, I. Katsuta), EP 0.552.809 **1992**; (g) H. Nagahara, M. Ono, Y. Fukuoka, *Stud. Surf. Sci. Catal.* **1994**, *92* 375; (g) R. R. Davda, J. W. Shabaker, G. W. Huber, R. D. Cortright, J. A. Dumesic, *Appl. Catal. B:* **2003**, *43*, 13; (h) J. R. Anderson; E. M. Campi, W. R. Jackson, Z. P. Yang, *J. Mol. Catal. A:* **1997**, *116*, 109; (i) J. W. Shabaker; J. A. Dumesic, *Ind. Eng. Chem. Res.* **2004**, *43*, 3105; (j) M. Gruber, M. Wagner, R. Heidenreich, J. G. E. Krauter, N. Coskun, K. Köhler, *Catal. Lett.* **2004**, *94*, 177; (k) S. S. Pröckl, W. Kleist, M. A. Gruber, K. Köhler, *Angew. Chem. Int. Ed.* **2004**, *43*, 1881; (l) Asahi Kasei KKK (K. Yamashita, H. Obana, I. Katsuta), US 5.457.251 **1996**; (m) Mitsubisgi CC (T. Suzuki, T. Ezaki), US 5.639.927 **1997**; (n) W. Keim, *Green Chem.* **2003**, *5*, 105.

23 (a) M. McCarthy, H. Stemmer, W. Leitner, *Green Chem.* **2002**, *4*, 501; (b) A. A. Galkin, B. G. Kostyik, V. V. Lunin, M. Poliakoff, *Angew. Chem. Int. Ed.* **2000**, *39*, 2738; (c) Y. Kayaki, T. Suzuki, T. Ikariya, *Chem. Lett.* **2001**, 1016; (d) University of California, US 6.479.708 **1999**; (e) T. Mathivet, E. Monflier, Y. Castanet, A. Mortreux, *Tetrahedron* **2002**, *58*, 3877; (f) B. Wang, Y.-R. Kang, L.-M. Yang, J.-S. Suo, *J. Mol. Catal. A:* **2003**, *203*, 29; (g) P. J. Dyson, D. J. Ellis, W. Henderson, G. Laurenczy, *Adv. Synth. Catal.* **2003**, *345*, 216; (h) G.-T. Wei, Z. Yang, C.-Y. Lee, H.-Y. Yang, C. R. C. Wang, *J. Am. Chem. Soc.* **2004**, *126*, 5036; (i) P. B. Webb, D. J. Cole-Hamilton, *Chem. Commun.* **2004**, 612; (j) J. Zhu, A. Robertson, S. C. Tsang, *Chem. Commun.* **2002**, 2044; (k) Y. Ouzumi, H. Tanaka, K. Shibatomi, *Org. Lett.* **2004**, *6*, 281; (l) K. Manabe, S. Ishikawa, T. Hamada, S. Kobayashi, *Tetrahedron* **2003**, *59*, 10439; (m) A. V. Narsalah, K. Ngaiah, *Synth. Commun.* **2003**, *33*, 3825; (n) U. Schuchardt, R. Rinaldi, 14th ISHC (Munich 2004), Abstracts, p. 320; (o) M. Lautens, J. Mancuso, H. Grover, *Synthesis* **2004**, 2006; (p) A. Mortreux, Y. Castanet, E. Monflier et al., *J. Mol. Catal. A:* **2001**, *176*, 105 and *Adv. Synth. Catal.* **2004**, *346*, 83; (q) S. Tilloy, H. Bricout, E. Monflier, *Green Chem.* **2003**, *4*, 188; (r) J. A. Pool, E. Lobkovsky, P. J. Chirik, *Nature* **2004**, *427*, 527; (s) L. Chen, C.-J. Li, *Org. Lett.* **2004**, *6*, 3151; (t) C. Torque, H. Bricout, F. Hapiot, E. Monflier, *Tetrahedron* **2004**, *60*, 6487; (u) M. de Champdore et al., *Tetrahedron* **2004**, *60*, 6555; (v) K. H. Shaughnessy, R. S. Booth, L. R. Moore, 14th ISHC (Munich **2004**), Abstracts, p. 169; (w) H. Klein, R. Jackstell, M. Beller, 14th ISHC (Munich **2004**), Abstracts, p. 181; (x) P. Csabai, F. Joó, 14th ISHC (Munich **2004**), Abstracts, p. 201; (y) W. Oberhauser et al., 14th ISHC (Munich **2004**), Abstracts, p. 236; (z) U. Schuchardt, R. Rinaldi, 14th ISHC (Munich **2004**), Abstracts, p. 320.

24 See references in [3].

2.2
State-of-the-Art

2.2.1
Organic Chemistry in Water

André Lubineau, Jacques Augé, and Marie-Christine Scherrmann

2.2.1.1
Introduction

Since the seminal contributions in the 1980s of Breslow [1a] for the Diels–Alder reaction and Kuntz and Ruhrchemie for hydroformylation reactions [1b], there has been an upsurge in interest in using water as the solvent, not only to enhance the reaction rates, but also to perform organic reactions that would otherwise be impossible, or to elicit new selectivities. Several reviews have been devoted specially to such a use [2], which nevertheless does not exclude the possibility of further catalyzing the reactions with Lewis acids [3] or the use of Lewis acid–surfactant-combined catalysis (LASC) [4].

As the most abundant liquid that occurs on Earth, water is very cheap and, more importantly, not toxic, so it can be used in large amounts without any associated hazard. In this medium, reactions can be carried out under mild conditions and yields and selectivities can therefore be largely improved. Furthermore, water-soluble compounds, such as carbohydrates, can be used directly without the need for the tedious protection–deprotection sequences. Finally, water-soluble catalysts can be reused after filtration, decantation, or extraction of the water-insoluble products [1b].

A remarkable feature of water-promoted reactions is that the reactants only need to be sparingly soluble in water and, most of the time, the effects of water occur under *biphasic* conditions (cf. Sections 2.1 and 2.3.1). If the reactants are not soluble enough, co-solvents can be used as well as surfactants. Another possibility for inducing water solubility lies in grafting a hydrophilic moiety (a sugar residue or carboxylate, for instance) onto the hydrophobic reactant.

This contribution encompasses the main concepts supporting the origin of the reactivity in water, along with some applications in organic synthesis.

2.2.1.2
Origin of the Reactivity in Water

The combination of a small size and a three-dimensional hydrogen-bonded network system is responsible for the complexity of the structure of water, which results in a large cohesive energy density (550 cal mL^{-1} or 2200 MPa), a high surface tension and a high heat capacity. These three attributes give water its unique structure as a liquid, and give rise to the special properties known as hydrophobic effects, which

Multiphase Homogeneous Catalysis
Edited by Boy Cornils and Wolfgang A. Herrmann et al.
Copyright © 2005 Wiley-VCH Verlag GmbH & Co. KGaA, Weinheim
ISBN: 3-527-30721-4

play a critical role in the folding of biological macromolecules, in the formation and stabilization of membranes and micelles, or in the molecular recognition processes, such as antibody–antigen, enzyme–substrate, and receptor–hormone binding. Since Breslow's discovery that the Diels–Alder reaction, which is known as insensitive to solvent effects, can be dramatically accelerated in aqueous solution, special attention was focused on the origin of the aqueous acceleration. Of interest was the observation of good correlations between solubilities of the reactants and Diels–Alder rate constants [5a]. The influence of hydrophobic effects on solubilities, reaction rates, and selectivities could be interpreted by the use of prohydrophobic and antihydrophobic additives [5b]. In 1986, Lubineau assumed that a fundamental issue is the high cohesive energy of water; he postulated then that a kinetically controlled reaction between two nonpolar molecules for which $\Delta V^{\neq}$ is negative must be accelerated in water [6]. The importance of the cohesive energy density to Diels–Alder reactions and Claisen rearrangements, both displaying a negative activation volume, was demonstrated by Gajewski [7]. By measuring standard Gibbs energy of transfer from organic to aqueous solvents, Engberts and co-workers showed that enforced hydrophobic interaction due to a decrease in the overall hydrophobic surface area during the activation process plays an important role in the rate acceleration in water [8]. This effect was considered to be a consequence of the high cohesive energy density of water [2a] and should be expressed in terms of pressure (cohesive pressure), but one must avoid any confusion with the internal pressure of water, which is small compared with other solvents [9]. The importance of hydrophobic effects was emphasized in Monte Carlo simulated Diels–Alder reactions especially when both reactants are nonpolar; when one of the reactants is a hydrogen-bond acceptor, enhanced hydrogen bonding interaction and hydrophobic effects contribute equally to the rate enhancement [10]. With methyl vinyl ketone as a dienophile model in Diels–Alder reactions, computed partial charges displayed greater polarization of the carbonyl bond in the transition state and consequently enhanced hydrogen bonding to the transition state; on the basis of Monte Carlo simulations [11a] and molecular orbital calculations [11b], hydrogen bonding was proposed as the key factor controlling the variation of the acceleration for Diels–Alder reactions in water. Monte Carlo simulations showed enhanced hydrogen bonding to the oxygen in the transition-state envelope of water molecules for Claisen rearrangements as well [12]. Such an enhanced hydrogen bonding effect was invoked to explain the experimental differences of reactivity between dienophiles in some Diels–Alder reactions [13a–c] and to understand the acceleration in water of the retro Diels–Alder reaction, a reaction with a slightly negative activation volume [13d].

In summary, the acceleration in water of reactions between neutral molecules arises from an enforced hydrophobic effect (especially when apolar reactants are involved), and a charge development in transition states in particular when one of the reactants is a hydrogen donor or acceptor. In both cases a negative volume of activation is expected. The two contributions (hydrophobic effects and polarity) could be active in the same reaction, which means a greater destabilization of the hydrophobic reactants in the initial state than in the transition state, and a greater stabilization of a more polar transition state.

2.2.1.3
Pericyclic Reactions

In his pioneering work [1a], Breslow studied the kinetics of the cycloaddition between cyclopentadiene and methyl vinyl ketone. The implication of the hydrophobic effect in Diels–Alder reactions was extensively supported by the effect of cyclodextrins [14] and additives, such as lithium chloride (salting-out agent) or guanidinium chloride (salting-in agent), which respectively increases or decreases the rate of the reaction [15]. By measuring activation parameters, it has later been shown that the acceleration arises from a favorable change of activation entropy, which is an indication of the implications of the hydrophobic effect [16]. Another aspect of the influence of water as the solvent in Diels–Alder reactions is the higher *endo* selectivity observed by comparison with organic solvents. Indeed, the Diels–Alder reaction has a negative activation volume and of the two possible transition states, the *endo* one is the more compact and therefore should be favored. An improvement in the rate of the aqueous Diels–Alder reaction came with the use of Lewis acid in aqueous media [17] also in the presence of micelles [18]. Use of metallo-vesicles or metallo-micelles was found to be very efficient [19].

The hetero Diels–Alder reaction has also been carried out efficiently in water. For instance, glyoxylic acid undergoes cycloadditions with various dienes [20] although the carbonyl function is almost exclusively present as its hydrate form. Other pericyclic reactions such as 1,3 dipolar [21] or [4+3] cycloadditions [Eq. (1), Table 1], and Claisen rearrangement [22] gave better results when conducted in aqueous media than in organic solvents.

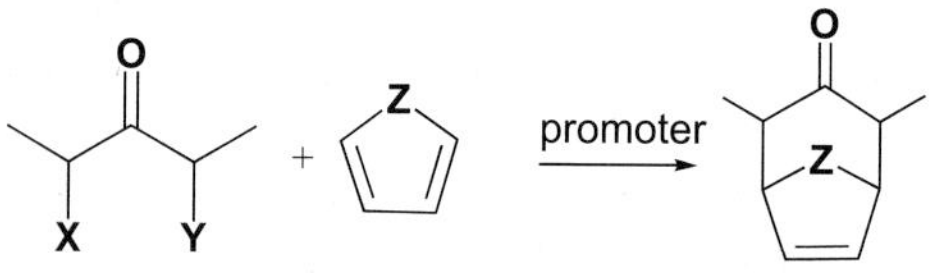

$$\text{(1)}$$

Table 1 Conditions and results for Eq. (1).

X	Y	Z	Solvent	Promoter	Time	Temp. [°C]	Yield [%]	Ref.
Br	Br	CH_2	Benzene	$Fe_2(CO)_9$	25 h	25	81	[23a]
Br	Br	CH_2	H_2O	Fe (powder)	13 h	20	60	[23b]
Br	Br	O	H_2O	Fe (powder)	34 h	20	82	[23b]
Cl	H	O	EtOH	Et_3N	330 d	20	70	[23c]
Cl	H	O	H_2O	Et_3N	5 h	20	88	[23b]
Cl	H	CH_2	H_2O	Et_3N	12 h	20	76	[23b]

2.2.1.4
Carbonyl Additions

Surprisingly enough, the reaction between a silyl enol ether, a potentially hydro-lyzable compound, and an aldehyde can be performed in water without catalyst. In this solvent, contrary to what is obtained in $TiCl_4$-catalyzed reactions, the formation of the *syn* adduct is favored, as observed for the aldolization in CH_2Cl_2 under high pressure (10 000 atm). The *syn* selectivity is due to the smaller activation volume of the corresponding transition state ($\Delta V^{\neq}_{syn} < \Delta V^{\neq}_{anti}$). This is an indication of the implication of hydrophobic effects during the activation process [6, 24]. The use of lanthanide triflates [3a, 25] as water-tolerant catalysts substantially improved the yields of the addition and allowed a good enantioselective control [26]. Similarly, the Henry [27], Reformatsky [28], Mannich [29], and Mannich-like reactions [30] were successfully performed in aqueous media.

The use of water as a solvent in the conjugate addition was first reported for 1,3-diketones [31] and then applied to β-unsubstituted enones [32], other conjugated enones [33], methyl vinyl ketone [34], α,β-unsaturated nitriles [35], and dehydro-alanine amides [36]. Also a related reaction, the Baylis–Hillman coupling, was found to be greatly accelerated in water compared with usual organic solvents [37].

The allylation of aldehydes via organotin reagents displays a negative activation volume [38], therefore it is accelerated by addition of water [39]. The reaction was extended to various aldehydes and ketones and to various allylic organotin dichlorides [40] or tetra-allyltin in acidic aqueous medium [41]. With scandium triflate as a catalyst, tetra-allyltin [42] or tetra-allylgermane [43] react smoothly in mixed aqueous solvents, providing high yields of the corresponding homoallylic alcohols. The aqueous Barbier-type allylation reaction was found to be mediated by zinc [44a–b] and tin [39, 44c, 45a], and also applied to unprotected carbohydrates to prepare diastereoselectively higher-carbon sugars [45b]. A major improvement was realized with the use of indium, a metal with a very low first ionization potential (5.8 eV) which works without ultrasonic radiation even at room temperature [46]. The reactivity and the diastereoselectivity are compatible with a chelation-controlled reaction [45a, 47]. Indeed, the methodology was used to prepare 3-deoxynonulosonic acids such as KDN [48], Neu5Ac [48b, 49a] and analogues [49b], and *C*-disaccharides [50].

2.2.1.5
Radical Reactions

Although significant solvent effects have been observed for radical reactions in water, the use of an aqueous medium for this type of chemistry remains, nowadays, rare. Oshima's group has reported a remarkable improvement for the triethylborane-mediated radical cyclization reactions of allyliodoacetates in butyrolactones [51a,b] [Eq. (2)]. Some other studies by the same group have recently been reviewed [51c]. Miyabe et al. described the indium- [52a] or triethylborane-mediated [52b] radical addition of alkyl iodide to electron-deficient C=N bonds and C=C bonds in water.

The same group reported indium-mediated tandem radical addition–cyclization–trap reactions in aqueous media [52c]. The use of water-soluble radical initiator 2,2′-azobis[2-(2-imidazolin-2-yl)propane], water-soluble chain carrier 1-ethyl-piperidinium hypophosphite (EPHP) and surfactant cetylammonium bromide (CTAB) allowed the C–C bond-forming radical reactions of highly hydrophobic substrates in water [53]. Similarly, the use of CTAB and EPHP in presence of 4,4′-azobis(4-cyanovaleric acid) promoted the indium-mediated radical addition to β-substituted conjugated alkenes in water [54].

$$\xrightarrow{\text{Et}_3\text{B/O}_2}$$

(2)

Solvent/yield (%): Water/78; DMSO/37;
MeCN/13; EtOH/13; CH$_2$Cl$_2$/<1; THF/<1

References

1 (a) D. C. Rideout, R. Breslow, *J. Am. Chem. Soc.* **1980**, *102*, 7816; (b) E. G. Kuntz, *CHEMTECH* **1987**, *17(a)*, 570.

2 (a) A. Lubineau, J. Augé, Y. Queneau, *Synthesis* **1994**, 741; (b) C. J. Li, T. H. Chan (Eds.), *Organic Reactions in Aqueous Media*, **1997**, Wiley-Interscience, New York; (c) *Organic Synthesis in Water* (Ed.: P. A. Grieco), Blackie, London **1998**; (d) A. Lubineau, J. Augé, in *Modern Solvents in Organic Synthesis, Top. Curr. Chem.* **1999**, *206*, 1; (e) U. M. Linström, *Chem. Rev.* **2002**, *102*, 2751; (f) S. Otto, J. B. F. N. Engberts, *Org. Biomol. Chem.* **2003**, *1*, 2809; (g) *Aqueous-Phase Organometallic Catalysis*, 2nd edition (Eds.: B. Cornils, W. A. Herrmann) **2004**, Wiley-VCH, Weinheim.

3 (a) S. Kobayashi, *Synlett* **1994**, 689; (b) J. B. F. N. Engberts, B. L. Feringa, E. Keller, S. Otto, *Recl. Trav. Chim. Pays-Bas* **1996**, *115*, 457.

4 (a) S. Kobayashi, T. Wakabayashi, *Tetrahedron Lett.* **1998**, *39*, 5389; (b) S. Kobayashi, K. Manabe, *Acc. Chem. Res.* **2002**, *35*, 209.

5 (a) R. Breslow, Z. Zhu, *J. Am. Chem. Soc.* **1995**, *117*, 9923; (b) R. Breslow, *Structure and Reactivity in Aqueous Solution* (Eds.: C. J. Cramer, D. G. Truhlar) **1994**, ACS Symposium Series 568, Chapter 20.

6 A. Lubineau, *J. Org. Chem.* **1986**, *51*, 2142.

7 (a) J. J. Gajewski, *J. Org. Chem.* **1992**, *57*, 5500; (b) J. J. Gajewski, N. L. Brichford, *Structure and Reactivity in Aqueous Solution* (Eds.: C. J. Cramer, D. G. Truhlar) **1994**, ACS Symposium Series 568, Chapter 16.

8 W. Blokzijl, M. J. Blandamer, J. B. F. N. Engberts, *J. Am. Chem. Soc.* **1991**, *113*, 4241.

9 M. R. J. Dack, *Chem. Soc. Rev.* **1975**, *4*, 211.

10 T. R. Furlani, J. Gao *J. Org. Chem.* **1996**, *61*, 5492.

11 (a) J. F. Blake, W. L. Jorgensen, *J. Am. Chem. Soc.* **1991**, *113*, 7430; (b) J. F. Blake, D. Lim, W. L. Jorgensen, *J. Org. Chem.* **1994**, *59*, 803.

12 D. L. Severance, W. L. Jorgensen, *J. Am. Chem. Soc.* **1992**, *114*, 10966.

13 (a) S. Otto, W. Blokzijl, J. B. F. N. Engberts, *J. Org. Chem.* **1994**, *59*, 5372;
(b) G. K. van der Wel, J. W. Wijnen, J. B. F. N. Engberts, *J. Org. Chem.* **1996**, *61*, 9001;
(c) J. B. F. N. Engberts, *Pure Appl. Chem.* **1995**, *67*, 823; (d) J. W. Wijnen,
J. B. F. N. Engberts, *J. Org. Chem.* **1997**, *62*, 2039.

14 H.-J. Schneider, N. K. Sangwan, *Angew. Chem. Int. Ed. Engl.* **1987**, *26*, 896.

15 R. Breslow, C. J. Rizzo, *J. Am. Chem. Soc.* **1991**, *113*, 4340.

16 A. Lubineau, H. Bienhaymé, Y. Queneau, M.-C. Scherrmann, *New J. Chem.* **1994**, *18*, 279.

17 F. Fringuelli, O. Piermatti, F. Pizzo, L. Vaccaro, *Eur. J. Org. Chem.* **2001**, 439.

18 (a) S. Otto, J. B. F. N. Engberts, J. C. T. Kwak, *J. Am. Chem. Soc.* **1998**, *120*, 9517; (b) K. Manabe, Y. Mori, S. Kobayashi, *Tetrahedron* **1999**, *55*, 11203.

19 T. Rispens, J. B. F. N. Engberts, *Org. Lett.* **2001**, *3*, 941.

20 (a) A. Lubineau, J. Augé, N. Lubin, *Tetrahedron Lett.* **1991**, *32*, 7529; (b) A. Lubineau, J. Augé, E. Grand, N. Lubin, *Tetrahedron* **1994**, *50*, 10265; (c) A. Lubineau, J. Augé, N. Lubin, *Tetrahedron* **1993**, *49*, 4639; (d) A. Lubineau, Y. Queneau, *J. Carbohydr. Chem.* **1995**, *14*, 1295; (e) A. Lubineau, E. Grand, M.-C. Scherrmann, *Carbohydr. Res.* **1997**, *297*, 169.

21 (a) A. Lubineau, G. Bouchain, Y. Queneau, *J. Chem. Soc., Perkin Trans. 1* **1995**, 2433; (b) G. Molteni, M. Orlandi, G. Broggini, *J. Chem. Soc., Perkin Trans. 1* **2000**, 3742.

22 (a) E. Brandes, P. A. Grieco, J. J. Gajewski, *J. Org. Chem.* **1989**, *54*, 515; (b) P. A. Grieco, E. B. Brandes, S. McCann, J. D. Clark, *J. Org. Chem.* **1989**, *54*, 5849; (c) A. Lubineau, J. Augé, N. Bellanger, S. Caillebourdin, *Tetrahedron Lett.* **1990**, *31*, 4147; (d) A. Lubineau, J. Augé, N. Bellanger, S. Caillebourdin, *J. Chem. Soc., Perkin Trans. 1* **1992**, 1631.

23 (a) H. Takaya, S. Makino, R. Noyori, *J. Am. Chem. Soc.* **1978**, *100*, 1765; (b) A. Lubineau, G. Bouchain, *Tetrahedron Lett.* **1997**, *38*, 8031; (c) B. Föhlisch, E. Gehrlach, J. J. Stezowski, P. Kollat, E. Martin, W. Gottstein, *Chem. Ber.* **1986**, *119*, 1661.

24 A. Lubineau, E. Meyer, *Tetrahedron* **1988**, *44*, 6065.

25 S. Kobayashi, K. Manabe, S. Nagayama, in *Modern Carbonyl Chemistry* (Ed.: J. Otera) **2000**, Wiley-VCH, Weinheim.

26 S. Kobayashi, T. Hamada, S. Nagayama, K. Manabe, *Org. Lett.* **2001**, *3*, 165.

27 R. Ballini, G. Bosica, *J. Org. Chem.* **1997**, *62*, 425.

28 T. H. Chan, C.-J. Li, M. C. Lee, Z. Y. Wei, *Canad. J. Chem.* **1994**, *72*, 1181.

29 V. Tychopoulos, J. H. P. Tyman, *Synth. Commun.* **1986**, *16*, 1401.

30 (a) S. D. Larsen, P. A. Grieco, W. F. Fobare, *J. Am. Chem. Soc.* **1986**, *108*, 3512; (b) P. A. Grieco, A. Bahsas, *J. Org. Chem.* **1987**, *52*, 1378.

31 (a) U. Eder, G. Sauer, R. Wiechert, *Angew. Chem. Int. Ed. Engl.* **1971**, *10*, 496; (b) Z. G. Hajos, D. R. Parrish, *J. Org. Chem.* **1974**, *39*, 1612.

32 E. Keller, B. L. Feringa, *Tetrahedron Lett.* **1996**, *37*, 1879.

33 R. Ballini, G. Bosica, *Tetrahedron Lett.* **1996**, *37*, 8027.

34 A. Lubineau, J. Augé, *Tetrahedron Lett.* **1992**, *33*, 8073.

35 G. Jenner, *Tetrahedron* **1996**, *52*, 13557.

36 B. N. Naidu, M. E. Sorenson, T. P. Connoly, Y. Ueda, *J. Org. Chem.* **2003**, *68*, 10098.

37 J. Augé, N. Lubin, A. Lubineau, *Tetrahedron Lett.* **1994**, *35*, 7947.

38 Y. Yamamoto, K. Maruyama, K. Matsumoto, *J. Chem. Soc., Chem. Commun.* **1983**, 489.

39 J. Nokami, J. Otera, T. Sudo, R. Okawara, *Organometallics* **1983**, *2*, 191.

40 D. Furlani, D. Marton, G. Tagliavini, M. Zordan, *J. Organomet. Chem.* **1988**, *341*, 345.

41 A. Yanagisawa, H. Inoue, M. Morodome, H. Yamamoto, *J. Am. Chem. Soc.* **1993**, *115*, 10356.

42 I. Hachiya, S. Kobayashi, *J. Org. Chem.* **1993**, *58*, 6958.

43 T. Akiyama, J. Iwai, *Tetrahedron Lett.* **1997**, *38*, 853.

44 (a) C. Einhorn, J.-L. Luche, *J. Organomet. Chem.* **1987**, *322*, 177; (b) C. Pétrier, J.-L. Luche, *J. Org. Chem.* **1985**, *50*, 910; (c) C. Pétrier, J. Einhorn, J.-L. Luche, *Tetrahedron Lett.* **1985**, *26*, 1449.

45 (a) E. Kim, D. M. Gordon, W. Schmid, G. M. Whitesides, *J. Org. Chem.* **1993**, *58*, 5500; (b) W. Schmid, G. M. Whitesides, *J. Am. Chem. Soc.* **1991**, *113*, 6674.

46 (a) C. J. Li, T. H. Chan, *Tetrahedron Lett.* **1991**, *32*, 7017; (b) T. H. Chan, Y. Yang, *J. Am. Chem. Soc.* **1999**, *121*, 3228.

47 (a) L. A. PAQUETTE, T. M. MITZEL, *Tetrahedron Lett.* **1995**, *36*, 6863; (b) L. A. PAQUETTE, T. M. MITZEL, *J. Am. Chem. Soc.* **1996**, *118*, 1931.

48 (a) T. H. CHAN, C. J. LI, *J. Chem. Soc, Chem. Commun.* **1992**, 747; (b) T. H. CHAN, M.-C. LEE, *J. Org. Chem.* **1995**, *60*, 4228.

49 (a) D. M. GORDON, G. M. WHITESIDES, *J. Org. Chem.* **1993**, *58*, 7937; (b) S.-Y. CHOI, S. LEE, G. M. WHITESIDES, *J. Org. Chem.* **1996**, *61*, 8739.

50 Y. CANAC, E. LEVOIRIER, A. LUBINEAU, *J. Org. Chem.* **2001**, *66*, 3206.

51 (a) H. YORIMITSU, T. NAKAMURA, H. SHINOKUBO, K. OSHIMA, *J. Org. Chem.* **1998**, *63*, 8604; (b) H. YORIMITSU, T. NAKAMURA, H. SHINOKUBO, K. OSHIMA, K. OMOTO, H. FUJIMOTO, *J. Am. Chem. Soc.* **2000**, *122*, 11041; (c) H. YORIMITSU, H. SHINOKUBO, K. OSHIMA, *Synlett* **2002**, *5*, 674.

52 (a) H. MIYABE, M. UEDA, A. NISHIMURA, T. NAITO, *Org. Lett.* **2002**, *4*, 131; (b) H. MIYABE, M. UEDA, T. NAITO, *J. Org. Chem.* **2000**, *65*, 5043; (c) M. UEDA, H. MIYABE, A. NISHIMURA, O. MIYATA, Y. TAKEMOTO, T. NAITO, *Org. Lett.* **2003**, *5*, 3835.

53 Y. KITA, H. NAMBU, N. G. RAMESH, G. ANILKUMAR, M. MATSUGI, *Org. Lett.* **2001**, *8*, 1157.

54 D. J. JANG, D. H. CHO, *Synlett* **2002**, *4*, 631.

2.2.2
Organometallic Chemistry of Water

Fritz E. Kühn and Wolfgang A. Herrmann

2.2.2.1
Introduction

Water plays a fundamental role in coordination chemistry. Not only do many, if not all, metals bind water molecules to fill their coordination sphere; it has been through the kinetics of water-exchange processes at metal ions that the basics of the theory of coordination chemistry have been conveyed [1]. By way of contrast, not so much is known about aqueous organometallic chemistry, because of the notorious lability of organometallics toward water. It has to be noted, however, that particularly in recent years the organometallic chemistry of water has gained increasing attention, particularly with water as the polar phase in multiphase systems, as can be seen in the appearance of an increasing number of original articles and several review articles in this field [2]. This section focuses on the key features of water in both aspects, i.e., coordination and organometallic chemistry.

2.2.2.2
Water as a Solvent and Ligand

Water as a solvent offers several opportunities as compared with organic solvents. It favors ionic reactions because of its high dielectric constant ($\varepsilon_{25\,°C} = 78.5$) and the ability to solvate cations as well as anions. Beyond that, water is the ideal solvent for radical reactions since the strong O–H bonds (enthalpy 436 kJ mol^{-1}) are not easily attacked [1g]. Furthermore water displays particularly high enthalpy changes during the solid–liquid (ΔH_{melt} = 6.003 kJ mol^{-1}) and particularly the liquid–gaseous (ΔH_{vap} = 40.656 kJ mol^{-1}) transfer, since the phase transfers are associated with the formation or breaking of hydrogen bridges. Additional advantages of water as a solvent are its high heat capacity, its strong pressure dependency of the viscosity, and the high cohesive energy density (c.e.d. = 2.303 kJ cm^{-3}) [2]. Classified as a ligand for metal ions, water has decent crystal field splitting properties, standing between oxygen-bound anions and nitrogen donors such as pyridines in the spectrochemical series. The water molecule is a good σ-donor ligand, while π back-bonding is negligible. For this reason, higher-valent transition metals form the more stable metal complexes, but the nature of the metal itself is important, too. For first-row metals (+2), an extra destabilization due to electrons in e$_g$ orbitals accounts for a ligand labilization. This effect is commonly referred to as "crystal field activation energy" (CFAE) [1a–c]. From these properties it is concluded that low-valent metals do not favor water in the ligand sphere. Note that $Cr(CO)_6$ is a stable compound because of the outstanding π back-bonding of carbon monoxide, while {$Cr(H_2O)_6$} does not exist – quite contrary to the common occurrence of $[Cr(H_2O)_6]^{3+}$. On the other hand, trivalent chromium does not form the (hypothetical) cationic carbonylchromium complex {$[Cr(CO)_6]^{3+}$}. This is, in short, one major reason why so little is known about typical organometallic water complexes. Comparatively few organometallic aquo complexes have been isolated in fact. Examples are the carbonylrhenium(I) and (π-benzene)ruthenium(II) complexes **1** and **2**, respectively, and their congeners **3–5** [3–6]. They can be used as convenient starting materials for complexes exhibiting the respective organometallic backbones, e.g., the $[Re(CO)_3]^+$ cation from **1** which is otherwise available only with difficulty [6]. The synthesis of aquo complexes from metal carbonyls proceeds via photolysis from the anhydrous parent compounds. The air-stable (!) rhenium complex **1** is conveniently generated from $[ReO_4]^-$ or $[ReOCl_4]^-$ and [$BH_3 \cdot$ solvent] in the presence of carbon monoxide [6a,b]. It is an outstanding precursor of products like $[L_3Re(CO)_3]^+$, with L = N- or P-donors [6].

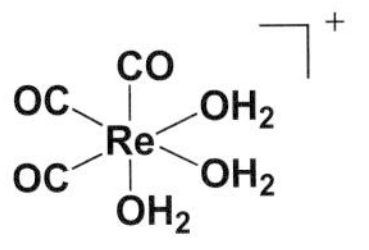

1

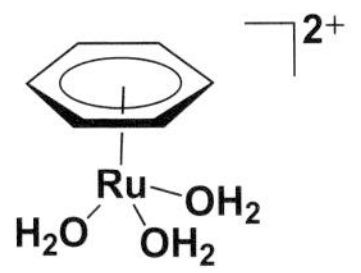

2

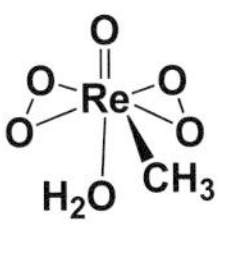

3

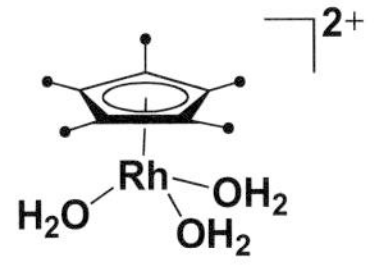

4

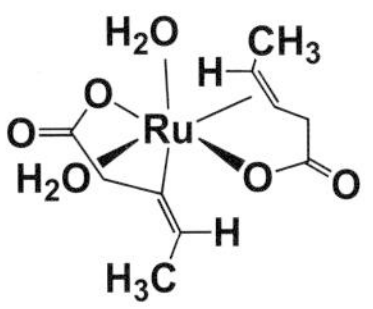

5

2.2.2.3
Organometallic Reactions of Water

Metal–carbon (M–C) bonds are thermodynamically unstable with regard to their hydrolysis products. Water can attack M–C bonds either by proton transfer (H^+, electrophilic reaction) or via the oxygen (OH_2 or OH^-, nucleophilic reaction). Examples are shown in Scheme 1. Ligands such as carbon monoxide and ethylene are activated toward nucleophilic attack upon coordination to (low-valent) metals, e.g., Pd^{2+}. A number of C-C bond-forming reactions derive from this activation. Allyl ligands are generated by proton attack at the terminal 1,3-diene carbon groups (Scheme 1). In other cases, protonation of heteroatoms of metal-attached ligands is followed by elimination steps; for example, the allyl alcohol ligand $H_2C=C(CH_3)CH_2OH$ (η^2) is converted by $H[BF_4]$ into the allyl cation $[H_2CC(CH_3)=CH_2]^+$, which is a standard route of making metal–allyl complexes [3]. High bond polarity yields increased reactivity with water. Thus, $Al(CH_3)_3$ and $In(CH_3)_3$ hydrolyze quickly. $Sb(CH_3)_3$ and $Sn(CH_3)_4$ are inert because of low bond polarity and efficient metal shielding (electron pair at Sb, coordination number 4 at Sn). $Si(CH_3)_4$ is water-stable (low bond polarity, good steric shielding), while SiH_4 hydrolyzes quickly due to inefficient shielding and nucleophilic attack, probably via 3d orbitals of the silicon. However, metal hydrides hydrolyze only if they are ionic or coordinatively unsaturated. For example, NaH (ionic) instantaneously extrudes hydrogen upon contact with water whereas the covalent hydrides $Mn(CO)_5H$, $(\eta^5\text{-}C_5H_5)Fe(CO)_2H$, and $(\eta^5\text{-}C_5Me_5)ReH_6$ remain unchanged.

Since the M–R bond is normally polarized toward an anionic organyl group ($R^{\delta-}$), a metal hydroxide forms along with the respective hydrocarbon from metal alkyls [Eq. (1)].

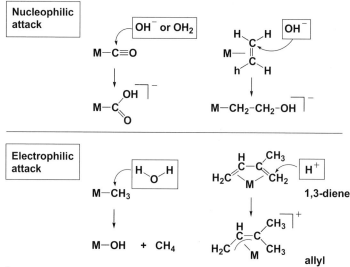

Scheme 1

$$\overset{\delta^+ \quad \delta}{M-R} + H_2O \rightarrow M-OH + R-H \tag{1}$$

$$2\ M-OH \rightarrow M=O + H_2O-M \tag{2}$$

$$2\ M-OH \rightarrow M-O-M + H_2O \tag{3}$$

$$2\ Re_2(CO)_{10} + 4\ H_2O \xrightarrow[-8\ CO]{h\nu} [(CO)_3ReOH]_4 + 2\ H_2 \tag{4}$$

Follow-up products may include metal oxides, be it in a mononuclear or in a di- or oligonuclear form [Eqs. (2)–(4)]. The tetranuclear rhenium(I) complex formed by photolysis [Eq. (4)] has a cubane-type structure [7]. Related complexes **6a–d** were made according to Scheme 2 [8].

Since many organometallics behave as Lewis bases due to electron-rich metals, protonation is a common reaction. For example, the tungsten hydride **7** undergoes reversible protonation at the metal, forming the water-soluble cationic hydride **8** [Eq. (5)]. In nickelocene **9**, a 20e⁻ complex, protonation occurs at the π-bonded cyclopentadienyl ligand; the intermediate **10** has a stable, isolable counterpart in

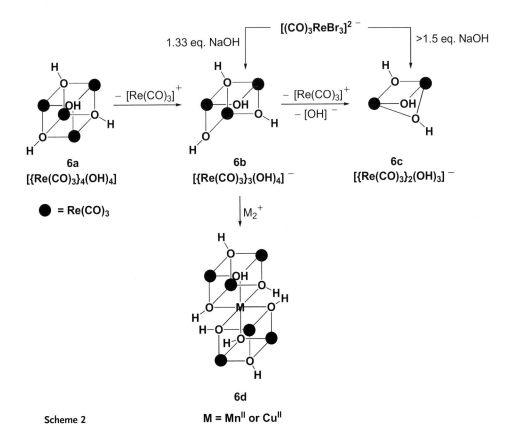

Scheme 2

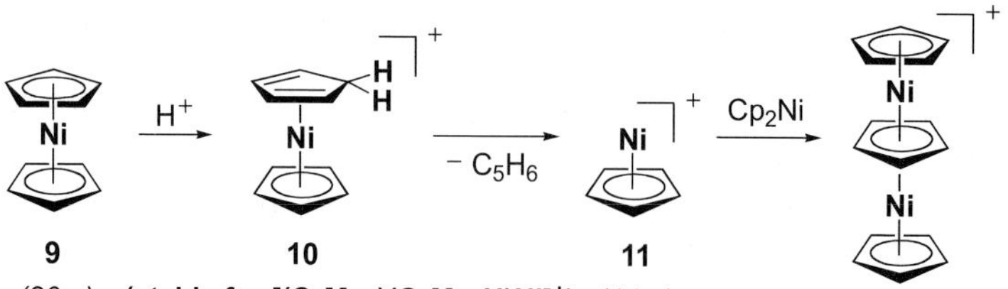

9

10

11

(20 e) (stable for [(C$_5$Me$_5$)(C$_5$Me$_5$H)Ni]$^+$) *(14 e)*

12

(34 e)

Scheme 3

the fully methylated derivative. Consecutive loss of cyclopentadiene forms the cation **11**, which adds to unchanged nickelocene forming the tripledecker sandwich **12** (Scheme 3). Pronouncedly oxophilic metals such as the rare-earth elements are particularly sensitive to water. There are numerous cases where an oxo ligand has been introduced by accidental moisture present under the conditions of reaction [9]. Chemistry of this type of metals has therefore to be performed under rigorous glove-box conditions.

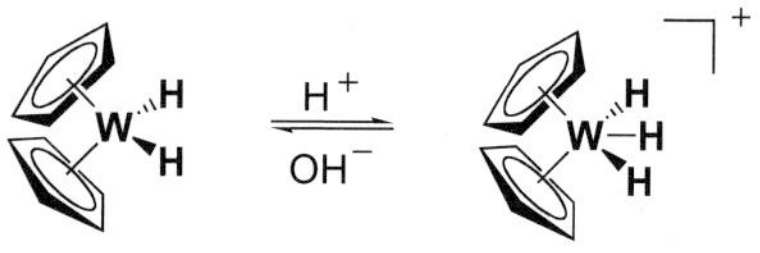

7

8

$$(5)$$

$$L_nRh-P(C_6H_5)_3 + H_2O \rightarrow O=P(C_6H_5)_3 + ... \tag{6}$$

$$(CO)_4CO-H + H_2O \rightarrow [(CO)_4)Co]^- + H_3O^+ \tag{7}$$

There are also cases where the *ligand* reacts with water. Organic phosphanes, for example, may show up as phosphane oxides under certain circumstances [cf. Eq. (6)]. The oxidant is not always obvious in these cases, but it can be water. Strong metalla-acids undergo protonation of water to form the corresponding anion; HCo(CO)$_4$ is such an example [Eq. (7)]. There are cases where the M–C bonds withstand cleavage. The organorhenium(VII) oxide **13** (pentamethyl(cyclopenta-dienyl)trioxorhenium) does not exchange its terminal oxo ligands with water, while the sterically less hindered and more electron-deficient cyclopentadienyltrioxo-rhenium **14** exchanges its oxo ligands slowly [10].

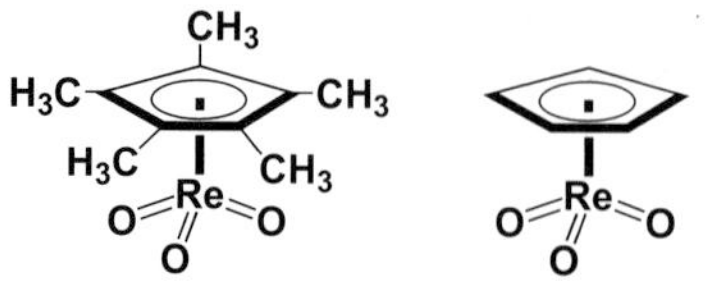

13

14

The water-soluble (ca. 30 g L^{-1}) methyltrioxorhenium(VII) (MTO) **15** [5] exchanges its oxo ligands quickly with water [Eq. (8)] [10] and it has been assumed that it possibly forms octahedral water adducts prior to undergoing aggregation to "polymeric MTO". Mainly steric but also electronic reasons are likely to account for this difference between the organorhenium(VII) oxides **13** and **14**, and **15**. MTO is also an example of a water-stable metal organyl: even in boiling water, only a few of the methyl groups are lost (as methane) after several hours [5, 11]! The reaction product of MTO with excess hydrogen peroxide, a highly reactive and catalytically active organometallic bisperoxo complex (**16**) can be stabilized – among other possibilities – and isolated with a coordinating water molecule [Eq. (9)] [6e, 12]. Water can also oxidize organometallic complexes. For example, the platinum(IV) complex cation $[(C_6H_5)_2Pt(OH)]^+$ is generated from divalent platinum by reaction with water [13].

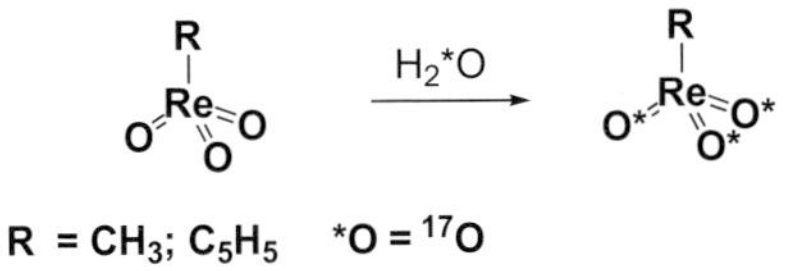

$$\text{(8)}$$

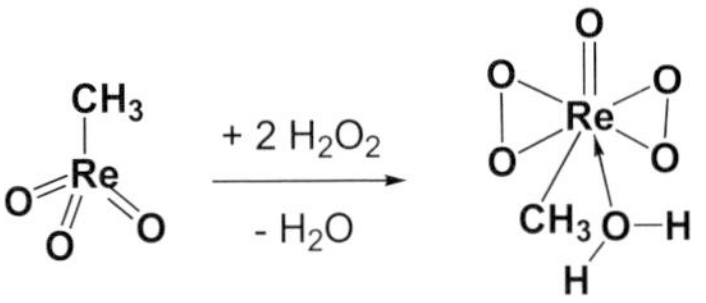

R = CH₃; C₅H₅   *O = ¹⁷O

$$\text{(9)}$$

15   16

2.2.2.4
Water-Soluble Metal Complexes

Following the pioneering work in the area of biphasic catalysis [14, 15], a steady demand for water-soluble and, at the same time, water-compatible metal complexes has been recognized. Despite a broad array of solubilizing ligands, sulfonated derivatives of ligands containing aryl groups have proven most successful, mostly because of the outstanding solubility in water. Notably, the standard tris(*m*-sulfonato-phenyl)phosphine $[P(C_6H_4$-*m*-SO_3-$Na^+)_3$; TPPTS] has a solubility of ca. 1.1 kg L^{-1} upon which the success of the catalyst system Rh/TPPTS depends in biphasic aqueous hydroformylation [15] (cf. Section 2.4.1.1). Numerous ionic organo-metallics, e.g., $Na[Re(CO)_5]$ or $[(\eta^5$-$C_5H_5)Fe(CO)_2$-$\{P(C_6H_5)_3\}]^+I^-$, are soluble in water and can be precipitated by large counter-ions (e.g., $[P(C_6H_5)_4]^+$, $[B(C_6H_5)_4]^-$, $[(C_6H_5)_3P=N=P(C_6H_5)_3]^+$). Basic ions such as $[Re(CO)_5]^-$ give hydrido complexes upon protonation, e.g., $HRe(CO)_5$.

2.2.2.5
Perspectives

Organometallic entities display a variety of reactions with water, following their commonly observed thermodynamic instability. However, in many cases, kinetic barriers prevent these reactions from occurring, making organometallics seemingly stable toward water and, often, even toward protic aqueous acids and hydroxides. Even in cases where highly charged metal ions coordinate water, further degradation of the (organic) ligand sphere does not necessarily occur. Therefore, water is much more compatible with organometallic compounds than has previously been assumed. The organic chemistry in water is just beginning to gain wider attention, too [2, 16]. Specific reaction modes arising from the reactivity of water have to be taken into account. Consider, for example, the sulfonated hydroformylation catalyst **17**, which undergoes slow but significant P–C bond cleavage (Scheme 4). The resulting phosphinidene intermediate **18** can react with water to give the phosphinous acid **20** via tautomerization of the hydroxyphosphine **19** [15a]. Thus, one must be aware of oxygenated (side) products when working with aqueous organometallics. The field is far from being fully explored.

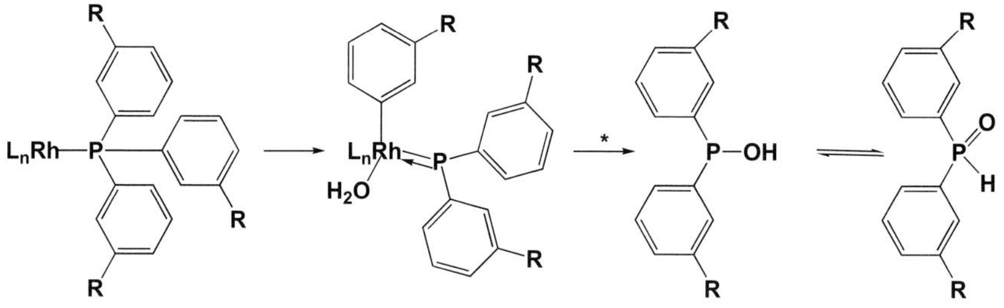

17   18   19   20

R = SO₃Na

Scheme 4

Supercritical water (the critical point of water is reached at $T = 374.2\,°C$ and $p = 22.1\,MPa$) offers opportunities in selectivity for organometallic reactions. The behavior of supercritical water is in several aspects (e.g.) heat capacity, viscosity, dissociation constant, quite different from water under normal conditions. This behavior change is due to the loss of hydrogen bridging. The dielectric constant, for example, changes from 78.5 to 6.0 in supercritical water and becomes quite similar to the dielectric constants of some organic solvents, such as tetrahydrofuran (THF) ($\varepsilon_{THF} = 7.4$). Several applications of supercritical water have been reported [17] during recent years. An example is the cyclotrimerization of certain alkynes, which is very selective for benzene derivatives in supercritical water at 374 °C [18]. Organometallic, water-soluble anticancer reagents, normally in the context of amino acid ligands, are also appearing more frequently in the literature now [19]. Important C–C coupling reactions, such as the Heck and Suzuki reactions, have been successfully transferred to aqueous media recently [20], the use of polymer-fixed homogeneous catalysts which can be dispersed in the aqueous phase is also becoming more widespread, and a particularly interesting development is the use of surface- or polymer-fixed catalysts which are reversibly released into the aqueous solution only in certain defined temperature intervals [21].

References

1 (a) J. D. Atwood, *Inorganic and Organometallic Reaction Mechanisms*, Brooks/Cole, Monterey, Canada **1985**; (b) F. Basolo, R. G. Pearson. *Mechanisms of Inorganic Reactions*. Wiley, New York. **1967**; (c) M. Gerloch, E. C. Constable, *Transition Metal Chemistry*, VCH, Weinheim **1994**; (d) C. F. Baer, R. E. Mesmer, *The Hydrolysis of Cations*, Wiley, New York **1976**; (e) D. M. Roundhill, *Advan. Organomet. Chem.* **1995**, *38*, 155; (f) F. Joó, A. Katho, *J. Mol. Catal.* **1997**, *116*, 3; (g) B. Cornils, C. W. Kohlpaintner, E. Wiebus, *Encyclopedia of Chemicals Processing and Design* (Eds.: J. McKetta, G. E. Weismantel), Dekker, New York **1998**.

2 (a) F. Joó, *Aqueous Organometallic Catalysis*, Kluwer, Dordrecht **2001**; (b) B. Auch-Schwelk, C. Kohlpaintner, *Chemie Unserer Zeit* **2001**, *35*, 306; (c) P. Wipf, R. L. Nunes, S. Ribe, *Helv. Chim. Acta* **2002**, *85*, 3478; (d) C. J. Bertole, C. A. Mims, G. Kiss, *J. Catal.* **2002**, *210*, 84; (e) D. Sinou, *Adv. Synth. Catal.* **2002**, *344*, 221; (f) F. Joo, *Acc. Chem. Res.* **2002**, *35*, 738; (g) J. Yoshida, K. Itami, *Chem. Rev.* **2002**, *102*, 3693; (h) H. Fuhrmann, T. Dwars, G. Oehme, *Chem. Unserer Zeit* **2003**, *37*, 40; (i) C. Baillie, J. L. Xiao, *Curr. Org. Chem.* **2003**, *7*, 477.

3 (a) J. P. Collman, L. S. Hegedus, J. R. Norton, R. G. Finke, *Principles and Applications of Organotransition Metal Chemistry*, University Science Books, Mill Valley, CA **1987**; (b) Ch. Elschenbroich, A. Salzer, *Organometallics, A Concise Introduction*, 2nd ed., VCH, Weinheim **1992**.

4 M. Herberhold, G. Süss, J. Ellermann. H. Gabelein, *Chem. Ber.* **1987**, *111*, 2931.

5 (a) C. Romão, F. E. Kühn, W. A. Herrmann, *Chem. Rev.* **1997**, *97*, 3197; (b) F. E. Kühn, M. Groarke, in *Applied Homogeneous Catalysis with Organometallic Compounds – A Comprehensive Handbook*, 2nd ed., Vol. 3 (Eds.: B. Cornils, W. A. Herrmann), Wiley-VCH, Weinheim **2002**, p. 1304.

6 (a) R. Alberto, A. Egli, V Abram, K. Hegetsweiler, V Gramlich, P A Schubiger, *J. Chem. Soc., Dalton Trans.* **1994**, 2815; (b) R. Alberto, R. Schibli, A. Egli, P. A. Schubiger, W. A. Herrmann, G. Artus, V. Abram, T. A. Kade, *J. Organomet. Chem.* **1995**, *492*, 217; (c) W A. Herrmann, R. Alberto, J. C. Bryan, A. P. Sattelberger, *Chem. Ber.* **1991**, *124*, 1107; (d) W A. Howard, G. Parkin, *Polyhedron* **1993**, *12*, 1253; (e) W. A. Herrmann, R. W. Fischer, W. Scherer, M. U. Rauch, *Angew. Chem. Int. Ed. Engl.* **1993**, *32*, 1157; (f) J. W. Steed, D. A. Tocker, *J. Chem. Soc., Chem. Commun.* **1991**, *22*, 1609; (g) G. Erker. C. Krüger. C. Starter, S. Werner, *J. Organomet. Chem.* **1989**, *377*, C55; (h) D. V. McGrath, R. H. Grubbs, J. W. Ziller, *J. Am. Chem. Soc.* **1991**, *113*, 3611; (i) D. M. Lynn, S. Kanaoka, R. H. Grubbs, *J. Am. Chem. Soc.* **1996**, *118*, 784; (j) S. Wache, *J. Organomet. Chem.* **1995**, *494*, 235; (k) M. Stehler-Rothlisberger, W. Hummel, P. Pittet, H. Burgi, A. Ludi, M. E. Merbach, *Inorg. Chem.* **1988**, *27*, 1358; (l) M. S. Eisen, A. Haskel, H. Chen, M. M. Olmstead, D. P. Smith, M. F. Maestre, R. H. Fish, *Organometallics* **1995**, *14*, 2806; (m) S. Ogo, H. Chen, M. M. Olmstead, R. H. Fish, *Organometallics* **1996**, *15*, 2009; (n) F. E. Kühn, C. C. Romão, W. A. Herrmann, in *Houben-Weyl – Science of Synthesis*, Category 1, Vol. 2 (Ed.: T. Imamoto) **2002**, 111; (o) S. Alves, A. Paulo, J. D. G. Correia, A. Domingos, I. Santos, *J. Chem. Soc., Dalton Trans.* **2002**, 4714; (p) J. Bernard, K. Ortner, B. Spingler, H. J. Pietzsch, R. Alberto, *Inorg. Chem.* **2003**, *42*, 1014.

7 M. Herberhold, G. Süss-Fink, *Angew. Chem. Int. Ed. Engl.* **1975**, *14*, 700.

8 W. A. Herrmann, A. Egli, E. Herdtweck, R. Alberto, F. Baumgärtner, *Angew. Chem. Int. Ed. Engl.* **1996**, *35*, 432.

9 W. A. Herrmann, N. W. Huber, O. Runte, *Angew. Chem. Int. Ed. Engl.* **1995**, *34*, 2187.

10 (a) W. A. Herrmann, R. Serrano, H. Bock, *Angew. Chem. Int. Ed. Engl.* **1984**, *23*, 383; (b) F. E. Kühn, W. A. Herrmann, R. Hahn, M. Elison, J. Blümel, E. Herdtweck, *Organometallics* **1994**, *13*, 1601.

11 (a) W. A. Herrmann, R. W. Fischer, *J. Am. Chem. Soc.* **1995**, *117*, 3223; (b) P. Gisdakis, N. Rösch, J. Mink, E. Bencze, I. S. Gonçalves, F. E. Kühn, *Eur. J. Inorg. Chem.* **2001**, 981.

12 W. A. Herrmann, J. D. G. Correia, G. R. J. Artus, R. W. Fischer, C. C. Romão, *J. Organomet. Chem.* **1996**, *520*, 139.

13 (a) A. J. Canty, H. Jin, A. S. Roberts, B. W. Skelton, A. H. White, *Organometallics* **1996**, *15*, 5713; (b) A. J. Canty, S. D. Fritsche, H. Jin, J. Patel, B. W. Skelton, A. H. White, *Organometallics* **1997**, *16*, 2175; (c) A. Milet, A. Dedieu, A. J. Canty, *Organometallics* **1997**, *16*, 5331; (d) A. Haskel, E. Keinan, *Organometallics* **1999**, *18*, 4677; (e) D. W. Lucey, D. S. Helfer, J. D. Atwood, *Organometallics* **2003**, *22*, 826.

14 (a) W. A. Herrmann, Ch. W. Kohlpaintner, *Angew. Chem., Int. Ed. Engl.* **1993**, *32*, 1524; (b) C. D. Frohning, Ch. W. Kohlpaintner, H. W. Bohnen, in *Applied Homogeneous Catalysis with Organometallic Compounds*, 2nd ed., Vol. 1 (Eds.: B. Cornils, W. A. Herrmann), Wiley-VCH, Weinheim **2002**, 31.

15 (a) B. Cornils, E. Wiebus, *Chem. Ing. Techn.* **1994**, *66*, 916; (b) B. Cornils, W. A. Herrmann, R. W. Eckl, *J. Mol. Catal.* **1997**, *116*, 27.

16 (a) A. Lubineau, J. Augé, *Modern Solvents in Organic Synthesis, Topics in Current Chemistry* **1999**, *206*, 1; (b) P. A. Grieco, *Organic Synthesis in Water*, Blackie, London **1998**; (c) U. M. Lindström, *Chem. Rev.* **2002**, *102*, 2751.

17 (a) P. E. Savage, *Chem. Rev.* **1999**, *99*, 603; (b) A. R. Katritzky, D. A. Nichols, M. Siskin, R. Murungan, M. Balasubramanian, *Chem. Rev.* **2001**, *101*, 837; (c) N. Akiya, P. E. Savage, *Chem. Rev.* **2002**, *102*, 2725.

18 K. S. Jerome, E. J. Parsons, *Organometallics* **1993**, *12*, 2991.

19 (a) P. Yang, M. L. Guo, *Coord. Chem. Rev.* **1999**, *186*, 189; (b) G. Mokdsi, M. M. Harding, *J. Inorg. Biochem.* **2001**, *83*, 205; (c) M. L. Guo, Z. J. Guo, P. J. Sadler, *J. Biol. Inorg. Chem.* **2001**, *6*, 698; (d) W. Beck, K. Severin, *Chem. Unserer Zeit* **2002**, *36*, 356; (e) J. Y. Zhang, X. K. Ke, C. Tu, J. Lin, J. Ding, L. P. Lin, H. K. Fun, X. Y. You, Z. J. Guo, *Biometals* **2003**, *16*, 485.

20 I. P. Beletskaya, A. V. Cheprakov, *Organic Synthesis in Water* (Ed.: P. A. Grieco), Blackie, London **1998**, p. 141.

21 D. Bergbreiter, *Chiral Catalyst Immobilization and Recycling* (Eds.: D. E. DeVos, I. F. J. Vankelekom, P. A. Jacobs), Wiley-VCH, Weinheim **2000**, p. 43.

2.2.3
Catalysts for an Aqueous Biphasic Catalysis

2.2.3.1
Central Atoms

Ana M. Santos and Fritz E. Kühn

2.2.3.1.1
Introduction

As has been discussed in some detail in the previous section, organometallic complexes are usually regarded as water-sensitive and several of these compounds are indeed very instable toward moisture and decompose readily in the presence of even minor amounts of water. Nevertheless, the use of water as a solvent in many catalytic reactions is of the utmost interest, due not only to its unique properties (high dielectric constant, ability to solvate both cations and anions, etc.) but also to environmental and cost–benefit reasons, when it replaces toxic, explosive, or flammable organic solvents on an industrial scale (see Sections 2.1 and 2.3). Because of these advantages, catalysis by transition metals in aqueous media has become increasingly attractive, it has received much attention, and much effort has been put into research on suitable active complexes [1]. The solubility of the transition metal complexes can be brought about by using water-solubilizing ligands or it can be due to the direct interaction of the metal center with water. Organometallic complexes are usually viewed with respect to the fields of their (potential) applications in catalysis or from the point of view of their ligands, enabling solubility and providing the right environment for particular reaction pathways. In this section a different perspective is chosen. Here, we treat the organometallic complexes enabling catalytic reactions in the aqueous phase with respect to their metal atom, giving a very brief but general overview of the most recent developments; we then turn in more detail to organometallic catalysts in a stricter sense (compounds with an M–C bond) and discuss one particular example with respect to the behavioral changes of potential catalysts if the metal atoms are changed within one group of transition metals.

2.2.3.1.2
Overview of the Metal Complexes Involved in Aqueous Catalysis

This section is not intended to give a detailed description of the applications of organometallic complexes in aqueous catalysis. In Table 1 some recent catalytic applications for different metals are presented as examples. The complexes here are considered to be organometallic in a broader context. The comparatively few examples described with organometallic compounds in a stricter sense (containing an M–C bond) will be dealt with in the following paragraph. Since only the most recent applications are mentioned, for further information see the References.

Table 1 Selected transition metal complexes described recently as catalysts in the aqueous phase.

Central atom	Catalyst	Type of reaction	Ref.
W	$\{PO_4[WO(O_2)_2]_4\}^{3-}$	Oxidation of fatty acids	[2]
Ru	$RuCl_2(PPh_3)_3$	Allylic C–H activation	[3]
Ru	$[RuCl_2(p\text{-cymene}]_2L^*{}^{a)}$	Cyclopropanation	[4]
Ru	$Na[\{O_3S(C_6H_4)CH_2C(CH_2PPh_2)_3Ru\}_2(\mu\text{-}Cl)_3]$	Hydrogenation of alkenes	[5]
Ru	$[RuCl_3(NO)(TPPMS)_2]/[RuCl_3(NO)(TPPTS)_2]^{b)}$	Hydrogenation of CO_2	[6]
Ru	$[RuCl_2(PTA)_4]^{c)}$	Hydrogenation of olefins	[7]
Os	$[Os(CO)_3Cl_2]_2$	Hydration of acetonitrile	[8]
Co	$Co(CO)_8{}^{d)}$	Pauson–Khand reaction	[9]
Rh	$Rh_2(COD)_2Cl_2$	Carbonylation, conjugated addition	[3]
Rh	$Rh_2(COD)_2BF_4$	Carbonylation, conjugated addition	[3]
Rh	$Rh(acac)(CO)_2/PNS^{e)}$	Hydroformylation	[10]
Rh	$Rh(COD)BF_4/\alpha,\alpha\text{-}\beta,\beta\text{-trehalose}$	Asymmetric hydrogenation	[4]
Rh	$RhCl(CO)(PPh_2\text{-PS-PEG-PPh}_2)^{f)}$	Hydroformylation of alkenes	[11]
Rh	$[RhCl(m\text{-TPPMS})_3]$	Hydrogenation	[7]
Ni	$Ni(COD)_2/CF_3COOH/L^{g)}$	Isomerization of allylic alcohols	[12]
Pd	Pd/C	Synthesis of asymmetric biaryls	[3]
Pd	$Pd(TTPTS)_3$	Copolymerization CO/olefins	[13]
Pd	$PdCl_2/TPPTS$	Hydrocarboxylation of styrenes	[14]
Pd	$Pd(OAc)_2$ or $PdCl_2/(bpy)^{h)}$	Oxidation of alcohols	[15]
Pd	$Pd(OAc)_2/m\text{-TPPTC}^{i)}$	Heck reaction	[16]
Pd	$PdCl_2(PhCN)_2$	Kharasch reaction	[17]
Pd	$Pd(OAc)_2/BINAS^{j)}$	Amination of aromatic halides	[18]
Pt	$cis\text{-PtCl}_2(TPPTS)_2{}^{k)}$	Alkyne hydration	[19]
Cu	$Cu(NO_3)_2/L\text{-abrine}^{l)}$	Chiral Diels–Alder reaction	[20]
Ag	$AgNO_3/(S)\text{-Tol-BINAP}^{m)}$	Allylation	[4]
Ce	$Ce(Otf)_4L^*{}^{n)}$	Allylation	[4]

Even a short look at Table 1 reveals that Ru, Rh, and Pd, the "classical" catalyst metals, are still more frequently examined than other transition metals in catalysis. These "classical" catalyst metals, however, are quite expensive, so the use of cheaper metals seems desirable if this is possible without a significant loss of activity. Nevertheless, in cases where the (organic) starting materials for the reactions to be catalyzed are relatively expensive or where significant excesses of co-ligands are used, the price of the catalyst metal may not be decisive. Anyway, changing the catalyst metal is usually only desirable when significantly higher selectivities and/ or turnovers are reached or the new catalyst is – with similar catalytic results – much more stable than its alternative/established competitors. Such general rules, however, are not strictly applicable for nonindustrial research where mechanistic or fundamental research may justify the use of complexes which are not applicable for industrial purposes.

It is also evident from Table 1 that seemingly a variety of ligands and even metal oxidation states may be applied for similar catalytic purposes. However, changes of both the surrounding ligands and the metal center make comparisons often difficult and generalizations dangerous.

2.2.3.1.3
Complexes Containing an M–C Bond in Aqueous Catalysis

Compounds containing one or more metal–carbon σ- or π- bonds (organometallic complexes in a strict sense) are still not so frequently used as catalysts in aqueous phase, despite the fact that their applications in aqueous systems are increasing. A more general account of organometallic complexes in water is given in Section 2.2.2, referring particularly to the role of water. However, the limited number of truly organometallic complexes applied in water containing organometallic catalytic systems are usually not systematically examined with respect to the influence of a change of the central atom (metal) on the behavior of the catalytic system. Ligand changes usually seem to provide an easier and subtler means of change in catalyst behavior than a change of the metal. In low oxidation state transition metal com-

---

Footnotes to Table 1
a) L* = bis(hydroxymethyldihydroxyoxazolyl)pyridine.
b) TPPMS = P(*m*-C$_6$H$_4$SO$_3$Na)Ph$_2$.
c) PTA = 1,3,5-triaza-7-phospha-adamantane.
d) Water was used as reaction promoter; worse yields were obtained with water as solvent.
e) PNS = Ph$_2$PCH$_2$CH$_2$C(O)NHC(CH$_3$)$_2$CH$_2$SO$_3$Li.
f) PS-PEG = polystyrene-poly(ethyleneglycol).
g) 1,4-bis(diphenylphosphanyl)butane.
h) bpy = substituted 2,2'-bipyridine.
i) *m*-TPPTC = P(*m*-C$_6$H$_4$CO$_2$Li)$_3$.
j) BINAS = sulfonated 2,2'-bis(diphenylphosphinomethyl)-1,1'-binaphthyl.
k) TPPTS = P(*m*-C$_6$H$_4$SO$_3$Na)$_3$.
l) N-(α)-methyl-L-tryptophan.
m) BINAP = 2,2'-bis(diphenylphosphino)-1.1'-binaphthyl.
n) L* = (*S*,*S*)-2,6-bis(4-isopropyl-2-oxazolin-2-yl)pyridine.

plexes, changes of the metal are usually regarded as of less significant influence than in high oxidation states, but such general statements should be regarded with great caution. Changes of both metal (often including its formal oxidation state) and (at least some of the) ligands are of course applied as well. In this case it is very hard to ascribe reliably and accurately the observed behavior changes to a defined source (such as the changed metal alone). In this section recent applications of complexes containing a metal–carbon bond in aqueous-phase catalysis are described briefly.

Molybdocene compounds have been known since the 1960s and have a very rich reaction chemistry. Recently, detailed investigations dealing with $[Cp_2MoCl_2]$ $(Cp = \eta^5\text{-}C_5H_5)$ have been reported, mainly due to its potential as anti-tumour agent [21]. This has led to a particular interest in the study of the aqueous chemistry of molybdocenes and to the discovery of new catalytic properties of these compounds. $[Cp'_2Mo(\mu\text{-}OH)_2MoCp'_2](OTs)_2$ **(1)** $(Cp' = \eta^5\text{-}C_5H_4CH_3$, OTs = p-toluenosulfonate) was prepared from $[Cp'_2MoCl_2]$ and HOTs in aqueous acetone [22]. This compound was found to catalyze intra- and intermolecular H/D exchange with alcoholic substrates, through the intermediacy of a π-carbonyl complex, and was therefore capable of activating C–H bonds in aqueous solution.

The reaction of nitriles with water to form amides was also found to be catalyzed by complex **1** [23]. This is a commercially relevant reaction since amides are used as lubricants, detergent additives, drug stabilizers, etc. [24]. In aqueous solution this complex was found to be in equilibrium with $[Cp'_2Mo(OH)(H_2O]^+$ **(2)**, and this monomer was proposed to be the active hydration catalyst. The hydration was proposed to occur by an intramolecular attack of a hydroxide ligand on a coordinating nitrile. The reaction occurred under mild conditions, preventing the autocatalytic formation of acrylamide. Although this system allows the hydration of simple and functionalized nitriles, including acrylonitrile, which is hydrated exclusively at the C≡N position, its major drawback resides in the fact that the reactions are subject to product inhibition.

With the work by Grubbs et al. [27] and Herrmann et al. [28], the use of ruthenium carbene complexes as homogeneous catalysts for the ROMP (Ring-Opening Metathesis Polymerization) of olefins was established (see Section 2.4.4.3). The development of catalysts that can catalyze living polymerization in water was an important goal to achieve, especially for applications in biomedicine. In this context, two water-soluble ruthenium carbene complexes (**3** and **4**) have been reported that act as initiators for the living polymerization of water-soluble monomers in a quick and quantitative manner [29].

The polymerization must be conducted in the presence of an acid (that leaves the ruthenium–alkylidene bond intact) that eliminates any hydroxide ion present that would lead to catalyst decomposition. In solution under acid conditions, one of the phosphines is protonated; the monophosphine species formed is surprisingly stable. This species is not formed stoichiometrically, but always coexists with the biphosphine species, which is less reactive. No surfactants or organic solvents are needed. The same research group also presented a Ni(II) system [30] capable of producing high molecular weight polyethylene, polymerizing functionalized olefins, and requiring no co-catalyst. One of the compounds was tested in a toluene/water

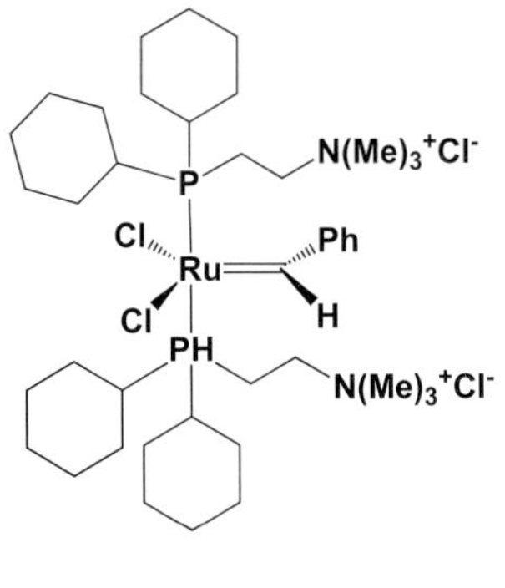

3

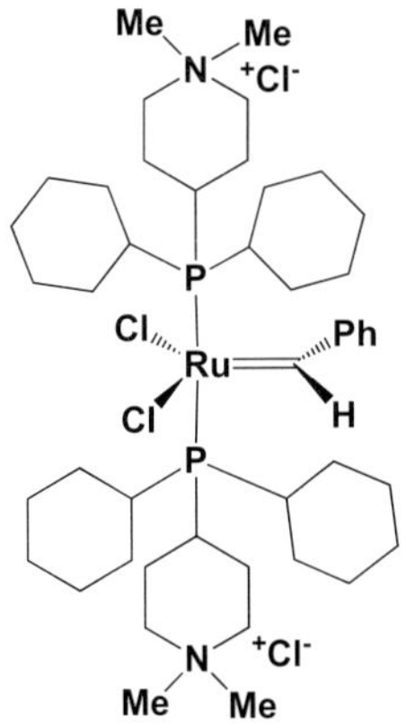

4

biphasic system, and the presence of water was found not to decrease the catalytic performance.

The compound $[Cp^*Ir(H_2O)_3]^{2+}$ ($Cp^* = \eta^5\text{-}C_5Me_5$) was found to be an efficient catalyst precursor for the hydrogenation of water-soluble compounds [25]. Such organometallic aqua complexes are very interesting, since their properties change drastically with variations of the solution pH, due to deprotonation of the aqua ligands [26]. The active catalyst was found to be a binuclear μ-hydride complex $[(Cp^*Ir)_2(\mu\text{-}H)(\mu\text{-}OH)(\mu\text{-}HCOO)]^+$ (5) that could be isolated and characterized.

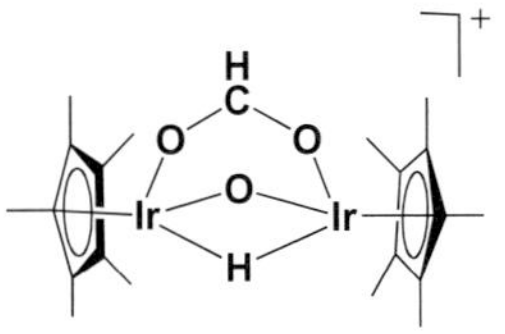

5

The pH-dependent transfer hydrogenation of water-soluble carbonyl compounds in the presence of a hydrogen donor (HCOONa) was studied for a different series of carbonyl compounds, namely straight-chain aldehydes, cyclic aldehydes, ketones, aldehyde-acids and keto-acids. The reaction rates were found to be higher for aldehydes than for ketones. Furthermore there was a strong pH dependence of the reaction, the best results being obtained at pH 3.2. This value corresponds to the higher yield of the catalyst formed from the catalyst precursor and HCOONa.

A Pd(II) complex bearing a diimine ligand (6) [31] was also found to catalyze the polymerization of ethylene in aqueous conditions. Highly branched or linear polyethylenes can be obtained at high rates. It is interesting to note that an aqueous suspension of the precursor complex displays a high activity and stability, but solutions in acetone/water mixtures are inactive toward ethylene. The polymers obtained have a higher molecular weight and are in general less branched than the ones obtained in organic solvents under otherwise identical conditions.

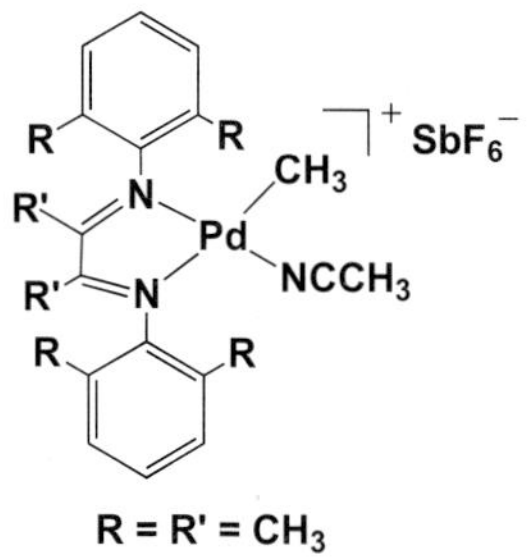

$$R = R' = CH_3$$

6

A further example of the application of an organometallic palladium complex as polymerization catalyst was presented by Sheldon et al. [32]. The active complex can be obtained by reacting [PdCl(CH$_3$)COD] **(7)** in methanol with a tetrasulfonated diphosphane, under abstraction of a chloro ligand by AgOTf. The complex can be isolated as an air-stable white solid and shows a remarkable activity on the copolymerization of ethylene and CO in water. In fact, activities of 32.2 kg copolymer per gram Pd are reached, corresponding to TOFs > 61 000 h^{-1}.

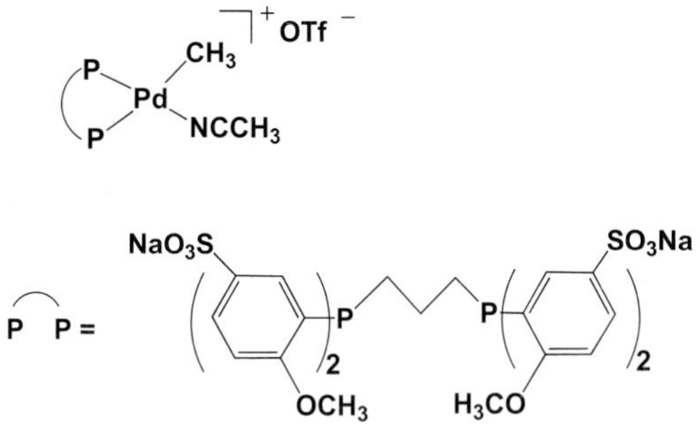

7

2.2.3.1.4
Mn, Tc, Re Complexes H$_3$C–MO$_3$ as an Example of Metal Variation in Potential Catalysts for Aqueous Systems

Methyltrioxorhenium(VII) (MTO, **8**) has been established as a powerful catalyst for a broad variety of organic transformations [33]. MTO has also found several applications in catalytic reactions performed in the aqueous phase, due to its pronounced stability to both water and temperature. A more detailed description of the application of MTO in aqueous systems is given in Section 2.4.3.3. When the particular importance of the presence of the Re–C bond in MTO was noted, the question arose of whether it would be possible to synthesize similar transition metal complexes derived from metals cheaper than Re, which would be applicable for the same purposes and would have comparable stabilities **(8–10)** [34]. Furthermore, would the lower homologues of Re parallel the chemistry of MTO and its derivatives, or differ principally as found within group 8 transition metals? Not much was known of the organometallic chemistry of technetium in the high valencies in the late 1980s [35].

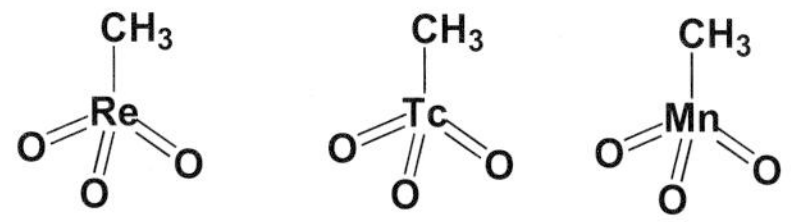

MTO   (unknown)

8   9   10

The reasons for this fact were manifold. Technetium is an artificial element and accordingly rare and not readily available. Therefore it was not examined in great detail with respect to its organometallic chemistry in the early days of this research field. However, since one of the isotopes, Tc-99(m), possesses interesting decay characteristics for application in nuclear medicine coordination chemistry in aqueous systems, it was investigated thoroughly with the goal of yielding compounds to be applied in life sciences. The main focus during this exploration was originally put on classical Werner-type complexes rather than on organometallics. During the late 1980s, growing interest arose in the high oxidation state chemistry of Tc, following the exploration of the chemistry of $(C_5(CH_3)_5))ReO_3$ and the finding of the catalytic usefulness of MTO [36]. In 1990, the synthesis of H_3C-TcO_3 was reported, starting from Tc_2O_7 and $Sn(CH_3)_4$, a pathway seemingly similar to the preparation of MTO [Eqs. (1) and (2)] [37]. However, while the reaction of Re_2O_7 with $Sn(CH_3)_4$ takes place in boiling THF and is completed under these conditions after ca. 4 h, the synthesis of CH_3TcO_3 (9) has to be performed below room temperature to avoid decomposition of the product, which is quite sensitive to both temperature (decomposition above ca. 20 °C) and moisture (formation of pertechnetium acid). Besides the desired product and the synthetically (in this particular procedure) unavoidable byproduct $(CH_3)_3Sn-O-TcO_3$, a dimer of composition $(CH_3)_2OTc(\mu-O)_2TcO(CH_3)_2$ is formed, due to reduction of Tc(VII).

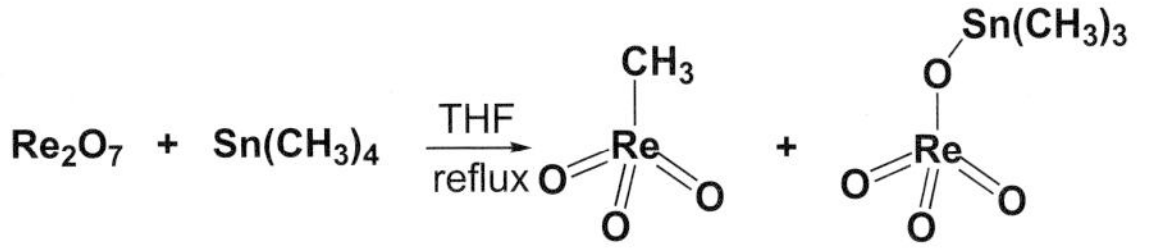

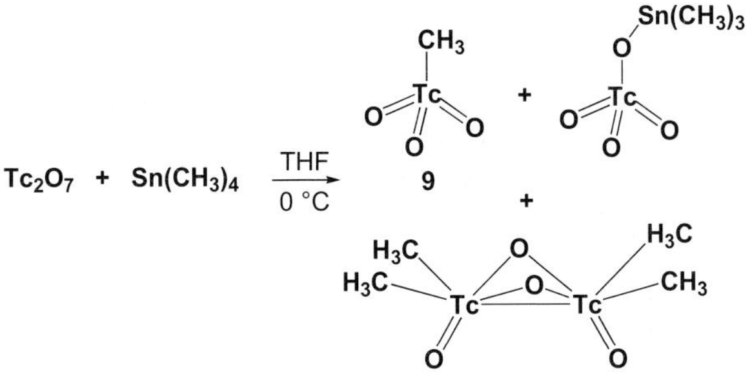

An analogous by-product is not formed during the synthesis of MTO. CH_3TcO_3 does not form stable adducts with N-donor Lewis bases, such as bipyridine and quinuclidine, a reaction that is quite characteristic for MTO and also of importance for its application in olefin epoxidation catalysis [38]. Vibrational spectroscopic investigations of CH_3TcO_3 indicate much weaker M=O interactions than in the case of its Re congener. In other reactions, MTO and CH_3TcO_3 differ not only marginally but also fundamentally. Olefins such as cyclohexene add to the Tc complex to form a glycolate Tc(V) compound. In the presence of water and acids this complex decomposes to yield the 1,2-diol and the disproportionation products $TcO_2(H_2O)_x$ and $[TcO_4]^-$ [Eq. (3)]. As observed for OsO_4, this reaction sequence occurs catalytically [39]. Based on these results it has been concluded that the even more powerful olefin-cleaving oxidizing agent RuO_4 might correspond to the still unknown $H_3C–MnO_3$ [37a].

$$\text{(3)}$$

It is interesting to note the attempts to synthesize $(C_5(CH_3)_5)TcO_3$ in this context. The Re homologue was originally prepared from $(C_5(CH_3)_5)Re(CO)_3$ and H_2O_2 [40]. The synthesis of "$(C_5(CH_3)_5)TcO_3$" was reported to work likewise; the reported structure of the product, however, was unparalleled in Re chemistry: "$(C_5(CH_3)_5)TcO_3$" was described as a linear polymer of formal composition $[(C_5(CH_3)_5)Tc(\mu\text{-}O)_3Tc]_n$, having an unusually short Tc–Tc distance [41]. Despite the fact that the structure has been rationalized theoretically [42] after it was described, its validity has been questioned, particularly since the published synthesis proved to be irreproducible and problems with the published structure became evident [37a]. A subsequent re-examination of the structural data of the Tc compound and a detailed examination of the structure of its Re congener (which suffers from both twinning and disorder phenomena) by Cotton et al. revealed that the compound of composition $[(C_5(CH_3)_5)Tc(\mu\text{-}O)_3Tc]_n$ was very likely in reality to be its Re congener (!) and "$(C_5(CH_3)_5)TcO_3$" possibly never existed in fact [43]. Nevertheless, the existence of $(C_5(CH_3)_5)TcO_3$ has been predicted on the basis of theoretical calculations [44].

Theoretical investigations (density functional investigations studying the influence of relativistic effects on the reactivity of the metal center) have been performed to account for the differences in chemical behavior of MTO and its Tc congener, which should be isostructural with their different behavior therefore originating solely or at least overwhelmingly from electronic reasons [45]. In order to quantify the Lewis acidity of the central metal atom, adduct formation with NH_3 as a probe molecule has been examined. MTO was calculated to form stable base adducts, which is in accordance with experimental results. The calculation for the

Tc complex yielded longer M–N bond distances and lower M–N stretching frequencies indicating fairly small association energies, also in accord with the above-mentioned experimental result that CH_3TcO_3 does not form Lewis adducts. From the calculated NH_3 adduct formation energies supported by an analysis of the charge distribution it was concluded that Re is a somewhat stronger Lewis acidic center than Tc in the MO_3 moiety. According to an analysis of Mulliken populations and of core and valence level shifts, charge transfer from NH_3 occurs preferentially to the Re center in the case of MTO and to the oxo ligands in CH_3TcO_3. From the charge distribution in CH_3MO_3 and in the corresponding NH_3 adducts it can be concluded that Re is the stronger and more reactive Lewis acidic metal center, as it withdraws electronic charge from all its ligands, including the electronegative oxygens. The hardness, calculated from the HOMO–LUMO splitting and from the M–O displacement derivatives of the charges as well as of the dipole moment, increases in the order $CH_3TcO_3 < OsO_4 < CH_3ReO_3$. This result rationalizes the fact that CH_3TcO_3 and OsO_4 preferentially react with softer bases, whereas MTO prefers to react with hard bases. Accordingly, MTO reacts with the harder Lewis base H_2O_2 to form mono- and bisperoxo complexes. Instead of oxidizing olefins directly, it even exhibits reactivity with epoxides to react further to glycolate complexes if no H_2O_2 is present. On the other hand, the MO_3 moiety of CH_3TcO_3 and OsO_4 has been calculated to be more polarizable. Therefore, the preferred reaction partners are soft Lewis bases, e.g., olefins through their π-bonds [45]. In fact, CH_3TcO_3 takes a place between MTO and the dihydroxylating agent MnO_4^- [46]. Based on Hartree–Fock results, H_3C-MnO_3 has been predicted to be unstable [44]. Considering the stability discussed above and sensitivity differences observed between MTO and CH_3TcO_3 this might really be the case. When considering $CpMnO_3$, it turned out by a DF theory approach [47] that the Cp ligand should be bound in a $η^1$ fashion, not in a $η^5$ fashion as observed for $CpReO_3$ [48]. It has to be noted, however, that $CpReO_3$ shares neither the water solubility nor the temperature stability of MTO and seems also to be much less applicable to catalytic reactions. However, despite the fact that in theoretical examinations it is usually regarded as an more easily calculated substitute for the experimentally much better examined $Cp*ReO_3$, the chemical behavior of these two complexes seem to be surprisingly different [48]. $CpReO_3$ seems to decompose under formation of $η^3$- and $η^1$-intermediates at elevated temperatures [49]. Accordingly, $CpTcO_3$ and $CpMnO_3$ might be even more unstable than their CH_3MO_3 congeners and should therefore not be applicable as catalysts in an aqueous medium in a comparable fashion to MTO.

In summary, the theoretical and experimental results obtained for CH_3MO_3 (M = Mn, Tc, Re) clearly show that the Tc complex is much more prone to reduction than the Re complex MTO. While the Tc(VI) complex $(CH_3)_3OTc(μ-O)_2TcO(CH_3)_2$ forms as a byproduct of the synthesis of CH_3TcO_3 from Tc_2O_7 and $Sn(CH_3)_4$ even below room temperature, the Re(VI) congener $(CH_3)_3ORe(μ-O)_2ReO(CH_3)_2$ is only made available by means of the much stronger methylation/reducing agent $Zn(CH_3)_2$ [50]. CH_3TcO_3 is very sensitive to water, while MTO is water-stable and only sensitive to basic reaction conditions, forming perrhenate as a reaction product [51]. While MTO forms peroxo complexes with H_2O_2, which are also water-stable and react with olefins under epoxide formation, CH_3TcO_3 reacts with olefins directly under formation of glycolate, Tc being a softer Lewis acid than Re in the MO_3 moiety. CH_3MnO_3 has been predicted to be far more sensitive than the Tc compound and has accordingly not been isolated to date.

This subsection shows that at least for certain transition metal complexes, such as high oxidation state compounds of group VII, a change of the metal in otherwise identical complexes may have severe consequences for their behavior in both catalysis and reaction chemistry, turning complexes from stable, water-soluble and catalytically very versatile, e.g., MTO- to temperature- and moisture-sensitive ones with very different chemical behavior, such as (CH_3TcO_3), or even to ones that do not exist, such as (CH_3MnO_3). When changing from CH_3TcO_3 to OsO_4 the differences in reaction chemistry are seemingly less pronounced, but a change of the metal alone is also not straightforwardly feasible. Cases where the changes from metal to metal cause less pronounced changes in the chemistry, however, are also known. Examples are some cases of low oxidation state transition metal chemistry and several aspects of the chemistry of the rare-earth complexes.

References

1 I. T. HORVÁTH, F. JOÓ, *Aqueous Organometallic Chemistry and Catalysis* (Eds.), Kluwer, Dordrecht **1995**.

2 I. V. KOZHEVNIKOV, G. P. MULDER, M. C. STEVERINK-DE ZOETE, M. G. OOSTWALD, *J. Mol. Catal. A:* **1998**, *134*, 223.

3 C.-J. LI, *Acc. Chem. Res.* **2002**, *35*, 533.

4 U. M. LINDSTRÖM, *Chem. Rev.* **2002**, *102*, 2751.

5 I. ROJAS, F. K. LINARES, N. VALENCIA, C. BIANCHINI, *J. Mol. Catal. A:* **1999**, *144*, 1.

6 A. KATHO, Z. OPRE, G. LAURENCZY, F. JOO, *J. Mol. Catal. A:* **2003**, *204*, 143.

7 F. JOÓ, *Acc. Chem. Res.* **2002**, *35*, 738.

8 E. CARIATI, C. DRAGONETTI, L. MANASSERO, D. ROBERTO, F. TESSORE, E. LUCENTI, *J. Mol. Catal. A:* **2003**, *204*, 279.

9 S. RIBE, P. WIPF, *Chem. Commun.* **2001**, 299.

10 A. M. TRZECIAK, J. J. ZIÓLKOWSKI, *Coord. Chem. Rev.* **1999**, *190–192*, 883.

11 Y. UOZUMI, M. NAKAZONO, *Adv. Synth. Catal.* **2002**, *344*, 274.

12 H. BRICOUT, E. MONFLIER, J. F. CARPENTIER, A. MORTREUX, *Eur. J. Inorg. Chem.* **1998**, *11*, 1739.

13 G. VERSPUI, J. FEIKEN, G. PAPADOGIANAKIS, R. A. SHELDON, *J. Mol. Catal. A:* **1999**, *146*, 299.

14 F. BERTOUX, S. TILLOY, E. MONFLIER, Y. CASTANET, A. MORTREUX, *J. Mol. Catal. A:* **1999**, *138*, 53.

15 G. J. TEN BRINK, I. W. C. E. ARENDS, M. HOOGENRAAD, G. VERSPUI, R. A. SHELDON, *Adv. Synth. Catal.* **2003**, *345*, 497.

16 R. AMENGUAL, E. GENIN, V. MICHELET, M. SAVIGNAC, J. P. GENET, *Adv. Synth. Catal.* **2002**, *344*, 393.

17 D. MOTODA, H. KINOSHITA, H. SHINOKUBO, K. OSHIMA, *Adv. Synth. Catal.* **2002**, *344*, 261.

18 G. WULLNER, H. JANSCH, S. KANNENBERG, F. SCHUBERT, G. BOCHE, *Chem. Commun.* **1998**, 1509.

19 L. W. FRANCISCO, D. A. MORENO, J. D. ATWOOD, *Organometallics* **2001**, *20*, 4237.

20 S. OTTO, G. BOCCALETTI, J. B. F. N. ENGBERTS, *J. Am. Chem. Soc.* **1998**, *120*, 4238.

21 (a) M. M. HARDING, G. MOKDSI, J. P. MACKAY, M. PRODIGALIDAD, S. W. LUCAS, *Inorg. Chem.* **1998**, *37*, 2432; (b) L. Y. KUO, M. G. KANATZIDIS, M. SABAT, A. L. TIPTON, T. J. MARKS, *J. Am. Chem. Soc.* **1991**, *113*, 9027.

22 C. Balzarek, D. R. Tyler, *Angew. Chem. Int. Ed.* **1999**, *38*, 2406.

23 K. L. Breno, M. D. Pluth, D. R. Tyler, *Organometallics* **2003**, *22*, 1203.

24 (a) F. Matsuda, *Chemtech.* **1977**, *7*, 306; (b) C. E. Haberman, in J. J. Kroschwitz, M. Howe-Grand (Eds.), *Encyclopedia of Chemical Technology*, John Wiley and Sons, New York **1991**, Vol. 1, p. 346.

25 S. Ogo, N. Makihara, Y. Watanabe, *Organometallics* **1999**, *18*, 5470.

26 See, for example: (a) U. Koelle, *Coord. Chem. Rev.* **1994**, *135/136*, 623 and references therein; (b) D. T. Richens, *The Chemistry of Aqua Ions*, John Wiley, Chichester **1997**, p. 592 and references therein.

27 (a) P. Schwab, M. B. France, J. W. Ziller, R. H. Grubbs, *Angew. Chem. Int. Ed. Engl.* **1995**, *34*, 2039; (b) E. L. Dias, S. T. Nugyen, R. H. Grubbs, *J. Am. Chem. Soc.* **1997**, *119*, 3887.

28 T. Weskamp, W. C. Schattenmann, M. Spiegler, W. A. Herrmann, *Angew. Chem. Int. Ed.* **1998**, *37*, 2490.

29 D. M. Lynn, B. Mohr, R. H. Grubbs, *J. Am. Chem. Soc.* **1998**, *120*, 1627.

30 T. R. Younkin, E. F. Connor, J. I. Henderson, S. K. Friedrich, R. H. Grubbs, D. A. Bansleben, *Science* **2000**, *287*, 460.

31 A. Held, S. Mecking, *Chem. Eur. J.* **2000**, *6*, 4623.

32 G. Verspui, F. Schanssema, R. A. Sheldon, *Angew. Chem. Int. Ed.* **2000**, *39*, 804.

33 (a) C. C. Romão, F. E. Kühn, W. A. Herrmann, *Chem. Rev.* **1997**, *97*, 3197; (b) K. P. Gable, *Adv. Organomet. Chem.* **1997**, *41*, 127.

34 W. A. Herrmann, *J. Organomet. Chem.* **1995**, *500*, 149.

35 R. Alberto, R. Schibli, A. Egli, U. Abram, S. Abram, T. A. Kaden, P. A. Schubiger, *Polyhedron* **1998**, *17*, 1133.

36 W. A. Herrmann, F. E. Kühn, *Acc. Chem. Res.* **1997**, *30*, 169.

37 (a) W. A. Herrmann, R. Alberto, P. Kiprof, F. Baumgärtner, *Angew. Chem. Int. Ed. Engl.* **1990**, *29*, 189; (b) W. A. Herrmann, J. G. Kuchler, J. K. Felixberger, E. Herdtweck, W. Wagner, *Angew. Chem. Int. Ed. Engl.* **1988**, *27*, 394.

38 (a) F. E. Kühn, A. M. Santos, P. W. Roesky, E. Herdtweck, W. Scherer, P. Gisdakis, I. V. Yudanov, C. di Valentin, N. Rösch, *Chem. Eur. J.* **1999**, *5*, 3603; (b) P. Ferreira, W. M. Xue, E. Bencze, E. Herdtweck, *Inorg. Chem.* **2001**, *40*, 5834.

39 (a) A. Veldkamp, G. Frenking, *J. Am. Chem. Soc.* **1994**, *116*, 4937; (b) P. O. Norrby, H. C. Kolb, K. B. Sharpless, *J. Am. Chem. Soc.* **1994**, *116*, 8470.

40 (a) W. A. Herrmann, R. Serrano, H. Bock, *Angew. Chem. Int. Ed. Engl.* **1984**, *23*, 383; (b) W. A. Herrmann, J. D. G. Correia, F. E. Kühn, G. R. J. Artus, C. C. Romão, *Chem. Eur. J.* **1996**, *2*, 168.

41 B. Kanellakopoulos, B. Nuber, K. Raptis, M. Ziegler, *Angew. Chem. Int. Ed. Engl.* **1989**, *28*, 1055.

42 A. W. E. Chan, R. Hoffmann, *Inorg. Chem.* **1991**, *30*, 1086.

43 A. K. Burell, F. A. Cotton, L. M. Daniels, V. Petricek, *Inorg. Chem.* **1995**, *34*, 4253.

44 T. Szyperski, P. Schwerdtfeger, *Angew. Chem. Int. Ed. Engl.* **1989**, *28*, 1228.

45 S. Köstlmeier, V. A. Nasluzov, W. A. Herrmann, N. Rösch, *Organometallics* **1997**, *16*, 1786.

46 J. March, *Advanced Organic Chemistry*, 3rd ed., Wiley, New York **1985**, p. 732.

47 P. Gisdakis, N. Rösch, *J. Am. Chem. Soc.* **2001**, *123*, 697.

48 F. E. Kühn, W. A. Herrmann, R. Hahn, M. Elison, J. Blümel, E. Herdtweck, *Organometallics* **1994**, *13*, 1601.

49 (a) W. A. Herrmann, F. E. Kühn, C. C. Romão, *J. Organomet. Chem.* **1995**, *489*, C56; (b) W. A. Herrmann, F. E. Kühn, D. A. Fiedler, M. Mattner, M. Geisberger, H. Kunkeley, A. Vogler, S. Steenken, *Organometallics* **1995**, *14*, 5377.

50 (a) W. A. Herrmann, C. C. Romão, P. Kiprof, J. Behm, M. R. Cook, M. Taillefer, *J. Organomet. Chem.* **1991**, *413*, 11; (b) F. E. Kühn, J. Mink, W. A. Herrmann, *Chem. Ber.* **1997**, *130*, 295.

51 (a) M. M. Abu-Omar, J. P. Hansen, J. H. Espenson, *J. Am. Chem. Soc.* **1996**, *118*, 4966; (b) K. A. Brittingham, J. P. Espenson, *J. Chem. Soc., Dalton Trans.* **1999**, *38*, 744.

2.2.3.2
Ligands

2.2.3.2.1
Phosphorus-Containing Ligands

Othmar Stelzer (†), Stefan Rossenbach, Dietmar Hoff, Marcel Schreuder-Goedheijt, Paul C. J. Kamer, Joost N. H. Reek, and Piet W. N. M. van Leeuwen

a) Monophosphines

General Features, Scope, and Limitations

The water solubility of the catalysts in aqueous two-phase catalysis can be achieved by appropriate modification of the phosphine ligands with polar groups such as SO_3^-, COO^-, NMe_3^+, OH, etc. [1a]. First attempts to carry out transition metal catalyzed reactions using water-soluble phosphine ligands date back to 1973 [1b]. The monosulfonated derivative of Ph_3P had already been synthesized in 1958 [1c], and the well-known standard ligand TPPTS (trisulfonated triphenylphosphine) was reported by Kuntz in 1975 [2]. The development of new types of hydrosoluble phosphine ligands with "tailor-made" structures for highly active and selective two-phase catalysts is an ongoing challenge to chemists working in this field. Aspects of the topic under review have been covered in the literature by review articles [3–6] and monographs [7].

Anionic Phosphines

Phosphines Containing Sulfonated Aromatic and Aliphatic Groups

Water-soluble phosphines of this type reported in the literature so far are collected as Structures **1–6**. Direct sulfonation of the neutral "mother phosphine" with oleum, introduced originally by Kuntz in 1975 [2] for the preparation of TPPTS (**1a**), is still the most important procedure; it has also been used for the syntheses of an extended series of TPPTS-type catalyst ligands, e.g. **1b** (R = Ph [8], C_6H_{13} [9]), **1c** (R = alkyl, CYHex; R' = Me, OMe [10]), **1d** (R = 4-F-C_6H_4 [11]), **1e** [12]. Sulfonated tris (ω-phenyl-alkyl)phosphines **5a** [13] and their *p*-phenylene analogues **5b** [14] as well as the bicyclic phosphine **4** [15] were obtained in an analogous manner. Triarylphosphines containing activated aryl groups, e.g., the dibenzofuran system, can be sulfonated with concentrated sulfuric acid under mild conditions (**1g**) [16c]. The kinetics of PPh_3 sulfonation have been investigated by Lecomte and Sinou [16a]. The selectivity has been studied by Chaudhari [16b].

The formation of phosphine oxides, which is a serious disadvantage inherent to this synthetic procedure [17], may be suppressed by addition of boric acid to the reaction mixtures [18].

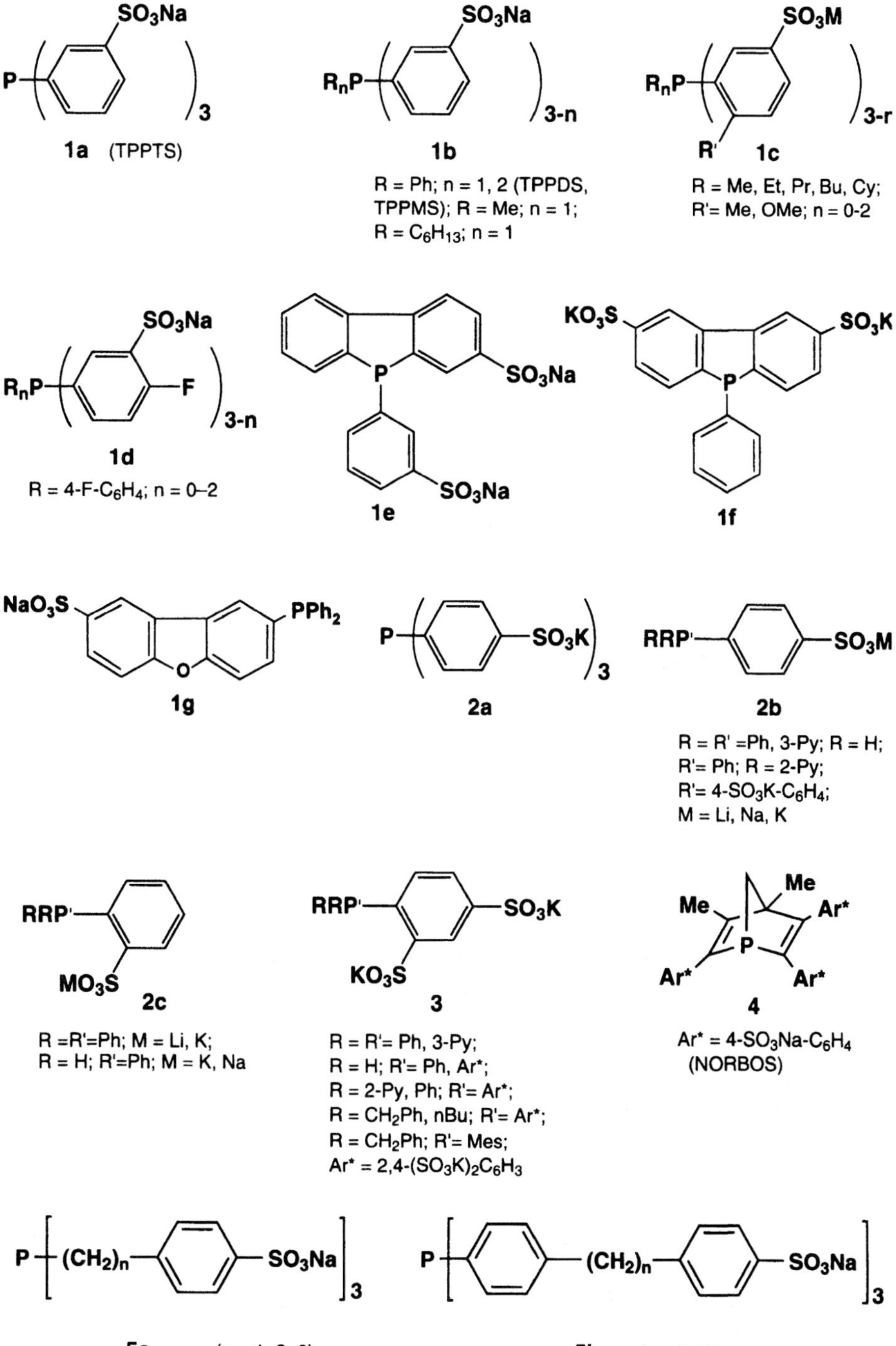

Figure 1 Water-soluble phosphines containing sulfonated aryl and alkyl side chains.

Ruthenium complexes of TPPTS and TPPMS [19–22] have been employed as catalysts or catalyst precursors for the hydrogenation of α,β-unsaturated carbonyl compounds in biphasic systems [19–23a]; cf. Section 2.4.2. Other ligands **7–13** are known and have been described [20, 23b, c].

The technically most important biphasic process in the Ruhrchemie/Rhône-Poulenc hydroformylation of propene uses the *in-situ* Rh(I) catalyst HRh(CO)-(TPPTS)$_3$ [6, 24]. Its formation from Rh(CO)$_2$(acac) and TPPTS in a syngas atmosphere has been studied in detail [25, 26]. The BINAS-Na (**10**)/Rh catalyst showed an outstanding performance in propene hydroformylation [15]. Binuclear thiolato-bridged rhodium complexes **11** have been used in 1-octene hydroformylation as precatalysts [41]. For details of the hydroformylation, see Section 2.4.1.1 [15, 27, 28].

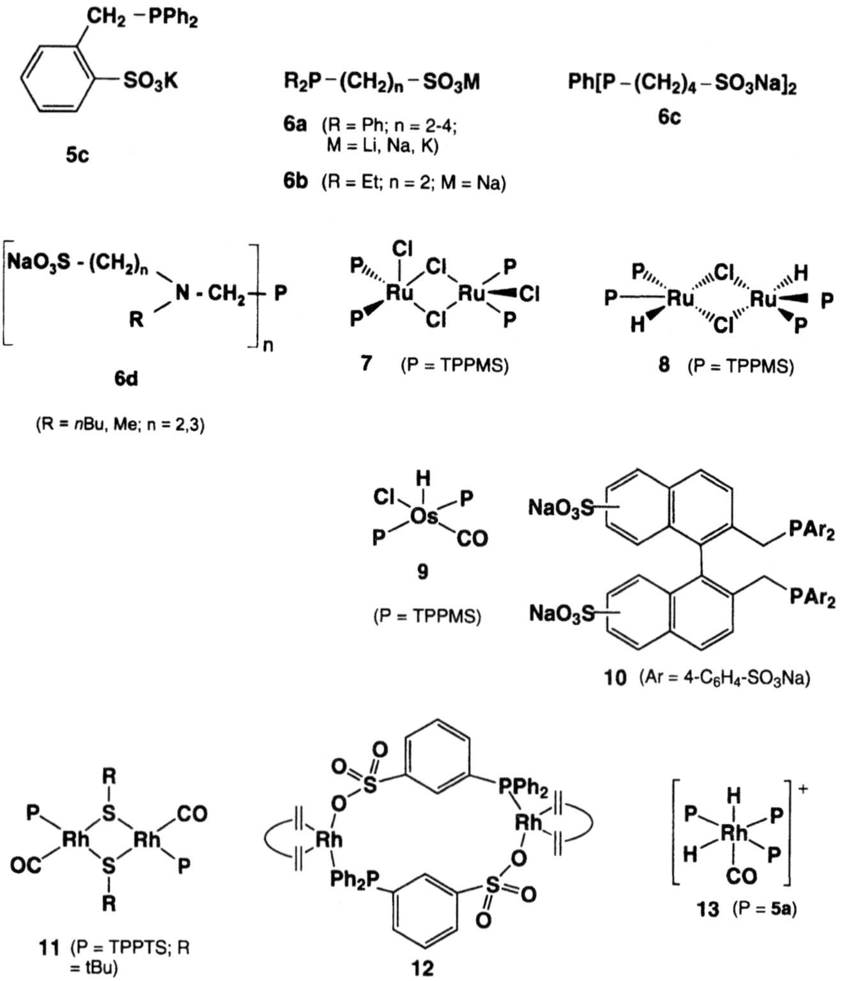

Figure 2 Structure of catalysts, ligands, and intermediates.

Rh(I) catalysts containing TPPTS-type ligands with electron-withdrawing groups in the aromatic rings show higher n/i selectivity in hydroformylation reactions. Thus *para*-fluorinated derivatives of TPPTS and TPPDS (**1d**), which are weaker bases and stronger π-acids than TPPTS and TPPDS [29], in 1-hexene hydroformylation gave n/i selectivities of 93 : 7, compared with 86 : 14 for the nonfluorinated ligands [11]. The application of aqueous two-phase catalysis in hydroformylation of longer chain olefins is hampered, however, in most cases by their low solubility in water. This problem has been overcome by using phosphine ligands like **5b**, showing a more pronounced surface-active character than TPPTS.

High ionic strengths stabilize the hydration sphere by minimizing the electrostatic repulsions between the sulfonate groups. As a consequence the dissociation energy for TPPTS with formation of the active species $\{HRh(CO)(TPPTS)_2\}$ will be increased, thus lowering the activity of the catalyst. In agreement with this reasoning a value of 30.6 or 22.4 kcal mol^{-1} has been calculated for the barrier of exchange at high or low ligands and complex concentration, respectively [30].

Palladium complexes of TPPTS and TPPMS have been employed extensively as catalysts for carbonylation, hydroxycarbonylation, and C–C cross-coupling reactions. Hydroxycarbonylation of bromobenzene in a biphasic medium using Pd(TPPTS)$_3$ as catalyst yields benzoic acid, which remains in the aqueous phase, thus avoiding the direct recycling of the catalyst [31]. The formation of Pd(TPPTS)$_3$ from PdCl$_2$ and TPPTS in aqueous solution has been studied in detail by ^{17}O, $\{^{1}H\}^{31}$P, and ^{35}Cl NMR spectroscopy. The complex [PdCl(TPPTS)$_3$]$^+$Cl$^-$ obtained initially is reduced by excess TPPTS, TPPTSO being formed. A more attractive synthesis of Pd(TPPTS)$_3$ involves the facile reduction of [PdCl(TPPTS)$_3$]$^+$Cl$^-$ with CO [32].

Ni(0) complexes of TPPTS have been employed as catalysts for the hydrocyanation of dienes and unsaturated nitriles. Product linearity and catalyst lifetimes can be improved if the catalysis is performed in a xylene/water biphasic system by using TPPTS as co-catalyst [33]. The Ni(0)/TPPTS complexes employed may be obtained by electrochemical reduction of Ni(CN)$_2$ in water in presence of TPPTS [34].

Dual metal Rh/Ir/TPPTS catalysts have been employed in the hydroaminomethylation of 1-pentene with synthesis gas (CO/H$_2$ = 1 : 5) and ammonia in an aqueous two-phase system [35]. With this catalyst system good n/i selectivities and a rapid hydrogenation of the imine to amine have been observed. Both primary and secondary amines can be formed with high selectivity by variation of the olefin/NH$_3$ ratio.

Rhodium(I) complexes HRh(CO)L$_3$ of alkali metal phenylphosphinoalkylsulfonates (L = **6a**, n = 3, 4; **6c**) have been used as catalysts for hydroformylation of higher olefins (e.g., *n*-1-tetradecene) in methanolic solution. The catalyst could be recovered with loss of activity by extraction of the isolated product with water.

Phosphines Containing Carboxylated Aromatic Groups and Side Chains

Structures **14–22** (Figure 3) are hydrophilic phosphine ligands bearing carboxylic groups. In contrast to their alkali metal salts, the free acids show only moderate

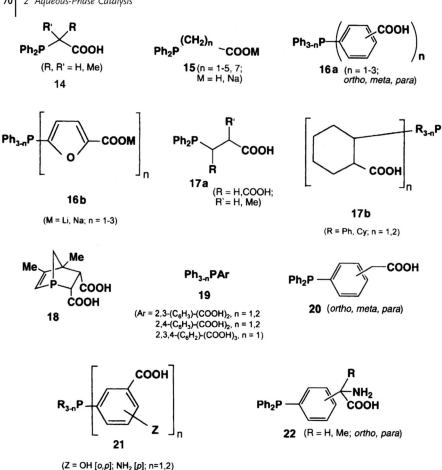

Figure 3 Carboxylated phosphines.

solubilities in water. The ligands **15–17a** have been obtained by standard preparative methods comprising alkylation and arylation of alkali metal organophosphides in organic solvents (THF, DME, dioxane) [36–38]. Improved synthetic procedures based on the nucleophilic phosphination of fluorobenzoic acids [39] or Pd-catalyzed P–C coupling reactions of bromo- and iodobenzoic acid with primary or secondary phosphines [40] are known. These methods are of broad applicability and can be used also for the syntheses of multiply functionalized phosphinocarboxylic acids (**19, 21**) including the diphenylphosphinophenylacetic acids (**20**) and the novel phosphine ligands containing amino acid moieties (**22**) [41].

Compared with their sulfonated analogues, phosphine ligands containing carboxylic acid moieties have been much less investigated as catalyst components, although some of them (e.g., Ph_2PCH_2COOH, **14**) have already been applied at an early stage in the Shell Higher-Olefin Process (SHOP; cf. Section 3.3) [42], the first large-scale industrial biphasic but nonaqueous catalytic process.

Phosphines Containing Phosphonated Aromatic Groups and Alkyl Side Chains

The water-soluble phosphonate-functionalized phosphines have been obtained by halogen–metal exchange on bromophenyldiphenylphosphines with *n*BuLi followed by reaction with diethyl chlorophosphate and subsequent hydrolysis [43].

Catalytic applications have been described [44–48].

Cationic Phosphines

The synthesis of AMPHOS, the prototype of cationic phosphines, requires intermediate protection of the phosphorus in $Ph_2P-(CH_2)_2-NMe_2$ by oxidation or coordination to a transition metal before N-quaternization [49, 50a]. In the case of the ligands **23a** and **23b** (Figure 4), borane was used as the protecting group.

Ligands **25** and **26a** containing guanidinium moieties constitute a novel type of cationic phosphine ligand showing extreme solubilities in water. The guanidinium groups were introduced into the corresponding aminoalkyl- and aminoaryl-phosphines by addition of cyanamides R_2N-CN (R = H, Me) or 1*H*-pyrazole-1-carboxamidine [50b,c].

Catalytic applications have been described [40b,c, 51–53].

Rhodium(I) and cobalt(0) complexes of AMPHOS, $[(nbd)Rh(AMPHOS)_2]^{3+}$, and $[Co(CO)_3(AMPHOS)]_2(PF_6)_2$, were already being used in the early 1980s for the hydrogenation and hydroformylation of maleic and crotonic acid or styrene and 1-hexene in water or aqueous biphasic systems [49]. The lower effectiveness of the Co catalyst was attributed to the lighter metal's proclivity to oxidation and phosphine

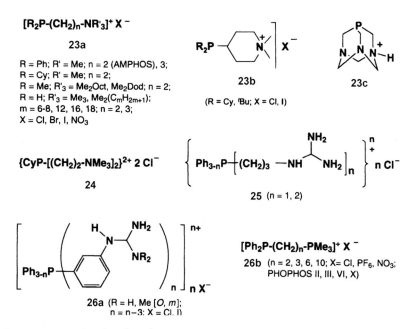

Figure 4 Cationic phosphine ligands.

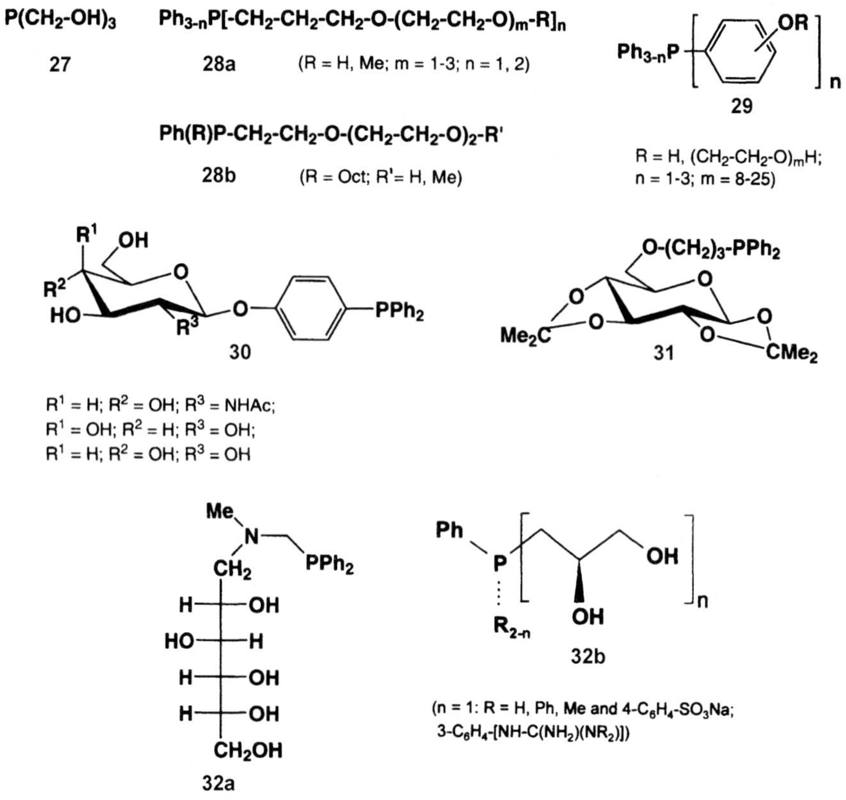
P(CH2-OH)3
27
Ph3-nP[-CH2-CH2-CH2-O-(CH2-CH2-O)m-R]n
28a
(R = H, Me; m = 1-3; n = 1, 2)
Ph3-nP
OR
n
29
R = H, (CH2-CH2-O)mH;
n = 1-3; m = 8-25)
Ph(R)P-CH2-CH2-O-(CH2-CH2-O)2-R'
28b
(R = Oct; R'= H, Me)
R1
OH
R2
HO
O
R3
PPh2
30
R1 = H; R2 = OH; R3 = NHAc;
R1 = OH; R2 = H; R3 = OH;
R1 = H; R2 = OH; R3 = OH
O-(CH2)3-PPh2
Me2C
CMe2
31
Me
N
CH2
PPh2
H
OH
HO
CH2OH
32a
Ph
P
R2-n
n
32b
(n = 1: R = H, Ph, Me and 4-C6H4-SO3Na;
3-C6H4-[NH-C(NH2)(NR2)])

Figure 5 Nonionic water-soluble phosphines.

dissociation. Binuclear thiolato-bridged Rh(I) complexes, $[Rh(CO)(AMPHOS)(tBuS)]_2^{2+}2[BPh_4]^-$ are highly active catalysts for the hydrogenation of unsaturated alcohols and acids [106]. Water-soluble Rh(I) complexes of monoprotonated 1,3,5-triaza-7-phosphaadamantane (PTA), $[Rh(PTA)_2(PTAH)Cl]Cl$ (PTAH = **24c**), have been used as catalysts for the regioselective reduction of unsaturated to saturated aldehydes with sodium formate in an aqueous biphasic system or dihydrogen in presence of ethanol as co-solvent.

Nonionic Water-Soluble Phosphines

The concept of attaining water solubility by incorporation of hydroxylic groups into the ligand periphery has either to attracted little attention. Large-scale syntheses for **27** [54] based on K_2PtCl_4-catalyzed addition of aqueous formaldehyde to PH_3 or decomposition of commercially available $[P(CH_2OH)_4]^+Cl^-$ have been published recently [55]. Ligands of type **28a** have been obtained by addition of $Ph_{3-n}PH_n$ to ethylene glycol monoallyl ethers [56]. Chiral ligands of this type, e.g., **28b**, showing unprecented η^3-mode bonding to transition metals, were prepared by Mathieu and co-workers [57]. The hydroxyphenylphosphines **29** are accessible by multistage syntheses [58] or by making use of the Pd-catalyzed P–C coupling reaction employing iodophenols as starting materials [40]. Ethoxylation of mono-, di- and tris-*p*-hydroxy-triphenylphosphines with ethylene oxide yields polyether-substituted triphenyl-phosphines (PETPPs) designed for use in micellar [59, 60] and thermally regulated [61] phase-transfer catalysis. Ligands whose water solubility is inversely dependent on temperature were first reported by Bergbreiter et al. [62]. This subject will be discussed in more detail in Chapter 7.

Ligands incorporating the diphenylphosphine moiety into sugar structures have been reported by a number of authors [68, 64]. In most cases, however, the hydroxyl functions were fully (e.g., **31**) or at least in part protected, the solubilities of these ligands in water being low [56]. By two-phase glycosidation of acetyl-protected halopyranoses with *p*-hydroxyphenyldiphenylphosphine at ambient temperature and subsequent O–Ac deprotection the aryl-*β-O*-glycosides of glucose, galactose, and glucosamine (**30**) could be obtained [65]. There are scattered examples of applications in catalysis [66, 67]. The same is true for **32**.

b) Diphosphines and Other Phosphines

Diphosphines: Introduction of Sulfonate Groups by Direct Sulfonation

Water-soluble diphosphines have been synthesized by controlled sulfonation in oleum (**33–37**). Although the number of chiral diphosphines reported is ever increasing, the number of sulfonated achiral diphosphines remains limited and includes BINAS-8 (**33**) [68] and BISBIS (**34**) [69], both of which contain an aromatic bridge between the two phosphino moieties, and the alkyl-bridged bidentates 1,2-bis[di(3-sulfonatophenyl)phosphino]ethane (**35**) [70] and 1,3-bis[di(3-sulfonato-phenyl)phosphino]propane (**36**) and some analogues [71]. Sulfonated analogs of Xantphos have also been prepared using this strategy and these water-soluble ligands from rhodium complexes that are very selective in the hydroformylation of alkenes [72].

The precise control of the number and position of the sulfonate groups remains a challenge. The concentration of SO_3 and reaction temperature have a major effect on the degree of sulfonation [73]. Oxidation of phosphorus may be avoided using a method developed by Herrmann et al., which makes use of a superacidic medium derived from orthoboric acid and anhydrous sulfuric acid [74a]. In that manner, BINAP was sulfonated almost without formation of phosphine oxides ($\leq$ 3 mol%).

The diphosphines **33** and **34** were tested as ligands in the rhodium-catalyzed *biphasic* hydroformylation of propene. Both catalysts were found to exhibit higher activities and gave rise to higher *l/b* ratios [68] than TPPTS. Furthermore, it was shown that displacement of the *biphenyl* unit of **34** by a *binaphthyl* unit in **33** leads to an increase of the catalytic activity which was ascribed to electronic effects.

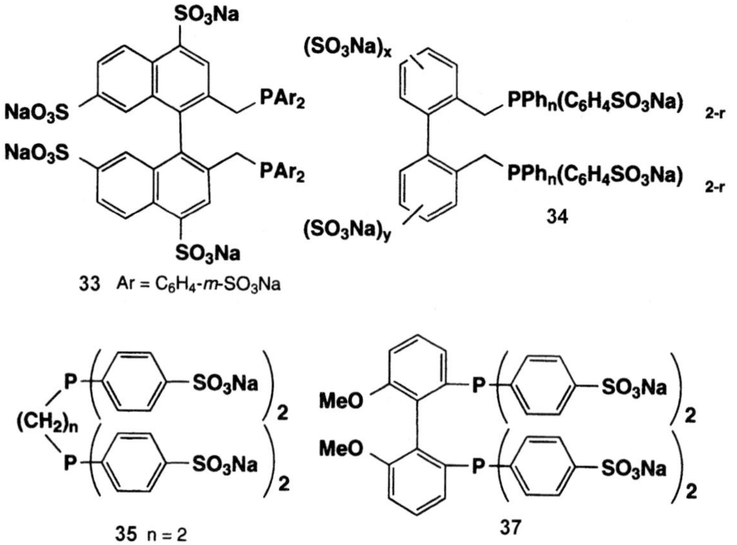

33 Ar = C_6H_4-*m*-SO_3Na; 34; 35 n = 2; 36 n = 3; 37

Figure 6 Directly sulfonated diphosphines.

Diphosphines with Quaternized Aminoalkyl or Aminoaryl Groups

Quaternization of nitrogen atoms of aminoalkyl or aminoaryl diphosphines opens up another route to water-soluble diphosphines. One of the first (chiral) diphosphines that has been synthesized by this method was shown to be very soluble in water [75], which stimulated several others to use similar procedures to prepare chiral water-soluble diphosphines [76]. Before quaternization, the phosphorus atom has to be protected either by oxidation (e.g., with hydrogen peroxide) or by coordination to a metal. Subsequent reduction or decomplexation then affords the water-soluble diphosphine; for example, the *chiral* diphosphine **38** (Figure 7) [77]. Stelzer and co-workers used a strategy involving a palladium-catalyzed P–C coupling reaction to introduce water-soluble groups in (di)phosphine ligands [78], which can be used to prepare guanidinium-modified (di)phosphines [79].

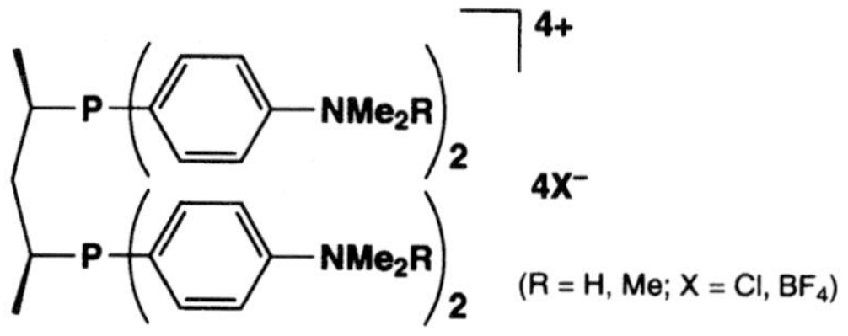

38

Figure 7 Chiral water-soluble quaternized diphosphine.

Diphosphines with Hydroxyalkyl or Polyether Substituents

After initial prelineary reports [80–82], the synthesis of 1,2-bis(di(hydroxyalkyl)-phosphino)ethans (**39**) was described [83a] [Eq. (1)].

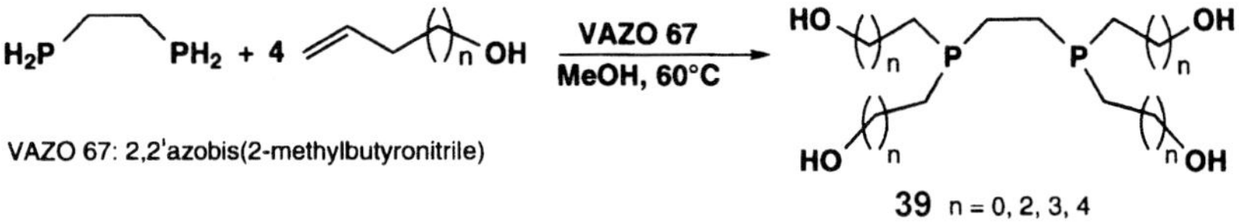

VAZO 67: 2,2'azobis(2-methylbutyronitrile)

39 n = 0, 2, 3, 4

Several metal complexes were prepared and characterized. For example, the reaction of **39** with NiCl$_2$ in a 2 : 1 ratio in methanol gave an orange product identified as Ni(DHPrPE)$_2$Cl [**40**; Eq. (2)]. The solubility of the metal complexes in water can be attributed to the hydrophilic hydroxyl groups, which surround the outside of the complexes, but even more to the charges on the molecules [84].

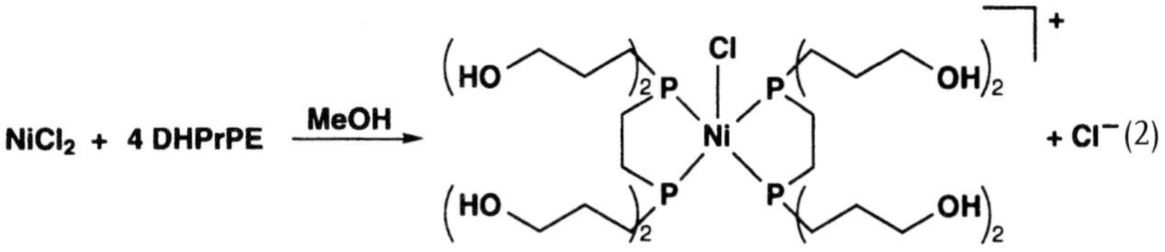

40

Substituting a phosphine with a polyether chain may also make the phosphine water-soluble. However, diphosphines of the type **41** (**41–43**) are only soluble in water when n > 15 [85]. This type of material can also be used to prepare thermally responsive catalysts [86].

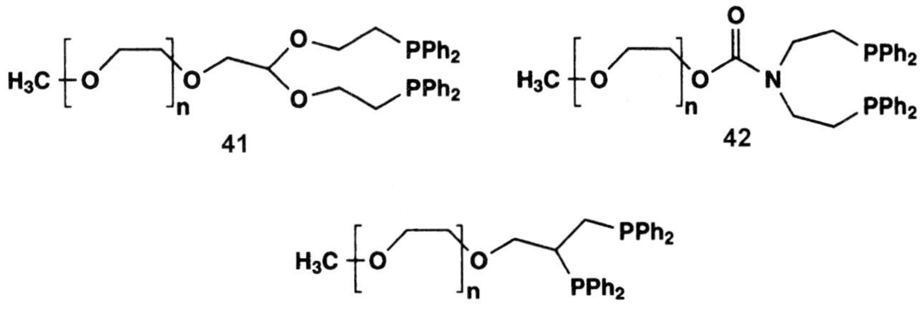

41   **42**

43

Figure 8 Diphosphines containing polyether chains.

Carboxylated Diphosphines

Phosphines with carboxylic groups were some of the earliest water-soluble phosphines investigated (see also p. 70) [87]. The Podlahová's group prepared a diphosphine, the phosphine analogue of ethylenediaminetetraacetic acid, which is obtained as a monohydrate of the tetrasodium salt (**44**) [87]. Jegorov and Podlahová published a short review on the catalytic uses of these carboxyalkylphosphines [88].

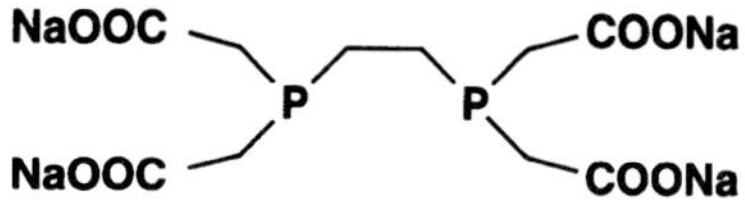

Van Doorn developed a water-soluble diphosphine based on 2,3-bis(diphenyl-phosphino)maleic anhydride, which was converted into the biscarboxylic acid **45** with sodium hydroxide (Scheme 1) [89a]. The compound was also described by Tyler and co-workers in 1993 [89b].

An advantage over other methods for the preparation of water-soluble phosphines is that diphosphine **45** could easily be purified by extraction followed by crystallization from diethyl ether. The structure of **35** was confirmed by X-ray crystallography. The ligand is readily soluble in aqueous solution at pH 5 or higher. For example, the solubility is > 1 M at pH 7. At this pH the ligand is deprotonated.

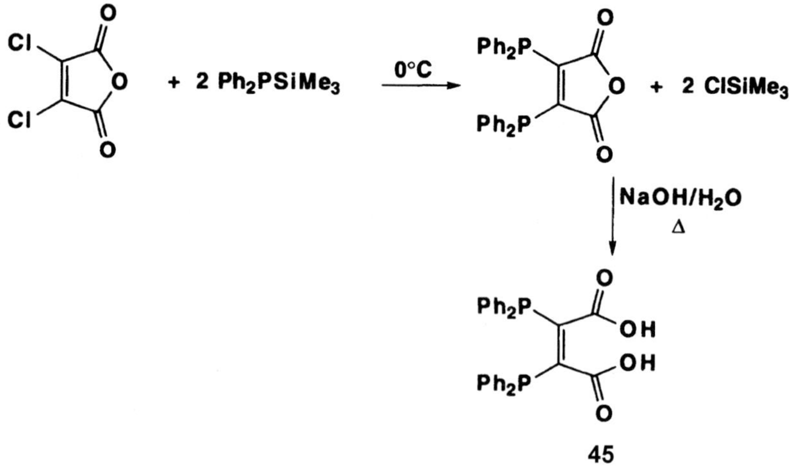

Scheme 1 Synthesis of the water-soluble diphosphine **45** based on 2,3-bis(diphenylphosphino)maleic anhydride.

Amphiphilic Diphosphines

Another approach to water-soluble phosphines with the emphasis on metal recycling has been reported [90]. The ligands synthesized are based on BISBI and **46–48**, and hydroformylation (for example) can be conducted in a homogeneous (organic) phase [90a]. After it has been used in the hydroformylation of olefins the catalyst can be removed by acidic extraction. It has been established that these novel diphosphines form active and highly selective catalysts.

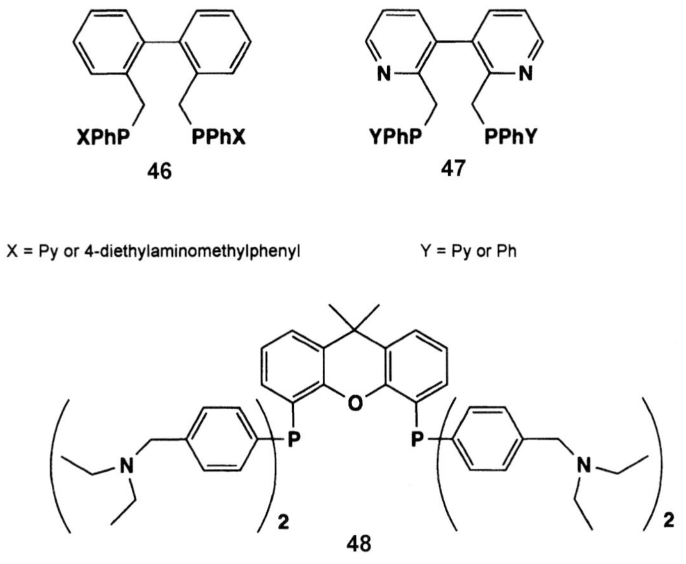

XPhP
PPhX
46
YPhP
PPhY
47
X = Py or 4-diethylaminomethylphenyl
Y = Py or Ph
48

Figure 9 Amphiphilic diphosphine ligands.

Other Phosphines

Water-soluble mono- and diphosphines represent the major class of phosphines used in aqueous-phase homogeneous catalysis. However, some new types of water-soluble phosphines have been developed, including phosphines containing sugar substructures [91] or phosphonate chains [92], and chiral sulfonated phosphines for the asymmetric hydrogenation of dehydropeptides [93] and phosphines with amino acid moieties [94].

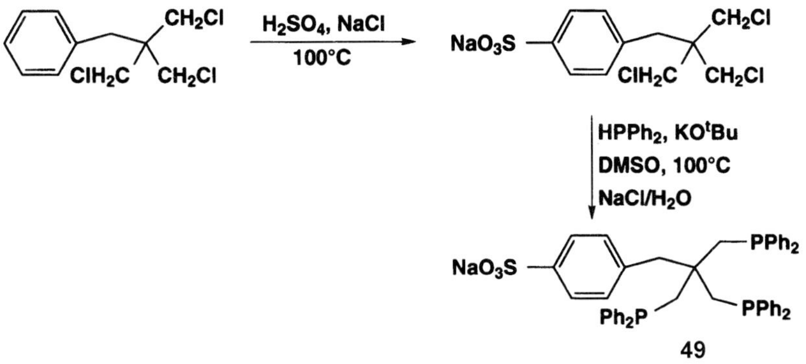

CH_2Cl
ClH_2C
CH_2Cl
H_2SO_4, NaCl
100°C
NaO_3S
$HPPh_2$, KO^tBu
DMSO, 100°C
NaCl/H_2O
PPh_2
Ph_2P
49

Scheme 2 Synthesis of SULPHOS (**54**).

Bianchini et al. [95] synthesized another water-soluble triphosphine, $NaO_3S(C_6H_4)CH_2C(CH_2PPh_2)_3$, the so-called SULPHOS (**49**; Scheme 2). This ligand is a TRIPHOS ligand [96] with a hydrophilic tail attached to the bridgehead carbon atom. It has been developed to facilitate catalyst separation in biphasic but aqueous systems. The synthesis involves the treatment of benzyltris(chloroethyl)-methane with concentrated sulfuric acid at 100 °C, which results in the regioselective *para* sulfonation of the phenyl ring. Reaction of $NaO_3S(C_6H_4)CH_2C(CH_2Cl)_3$ with $KPPh_2$ in DMSO at 100 °C gives **49**. The complexes **(49)**Rh(cod) and **(49)**Rh(CO)$_2$ were used in biphasic catalysis as hydrogenation and hydroformylation catalysts respectively.

Katti et al. have developed the new water-soluble triphosphine $PhP[CH_2CH_2P(CH_2OH)_2]$ **(50)** [97]. This triphosphine, upon interaction with $[Rh(cod)Cl]_2$ under biphasic (water/dichloromethane) conditions, produces a water-soluble rhodium(I) complex in which the rhodium center is tripodally coordinate via the PPh and $P(CH_2OH)_2$ functionalities, as established by NMR spectroscopy. The presence of PPh and $P(CH_2OH)_2$ groups of disparate basicities makes it unusual in comparison with the traditional triphosphines (e.g., TRIPHOS: $PhP(CH_2CH_2PPh_2)_2$). They suggested that the different basicities in **50** may aid the development of catalytically useful transition metal complexes in which the weaker of the two different M–P bonds may be reversibly cleaved in the presence of a substrate molecule.

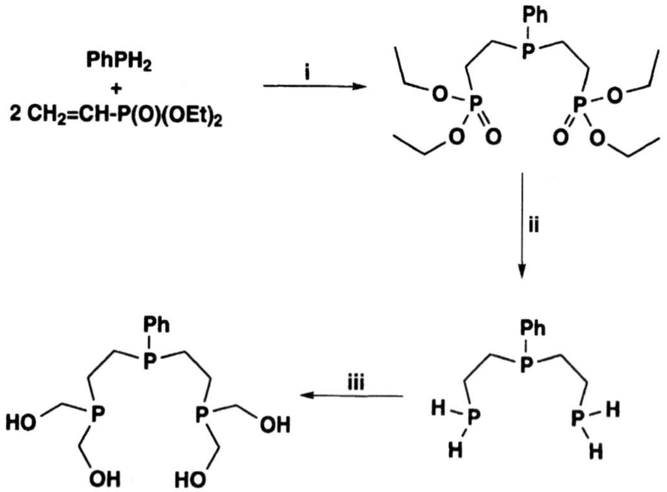

Scheme 3 Synthesis of water-soluble triphosphine **50**.
Reagents: (i) KOBu^t, THF; (ii) LiAlH$_4$, Et$_2$O; (iii) HCHO, EtOH.

References

1 (a) J. MANASSEN, *Catalysis: Progress in Research* (Eds.: F. BASOLO, R. L. BURWELL), Plenum Press, London **1973**, pp. 177, 183; (b) J. CHATT, G. J. LEIGH, R. M. SLADE, *J. Chem. Soc., Dalton Trans.* **1973**, 2021; (c) S. AHRLAND, J. CHATT, N. R. DAVIES, A. A. WILLIAMS, *J. Chem. Soc.* **1958**, 276.

2 Rhône-Poulenc Recherche (E. KUNTZ), FR 2.314.910 (**1975**).

3 M. Barton, J. D. Atwood, *J. Coord. Chem.* **1991**, *24*, 43; D. Sinou, *Bull. Soc. Chim. Fr.* **1987**, *3*, 480.

4 W. A. Herrmann, C. W. Kohlpaintner, *Angew. Chem.* **1993**, *105*, 1588; *Angew. Chem., Int. Ed. Engl.* **1993**, *32*, 1524.

5 B. Cornils, *Angew. Chem.* **1995**, *107*, 1709; *Angew. Chem., Int. Ed. Engl.* **1995**, *34*, 1574.

6 B. Cornils, E. G. Kuntz, *J. Organomet. Chem.* **1995**, *502*, 177; B. Cornils, W. A. Herrmann, R. W. Eckl, *J. Mol. Catal. A: Chem.* **1997**, *116*, 27.

7 B. Cornils, W. A. Herrmann, *Applied Homogeneous Catalysis with Organometallic Compounds* (Eds.: B. Cornils, W. A. Herrmann), VCH, Weinheim **1996**.

8 Ruhrchemie AG (H. Bahrmann, B. Cornils, W. Lipps, P. Lappe, H. Springer), CA 1.247.642 (**1988**), *Chem. Abstr.* *111*, 97 503; Ruhrchemie AG (L. Bexten, B. Cornils, D. Kupies), DE 3.431.643 (**1986**).

9 I. Tóth, I. Guo, B. E. Hanson, *J. Mol. Catal. A: Chem.* **1997**, *116*, 217.

10 Hoechst AG (G. Albanese, R. Manetsberger, W. A. Herrmann), EP 704.450 (**1996**).

11 B. Fell, G. Papadogianakis, *J. Prakt. Chem.* **1994**, *336*, 591.

12 Hoechst AG (G. Albanese, R. Manetsberger, W. A. Herrmann, R. Schmid), EP 704.452 (**1996**).

13 T. Bartik, B. Bartik, B. E. Hanson, I. Guo, I. Tóth, *Organometallics* **1993**, *12*, 164.

14 H. Ding, B. E. Hanson, J. Bakos, *Angew. Chem.* **1995**, *107*, 1728; *Angew. Chem., Int. Ed. Engl.* **1995**, *34*, 1645; H. Ding, B. E. Hanson, T. Bartik, B. Bartik, *Organometallics* **1994**, *13*, 3761.

15 W. A. Herrmann, C. W. Kohlpaintner, R. B. Manetsberger, H. Bahrmann, H. Kottmann, *J. Mol. Catal. A: Chem.* **1995**, *97*, 65.

16 (a) L. Lecomte, D. Sinou, *Phosphorus, Sulfur, Silicon* **1990**, *53*, 239; (b) B. M. Bhanage, S. S. Divekar, R. M. Deshpande, R. V. Chaudhari, *Org. Proc. Res. Dev.* **2000**, *3*, 342; (c) A. E. Sollewijn Gelpe, J. J. N. Veerman, M. Schreuder Goedheijt, P. C. J. Kamer, P. W. N. M. van Leeuwen, H. Hiemstra, *Tetrahedron* **1999**, *55*, 6657.

17 T. Bartik, B. Bartik, B. E. Hanson, T. Glass, W. Bebout, *Inorg. Chem.* **1992**, *31*, 2667.

18 W. A. Herrmann, G. P. Albanese, R. B. Manetsberger, P. Lappe, H. Bahrmann, *Angew. Chem.* **1995**, *107*, 893; *Angew. Chem., Int. Ed. Engl.* **1995**, *34*, 811; Hoechst AG (G. Albanese, R. Manetsberger, W. A. Herrmann, C. Schwer), EP 704.451 (**1996**).

19 A. Andriollo, J. Carrasquel, J. Mariño, F. A. López, D. E. Páez, I. Rojas, N. Valencia, *J. Mol. Catal. A: Chem.* **1997**, *116*, 157.

20 R. A. Sánchez-Delgado, M. Medina, F. López-Linares, A. Fuentes, *J. Mol. Catal. A: Chem.* **197**, *116*, 167.

21 M. Hernandez, P. Kalck, *J. Mol. Catal. A: Chem.* **1997**, *116*, 117.

22 M. Hernandez, P. Kalck, *J. Mol. Catal. A: Chem.* **1997**, *116*, 131.

23 (a) J. M. Grosselin, C. Mercier, G. Allmang, F. Grass, *Organometallics* **1991**, *10*, 2126; (b) A. Bényei, J. N. W. Stafford, A. Kathó, D. J. Darensbourg, F. Joó, *J. Mol. Catal.* **1993**, *84*, 157; (c) D. J. Darensbourg, F. Joó, A. Kathó, J. N. White-Stafford, A. Bényei, J. H. Reibenspies, *Inorg. Chem.* **1994**, *33*, 175.

24 Hoechst AG (W. A. Herrmann, J. Kulpe, J. Kellner, H. Riepl), DE 3.840.600 (**1990**); DE 3.921.295 (**1991**); Rhône-Poulenc SA (E. Kuntz), DE 2.627.354 (**1976**).

25 J. P. Arhancet, M. E. Davis, J. S. Merola, B. E. Hanson, *J. Catal.* **1990**, *121*, 327.

26 I. T. Horvath, R. V. Kastrup, A. A. Oswald, E. J. Mozeleski, *Catal. Lett.* **1989**, *2*, 85.

27 (a) H. Bahrmann, C. D. Frohning, P. Heymanns, H. Kalbfell, P. Lappe, D. Peters, E. Wiebus, *J. Mol. Catal. A: Chem.* **1997**, *116*, 35; (b) Hoechst AG (H. Bahrmann, P. Lappe, E. Wiebus, B. Fell, P. Herrmanns), DE 19.532.394 (**1997**).

28 F. Monteil, R. Queau, P. Kalck, *J. Organomet. Chem.* **1994**, *480*, 177.

29 T. Allman, R. G. Goel, *Can. J. Chem.* **1982**, *60*, 716.

30 H. Ding, B. E. Hanson, T. E. Glass, *Inorg. Chim. Acta* **1995**, *229*, 329; H. Ding, *Dissert. Abstr. Int.* **1996**, *56*, 6735.

31 F. Monteil, P. Kalck, *J. Organomet. Chem.* **1994**, *482*, 45.

32 G. Papadogianakis, J. A. Peters, L. Maat, R. A. Sheldon, *J. Chem. Soc., Chem. Commun.* **1995**, 1105.

33 Rhône-Poulenc Chimie (M. Huser, R. Perron), EP 650.959 (**1995**).

34 Rhône-Poulenc Fiber et Resin Intermediates (A. Chamard, D. Horbez, M. Huser, R. Perron), EP 715.890 (**1996**).

35 B. Zimmermann, J. Herwig, M. Beller, *Angew. Chem., Int. Ed.* **1999**, *38*, 2372.

36 W. Keim, R. P. Schulz, *J. Mol. Catal.* **1994**, *92*, 21.

37 J. A. van Doorn, N. Meijboom, *Phosphorus, Sulfur, Silicon* **1989**, *42*, 211.

38 J. E. Hoots, T. B. Rauchfuss, D. A. Wrobleski, *Inorg. Synth.* **1982**, *21*, 175.

39 M. Hingst, M. Tepper, O. Stelzer, *Eur. J. Inorg. Chem.* **1998**, 73.

40 (a) O. Herd, A. Hessler, M. Hingst, M. Topper, O. Stelzer, *J. Organomet. Chem.* **1996**, *522*, 69; (b) O. Herd, A. Hessler, M. Hingst, P. Machnitzki, M. Tepper, *Catal. Today* **1998**, *42*, 413; (c) P. Machnitzki, M. Tepper, K. Wenz, O. Stelzer, E. Herdtweck, *J. Organomet. Chem.* **2000**, *602*, 158.

41 (a) M. Tepper, O. Stelzer, T. Häusler, W. S. Sheldrick, *Tetrahedron Lett.* **1997**, *38*, 2257; (b) D. J. Brauer, S. Rossenbach, S. Schenk, M. Tepper, O. Stelzer, T. Häusler, W. s. Sheldrick, *J. Organomet. Chem.* **2000**, *598*, 116.

42 (a) M. Peuckert, W. Keim, *Organometallics* **1983**, *2*, 594; (b) Shell Int. Res. (W. Keim, T. M. Shryne, R. S. Bauer, H. Chung, P. W. Glockner, H. van Zwet), DE 2.054.009 (**1969**); W. Heim, *Chem. Ing. Techn.* **1984**, *56*, 850; (c) A. Behr, W. Keim, *Arab. J. Science Eng.* **1985**, *10*, 377.

43 (a) T. L. Schull, J. C. Feitinger, D. A. Knight, *Inorg. Chem.* **1996**, *35*, 6717; (b) T. L. Schull, J. C. Feitinger, D. A. Knight, *J. Chem. Soc., Chem. Commun.* **1995**, 1487.

44 (a) P. Machnitzki, T. Nickel, O. Stelzer, C. Landgrafe, *Eur. J. Inorg. Chem.* **1998**, 1029; (b) W. J. Dressick, C. George, S. L. Brandow, T. L. Schull, D. A. Knight, *J. Org. Chem.* **2000**, *65*, 5059.

45 S. Bischoff, A. Weigt, H. Miessner, B. Lücke, *National Meeting, Am. Chem. Soc., Div. Fuel Chem., Washington* **1995**, *40*, 114.

46 S. Lelièvre, F. Mercier, F. Mathey, *J. Org. Chem.* **1996**, *61*, 3531.

47 Hoechst AG (S. Haber, H. J. Kleiner), WO 97/05104 (**1997**).

48 SNPE (F. Mathey, D. Neibecker, A. Brèque), FR 2.588.197 (**1985**).

49 R. T. Smith, R. K. Ungar, L. J. Sanderson, M. C. Baird, *Organometallics* **1983**, *2*, 1138.

50 (a) R. T. Smith, M. C. Baird, *Transition Met. Chem.* **1981**, *6*, 197; (b) A. Hessler, O. Stelzer, H. Dibowski, K. Worm, F. P. Schmidtchen, *J. Org. Chem.* **1997**, *62*, 2362; (c) H. Dibowski, F. P. Schmidtchen, *Angew. Chem.* **1998**, *110*, 487.

51 (a) B. Mohr, D. M. Lynn, R. H. Grubbs, *Organometallics* **1996**, *15*, 4317; (b) K. H. Shaughnessy, R. A. Booth, *Org. Lett.* **2001**, *3*, 2757; (c) D. M. Lynn, B. Mohr, R. H. Grubbs, L. M. Henling, M. W. Day, *J. Am. Chem. Soc.* **2000**, *122*, 6601; (d) D. M. Lynn, B. Mohr, R. H. Grubbs, *J. Am. Chem. Soc.* **1998**, *120*, 1627; (e) T. A. Kirkland, D. M. Lynn, R. H. Grubbs, *J. Org. Chem.* **1998**, *63*, 9904.

52 F. Bitterer, S. Kucken, O. Stelzer, *Chem. Ber.* **1995**, *128*, 275; D. J. Brauer, J. Fischer, S. Kucken, K. P. Langhans, O. Stelzer, N. Weferling, *Z. Naturforsch. Teil B* **1994**, *49*, 1511.

53 E. Renaud, R. B. Russell, S. Fortier, S. J. Brown, M. C. Baird, *J. Organomet. Chem.* **1991**, *419*, 403.

54 P. G. Pringle, M. B. Smith, *Platinum Met. Rev.* **1990**, *34*, 74.

55 J. W. Ellis, K. N. Harrison, P. A. T. Hoye, A. G. Orpen, P. G. Pringle, M. B. Smith, *Inorg. Chem.* **1992**, *31*, 3026.

56 T. N. Mitchell, K. Heesche-Wagner, *J. Organomet. Chem.* **1992**, *436*, 43.

57 E. Valls, J. Suades, B. Donadieu, R. Mathieu, *J. Chem. Soc., Chem. Commun.* **1996**, 771.

58 L. Maier, *Organic Phosphorus Compounds* (Eds.: G. M. Kosolapoff, L. Maier), Wiley Interscience, New York **1972**, p. 1.

59 Z. Jin, Y. Yan, H. Zuo, B. Fell, *J. Prakt. Chem.* **1996**, *338*, 124.

60 (a) B. Fell, C. Schobben, G. Papadogianakis, *J. Mol. Catal. A: Chem.* **1995**, *101*, 179; (b) B. Fell, D. Leckel, C. Schobben, *Fat. Sci. Technol.* **1995**, *97*, 219.

61 Z. Jin, X. Zheng, B. Fell, *J. Mol. Catal. A: Chem.* **1997**, *116*, 55.

62 D. E. Bergbreiter, L. Zhang, V. M. Mariagnanam, *J. Am. Chem. Soc.* **1993**, *115*, 9295.

63 (a) M. Yamashita, M. Kobayashi, M. Sugiura, K. Tsunekawa, T. Oshikawa, S. Inokawa, H. Yamamoto, *Bull. Chem. Soc. Jpn.* **1986**, *59*, 175; (b) T. H. Johnson, G. Rangerjan, *J. Org. Chem.* **1980**, *45*, 62; (c) D. Lafont, D. Sinou, G. Descotes, *J. Organomet. Chem.* **1979**, *169*, 87.

64 H. U. Blaser, *Chem. Rev.* **1992**, *92*, 935.

65 M. Beller, J. G. E. Krauter, A. Zapf, *Angew. Chem.* **1997**, *109*, 793; *Angew. Chem., Int. Ed. Engl.* **1997**, *36*, 793.

66 G. M. Olsen, W. Henderson, *Proc. XVIIth Int. Conf. on Organometallic Chemistry*, Brisbane **1996**, p. 168.

67 D. J. Brauer, P. Machnitzki, T. Nickel, O. Stelzer, *Eur. J. Inorg. Chem.* **2000**, 65.

68 W. A. Herrmann, C. W. Kohlpaintner, R. B. Manetsberger, H. Bahrmann, H. Kottmann, *J. Mol. Catal.* **1995**, *97*, 65.

69 W. A. Herrmann, C. W. Kohlpaintner, H. Bahrmann, W. Konkol, *J. Mol. Catal.* **1992**, *73*, 191.

70 T. Bartik, B. Bunn, B. Bartik, B. E. Hanson, *Inorg. Chem.* **1994**, *33*, 164.

71 (a) G. Verspui, G. Papadogianakis, R. A. Sheldon, *Angew. Chem., Int. Ed.* **2000**, *39*, 804; (b) W. P. Mul, H. Dirkzwager, A. A. Broekhuis, H. J. Heeres, A. J. van der Linden, A. G. Orpen, *Inorg. Chim. Acta* **2002**, *327*, 147.

72 (a) M. Schreuder Goedheijt, P. C. J. Kamer, P. W. N. M. van Leeuwen, *J. Mol. Catal.* **1998**, *134*, 243; (b) M. Schreuder Goedheijt, B. J. Hanson, J. N. H. Reek, P. C. J. Kamer, P. W. N. M. van Leeuwen, *J. Am. Chem. Soc.* **2000**, *220*, 1650.

73 Hoechst AG (H. Bahrmann, P. Lappe, W. A. Herrmann, R. Manetsberger, G. Albanese), DE 4.321.512 (**1993**).

74 (a) W. A. Herrmann, G. P. Albanese, R. B. Maneetsberger, P. Lappe, H. Bahrmann, *Angew. Chem., Int. Ed. Engl.* **1995**, *34*, 811; (b) Ruhrchemie AG (R. Gärtner, B. Cornils et al.), DE 3.235.029 and 3.235.030 (**1982**).

75 U. Nagel, E. Kingel, *Chem. Ber.* **1986**, *119*, 1731.

76 For example: (a) T. Lamouille, C. Saluzzo, R. ter Halle, F. le Guyader, M. Lemaire, *Tetrahedron Lett.* **2001**, *52*,d 666; (b) P. Guerreiro, V. Ratovelomana-Vidal, J.-P. Gênet, P. Dellis, *Tetrahedron Lett.* **2001**, *52*, 3223.

77 I. Tóth, B. E. Hanson, *Tetrahedron: Asymmetry* **1990**, *1*, 895.

78 O. Herd, A. Hessler, M. Hingst, P. Machnitzki, M. Tepper, O. Stelzer, *Catal. Today* **1998**, *42*, 413.

79 (a) P. Machanitzki, M. Tepper, K. Wenz, O. Stelzer, E. Herdtweck, *J. Organomet. Chem.* **2000**, *602*, 158; (b) P. Wasserscheid, H. Waffenschmidt, P. Machnitzki, K. W. Kottsieper, O. Stelzer, *Chem. Commun.* **2001**, 451.

80 J. Chatt, G. L. Leigh, R. M. Slade, *J. Chem. Soc., Dalton Trans.* **1973**, 2021.

81 D. Klötzer, P. Mäding, R. Munze, *Z. Chem.* **1984**, *24*, 224.

82 J. Holz, A. Boerner, A. Kless, S. Borns, S. Trinkhaus, R. Seleke, D. Heller, *Tetrahedron: Asymmetry* **1995**, *6*, 1973.

83 (a) G. T. Baxley, W. K. Miller, D. K. Lyon, B. E. Miller, G. F. Nieckarz, T. J. R. Weakley, D. R. Tyler, *Inorg. Chem.* **1995**, *35*, 6688; (b) G. T. Baxley, T. J. R. Weakley, W. K. Miller, D. K. Lyon, D. R. Tyler, *J. Mol. Catal. A: Chem.* **1997**, *116*, 191.

84 P. G. Pringle, D. Brewin, M. B. Smith, K. Worboys, in *Aqueous Organometallic Chemistry and Catalysis* (Eds.: I. T. Horváth, F. Joó), Kluwer, Dordrecht **1995**, p. 111.

85 Y. Amrani, D. Sinou, *J. Mol. Catal.* **1984**, *24*, 231.

86 For a review see D. E. Bergbreiter, *Chem. Rev.* **2002**, *102*, 3345.

87 J. Podlahová, J. Podlahová, *Collect. Czech. Chem. Commun.* **1980**, *45*, 2049.

88 A. Jegorov, J. Podlahová, *Catal. Lett.* **1991**, *9*, 9.

89 (a) J. A. van Doorn, *Thesis*, University of Amsterdam **1991**, p. 31; (b) A. Avey, D. M. Schut, T. J. R. Weakley, D. R. Tyler, *Inorg. Chem.* **1993**, *32*, 233.

90 (a) A. Buhling, P. C. J. Kamer, P. W. N. M. van Leeuwen, *J. Mol. Catal. A:* **1995**, *98*, 69; (b) A. Buhling, J. W. Elgersma, S. Nkrumah, P. C. J. Kamer, P. W. N. M. van Leeuwen, *J. Chem. Soc., Dalton Trans.* **1996**, 2143; (c) A. Buhling, P. C. J. Kamer, P. W. N. M. van Leeuwen, J. W. Elgersma, K. Goubitz, J. Fraanje, *Organometallics* **1997**, *16*, 3027.

91 T. N. MITCHELL, K. HEESCHE-WAGNER, *J. Organomet. Chem.* **1992**, *436*, 43.

92 S. GANGULY, J. T. MAGUE, D. M. ROUNDHILL, *Inorg. Chem.* **1992**, *31*, 3500.

93 M. LAGHMARI, D. SINOU, A. MASDEU, C. CLAVER, *J. Organomet. Chem.* **1992**, *438*, 213.

94 (a) S. R. GILBERTSON, G. W. STARKEY, *J. Org. Chem.* **1996**, *61*, 2922; (b) D. J. BRAUER, S. SCHENK, S. ROSSENBACH, M. TEPPER, O. STELZER, T. HAUSLER, W. S. SHELDRICK, *J. Organomet. Chem.* **2000**, *598*, 116.

95 C. BIANCHINI, P. FREDIANI, V. SERNAU, *Organometallics* **1995**, *14*, 5458.

96 (a) C. BIANCHINI, A. MELI, M. PERUZZINI, F. VIZZA, P. FREDIANI, J. A. RAMIREZ, *Organometallics* **1990**, *9*, 226; (b) C. BIANCHINI, A. MELI, M. PERUZZINI, F. VIZZA, F. ZANOBINI, *Coord. Chem. Rev.* **1992**, *120*, 193; (c) V. SERNAU, G. HUTTNER, M. FRITZ, L. ZSOLNAI, O. WALTER, *J. Organomet. Chem.* **1993**, *453*, C23.

97 C. J. SMITH, V. S. REDDY, K. V. KATTI, *J. Chem. Soc., Chem. Commun.* **1996**, 2557.

2.2.3.2.2
Nitrogen- and Sulfur-Containing Water-Soluble Ligands

Philippe Kalck and Martine Urrutigoïty

Introduction

Biphasic catalysis involving the organometallic complex immobilized in the aqueous phase is concerned exclusively with transition metal complexes coordinated by water-soluble ligands which act as ancillary ligands to tune the coordination sphere of the metallic center and confer to it a good solubility in water. As catalysis is devoted to the functionalization of the substrate, the complex will contain ligands like hydride, carbonyl, allyl etc., arising from the activation of the substrate and the reactants. It is thus important that the water-soluble ligand maintains the various complexes and intermediates of the catalytic cycle in the aqueous phase. The various strategies that have been developed to modify a ligand with appropriate functionalities recall to some extent the methodology of incorporating protecting groups in organic syntheses. For instance, nitrogen-containing groups have been reviewed recently [1].

Nitrogen-Containing Ligands

Ethylenediaminetetraacetic acid or its sodium salt (EDTA) is a well-known complexing agent which can form an abundant variety of water-soluble complexes with all metals. EDTA is widely used in industry and environmental studies are concerned with the degradation of its complexes [2] or the bioaccumulation of platinum group metals [3]. Directly related to the aim of this book is the catalytic activity in various carbonylation reactions of the precursor [Ru(EDTA)(H$_2$O)], as reported by Taqui Khan and co-workers [4]. In fact the two nitrogen and three carboxylate functions bind the ruthenium(III) center, and under a CO atmosphere this complex is reduced to the ruthenium(II) species [Ru(EDTA)(CO)]$^-$ in which only one carboxylate group is not bonded to the metal. Apart the water gas shift reaction, which transforms CO/H_2O into CO_2 and H_2, this complex catalyzes carbonylation of terminal alkenes, cyclohexenes, primary or secondary amines, and aryl halides. Similarly triethanol-amine, $N(CH_2CH_2OH)_3$, can bind a metal center, as in $[MoO_4(TEA)]^{2-}$, and confer to it some solubility in water, although some compatibility with organic solvents has been noted [5].

Bidendate ligands containing the bipyridine framework have been synthesized by introduction of one or two sulfonate substituents onto the pyridine rings, or even of two carboxylate groups [6]. The corresponding water-soluble complexes have been prepared [7], and extension to carboxylato- or sulfonatopyridines [8], as well as to pyridineimine ligands, have been reported [9].

Chiral diamine ligands bearing hydrophilic substituents in the two *para* positions of (1R,2R)-(+)-N,N'-dimethyl-1,2-diphenylethylene diamine have been designed (hydroxy-, methoxy- or methoxytriethyleneglycol) and coordinated to a cationic iridium species to perform the enantioselective hydrogenation of phenylglyoxate methyl ester and acetophenone. Catalysis in water increases both the activity and the enantioselectivity (up to $ee = 68\%$) and an encouraging recycling efficiency has been obtained [10].

Initial studies by Baird et al. have shown that quaternization of the nitrogen atom of an aminophosphine induces an amphiphilic character: the prototype is thus $Ph_2P(CH_2)_2NMe_3^+$ (AMPHOS), which has been coordinated to rhodium and tested in hydrogenation or hydroformylation [11]. The two phenyl groups have been substituted with long carbon chains containing C_8 or C_{10} frames and the resulting ligands coordinated to palladium [12], or with trimethylammoniumphenyl groups in chiral diphosphines which have been introduced in the coordination sphere of rhodium in order to study the asymmetric hydrogenation of dehydroamino acid derivatives [13].

Interestingly, the synthesis of guanidiniumphenylphosphines has been accomplished by reaction of Ph_2PH with *meta*- or *para*-iodophenylguanidine in the presence of catalytic amounts of palladium acetate. This is prepared in a straightforward way from iodoaniline, as shown in Scheme 1.

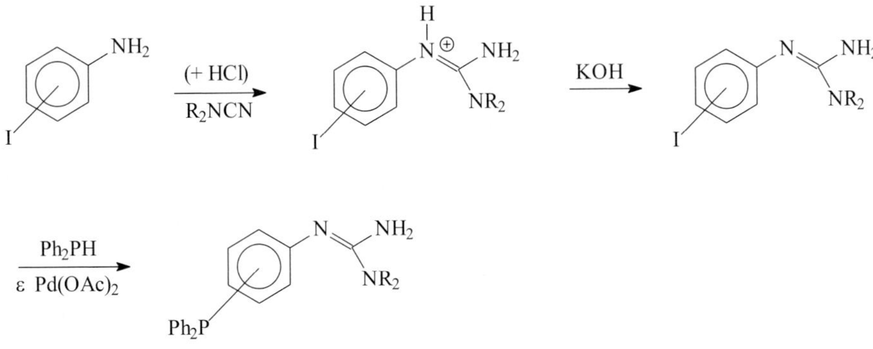

Scheme 1 Synthesis of guanidinium phenyl phosphines.

The corresponding complexes with molybdenum and palladium have been prepared; particularly, the Suzuki and Sonogashira reactions have been examined using the protonated form $[PhP(GuaH)_2]^{2+}$, and the results in the latter reaction show better yields than with monosulfonated triphenylphosphine [14].

Introduction in the *para* position on one phenyl group of triphenylphosphine of a $CH_2N(CH_2CH_2NEt_2)_2$ substituent containing three amino functions provides a ligand with a good solubility in water and particularly in water/toluene mixtures. The catalytic activity for the hydroformylation of 1-hexene with rhodium is rather modest [15]. Phosphatriazaadamantane (PTA); **1** shows some solubility in water and various ruthenium or rhodium complexes bearing the PTA and PTAH$^+$ ligands have been prepared by Darensbourg and co-workers and evaluated in catalysis, more particularly in regioselective hydrogenations [16].

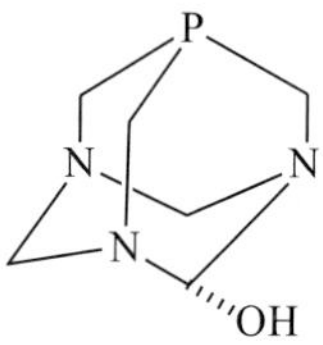

1

Structure 1 The phosphatriazaadamantane (PTA) molecule.

Concerning pyrazoles, substitution of a in the 4-position by an (ethylamino)methyl or an (isopropylamino)methyl group provides the ligand and the bridged dirhodium complexes with a good solubility, presumably because the ammonium function introduced does not interact with the rhodium centers as represented for the tetracarbonyl complex (**2**) [17].

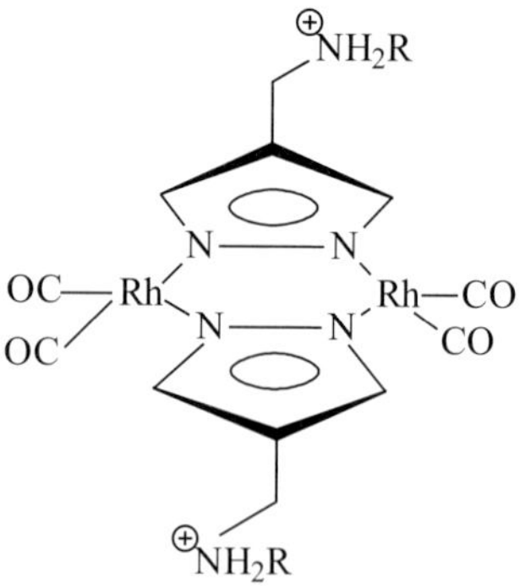

2

Structure 2 [Rh(Pz)CO$_2$]$_2$ complex (Pz = pyrazolate ammonium) (adapted from [17]).

Finally, simultaneous introduction of amines and ether functions allows diphosphine ligands to gain some solubility in water. The two ligands (**3** and **4**) have been described by van Leeuwen et al. and their behavior as ancillary ligands of rhodium has been examined in the hydroformylation reaction of terminal alkenes [18].

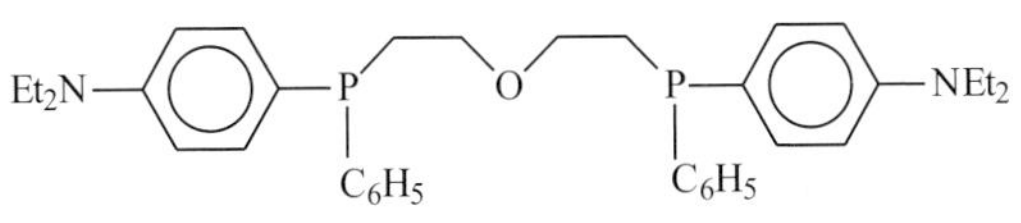

Structure 3 POPam ligand (adapted from [18]).

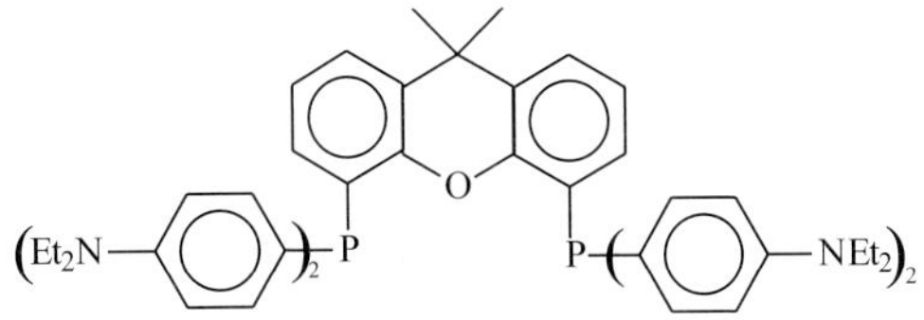

Structure 4 Xantham ligand (adapted from [18]).

Potential catalysts which would be immobilized in an aqueous phase have been built up from air-stable, water-soluble phosphino amino acids [19]. Heterobimetallic [Fe$_2$Pd] and [Fe$_2$Pt] complexes have been prepared with the assembling ligand 1,5-bis(*meta*-dicarboxyphenyl)-3,7-bis(ferrocenylmethyl)-1,5-diaza-3,7-diphosphacyclo-octane (**5**).

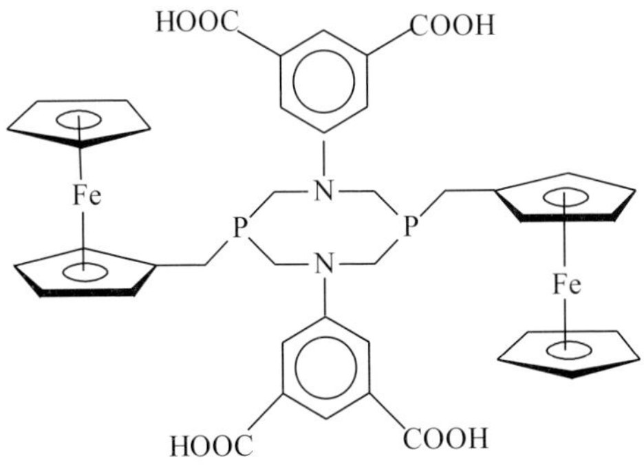

Structure 5 1,5-bis(meta-dicarboxyphenyl)-3,7-bis(ferrocenylmethyl)-1,5-diaza-3,7-diphosphacyclooctane ligand.

Sulfur-Containing Ligands

Sulfur plays a central role in enzymatic catalysis and metal sulfido enzymes can activate various small molecules in biological conditions [20]. Much work has been devoted to the isolation and understanding of the intimate mechanisms of activation of biomimetic sulfur-containing complexes [21]. However, these systems do not present significant solubility in water under the studied conditions. We can nevertheless refer to the Sellman group's work in which the authors isolated sulfur-containing representative [FeMo] complexes of the cofactor of [FeMo] nitrogenases [22, 23], and were able to produce water-soluble complexes by introduction of carboxylic acid functions. The corresponding iron and ruthenium compounds gain solubility in water by deprotonation of the carboxylic groups, as for instance in the anionic species **(6)** [23].

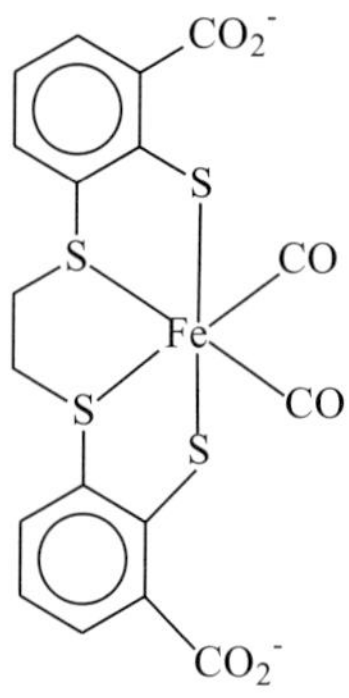

6

Structure 6 Sulfur-water-soluble iron complex (adapted from [23]).

More recently, various ligands have been designed to produce water-soluble iron complexes which can bind CO or NO as extra ligands [24]; e.g., **7**.

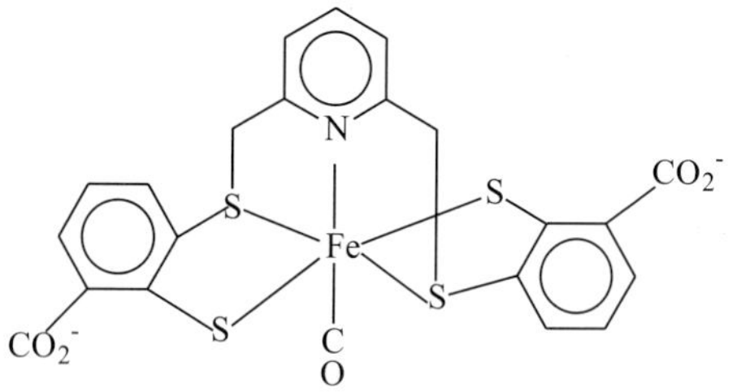

7

Structure 7 $[Fe(pyCO_2S_4)]^{2-}$ complex (adapted from [24]).

Thiolato ligands bearing a dimethylaminopropyl chain, which can be protonated by diluted sulfuric acid, induces water solubility to the dinuclear complexes $[Rh_2(\mu\text{-}S(CH_2)_3NMe_2)_2(CO)_2(PR_3)_2]$, especially after catalysis to separate the rhodium complex from the organic products [25]. The water solubility of the related rhodium–TPPTS complex ($PR_3 = $ TPPTS) can be raised by using the same bridging ligand in acidic media [26].

Chiral aminosulfonamide ligands containing a phenylsulfonic acid substituent display a significant solubility in water and have been engaged directly with a ruthenium precursor to reduce enantioselectively aromatic ketones to the corresponding alcohols [27].

Interestingly, although no catalytic activity has been examined, novel dithiabisphosphine ligands have been designed (one example is shown in **8**) and coordinated to ^{99m}Tc to produce water-soluble radiolabeling complexes [28]. The authors are extending their investigations on the capacity of coordination of such ligands to new systems containing nitrogen in place of the sulfur atoms [29].

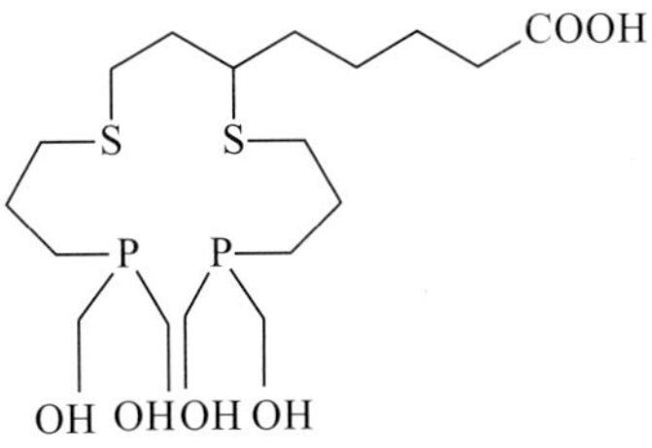

Structure 8 Structure of Dithia(bishydroxymethyl)bisphosphine "P_2S_2-COOH" (adapted from [28]).

Cyclodextrins can be modified by thiol substituents and further attachment to phosphine or aminophosphine ligands (see **9**) allows the coordination to rhodium catalysts [30]. The activity in the hydroformylation reaction of 1-octene is interesting since turnover frequencies near to 180 h^{-1} can be reached with a selectivity in C_9-aldehydes higher than 99% [31].

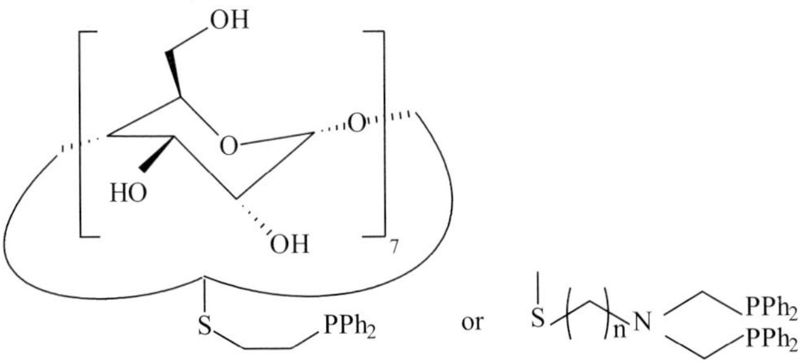

Structure 9 Cyclodextrin modified by thiophosphine or thioaminophosphine ligands (adapted from [30]).

Conclusion

In fact, various ligands which coordinate a metal center through a nitrogen-, a sulfur-, or a phosphorus-donating atom are made water-soluble by introduction of various functionalities. Apart from the sulfonated phosphines which have been largely studied, presumably due to their great capacity to dissolve in water, especially TPPTS (cf. Section 2.2.3.2.1), these ligands, bearing various hydrophilic groups, can provide a good way to tune the coordination sphere of the metal center while still being able to control water solubility.

References

1 G. THEODORIDIS, *Tetrahedron* **2000**, *56*, 2339.

2 M. FUERHACKER, G. LORBEER, R. HABERL, *Chemosphere* **2003**, *52*, 253.

3 S. ZIMMERMANN, C. M. MENZEL, D. STÜBEN, H. TARASCHEWSKI, B. SURES, *Environmental Pollution* **2003**, *124*, 1.

4 M. M. TAQUI KHAN, *Platinum Metals Rev.* **1991**, *35*, 70 and references cited herein.

5 S. SZALONTAI, G. KISS, L. BARTHA, *Spectrochimica Acta Part A.* **2003**, *59*, 1995.

6 S. ANDERSON, E. C. CONSTABLE, K. R. SEDDON; J. E. TURP, J. E. BAGGOT, M. J. PILLING, *J. Chem. Soc. Dalton Trans.* **1985**, 2247.

7 W. A. HERRMANN, W. R. THIEL, J. G. KUCHLER, *Chem. Ber.* **1990**, *123*, 1953.

8 P. K. BAKER, A. E. JENKINS, *Polyhedron* **1997**, *16*, 2279.

9 A KUNDU, B. P. BAFFIN, *Organometallics* **2001**, *20*, 3635.

10 A. FERRAND, M. BRUNO, M. L. TOMMASINO, M. LEMAIRE, *Tetrahedron: Asymmetry* **2002**, *13*, 1379.

11 R. T. SMITH, R. K. UNGAR, L. J. SANDERSON, M. C. BAIRD, *Organometallics* **1983**, *2*, 1138.

12 A. HESSLER, S. KUCHEN, O. STELZER, J. BLOTEVOGEL-BALTRONAT, W. S. SHELDRICK, *J. Organomet. Chem.* **1995**, *501*, 293.

13 (a) I. TOTH, B. E. HANSON, *Tetrahedron: Asymmetry* **1990**, *1*, 885; (b) I. TOTH, B. E. HANSON, *Organometallics* **1993**, *12*, 1506; (c) I. TOTH, B. E. HANSON, M. E. DAVIES, *Tetrahedron: Asymmetry* **1990**, *1*, 895.

14 P. MACHNITZKI, M. TEPPER, K. WENZ, O. STELZER, E. HERDTWECK, *J. Organomet. Chem.* **2000**, *602*, 158.

15 M. KARLSSON, M. JOHANSSON, C. ANDERSSON, *J. Chem. Soc. Dalton Trans* **1999**, 4187.

16 D. J. DARENSBOURG, N. WHITE STAFFORD, F. JOÓ, J. H. REIBENSPIES, *J. Organomet. Chem.* **1995**, *488*, 99.

17 G. ESQUIUS, J. PONS, R. YANEZ, J. ROS, X. SOLANS, M. FONT-BARDIA, *J. Organomet. Chem.* **2000**, *605*, 226.

18 A. BUHLING, P. C. J. KAMER, P. W. N. M. VAN LEEUWEN, J. W. ELGERSMA, K. GOUBITZ, J. FRANJE, *Organometallics.* **1997**, *16*, 3027.

19 A. A. KARAZIK, R. N. NAUMOV, R. SOMMER, O. G. SINYASHIN, E. HEY-HAWKINS, *Polyhedron* **2002**, *21*, 2251.

20 D. C. REES, *Annu. Rev. Biochem.* **2002**, *71*, 221.

21 T. B. RAUCHFUSS, *Inorg. Chem.* **2004**, *43*, 14.

22 D. SELLMAN, T. BECKER, F. KNOCH, *Chem. Ber.* **1996**, *129*, 509.

23 D. SELLMAN, W. SOGLOWEK, F. KNOCH, G. RITTER, J. DENGLER, *Inorg. Chem.* **1992**, *31*, 3711.

24 D. SELLMAN, K. P. PETERS, F. W. HEINEMANN, *Eur. J. Inorg. Chem.* **2004**, in press.

25 J. C. BAYÒN, J. REAL, C. CLAVER, A. POLO, A. RUIZ, *J. Chem. Soc., Chem. Commun.* **1989**, 1056.

26 F. MONTEIL, R. QUÉAU, PH. KALCK, *J. Organomet. Chem.* **1994**, *480*, 177.

27 C. BUBERT, J. BLAKER, S. M. BROWN, J. CROSBY, S. FITZJOHN, J. P. MUXWORTHY, T. THORPE, J. M. J. WILLIAMS, *Tetrahedron Lett.* **2001**, *42*, 4037.

28 R. SCHIBLI, S. R. KARRA, H. GALI, K. V. KAHI, C. HIGGINBOTHAM, W. A. VOLKERT, *Radiochim. Acta* **1998**, *83*, 211.

29 K. K. KOTHARI, K. RAGHURAMAN, N. K. PILLARSETTY, T. J. HOFFMAN, N. K. OWEN, K. V. KATTI, W. A. VOLKERT, *Applied Radiation and Isotopes* **2003**, *58*, 543.

30 M. T. REETZ, J. RUDOLPH, *Tetrahedron: Asymmetry* **1993**, *4*, 2405.

31 M. T. REETZ, S. R. WALDOGEL, *Angew. Chem., Int. Ed. Engl.* **1997**, *36*, 865.

2.2.3.2.3
Other Concepts: Hydroxyphosphines

Mandy-Nicole Gensow and Armin Börner

Apart from sulfonate, carboxylate and ammonium groups, hydroxy groups incorporated in phosphorus ligands also increase the solubility of the catalyst in water. Several members of this large class of ligands have been synthesized [1] and were tested in catalysis in organic solvents as well as in water [2]. Particular attention has been given to the special effect of the hydroxy group in acting as a hemilabile ligand or establishing secondary interactions with a suitable substrate [3].

Katti et al. synthesized polyhydroxyphosphines (Structures 1–5) by the nucleophilic addition of PH_3 or primary phosphines to formaldehyde or other carbonyl compounds [4]. The utility of such ligands in the formation of water-soluble transition metal complexes was evidenced in several cases [5]. It should be noted that besides the phosphorus(III) atom, hydroxy groups also can coordinate with "soft" transitions metals [3, 6].

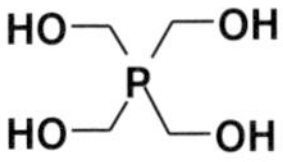

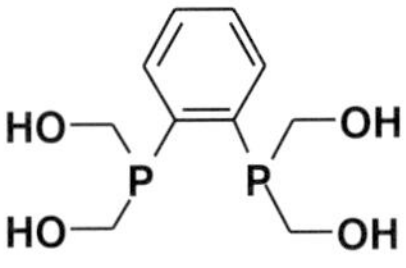

By addition of PH_3 or 1,2-ethylenediphosphine to allyl acetate [7] or allyl alcohol, 1,3-polyhydroxyalkylphosphines 6 and 7 can be prepared [8].

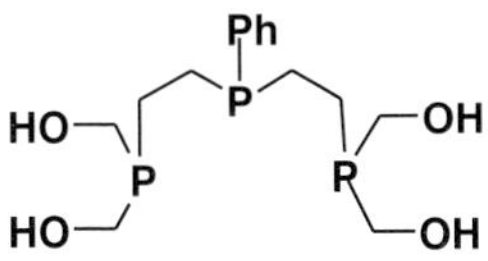

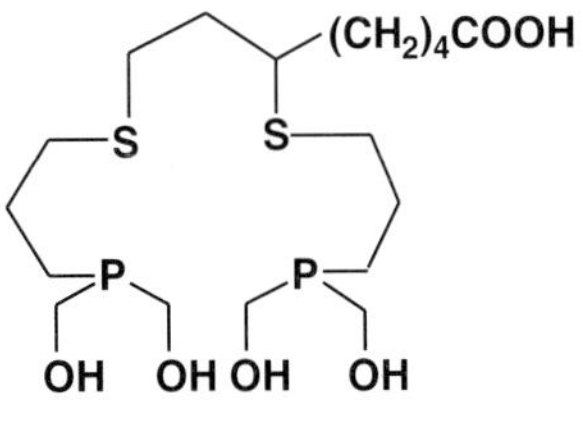

Related 1,3-polyhydroxyphosphines 8 and 9 were synthesized by Lindner et al. by photochemical addition of 1,3-propylenediphosphine to the relevant alkenes [9]. Pd(II) complexes of the new ligands were successfully employed for the copolymerization of CO and ethylene in water.

Carbohydrates represent one of the most versatile tools for the synthesis of *P*-ligands bearing hydroxy groups [1, 10]. Thus, Beller et al. used *β*-*O*-glycoside ligands 10 bearing a phenylphosphine group as aglycones for the Pd-catalyzed Suzuki coupling and Heck reaction in an aqueous two-phase system [11].

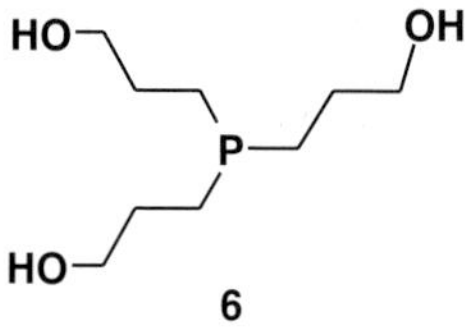

HO(CH$_2$)$_n$P⌒⌒P(CH$_2$)$_n$OH   [(HOCH$_2$)$_2$CH(CH$_2$)$_n$]$_2$P⌒⌒P[(CH$_2$)$_n$CH(CH$_2$OH)]$_2$

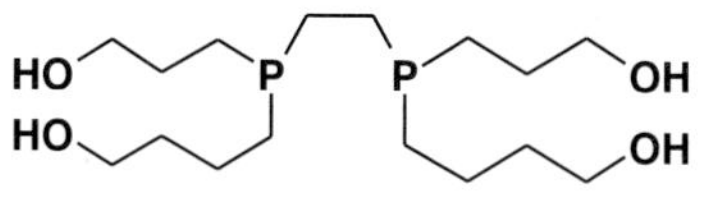

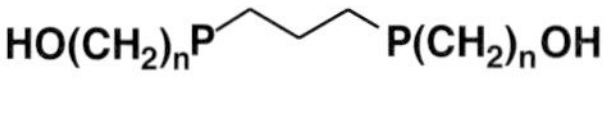

8 (n = 1-8) 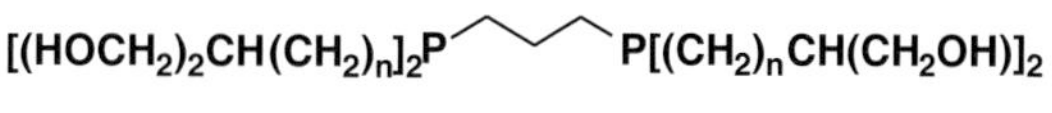   **9** (n = 3-6)

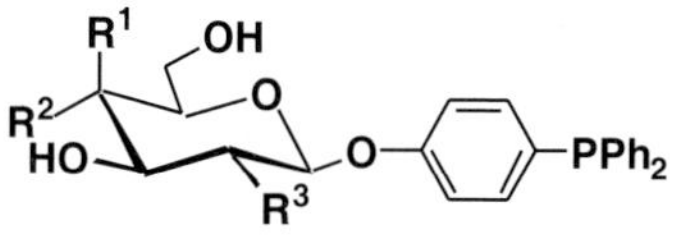

R^1 = H, R^2 = OH, R^3 = NHAc
R^1 = OH, R^2 = H, R^3 = OH
R^1 = H, R^2 = OH, R^3 = OH

10

In comparison to the application of PPTS, with ligands of type **10** improved catalyst activities and yields were noted. These effects were rationalized by the higher catalyst concentration in the nonpolar phase.

Similarly superior results in comparison to the application of sulfonated phosphine ligands were achieved in the Pd-catalyzed biaryl coupling of arylboronic acids in aqueous media by use of the ligand **11** (glcaPHOS) [12].

Selke et al. advantageously used rhodium(I) precatalysts **12a,b** based on Ph-β-glup-type ligands bearing two hydroxy groups in the sugar backbone for the asymmetric hydrogenation of Z-acetamidocinnamic acid derivatives in water [13]. Strong effects upon the enantioselectivity were noted which depended on the orientation of the hydroxy groups at the pyranose (besides D-glucose derivatives of D-galactose were used also). By addition of amphiphiles the enantioselectivity could be considerably enhanced in comparison to the reactions in a water blank.

Ohe and Uemura investigated the asymmetric hydrogenation of several enamides and itaconic acid in water in the presence of a Rh(I) catalyst based on disaccharide (e.g., trehalose) diphosphinite ligands **13a,b** [14]. When sodium dodecyl sulfate (SDS) was used as an additive the amount of catalyst could be significantly reduced. Simultaneously the enantioselectivity in the product was enhanced.

Evidence for a synergistic effect of internal hydroxy groups and amphiphiles on the Rh-catalyzed asymmetric hydrogenation in water was given by Selke and Börner. Thus, incorporation of a single hydroxy group in the DIOP-analogue ligand **14** increased the enantioselectivity by up to 70% *ee* in comparison to the blank experiment [15].

Such an effect was absent in the trial with the parent ligand (DIOP). It is noteworthy that the enantioselectivity achieved under optimized conditions distinctly exceeded that in pure methanol.

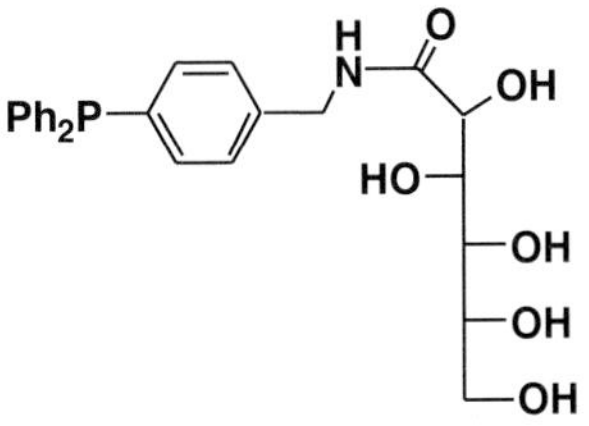

11

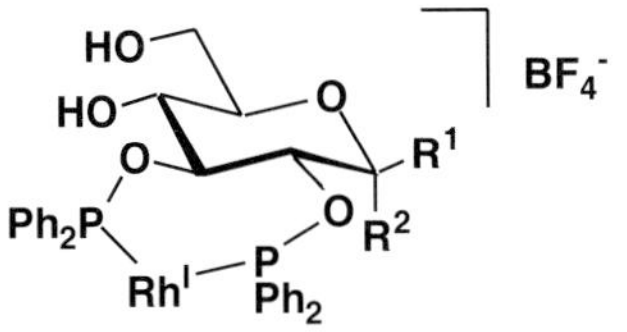

12a: R^1 = H; R^2 = OPh, OMe
12b: R^1 = OPh, OMe; R^2 = H
13a: R^1 = H; R^2 = 1-α-D-glucopyranose
13b: R^1 = 1-β-D-glucopyranose; R^2 = H

The chiral Rh(I) complex **15** bearing the tetrahydroxydiphospholane $(HO)_4$-BASPHOS was prepared by Holz and Börner [16]. In comparison to the strongly related Et-DuPHOS-Rh precatalyst, the four hydroxy groups increased the water solubility of the complex by a factor of four [17]. In the asymmetric hydrogenation of acetamidoacrylate in water with this cationic $(HO)_4$-BASPHOS catalyst, (*S*)-*N*-acetylalanine was obtained in quantitative yield and in more than 99% *ee*. Noteworthy is the unusually short time necessary in order to get complete conversion of the substrate. Later, RajanBabu and co-workers succeeded in the preparation of the free ligand, $(HO)_4$-BASPHOS [18]. This method was revealed to be more advantageous than the cleavage of HO-protective groups with acids in aqueous solutions affording mainly phosphonium salts [19].

Polyhydroxyphospholanes of the RoPHOS type [20] were prepared by Zhang and co-workers [19, 21]. With the ethyl-substituted ligand **16b** in the Rh(I)-catalyzed asymmetric hydrogenation of itaconic acid in a water–methanol mixture as solvent, 100% conversion and > 99% *ee* were achieved.

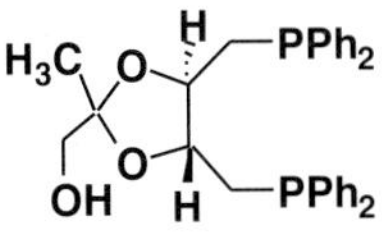

14 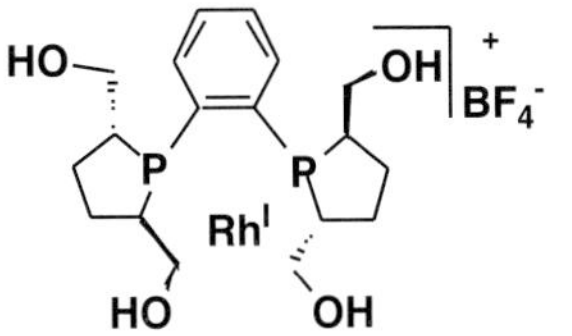   **15** 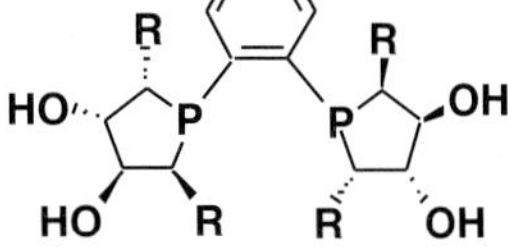   **16a:** R = Me
16b: R = Et

It should finally be noted that hydroxyphosphines can be converted under very smooth conditions into sulfonated phosphines by acylation with *o*-sulfobenzoic anhydride, as shown by Börner et al. [Eq. (1)] [22]. With this methodology in hand the severe conditions commonly used for the incorporation of sulfonate groups in phosphines can be avoided. Acid-labile functional groups like acetals survive under these conditions.

In comparison to the parent hydroxyphosphines the water solubility of the relevant Rh catalysts was strongly enhanced [23]. In the asymmetric hydrogenation of prochiral olefins, moderate enantioselectivities were achieved.

$$R\text{-OH} \xrightarrow[\text{Na}_2\text{CO}_3]{n\text{-BuLi or}} \text{(product)}$$

M = Na, Li
R =

(1)

References

1 For a review, see: J. Holz, M. Quirmbach, A. Börner, *Synthesis* **1997**, 983.

2 For a recent general review of water-soluble phosphines, see: N. Pinault, D. W. Bruce, *Coord. Chem. Rev.* **2003**, *241*, 1.

3 For reviews, see: (a) A. Börner, *Eur. J. Inorg. Chem.* **2001**, 327; (b) A. Börner, *Chirality* **2001**, *13*, 625.

4 (a) H. Gali, K. R. Prabhu, S. R. Karra, K. V. Katti, *J. Org. Chem.* **2000**, *65*, 676; (b) K. V. Katti, *Curr. Sci.* **1996**, *70*, 219; (c) K. V. Katti, H. Gali, C. J. Smith, D. E. Berning, *Acc. Chem. Res.* **1999**, *32*, 9.

5 (a) D. E. Berning, K. V. Katti, C. L. Barnes, W. A. Volkert, *Chem. Ber./Recueil* **1997**, *130*, 907; (b) C. J. Smith, V. Sreenivasa Reddy, S. R. Karra, K. V. Katti, L. Barbour, *Inorg. Chem.* **1997**, *36*, 1786; (c) C. J. Smith, V. Sreenivasa Reddy, K. V. Katti, *Chem. Commun.* **1996**, 2557; (d) J. W. Ellis, K. N. Harrison, P. A. T. Hoye, A. G. Orpen, P. G. Pringle, M. B. Smith, *Inorg. Chem.* **1992**, *31*, 3026; (e) K. N. Harrison, P. A. T. Hoye, A. G. Orpen, P. G. Pringle, M. B. Smith, *J. Chem. Soc., Chem. Commun.* **1989**, 1096.

6 a) D. J. Brauer, P. Machnitzki, T. Nickel, O. Stelzer, *Eur. J. Inorg. Chem.* **2000**, 65;(b) S. Borns, R. Kadyrov, D. Heller, W. Baumann, J. Holz, A. Börner, *Tetrahedron: Asymmetry* **1999**, *10*, 1425; (c) S. Borns, R. Kadyrov, D. Heller, W. Baumann, A. Spannenberg, R. Kempe, J. Holz, A. Börner, *Eur. J. Inorg. Chem.* **1998**, 1291; (d) A. Börner, A. Kless, R. Kempe, D. Heller, J. Holz, W. Baumann, *Chem. Ber.* **1995**, *128*, 767.

7 B. Driessen-Hölscher, J. Heinen, *J. Organomet. Chem.* **1998**, *570*, 139.

8 (a) G. T. Baxley, T. J. R. Weakley, W. K. Miller, D. K. Lyon, D. R. Tyler, *J. Mol. Catal. A:* **1997**, *116*, 191; (b) G. T. Baxley, W. K. Miller, D. K. Lyon, B. E. Miller, G. F. Nieckarz, T. J. R. Weakley, D. R. Tyler, *Inorg. Chem.* **1996**, *35*, 668.

9 E. Lindner, M. Schmid, J. Wald, J. A. Queisser, M. Geprägs, P. Wegner, C. Nachtigal, *J. Organomet. Chem.* **2000**, *602*, 173.

10 D. Steinborn, H. Junicke, *Chem. Rev.* **2000**, *100*, 4283.

11 M. Beller, J. G. E. Krauter, A. Zapf, *Angew. Chem.* **1997**, *109*, 793; *Angew. Chem., Int. Ed. Engl.* **1997**, *36*, 772.

12 M. Ueda, M. Nishimura, N. Miyaura, *Synlett* **2000**, 856.

13 R. Selke, M. Ohff, A. Riepe, *Tetrahedron* **1996**, *52*, 15 079.

14 (a) K. Yonehara, T. Hashizume, K. Ohe, S. Uemura, *Tetrahedron: Asymmetry* **1999**, *10*, 4029; (b) K. Yonehara, K. Ohe, S. Uemura, *J. Org. Chem.* **1999**, *64*, 9381.

15 R. Selke, J. Holz, A. Riepe, A. Börner, *Chem. Eur. J.* **1998**, *4*, 769.

16 (a) J. Holz, D. Heller, R. Stürmer, A. Börner, *Tetrahedron Lett.* **1999**, *40*, 7059; (b) J. Holz, R. Stürmer, U. Schmidt, H.-J. Drexler, D. Heller, H.-P. Krimmer, A. Börner, *Eur. J. Org. Chem.* **2001**, 4615.

17 R. SELKE, A. BÖRNER, unpublished results.

18 T. V. RAJANBABU, Y.-Y. YAN, S. SHIN, *J. Am. Chem. Soc.* **2001**, *123*, 10 207.

19 W. LI, Z. ZHANG, D. XIAO, X. ZHANG, *Tetrahedron Lett.* **1999**, *40*, 6701.

20 J. HOLZ, M. QUIRMBACH, U. SCHMIDT, D. HELLER, R. STÜRMER, A. BÖRNER, *J. Org. Chem.* **1998**, *63*, 8031.

21 W. LI, Z. ZHANG, D. XIAO, X. ZHANG, *J. Org. Chem.* **2000**, *65*, 3489.

22 S. TRINKHAUS, J. HOLZ, R. SELKE, A. BÖRNER, *Tetrahedron Lett.* **1997**, *38*, 807.

23 S. TRINKHAUS, R. KADYROV, J. HOLZ, R. SELKE, L. GÖTZE, A. BÖRNER, *J. Mol. Catal. A:* **1999**, *144*, 15.

2.2.3.3
Water-soluble Receptor as Mass Transfer Promoter

Eric Monflier

Solubilization of water-insoluble substrate in the aqueous phase containing the organometallic catalyst can be achieved by using catalytic amounts of water-soluble receptors such as cyclodextrins [1] or calixarenes [2]. The beneficial effect of these water-soluble host compounds on the mass transfer is ascribed to their complexing properties and it is postulated that these compounds operate like inverse phase-transfer catalysts according to Figure 1.

In this mechanism the receptor, which is represented by a truncated cone, forms a host/guest complex with the substrate at the liquid/liquid interface and transfers the water-insoluble substrate (**S**) into the aqueous phase where it reacts with the water-soluble organometallic catalyst. After reaction, the product (**P**) is released in the organic phase and the transfer cycle can go on.

Cyclodextrins are the most used host compounds so far. Native cyclodextrins are effective inverse phase-transfer catalysts for the deoxygenation of allylic alcohols [3], epoxidation [4], oxidation [5], or hydrosilylation [6] of olefins, reduction of α,β-unsaturated acids [7], α-keto ester [8], conjugated dienes [9] or aryl alkyl ketones [10]. Interestingly, chemically modified cyclodextrins like the partially *O*-methylated β-cyclodextrin (RAME-β-CD) show a better catalytic activity than native cyclodextrins in numerous reactions such as the Wacker oxidation [11], hydrogenation of

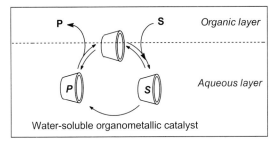

Figure 1 Cyclodextrins as water-soluble host compounds.

aldehydes [12], the Suzuki cross-coupling reaction [13], hydroformylation [14] or hydrocarboxylation [15] of olefins. This outstanding effect of RAME-β-CD on reaction rate was attributed to its slight surface activity and to the presence of a deep hydrophobic host cavity that accommodates the substrate properly. RAME-β-CD was also used successfully to perform substrate-selective reactions in a two-phase system [16]. Thus, in the presence of a mixture of isomers, the water-soluble catalyst reacts with the isomer that preferentially interacts with the cyclodextrin cavity, inducing substrate selectivity. For instance, a 97 : 3 product ratio was observed during the palladium-catalyzed cleavage of a 50 : 50 mixture of *N*-dodecyl-*O*-allylurethane and *N,N*-dihexyl-*O*-allylurethane [17]. Surprisingly, it was found that the cyclodextrins can form inclusion complexes with the hydrosoluble phosphine used to dissolve the catalyst in the aqueous phase [18]. These complexes were responsible for the decrease in the linear/branched aldehydes ratio during the rhodium-catalyzed hydroformylation reaction [19], the modification of the catalyst structure [20], and the drop in cyclodextrin activity in some experimental conditions [21]. Interestingly, more sophisticated approaches involving covalent attachment of the phosphane ligand to the cyclodextrin through a spacer have also been performed to combine molecular recognition, phase-transfer properties and aqueous organometallic catalysis [5d, 22, 23]. Reetz et al. have reported the synthesis of rhodium complexes with β-cyclodextrin modified by diphosphanes as ligands [22]. Numerous substrates including internal olefins have been hydrogenated or hydroformylated successfully in a biphasic medium. Unfortunately, the catalytic systems cannot be recovered quantitatively due to partial transfer of the catalyst into the organic phase during the reaction [22, 23].

Recently, the use of water-soluble calixarenes as inverse phase-transfer catalysts has also been reported to overcome mass transfer limitations in aqueous organo-metallic catalysis. In the Suzuki cross-coupling reaction, the efficiency of these synthetic macrocycles was higher than that observed for β-cyclodextrins [13, 24]. Water-soluble sulfonated calixarenes modified by phosphine [25]- or nitrile [26]-containing groups are efficient ligands for the hydroformylation or Wacker oxidation of olefins, respectively. Contrary to the cyclodextrins modified by phosphine groups, no decrease in the catalytic activity was observed during recycling experiments, suggesting that the rhodium catalyst is stable and quantitatively immobilized in the aqueous phase. Unfortunately, the selectivity in aldehydes (55–80%) was notably lower than selectivities observed with systems described by Monflier (95%) [14] or Reetz (99%) [22].

References

1   (a) J. Szejtli, *Chem. Rev.* **1998**, *98*, 1743; (b) G. Wenz, *Angew. Chem., Int. Ed. Engl.* **1994**, *33*, 803.
2   (a) L. Mandolini, R. Ungaro (Eds.), *Calixarenes in Action*, Imperial College Press, London **2000**; (b) V. Böhner, *Angew. Chem., Int. Ed.* **1995**, *34*, 713.
3   J. T. Lee, H. Alper, *Tetrahedron Lett.* **1990**, *31*, 4101.
4   P. A. Ganeshpure, S. Satish, *J. Chem. Soc., Chem. Commun.* **1988**, 981.

5 (a) H. A. ZAHALKA, K. JANUSZKIEWICZ, H. ALPER, *J. Mol. Catal.* **1986**, *35*, 249; (b) A. HARADA, Y. HU, S. TAKAHASHI, *Chem. Lett.* **1986**, 2083; (c) E. A. KARAKHANOV, T. Y. FILIPPOVA, S. A. MARTYNOVA, A. L. MAXIMOV, V. V. PREDEINA, I. N. TOPCHIEVA, *Catal. Today* **1998**, *44*, 189; (d) E. KARAKHANOV, A. MAXIMOV, A. KIRILLOV, *J. Mol. Catal. A:* **2000**, *157*, 25.

6 (a) L. N. LEWIS, C. A. SUMPTER, *J. Mol. Catal. A:* **1993**, *104*, 293; (b) L. N. LEWIS, C. A. SUMPTER, J. STEIN, *J. Inorg. Organomet. Polymers* **1996**, *6*, 123.

7 (a) J. T. LEE, H. ALPER, *Tetrahedron Lett.* **1990**, *31*, 1941; (b) H. ARZOUMANIAN, D. NUEL, *C.R. Acad. Sci. Paris* **1999**, *Série IIc*, 289.

8 C. PINEL, N. GENDREAU-DIAZ, A. BRÉHÉRET, M. LEMAIRE, *J. Mol. Catal. A.* **1996**, *112*, L157.

9 J. T. LEE, H. ALPER, *J. Org. Chem.* **1990**, *55*, 1854.

10 H. ZAHALKA, H. ALPER, *Organometallics* **1986**, *5*, 1909.

11 (a) E. MONFLIER, E. BLOUET, Y. BARBAUX, A. MORTREUX, *Angew. Chem., Int. Ed. Engl.* **1994**, *33*, 2100; (b) E. MONFLIER, S. TILLOY, E. BLOUET, Y. BARBAUX, A. MORTREUX *J. Mol. Catal. A:* **1996**, *109*, 27.

12 (a) S. TILLOY, H. BRICOUT, E. MONFLIER, *Green Chem.* **2002**, *4*, 188; (b) E. MONFLIER, S. TILLOY, Y. CASTANET, A. MORTREUX, *Tetrahedron Lett.* **1998**, *39*, 2959.

13 F. HAPIOT, J. LYSKAWA, S. TILLOY, H. BRICOUT, E. MONFLIER, *Adv. Synth. Catal.* **2004**, *346*, 83.

14 (a) E. MONFLIER, G. FREMY, Y. CASTANET, A. MORTREUX, *Angew. Chem., Int. Ed. Engl.* **1995**, *34*, 2269; (b) E. MONFLIER, S. TILLOY, G. FREMY, Y. CASTANET, A. MORTREUX, *Tetrahedron Lett.* **1995**, *52*, 9481; (c) P. KALCK, L. MOQUEL, M. DESSOUDEIX, *Catal. Today* **1998**, *42*, 431; (d) M. DESSOUDEIX, M. URRUTIGOÏTY, P. KALCK, *Eur. J. Inorg. Chem.* **2001**, 1797.

15 (a) E. MONFLIER, S. TILLOY, F. BERTOUX, Y. CASTANET, A. MORTREUX, *New J. Chem.* **1997**, *21*, 857; (b) S. TILLOY, F. BERTOUX, A. MORTREUX, E. MONFLIER, *Catal. Today* **1999**, *48*, 245.

16 T. LACROIX, H. BRICOUT, S. TILLOY, E. MONFLIER, *Eur. J. Org. Chem.* **1999**, 3127.

17 H. BRICOUT, C. CARON, D. BORMANN, E. MONFLIER, *Catal. Today* **2001**, *66*, 355.

18 (a) E. MONFLIER, S. TILLOY, C. MÉLIET, A. MORTREUX, S. FOURMENTIN, D. LANDY, G. SURPATEANU, *New J. Chem.* **1999**, *23*, 469; (b) E. MONFLIER, S. TILLOY, L. CARON, J. M. WIERUSZESKI, G. LIPPENS, S. FOURMENTIN D. LANDY, G. SURPATEANU, *J. Incl. Phenom.* **2000**, *38*, 6111; (c) L. CARON, C. CHRISTINE, S. TILLOY, E. MONFLIER, D. LANDY, S. FOURMENTIN, G. SURPATEANU, *Supramol. Chem.* **2002**, *14*, 11; (d) M. CANIPELLE, L. CARON, C. CALINE, S. TILLOY, E. MONFLIER, *Carbohyd. Res.* **2002**, *337*, 281; (e) M. CANIPELLE, L. CARON, H. BRICOUT, S. TILLOY, E. MONFLIER, *New J. Chem.* **2003**, *27*, 1603.

19 T. MATHIVET, C. MÉLIET, Y. CASTANET, A. MORTREUX, L. CARON, S. TILLOY, E. MONFLIER, *J. Mol. Catal. A:* **2001**, *176*, 105.

20 L. CARON, M. CANIPELLE, S. TILLOY, H. BRICOUT, E. MONFLIER, *Eur. J. Inorg. Chem.* **2003**, 595.

21 (a) C. BINKOWSKI, J. CABOU, H. BRICOUT, F. HAPIOT, E. MONFLIER, *J. Mol. Catal. A:* **2004**, *215*, 23; (b) J. R. ANDERSON, E. M. CAMPI, W. R. JACKSON, *Catal. Lett.* **1991**, *9*, 55; (c) A. BENYEI, F. JOO, *J. Mol. Catal.* **1990**, *58*, 151.

22 (a) M. T. REETZ, *J. Heterocyclic Chem.* **1998**, *35*, 1065; (b) M. T. REETZ, *Topics Catal.* **1997**, *4*, 187; (c) M. T. REETZ, S. R. WALDVOGEL, *Angew. Chem., Int. Ed. Engl.* **1997**, *36*, 865.

23 D. ARMSPACH, D. MATT, *Chem. Commun.* **1999**, 1073.

24 M. BAUR, M. FRANK, J. SCHATZ, F. SCHILDBACH, *Tetrahedron* **2001**, *57*, 6985.

25 (a) S. SHIMIZU, S. SHIRAKAWA, Y. SASAKI, C. HIRAI, *Angew. Chem., Int. Ed.* **2000**, *39*, 1256; (b) S. SHIRAKAWA, S. SHIMIZU, Y. SASAKI, *New J. Chem.* **2001**, *25*, 777.

26 E. KARAKHANOV, T. BUCHNEVA, A. MAXIMOV, M. ZAVERTYAEVA, *J. Mol. Catal. A:* **2002**, *184*, 11.

2.3
Homogeneous Catalysis in the Aqueous Phase as a Special Unit Operation

2.3.1
Fundamental Reaction Engineering of Aqueous Biphasic Catalysis

Zai-Sha Mao and Chao Yang

2.3.1.1
Introduction

Several considerations invoke the association, or incorporation, of the general catalytic study with the relevant chemical engineering study. Many chemists have been aware of such intrinsic connections between chemistry and chemical engineering, but there is still misunderstanding or neglect of chemical engineering factors in the study of catalysis. Among the most important aspects are the following ones.

Firstly, chemical engineering measures need to be taken to guarantee hydro-formylation, especially biphasic hydroformylation, in a well-defined and controlled physico-chemical environment. For example, the reaction rate and conversion of olefins to aldehydes is largely dependent on the mass transfer rate between the gas (syngas consisting of CO and H_2) and the liquid phase (aqueous phase containing a water-soluble transition metal complex catalyst, and organic phase of long-chain olefins and solvent), and the mass transfer rate itself is in turn the product of the interfacial mass transfer coefficient and the specific interfacial area. It is usually believed intuitively that the mass transfer rate or the specific interfacial area will reach an asymptotic value as the agitation speed is increased to a sufficiently high value, beyond which little further increase in the conversion or TOF of olefins would be observed, and that under such a condition the limitation of interphase mass transfer to chemical reaction is eliminated. For example, Zhang et al. [1] examined the effects of the stirring speed on the conversion of biphasic hydro-formylation of 1-dodecene catalyzed by a water-soluble rhodium complex and the normal/iso (n/i) ratio of the product aldehydes. The experimental results presented in Figure 1 were very typical, and the speed of $N = 15$ rps was selected for the rest of the hydroformylation experiments. However, rigorous analysis according to the principles of chemical engineering suggests that such selection is not fully justified. To reduce the mass transfer boundary layer thickness as much as possible so as to eliminate the mass transfer resistance, motion of the bulk phase relative to gas bubbles and droplets needs to be maximized, but this may not be accomplished by simply increasing the rotational speed of the stirrers. At a higher stirring speed, the bubbles and drops would follow closely the bulk liquid phase flow, and the relative motion would not be increased as effectively as expected. Therefore, achievement of the maximum or the asymptotic conversion is not an indication that conclusions on the intrinsic kinetics are guaranteed, and this may give an

Multiphase Homogeneous Catalysis
Edited by Boy Cornils and Wolfgang A. Herrmann et al.
Copyright © 2005 Wiley-VCH Verlag GmbH & Co. KGaA, Weinheim
ISBN: 3-527-30721-4

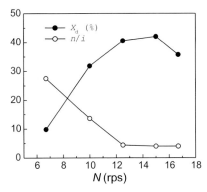

Figure 1 Variation of X_d and n/i of hydroformylation of 1-dodecene versus stirrer speed (N) [1].

inconsistent basis for comparison of similar studies. Chemical engineers would suggest using an autoclave with standard wall baffles to enhance the relative motion between phases and to standardize the reactor/agitator configuration to ensure a consistent basis for quantifying and comparing catalytic studies. In Figure 1 the decrease of conversion at above $N = 15$ rps suggests further complications caused by chemical and engineering factors, which have to be analyzed by the principles of two crosslinked disciplines, chemistry and chemical engineering.

Secondly, the theory incorporated with chemical engineering principles is necessary so as to offer means to elucidate the phenomena and data in catalytic research. It is now generally accepted that the surfactant added to the biphasic hydroformylation system causes the formation of micelles so that the associated solubilization, etc., promotes the reaction rate and influences the n/i ratio (see Section 2.3.4 [2, 3]. Chen et al. [2] suggested that the hydroformylation of 1-dodecene in biphasic systems occurred at the interface between aqueous and organic phases; the micelle structure formed was favorable for the n/i ratio; the key factor to the enhancement of reaction rate was the enrichment of catalytic species by the static electric attraction between the negatively charged active rhodium anion species and the positively charged cationic ends of surfactant solubilized in the organic phase. The data also supported the proposition of solubilization of olefin in micelles [4]. Van Vyve and Renken [3] also reported that the reaction rate and normal aldehyde selectivity were improved for the hydroformylation of C_6–C_{16} olefins in reverse micellar systems. To thoroughly and quantitatively understand the surfactant-enhanced hydroformylation, a chemical engineering study on the formation of micelles and the quantitative formulation of their physico-chemical properties is necessary. This should be incorporated into the chemical research and process development of hydroformylation of olefins, in particular the heterogeneous hydroformylation of long-chain olefins. To date no report is available about the hydroformylation of olefins based on quantitative mathematical models of micellar systems.

Thirdly, chemical engineering studies are very important to commercial exploitation of the results of homogeneous and biphasic hydroformylation, but it seems that chemical and chemical engineering studies constitute two separate periods of an originally indivisible research program. Thus, chemists should be alerted to plan their study and analyze the results with due consideration of the relevant chemical engineering factors; it is anticipated that close cooperation of chemists and chemical engineers would bring more benefits in efficiency and productivity in research and development of biphasic hydroformylation.

2.3.1.2
Chemical Engineering Factors

Many engineering factors have a profound influence on the performance of catalysts in hydroformylation, and their effects are demonstrated in terms of reaction indices such as conversion and selectivity. More concretely, these factors influence the mass transfer rate, phase dispersion, emulsification and demulsification, catalyst distribution, product separation, etc. (Table 1).

No comprehensive investigation is available to elucidate all the facets of the effects listed in Table 1. From the point of view of chemical engineering, some of them may be addressed only qualitatively, although some interesting experimental evidence has appeared. The mechanistic representation in Figure 2 [2] illustrates several factors listed in Table 1. When the biphasic system consisting of 1-dodecene (organic phase) and an aqueous solution of water-soluble rhodium catalyst precursor RhCl(CO)(TPPTS)$_2$ and surfactant cetyltrimethylammonium bromide (CTAB) is agitated sufficiently, the reduction in liquid–liquid surface tension due to the presence of CTAB makes the formation of fine gas bubbles and 1-dodecene droplets much easier, leading to higher specific gas–liquid and liquid–liquid interfacial areas. This would be favorable to the dissolution of gaseous reactants in the liquid phases and to a greater area for the hydroformylation reaction. If the concentration of CTAB is above the critical micelle concentration (CMC), micelles with organic kernels would be formed, and the interfacial liquid–liquid area would be increased

Table 1 Chemical engineering factors influencing the performance of biphasic hydroformylation.

Factor	Influencing mechanism	Target index	Ref.
Number of phases	Phase dispersion, dispersion uniformity, mass transfer resistance	Conversion	–
Phase ratio	Dispersion type (O/W, W/O), dispersion uniformity	Conversion *n/i* ratio	[5]
Stirring speed	Interphase mass transfer coefficient, specific interfacial area, uniformity of phase dispersion, emulsification	Conversion *n/i* ratio	[1, 2, 5, 6]
Stirrer type	Specific interfacial area, uniformity of phase dispersion, emulsification	Conversion *n/i* ratio	[5]
Surfactant	Formation of micelles, catalyst distribution, specific interfacial area	Conversion *n/i* ratio	[2]

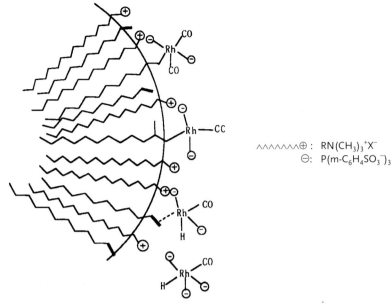

$$\wedge\wedge\wedge\wedge\wedge\wedge\wedge\oplus : \quad RN(CH_3)_3{}^+X^-$$
$$\ominus: \quad P(m\text{-}C_6H_4SO_3{}^-)_3$$

Figure 2 Catalytic active species in the interfacial layer of a cationic micelle [2].

further. Besides, the micelles probably solubilize the water-soluble catalyst species and this becomes the second factor in favor of a higher hydroformylation rate, in addition to a higher area for the reaction to occur. The orderly assembly formed by the long hydrophobic cetyl ends of CTAB at the liquid–liquid interface induces an orderly orientation of 1-dodecene molecules, thus contributing a higher selectivity of terminal aldehyde (higher n/i ratio). The contribution of static electric attraction between positive CTA ions and negative Rh complex ions, which results in enrichment of catalyst species at the interface, is also illustrated by Figure 2. Although the formation of a relatively stable emulsion makes the gravitational phase separation more difficult than just a simple decantation, high activity, high regioselectivity, and much lower ligand costs for the hydroformylation of higher olefins render it attractive and feasible for potential commercialization.

It has been found that the n/i ratio is significantly determined by the intensity of agitation [2, 5], and this was conjectured to be related to stability of the ordered assemblies of surfactant hydrophobic ends at the liquid–liquid interface. Violent disturbance of the interface and a less orderly surfactant arrangement make possible more diversified steric routes of hydroformylation of linear olefin molecules. This relationship of the n/i ratio with the intensity of agitation has not been elucidated in a quantitative and mechanistic sense so far.

Moreover, the above physico-chemical concept of multiphase reaction mechanisms has not been formulated into a quantitative mathematical model as a tool to predict better the performance of biphasic hydroformylation of long-chain olefins and explore efficiently the provisos of industrialization.

2.3.1.3
Gas–Liquid–Liquid Three-Phase Hydroformylation

Industrial homogeneous hydroformylation with a water-soluble complex catalyst has been commercialized successfully for propylene (Ruhrchemie/Rhône-Poulenc process) [7, 8] and butylene [9], because they have significant water solubility and the reaction proceeds at a reasonable rate in the bulk aqueous phase. For long-chain substrates such as 1-dodecene with low solubility in the aqueous phase, the rate of biphasic reaction is slow. In addition to finding novel more active catalysts and catalytic systems, many chemical engineering efforts have been devoted since the mid-1990s to improving the interphase mass transfer between the organic and aqueous phases; the proposed approaches depend on the chemists' and chemical engineers' understanding of the mechanisms of biphasic hydroformylation. Different chemical engineering measures were proposed to enhance the rate and regioselectivity of hydroformylation reactions.

Purwanto and Delmas [10] proposed the addition of co-solvent (ethanol) to enhance the solubility of 1-octene in the aqueous phase so that the overall reaction rate was increased, and their kinetic study led to a rate model similar to that in homogeneous liquid systems consistently from the point of view of bulk reaction mechanism. Chaudhari et al. [11] reported the improvement of the hydroformylation rate by addition of a small amount of PPh_3 to the biphasic system to enrich the effective catalyst species at the liquid–liquid interface. Kalck et al. [12] tested two more approaches to improve the mass transfer rate of biphasic hydroformylation of 1-octene and 1-decene with catalyst precursor $[Rh_2(\mu\text{-}S^tBu)_2(CO)_2(TPPTS)_3]$: use the phase-transfer agent β-cyclodextrin to transport the substrate into the aqueous phase to react there (see Section 2.2.3.2.2), and the supported aqueous-phase (SAP) catalyst to increase the reaction area due to the high specific surface area of porous silica (see Section 2.6). The improved conversion and TOF gave informative suggestions for the reaction mechanisms.

Wachsen et al. [13] presented a chemical engineering analysis of typical biphasic hydroformylation, the RCH/RP process using propylene, to demonstrate that the reaction occurred at the gas–liquid interface. With the comparison of the model of bulk liquid-phase reaction and that of reaction in the interphase region with experimental data, it was found that only the latter model elucidated the experimental measurements on the gas-phase pressure and the flux of reaction heat. This model is instructive in further efforts to improve the biphasic hydroformylation performance.

However, the analysis did not clearly state at which interface the reaction took place: gas–aqueous, or gas–organic phase. Also there exists a possibility that the hydroformylation takes place at the liquid–liquid interface, where gaseous propylene and syngas first dissolve largely in the organic phase and are then transported to the reaction location. When surfactants were added to accelerate the catalytic reaction, the performance indicated that the biphasic reaction might occur mainly in the interface of the aqueous–organic phases instead of in the bulk of the liquid mixture [2, 14].

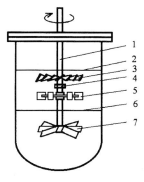

Figure 3 Hydroformylation autoclave with a patented composite surface aeration configuration: 1 Shaft; 2 Gas–liquid interface; 3 Self-rotating floating baffle (SRFB); 4 Annulus for position limiting; 5 Standard Rushton disk turbine (DT); 6 Liquid–liquid interface; 7 Pitched blade turbine upward (PBTU) [5, 18].

Previous studies [12, 15, 16] indicated that the intensity and mode of stirring and the reactor configuration affected dramatically the reaction rate and selectivity of biphasic hydroformylation of 1-dodecene catalyzed by water-soluble rhodium complexes. Yang et al. [5] further investigated several key engineering factors, including intensity of agitation, reactor configuration, and the technological conditions of reaction on hydroformylation of 1-dodecene as a typical higher olefin using $RhCl(CO)(TPPTS)_2$ complex catalyst and cationic surfactant. Hydroformylation of 1-dodecene was carried out in a 500 mL stirred stainless steel autoclave (Figure 3) in the $RhCl(CO)(TPPTS)_2/TPPTS/CTAB$ system. Both a six-blade Rushton disk turbine (DT) and a six-blade pitched blade turbine upward (PBTU) were used as surface aerators for entraining gaseous reactants into the liquid solution phases and for avoiding the recovery, recompression, and recycle of unreacted gases. A home-made baffle called a self-rotating floating baffle (SRFB) [17] was also applied for improvement of gas and liquid dispersion and pumping capability.

2.3.1.3.1
Effect of Agitation Speed

Agitation is certainly an important engineering factor that is strongly relevant to the performance of hydroformylation. In addition to the evidence presented in Figure 1, Yang et al. [5] also reported a significant influence of the agitation speed on the initial rate, conversion, and *normal/iso* aldehyde ratio of biphasic hydroformylation of 1-dodecene in a preliminary study. As shown in Figure 4, the initial rate and conversion increased with the agitation speed, but the regioselectivity to $n\text{-}C_{13}$ aldehyde decreased. These data demonstrated that violent stirring was unfavorable for the formation of linear aldehyde in the presence of cationic surfactant CTAB. The data presented by Chen identified a similar trend [2]. The opposite trends of the conversion and regioselectivity versus agitation make the conventional way of choosing the experimental agitation speed disputable, and it is suggested that a thorough understanding of the role of agitation is lacking, and that the laboratory experiments are not sufficient to advise on a proper compromise between the hydroformylation rate and the regioselectivity. However, the simple trend of a monotonic approach to an asymptotic level of hydroformylation

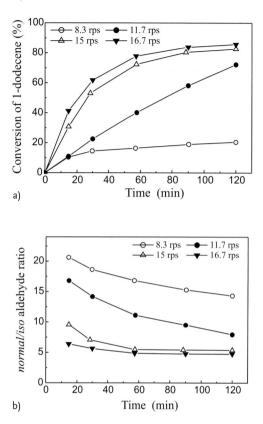

a)

b)

Figure 4 Influence of agitation speed on the conversion and regioselectivity of 1-dodecene hydroformylation [5].
(a) Conversion of 1-dodecene, (b) *n/i* aldehyde ratio.

conversion is frequently the case in simple biphasic systems without surfactant. For example, Lekhal et al. [6] observed that in the biphasic system with a co-solvent and without surfactant the reaction rate of 1-octene hydroformylation increased with the stirrer speed, but the selectivity of linear aldehyde was not affected and always close to 0.75. Only in such a case is it acceptable to select the stirrer speed corresponding to the stabilized hydroformylation conversion.

2.3.1.3.2
Orthogonal Experiment on Engineering Factors

Several chemical engineering factors affect the biphasic hydroformylation of 1-dodecene in a gas–liquid–liquid three-phase reaction system. In previous research [1], effects of temperature, total pressure, H_2/CO molar ratio, catalyst and ligand concentration, olefin concentration, surfactant concentration, and organic/aqueous-phase volume ratio on the hydroformylation kinetics were studied with

Table 2 Orthogonal table of L_{12} $(3^1 \cdot 2^4)$ for experimental design and the experimental results [5].

No.	N (rps)	V_O/V_W	C_{CTAB} [a] $\cdot 10^3$	Stirrer type	C_d [a]	r_L [b] $\cdot 10^3$	r_O [b] $\cdot 10^3$	n/i [c]	x [c] [%]	P [W]
1	1 (8.3)[d]	1 (3/7)	1 (11.0)	2 (PBTU)	2 (4.4)	0.0577	0.192	14.3	20.5	0.18
2	1	2 (1/1)	1	2	1 (2.2)[e]	0.0393	0.079	8.71	37.2	0.10
3	1	1	2 (12.3)	1 (DT)	1	0.0262	0.087	12.7	57.8	0.20
4	1	2	2	1	2	0.0865	0.173	9.13	33.3	0.39
5	2 (11.7)	1	1	1	2	0.135	0.451	5.17	77.3	0.48
6	2	2	1	2	1	0.131	0.263	6.37	58.7	0.42
7	2	1	2	2	2	0.208	0.695	9.08	68.5	0.35
8	2	2	2	1	1	0.134	0.268	4.87	54.3	0.43
9	3 (15.0)	1	1	1	1	0.192	0.640	4.37	77.0	1.05
10	3	2	1	1	2	0.233	0.467	4.22	68.1	1.04
11	3	1	2	2	1	0.176	0.586	4.75	78.6	0.91
12	3	2	2	2	2	0.357	0.713	4.85	72.5	1.45

Other reaction conditions: $T = 100\ °C$, $p = 1.1$ MPa, $y_{H2}/y_{CO} = 48.3/51.7$, [TPPTS]/[Rh] = 18, $C_{cat} = 1.5 \cdot 10^{-3}$ kmol m^{-3}, 2 h reaction time.

a) C_{CTAB} and C_d in kmol m^{-3}.
b) r_L and r_O in kmol m^{-3} s^{-1}; r_L was based upon total liquid volume, and r_O on the organic phase volume.
c) Observed values after 2 h reaction time.
d) Parameter values at different levels.
e) n-Decane was used as the organic diluent for adjusting the concentration of 1-dodecene to 2.2 kmol m^{-3}.

an orthogonal experiment of L_{18} $(2^1 \cdot 3^7)$. The preliminary optimal reaction conditions were found to be $T = 100\ °C$, $p = 1.1$ MPa, $V_O/V_W = 3/7$, $y_{H2}/y_{CO} = 1/1$, $C_{cat} = 1.5 \cdot 10^{-3}$ kmol m^{-3}, [TPPTS]/[Rh] = 18, $C_d = 2.2$ kmol/m^3 and $C_{CTAB} = 11.0 \cdot 10^{-3}$ kmol m^{-3}. Then Yang et al. [5] adopted another orthogonal experimental design to further investigate several key parameters: stirrer speed, CTAB concentration, 1-dodecene concentration, stirrer type, and the of organic/aqueous-phase volume ratio. The orthogonal design table of L_{12} $(3^1 \cdot 2^4)$ and the experimental results are tabulated in Table 2.

The margin analysis based on the experimental results in Table 2 suggests that the most important factor investigated is the agitation speed N and the second most important is the V_O/V_W ratio, as far as the maximum initial reaction rate, *normal/iso* aldehyde ratio and conversion are concerned (Table 3). The effect of surfactant concentration was the least important in the selected range of reaction

Table 3 Results of marginal analysis of orthogonal tests [16].

	N [rps]	V_O/V_W	C_{CTAB} [mol m^{-3}]	Stirrer type	C_d [kmol m^{-3}]
r_{O1a}	0.133	0.467	0.390	0.390	0.367
r_{O2a}	0.417	0.333	0.420	0.422	0.450
r_{O3a}	0.617				
R_{rO}	0.484	0.134	0.030	0.032	0.083
n/i_{1a}	11.22	8.40	6.79	6.75	6.60
n/i_{2a}	6.37	6.36	7.57	7.95	7.79
n/i_{3a}	4.51				
$R_{n/i}$	6.71	2.04	0.78	1.20	1.19
P_{1a}	0.26	0.71	0.726	0.78	0.70
P_{2a}	0.49	0.75	0.732	0.67	0.76
P_{3a}	1.29				
R_p	1.03	0.04	0.006	0.11	0.06
x_{1a}	37.17	65.24	59.38	63.56	62.93
x_{2a}	64.69	53.99	60.82	55.98	56.68
x_{3a}	74.64				
R_x	37.47	11.25	1.44	7.58	6.25
r_{L1a}	0.052	0.133	0.131	0.135	0.115
r_{L2a}	0.152	0.163	0.164	0.161	0.149
r_{L3a}	0.240				
R_{rL}	0.188	0.030	0.033	0.026	0.034

Note: Analysis was based on the results in Table 2.
r_{Oja} (j = 1, 2, 3) is the average value of r_O at level j of factor i (i = 1, 2, ..., 5),
R_{rO} = max (r_{Oja}) − min (r_{Oja}).
n/i_{ja} (j = 1, 2, 3) is the average value of n/i at level j of factor i (i = 1, 2, ..., 5),
$R_{n/i}$ = max (n/i_{ja}) − min (n/i_{ja}).
P_{ja} (j = 1, 2, 3) is the average value of P at level j of factor i (i = 1, 2, ..., 5),
R_p = max (P_{ja}) − min (P_{ja}).
x_{ja} (j = 1, 2, 3) is the average value of x at level j of factor i (i = 1, 2, ..., 5),
R_x = max (x_{ja}) − min (x_{ja}).

conditions [16]. The results of variance analysis were similar to the margin analysis, except for a minor difference in the sequence of importance of influence of two less important factors: the importance of the concentrations of CTAB and 1-do-decene to the index of conversion. The main result was that when the agitation speed was increased, the power consumption, conversion, and initial rate all increased, whereas the regioselectivity to *n*-tridecylic aldehyde decreased. It is somewhat surprising that a seemingly irrelevant factor, the V_O/V_W ratio, plays a significant role in the hydroformylation performance: the reaction rate based on organic phase volume and regioselectivity was enhanced by a decrease in V_O/V_W. This is possibly attributable to easy and stable formation of micelles and O/W emulsion as well as a greater amount of catalyst, but the exact mechanism awaits quantitative interpretation.

Although the optimal set of variable levels corresponding to each index was not completely coincident, the optimal reaction conditions for a compromise between high productivity and selectivity and low power consumption were suggested to be $N = 11.7$ rps, $V_O/V_W = 3/7$, $C_d = 4.4$ kmol m^{-3}, $C_{CTAB} = 12.4 \cdot 10^{-3}$ kmol m^{-3} and the stirrer type PBTU [5]. It is noticeable that the optimal value of the organic/aqueous-phase volume ratio agrees with that obtained by Zhang et al. [1]. It is also reported that low volume ratio (V_O/V_W) was in favor of conversion and TOF frequency in hydroformylation of 1-dodecene [19]. However, no chemical engineering model is available for a quantitative account of the influence of this factor.

2.3.1.3.3
Influence of Agitation Configuration

Biphasic hydroformylation of 1-dodecene indexed by the chemical conversion and regioselectivity can be optimized by exploration of the appropriate agitation configuration and intensity of agitation, which affects the hydrodynamics of mixing and dispersion correlated with interphase mass transfer. Five cases of agitation configuration were tested [5, 15]. Cases 1 and 2 were standard surface aeration configurations using single DT and PBTU, respectively. Novel composite impeller configurations were featured with a self-rotating floating baffle (SRFB) located between the gas–liquid interface and a DT impeller (Case 3) or a PBTU impeller (Case 4) to improve surface aeration. As shown in Figure 3, a dual impeller including an upper DT used as a surface aerator and a PBTU fixed in the middle of the reactor for circulating the reaction mixture was combined with an SRFB placed above the DT, and this composite agitation configuration was Case 5 [18].

As shown in Table 4 and Figure 5, the single PBTU impeller (Case 2) achieved a better performance with a higher initial rate and n/i aldehyde ratio accompanied by a lower power consumption than the single DT impeller (Case 1) under the same conditions of operation. This was possibly due to the better capability for overall circulation and lower agitation intensity of the PBTU impeller, which favored maintenance of the orderly micelle structure. When the novel SRFB was used (Cases 3 and 5), the initial rate of the biphasic hydroformylation increased remarkably from $0.630 \cdot 10^{-3}$ kmol m^{-3} (Case 1) to $1.24 \cdot 10^{-3}$ and $2.03 \cdot 10^{-3}$ kmol m^{-3} respectively, by a factor of almost 2–3. However, the conversion and regioselectivity after 2 h of reaction were almost the same. Therefore, it is clear that an SRFB can increase the initial rate of 1-dodecene hydroformylation, but fails to further enhance

Table 4 Effect of agitation configuration on hydroformylation of 1-dodecene [5].

Case	Stirrer type	$r_L \cdot 10^3$ [kmol m^{-3} s^{-1}]	$r_0 \cdot 10^3$ [kmol m^{-3} s^{-1}]	n/i [a]	x [a] [%]	P [W]
1	DT	0.189	0.630	6.39	84.0	0.42
2	PBTU	0.208	0.695	9.08	68.5	0.35
3	DT + SRFB	0.372	1.24	6.24	83.0	0.56
4	PBTU + SRFB	0.0620	0.207	10.5	42.8	0.41
5	DT + PBTU + SRFB	0.609	2.03	6.06	84.2	–

a) Observed values after 2 h reaction time.

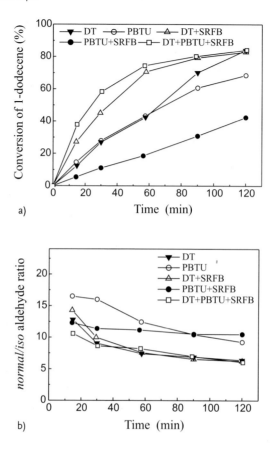

Figure 5 Hydroformylation conversion and regioselectivity versus
reaction time with different agitation configurations [5].
(a) Conversion of 1-dodecene, (b) *n/i* aldehyde ratio.

the conversion and selectivity over a full 2 h run. The poor performance in Case 4
was probably caused by the failure to self-rotate and float of the SRFB placed above
the PBTU, with the consequence that normal contact, circulation, and mixing of
gas and liquid streams were weakened.

Yu et al. [17] demonstrated that the composite surface aeration configurations
with an SRFB were effective in enhancing the gas–liquid volumetric mass transfer
coefficient by intensifying turbulence in gas–liquid–liquid flow and in generating
more small gas bubbles by stronger shear. Since the gas–liquid mass transfer of
biphasic hydroformylation systems may be limiting [6], the composite agitation
configuration displays a better performance in promoting the reaction rate and is
recommended for stirred autoclaves. From the above results, it can be concluded
that improving the agitation configuration and choosing a suitable agitation intensity
in gas–liquid and gas–liquid–liquid systems require as much attention as the
catalytic chemistry itself.

2.3.1.3.4
Emulsification of Reaction Mixture

It is observed from Figures 4 and 5 that the molar ratio of *normal* to *iso* aldehyde products decreased significantly with time, and it obviously varied with the agitation intensity and stirrer type. This phenomenon, closely related to the mechanisms of biphasic hydroformylation, is difficult to explain even in a qualitative manner. At the same time, significant variation of emulsification of the reaction mixture occurred in correlation with the hydroformylation system, surfactant addition, and agitation.

It is believed that the extent of emulsification is a reflection of the operating conditions and physicochemical properties of the reaction system. Because it is very difficult to characterize quantitatively, the extent of emulsification, a simple but somewhat subjective and fuzzy index was proposed to represent the extent of emulsification [15]. The defined fuzzy index ranges from 0 (significant phase settlement in less than 1 min after sample withdrawal from the reactor) to 5 (severe emulsification and no phase separation by gravity observed over 4 h at ambient temperature); for intermediate values, 1 is assigned to the situation with significant phase separation in 1–5 min, 2 for 5–20 min, 3 for 20–60 min and 4 for 1–4 h.

As shown in Figures 6 and 7, the extent of emulsification was observed to develop as the hydroformylation reaction proceeded, and to be intensified with the agitation intensity and CTAB concentration. The agitation configuration leading to higher olefin conversion resulted in more severe emulsification (Figure 8). It is obvious that emulsification favors the hydroformylation conversion and disfavors the regioselectivity (Figure 9). It may be conjectured that the oscillating microstructure of micelles accompanied by severe emulsification was not kept as orderly and compact as that at the beginning of the reaction, and this effect not only increased the difficulty of product separation at the end of hydroformylation, but was also unfavorable for the formation of linear aldehyde. It seems that the correlation of the extent of emulsification with hydroformylation indices must be interpreted from the chemical and chemical engineering point of view, and more work on

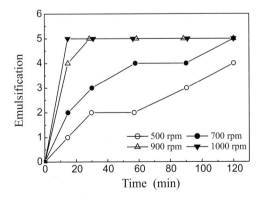

Figure 6 Influence of agitation speed on emulsification [5].

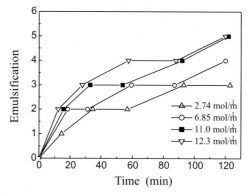

Figure 7 Influence of surfactant concentration on emulsification [5].

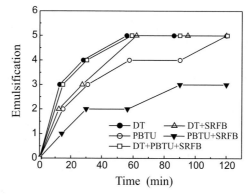

Figure 8 Influence of agitation configuration on emulsification [5].

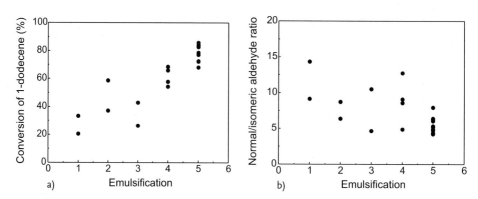

Figure 9 Conversion and regioselectivity versus emulsification extent of hydroformylation mixture in 2 h under different reaction conditions based on the experimental data in Table 2 and Figures 6–8 [5]. (a) Conversion of 1-dodecene, (b) *n/i* aldehyde ratio.

physicochemical properties of the emulsification, such as optical refraction, droplet size and interface features, micelle microstructure, etc., is required so as to facilitate further and in-depth understanding of the underlying mechanism of biphasic hydroformylation.

2.3.1.3.5
Kinetics of Biphasic Hydroformylation

Biphasic hydroformylation is a typical and complicated gas–liquid–liquid reaction. Although extensive studies on catalysts, ligands, and catalytic product distributions have appeared, the reaction mechanism has not been understood sufficiently and even contradictory concepts of the site of hydroformylation reaction were developed [11, 13, 20]. Studies on the kinetics of hydroformylation of olefins are not only instructive for improvement of the catalytic complexes and ligands but also provide the basic information for design and scale-up of novel commercial reactors. The kinetics of hydroformylation of different olefins, such as ethylene, propylene, 1-hexene, 1-octene, and 1-dodecene, using homogeneous or supported catalysts has been reported in the literature. However, the results on the kinetics of hydroformylation in aqueous biphasic systems are rather limited and up to now no universally accepted intrinsic biphasic kinetic model has been derived, because of the unelucidated reaction mechanism and complicated effects of multiphase mass transfer (see also Section 2.4.1.1.2).

The kinetics of low-carbon olefins, ethylene [21] and propylene [22], in aqueous systems was reported and different rate models proposed. Wachsen et al. [13] proved that aqueous biphasic hydroformylation of propylene took place at the interfacial region, in contrast to two preliminary kinetic models that incorporate mass transport.

The biphasic hydroformylation of 1-octene was studied in the presence of ethanol as a co-solvent and a proposed kinetic rate expression was nearly identical to that of the homogeneous system [10, 23]. A further kinetic study of this biphasic hydroformylation system was conducted by Lekhal et al. [6] to analyze the experimental data by coupling kinetics to a pseudo-homogeneous gas–liquid–liquid macroscopic conservation model; the authors proved that gas–liquid mass transfer was the only limitation.

Borrmann et al. [24] reported the kinetic data of the biphasic hydroformylation of several long-chain and branched olefins using a water-soluble rhodium polyethylene glycolate catalyst, with first-order reaction for 1-dodecene and 2,4,4-trimethylpent-1-ene and zero-order for styrene. The activation energy of the biphasic hydroformylation in a polyethylene glycol (PEG) solvent is higher than that in water as solvent (cf. Section 2.3.5). Based on some conjecture about the mechanism of 1-dodecene hydroformylation, Zhang et al. [1, 25] investigated the macro-kinetics of 1-dodecene hydroformylation catalyzed by a $RhCl(CO)(TPPTS)_2$ complex in the presence of CTAB, and analyzed the influence of interphase mass transfer by a computational fluid dynamics method. Several complicated kinetic and regioselectivity models, including the influence of concentrations of higher olefin and

Table 5 Activation energy of biphasic hydroformylation of olefins.

Olefin	Catalyst	E [kJ mol^{-1}]	Ref.
1-Octene	[Rh(COD)Cl]$_2$TPPTS/ethanol	66	[23]
1-Hexene	Supported Rh/P4VP	18–59	[26]
Cyclohexene		22–93	
1-Hexene	Rh(PEG$_x$)water/PEG	23–71	[24]
1-Octene		94–106	
1-Dodecene		95–105	
Diisobutene		30–51	
Propylene	RhCl(CO)(TPPTS)$_2$	77–83	[22]
1-Dodecene	RhCl(CO)(TPPTS)$_2$	62–73	[1]

catalyst, partial pressures of H$_2$ and CO, molar ratios of ligand to catalyst, volume ratios of organic/aqueous phases, weight percentages of surfactant and temperature, were presented. The macro-kinetic data reflecting the intrinsic chemical reaction and mass transfer under practical technological conditions provide the necessary basis for modeling and designing multiphase reactors.

Summarizing results reported in literature, the kinetic rate of biphasic hydroformylation was found to be influenced positively by increasing the concentration of catalyst, olefin, and hydrogen, whereas a higher carbon monoxide pressure exerted a negative effect. It seems that regioselectivity has not been approached by kinetic analysis to date. The apparent activation energy for hydroformylation was found to be in the range from 20 to 100 kJ mol^{-1} (Table 5), which covers the lower range of usual chemical reactions. Thus, the influence of mass transfer on the intrinsic kinetics of biphasic hydroformylation is not completely eliminated, though many precautions were taken to minimize the resistance of mass transfer in kinetic experiments. The mass transfer resistance depends largely on the solubility of the reactants, the thermodynamic phase equilibrium, and the interfacial properties, as well as the hydrodynamics of gas–liquid–liquid dispersion, agitation, and mixing. In practice, for improved performance of a gas–liquid–liquid hydroformylation reaction in a multiphase reactor, it is necessary to study and combine the intrinsic kinetics, mass and heat transfer, residence time distribution, and reactor configuration in an appropriate mathematical model.

2.3.1.4
Mathematical Model for Biphasic Hydroformylation Reactor

It is rather difficult to establish a reasonably accurate mathematical model of a gas–liquid–liquid three-phase reactor for biphasic hydroformylation, because of the complexity in formulating all the necessary mechanisms such as phase dispersion and distribution, multiphase flow, interphase mass transfer, micromixing, and the hydroformylation reaction. Besides, the task is further complicated by turbulence in multiphase flow and the complex domain of stirred-tank reactors.

Although a chemical engineering study is in rapid progress toward this goal, only a few simpler but still rather difficult models, aimed at part of the chemical engineering process taking place in the reactor, have been proposed for application in the biphasic hydroformylation of olefins.

Lekhal et al. [6] proposed a pseudo-homogeneous gas–liquid–liquid model based on the Higbie penetration theory to account for simultaneous absorption of two gases into the liquid phases. Because of the assumption of rapid liquid–liquid mass transfer of reactants leading to the equilibrium between two liquid phases, the model was simplified greatly and the detail of phase dispersion and distribution and multiphase flow was avoided. Reasonable success was achieved and the results of analysis suggested that the only limitation to the conversion of hydroformylation of 1-octene was the gas–liquid mass transfer of CO and H_2.

Van Elk et al. [27] used a similar mathematical model, based on the penetration model for three reactants in an ideally stirred reactor, to study the dynamic behavior of the gas–liquid homogeneous hydroformylation process. The influence of mass and heat transfer on the reactor stability in the kinetically controlled regime was analyzed and it brought to mind the existence of a dynamically unstable (limit circle) state under certain operating conditions. This model needs to be extended to account for the presence of a second liquid phase in biphasic hydroformylation.

In fact, a complete mathematical model of a hydroformylation reactor includes many sub-models. At present, the kinetics reported in the literature seems to contain uncertainty due to incomplete elimination of interphase mass transfer resistance in conventional experimental studies. In view of this, Zhang et al. [25] tried to use a reactor of simple geometry (a regular cylinder) in combination with a numerical approach to account accurately for the gas–liquid and liquid–liquid mass transfer in determining the interfacial hydroformylation kinetics and removing the possible uncertainty of estimating effective reactant concentrations. As for the multiphase flow in stirred reactors, great success in numerical simulation of gas–liquid flow has been achieved [28–31], but the turbulent interphase mass transfer is not resolved yet. Therefore, tremendous efforts are required before the mathematical model of a hydroformylation reactor as a predicting tool can be used for the purpose of commercial biphasic hydroformylation.

2.3.1.5
Prospect of Reaction Engineering Study on Hydroformylation

It is gradually being recognized by chemists that the engineering aspects indeed have important relevance to the chemical study of homogeneous and heterogeneous hydroformylation. In a series of regular reviews on yearly progress in hydroformylation research, a subtitle "Engineering aspects" started to appear in the contents [32] but the discussion was less voluminous. Meanwhile, several developments open new routes for effectively conducting hydroformylation of higher molecular weight olefins, in particular for facilitating the catalyst separation. Horváth et al. [33a,b] proposed fluorous biphasic systems (FBSs) which consist of a fluorous phase containing dissolved reagent or catalyst and a product phase with limited solubility in the fluorous phase (see Chapter 4). Based on nonionic phosphine ligands, the concept of thermoregulated phase transfer catalysis (TRPTC) was developed and applied in aqueous/organic biphasic hydroformylation of higher olefins [34, 33] (cf. Section 2.3.5). The separation of catalyst from substrate and product phase can also be achieved by using the special properties of ionic liquids, as suggested by Webb et al. [38], who conducted hydroformylation of 1-dodecene in a continuous flow system consisting of ionic liquid and supercritical CO_2 to effect downstream separation of products from reaction solvent and catalyst (cf. Chapters 5 and 6). The improved SLP-catalyst technology, combined with flow reversal and artificial unsteady operation, approaches steadily the same target along the same lines [39, 40]. Leitner [36] also devoted effort to understanding the role of engineering aspect factors on the performance and kinetics of hydroformylation of olefins. Here we use the conclusion from Leitner as the final remark, i.e., "Increasing research efforts are urgently needed to improve our fundamental knowledge and to broaden the scope of the basic methodology. At the same time, reaction engineering becomes more and more important to evaluate various process designs in terms of efficiency and costs. Interdisciplinary efforts involving preparative chemists, researchers in catalysts, physico-chemists, and chemical engineers appear to have the highest chances to meet this challenge."

Symbols

C	concentration of various component, kmol m^{-3}
E	activation energy, kJ mol^{-1}
N	agitation speed, rps
n/i	ratio of *normal/iso* aldehydes
P	power consumption for agitation, W
p	total pressure, MPa
r	initial rate of hydroformylation reaction, kmol m^{-3} s^{-1}
T	reaction temperature, °C
V_O/V_W	volume ratio of organic/aqueous phases
W	weight percentage in water, %
x	conversion of olefin
y	molar fraction of gas component

Subscripts

a	average value
cat	rhodium complex catalyst
CO	carbon monoxide
CTAB	cetyltrimethylammonium bromide
d	1-dodecene
H_2	hydrogen
L	total liquid phase based
O	organic phase volume based

References

1 Y. Q. ZHANG, Z.-S. MAO, J. Y. CHEN, *Catal. Today* **2002**, *74*, 23.

2 H. CHEN, Y. Z. LI, J. R. CHEN, P. M. CHENG, Y. E. HE, X. J. LI, *J. Mol. Catal. A: Chem.* **1999**, *149*, 1.

3 F. VAN VYVE, A. RENKEN, *Catal. Today* **1999**, *48*, 237.

4 L. B. WANG, H. CHEN, Y. E. HE, Y. Z. LI, M. LI, X. J. LI, *Appl. Catal. A:* **2003**, *242*, 85.

5 C. YANG, X. Y. BI, Z.-S. MAO, *J. Mol. Catal. A:* **2002**, *187*, 35.

6 A. LEKHAL, R. V. CHAUDHARI, A. M. WILHELM, H. DELMAS, *Catal. Today* **1999**, *48*, 265.

7 B. CORNILS, E. G. KUNTZ, *J. Organomet. Chem.* **1995**, *502*, 177.

8 B. CORNILS, W. A. HERRMANN, R. W. ECKL, *J. Mol. Catal. A:* **1997**, *116*, 27.

9 B. CORNILS, W. A. HERRMANN (Eds.), *Applied Homogeneous Catalysis with Organometallic Compounds*, Wiley-VCH, Weinheim **2000**.

10 P. PURWANTO, H. DELMAS, *Catal. Today* **1995**, *24*, 135.

11 R. V. CHAUDHARI, B. M. BHANAGE, R. M. DESHPANDE, H. DELMAS, *Nature* **1995**, *373*, 501.

12 P. KALCK, L. MIQUEL, M. DESSOUDEIX, *Catal. Today* **1998**, *42*, 431.

13 O. WACHSEN, K. HIMMLER, B. CORNILS, *Catal. Today* **1998**, *42*, 373.

14 H. DING, B. E. HANSON, T. BARTIK, B. BARTIK, *Organometallics* **1994**, *13*, 3761.

15 Z.-S. MAO, X. Y. BI, G. Z. YU, Y. Q. ZHANG, C. YANG, R. WANG, *Chinese J. Chem. Eng.* **2002**, *10*, 45.

16 X. Y. BI, C. YANG, Z.-S. MAO, *J. Chem. Ind. Eng. (China)* **2001**, *52*, 570 (in Chinese). X. Y. Bi, M. S., Thesis, Institute of Process Engineering, Chinese Academy of Sciences, Beijing, China **2001** (in Chinese).

17 G. Z. YU, Z.-S. MAO, R. WANG, *Chinese J. Chem. Eng.* **2002**, *10*, 39.

18 C. Yang, G. Z. Yu, Z.-S. Mao, Chinese Patent Application 02 104 088.5, **2002**.

19 Y. Z. LI, H. CHEN, J. R. CHEN, P. M. CHENG, J. Y. HU, X. J. LI, *Chinese J. Chem.* **2001**, *19*, 58.

20 P. KALCK, M. DESSOUDEIX, S. SCHWARZ, *J. Mol. Catal. A:* **1999**, *143*, 41.

21 R. M. DESHPANDE, B. M. BHANAGE, S. S. DIVEKAR, S. KANAGASABAPATHY, R. V. CHAUDHARI, *Ind. Eng. Chem. Res.* **1998**, *37*, 2391.

22 C. YANG, Z.-S. MAO, Y. F. WANG, J. Y. CHEN, *Catal. Today* **2002**, *74*, 111.

23 R. M. DESHPANDE, P. PURWANTO, H. DELMAS, R. V. CHAUDHARI, *Ind. Eng. Chem. Res.* **1996**, *35*, 3927.

24 T. BORRMANN, H. W. ROESKY, U. RITTER, *J. Mol. Cat. A:* **2000**, *153*, 31.

25 Y. Q. ZHANG, Z.-S. MAO, J. Y. CHEN, *Ind. Eng. Chem. Res.* **2001**, *40*, 4496.

26 M. M. MDLELENI, R. G. RINKER, P. C. FORD, *Inorg. Chim. Acta* **1998**, *270*, 345.

27 E. P. VAN ELK, P. C. BORMAN, J. A. M. KUIPERS, G. F. VERSTEEG, *Chem. Eng. Sci.* **2001**, *56*, 1491.

28 W. J. WANG, Z.-S. MAO, *Chinese J. Chem. Eng.* **2002**, *10*, 385.

29 K. E. MORUD, B. H. HJERTAGER, *Chem. Eng. Sci.* **1996**, *51*, 233.

30 A. D. GOSMAN, C. LEKAKOU, S. POLITIS, R. I. ISSA, M. K. LOONEY, *AIChE J.* **1992**, *38*, 1946.

31 V. V. RANADE, H. E. A. VAN DEN AKKER, *Chem. Eng. Sci.* **1994**, *49*, 5175.

32 F. UNGVARY, *Coord. Chem. Rev.* **2002**, *228*, 61.

33a I. T. HORVATH, J. RABAI, *Science* **1994**, *266*, 72.

33b I. T. HORVÁTH, G. KISS, R. A. COOK, J. E. BOND, P. A. STEVENS, J. RABAI, E. J. MOZELESKI, *J. Am. Chem. Soc.* **1998**, *120*, 3133.

34 Z. JIN, Z. L. ZHENG, B. FELL, *J. Mol. Catal. A:* **1997**, *116*, 55.

35 Y. WANG, J. JIANG, F. CHENG, Z. JIN, *J. Mol. Catal. A:* **2002**, *188*, 79.

36 W. LEITNER, *Acc. Chem. Res.* **2002**, *35*, 746.

38 P. B. WEBB, M. F. SELLIN, T. E. KUNENE, S. WILLIAMSON, A. M. Z. SLAWIN, D. J. COLE-HAMILTON, *J. Am. Chem. Soc.* **2003**, *125*, 15577.

39 A. BECKMANN, F. J. KEIL, *Chem. Eng. Sci.* **2003**, *58*, 841.

40 K. G. W. HUNG, D. PAPADIAS, P. BJORNBOM, M. ANDERLUND, B. AKERMARK, *AIChE J.* **2003**, *49*, 151.

2.3.2
Technical Solutions

Arno Behr and Joachim Seuster

2.3.2.1
Introduction

A question that applies generally in homogeneous catalysis using organometallic catalysts is how to separate the catalyst from the product. In homogeneous catalysis this problem can become the most critical point in the technical realization of a chemical process. One benefit of the aqueous-phase chemistry is the possible formation of a second nonpolar phase by organic products, thus enabling the separation of the catalyst from the products. The use of two liquid phases with the aim of separating the homogeneous catalyst is called the liquid–liquid two-phase technique (LLTP).

A further benefit of this LLTP technique is the suppression of consecutive reactions. This advantage is described by the general equation (1), in which starting components A and B usually yield the products C and D, but also an undesired consecutive product E. If the catalyst and the products C and D form only one phase, a further reaction to E cannot be avoided. If, however, the catalyst is dissolved in the aqueous phase and the products C and D in the second organic phase, no further reaction can proceed. Consequently, by using the LLTP-technique, the products C and D are formed with high selectivity.

$$A + B \xrightarrow{\text{[cat]}} C + D \xrightarrow{\text{[cat]}} E \tag{1}$$

2.3.2.2
Technical Realization

Table 1 gives a brief overview of the industrially realized polar-phase organometallic-catalyzed reactions together with some related reactions which are still under development.

The most important process worldwide is the "Shell higher-olefin process" (SHOP) in which ethylene is oligomerized to higher molecular mass, linear α-olefins. The nickel catalyst, containing a phosphorus/oxygen chelate ligand, is dissolved in the polar solvent 1,4-butanediol, which is not miscible with the α-olefins (see Chapter 3 [1–5]).

A very important process with the solvent water is the hydroformylation of propene to butyraldehydes, known as the Ruhrchemie/Rhône-Poulenc process. The reaction is catalyzed by a rhodium complex containing the water-soluble ligand triphenyl-phosphine trisulfonate (TPPTS) (see Section "Lower Alkenes" [6–11]).

All other industrial applications are still on a smaller scale. The Kuraray telomerization of butadiene with water is carried out at 5000 t y^{-1} (see Section 2.4.4.2

Table 1 Examples of organometallic-catalyzed reactions.

Company or University	Reaction	Catalyst	Solvent	Ref.
Shell (SHOP)	Oligomerization of ethene	Ni	1,4-Butanediol	[1–5]
Ruhrchemie/ Rhône-Poulenc	Hydroformylation of olefins	Rh	Water	[6–11]
Rhône-Poulenc	Co-oligomerization with myrcene	Rh	Water	[12–14]
Rhône-Poulenc	Hydrogenation of unsaturated aldehydes	Ru	Water	[15–18]
Wasserscheid/ Univ. Erlangen	Hydrogenation of sorbic acid	Ru	Ionic liquids	[19]
Kuraray	Telomerization of butadiene with water	Pd	Water/sulfolane	[20–28]
Behr/Univ. Dortmund	Telomerization of isoprene with methanol	Pd	Methanol	[29]
Behr/Univ. Dortmund	Telomerization of butadiene with glycol	Pd	Acetonitrile	[30–31]
Driessen-Hölscher/ Univ. Paderborn	Telomerization of butadiene with ammonia	Pd	Water	[32–33]
Keim/ Univ. Aachen	Telomerization of butadiene with phthalic acid	Pd	Acetonitrile	[34]
Keim/ Univ. Aachen	Dimerization of butadiene	Pd	DMSO	[34]

[20–28]); the Rhône-Poulenc reactions have been developed into bulk processes [12–18]. The investigations at the Universities of Dortmund, Aachen, Paderborn, and Erlangen (Germany) are still on laboratory or mini-plant scale.

After the reaction, the products have to be separated from the catalyst. This can be done by separation either of the product or of the catalyst from the residue. The separation methods can be varied: simple separation of two liquid phases, extraction with an additional solvent, or chemical treatment with additional bases or acids.

In all these cases a reaction is combined with a separation step. This can be done in the same unit (simultaneously) or in separate units (successively).

In total, by combining all variations there are 12 cases which will be discussed in more detail using the general reaction [Eq. (1)]. To simplify the discussion it is assumed that the expected products are organic nonpolar compounds. Of course, analogous considerations apply if the products are polar. In these cases the terms "polar" and "nonpolar" in the following sections have to be interchanged.

2.3.2.2.1
Reaction and Product Separation

The simplest way to combine catalysis in a polar medium and product separation is shown in Figure 1, where both operations proceed in one unit. The nonpolar phase can be separated at the top of the column while the polar catalyst phase remains in the reactor. This principle is realized in the SHOP process using 1,4-butanediol with a nickel catalyst as the catalytic phase.

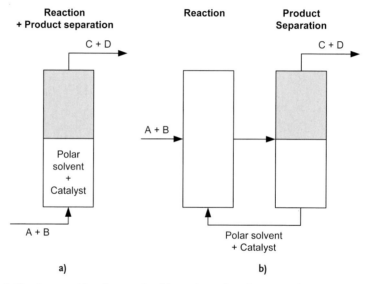

Figure 1 Simultaneous (a) and successive (b) reaction and product separation.

The simultaneous process shown in Figure 1 (a) can be divided into two sequential units, shown in Figure 1 (b), as realized in the Rhône-Poulenc/Ruhrchemie hydroformylation process of propene. The reaction takes place in a continuous stirred tank reactor (CSTR) while the phase separation is carried out in a decanter.

2.3.2.2.2
Reaction and Product Extraction

Unlike the processes described above, the processes in this section use an additional organic nonpolar solvent for the product extraction. Figure 2 shows the simultaneous reaction and product extraction also called "in-situ extraction". The nonpolar extractant absorbs the reaction product in the reactor and is separated from the polar catalyst phase in the following separation step. Due to the subsequent solvent distillation this process is somewhat more complex.

An application of this principle is in the telomerization of butadiene with ethylene glycol. One-phase reactions yield both mono- and ditelomers, while the two-phase system using a water/glycol phase containing a polar palladium catalyst avoids the

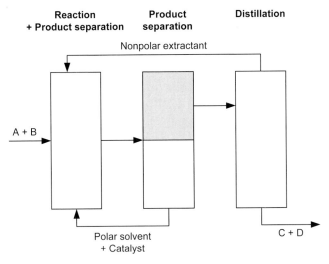

Figure 2 Simultaneous reaction and product extraction.

consecutive reaction to the ditelomer, thus yielding the desired monotelomers with high selectivity [30–31].

The successive process alternative, shown in Figure 3, divides the combined reaction/extraction step into a reaction unit with a single polar phase and the product extraction unit.

The practicability of this concept will be demonstrated by two examples. In the first, the dimerization of butadiene to 1,3,7-octatriene in the presence of a catalyst containingg a palladium precursor and triphenylphosphine as well as the alcohol $PhC(CF_3)_2OH$ is carried out favorably in the solvent acetonitrile. The following

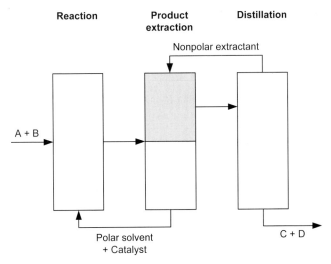

Figure 3 Successive reaction and product separation.

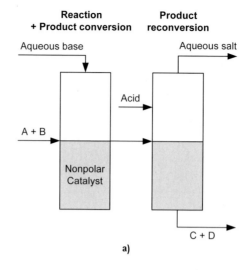

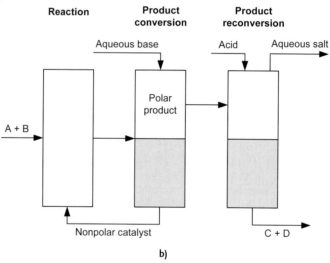

Figure 4 Simultaneous (a) and successive (b) reaction and product treatment.

extraction with isooctane separates the product from the acetonitrile phase, which still contains the catalyst and alcohol. Finally, the isooctane is distilled from the product and recycled into the extraction unit. In a continuously working miniplant 3 kg of octatriene per gram Pd could be produced.

The second example is the telomerization of phthalic acid with butadiene yielding bis(octadienyl) phthalates, which can be used as plastic softeners after hydrogenation. The polar solvent dimethyl sulfoxide contains the palladium catalyst formed from Pd(acac)$_2$ and tris(p-methoxyphenyl)phosphite. This extraction uses isooctane as well [34].

2.3.2.2.3
Reaction and Product Treatment

An interesting concept is the synthesis of an organic nonpolar product with a nonpolar catalyst, and the parallel conversion of the nonpolar product into a polar one and its separation with the polar phase from the catalytic nonpolar phase as shown in Figure 4 (a).

An example is the synthesis of nonpolar long-chain carboxylic acids, which are transformed into the polar carboxylic salts by addition of an aqueous base. In a second unit the carboxylic salt is reconverted into the nonpolar product and a salt solution which can easily be separated. The production of salt waste is one of the disadvantages of this and related chemical separation techniques.

The alternative process of successive reaction and product treatment, where the reaction and product conversion is split into two units, is shown in Figure 4 (b).

2.3.2.2.4
Reaction and Catalyst Separation

Another alternative for catalyst separation is the filtration after the catalyst has been precipitated. An interesting application is given by Fell and co-workers [35] for the hydroformylation of higher olefins by a methanol- and water-soluble rhodium catalyst. As shown in Figure 5, the reaction is carried out in a homogeneous methanol solution. After distillation of the methanol the catalyst is precipitated, filtered off, and again prepared with the distilled methanol. The products leave the process after the filtration step.

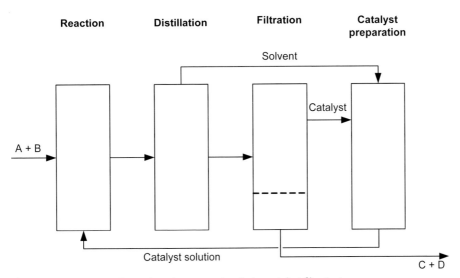

Figure 5 Successive reaction and catalyst separation (using catalyst filtration).

2.3.2.2.5
Reaction and Catalyst Extraction

The simultaneous reaction and catalyst extraction shown in Figure 6 (a) is not favorable, due to the fact that the catalyst leaves the reaction medium during the reaction. The successive reaction and catalyst extraction shown in Figure 6 (b), where reaction and extraction are carried out in two different units, is however much more favorable.

The most important point is the complete separation of the catalyst and the products. The Union Carbide process proposal for the hydroformylation of higher olefins solves this problem [36, 37]. The catalyst leaching is lower than 20 ppb of the rhodium catalyst.

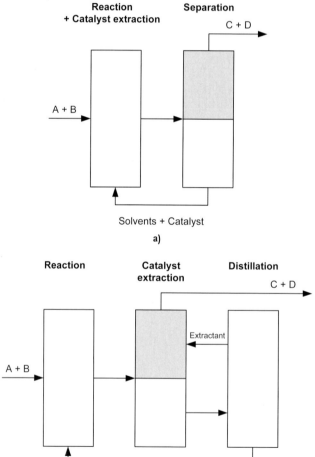

Figure 6 Simultaneous (a) and successive (b) reaction and catalyst extraction.

2.3.2.2.6
Reaction and Catalyst Treatment

Analogously to the process described in Section 2.3.2.2.3, where the product is treated with bases or acids, the catalyst also can be treated in this way. As discussed in the previous section, removal of the catalyst while the reaction is occurring is not favorable.

Figure 7 shows the successive reaction and catalyst treatment. After the single-phase reaction using a nonpolar solvent and catalyst, the nonpolar catalyst is converted into a polar one. In the third step, the catalyst is reconverted into a nonpolar one and recycled to the reaction unit. The conversions can be processed by using a catalyst containing anionic ligands with higher alkylammonium cations. Addition of sodium hydroxide will cause the alkylammonium cation to be exchanged by the sodium cation and leave the nonpolar phase. Of course, this process can be reversed by adding an acid and new ammonium ions.

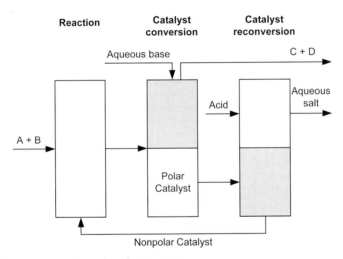

Figure 7 Successive reaction and catalyst treatment.

References

1 W. KEIM, *Chem. Ing. Tech.* **1984**, *56*, 850.
2 A. BEHR, W. KEIM, *Arabian J. Sci. Eng.* **1985**, *10*, 377.
3 W. KEIM, A. BEHR, B. GRUBER, B. HOFFMANN, F. H. KOWALDT, U. KÜRSCHNER, B. LIMBÄCKER, F. P. SISTIG, *Organometallics* **1986**, *5*, 2356.
4 W KEIM, *New J. Chem.* **1987**, *11*, 531.
5 W KEIM, *J. Mol. Catal.* **1989**, *52*, 19.
6 Rhône-Poulenc (E. G. KUNTZ), DE 2.627.354 (**1976**).
7 E. G. KUNTZ, *CHEMTECH* **1987**, 570.
8 E. WIEBUS, B. CORNILS, *Chem. Ing. Tech.* **1994**, *66*, 961.
9 B. CORNILS, E. G. KUNTZ, *J. Organomet. Chem.* **1995**, *502*, 177.

10 E. Wiebus, B. Cornils, *Hydrocarb. Proc.*, March **1996**, 63.
11 B. Cornils, W. A. Herrmann, R. W. Eckl, *J. Mol. Catal A:* **1997**, *116*, 27.
12 Rhône-Poulenc, EP 44.771 (**1980**), EP 441.708 (**1991**).
13 C. Mercier, G. Mignani, M. Aufrand, G. Allmang, *Tetrahedron Lett.* **1991**, *32*, 1433.
14 C. Mercier, P. Chabardes, *Pure Appl. Chem.* **1994**, *66*, 1509.
15 Rhône-Poulenc, BP 362.037 (**1987**), EP 320.339 (**1987**).
16 J. M. Grosselin, C. Mercier, *J. Mol. Catal.* **1990**, *63*, L26.
17 G. Allmang, F. Grass, J. M. Grosselin, C. Mercier, *J. Mol. Catal.* **1991**, *66*, L27.
18 J. M. Grosselin, C. Mercier, *Organometallics* **1991**, *10*, 2126.
19 S. Steines, B. Driessen-Hölscher, P. Wasserscheid, *J. Prakt. Chem.* **2000**, *342*, 348.
20 K. E. Atkins, W. E. Walker, R. M. Manyik, *Chem. Commun.* **1971**, 330.
21 Esso (M. G. Romanelli), US 3.670.032 (**1972**).
22 Kureha Chem. Ind. (S. Enomoto, H. Takita, S. Wada, Y. Mukaida, M. Yanaka), JP 49/35.603 B4 (**1974**).
23 Kuraray Ind. (N. Yoshimura, M. Tamura), US 4.356.333 (**1982**).
24 Kuraray Ind. (Y. Tokitoh, T. Higashi, K. Hino, M. Murasawa, N. Yoshimura), US 5.118.885 (**1992**).
25 E. Monflier, P. Bourdauducq, J.-L. Couturier, J. Kervennal, 1. Suisse, J. A. Mortreux, *Catal. Lett.* **1995**, *34*, 201.
26 E. Monflier, P. Bourdauducq, J.-L. Couturier, J. Kervennal, J. A. Mortreux, *Appl. Catal. A:* **1995**, *131*, 167.
27 E. Monflier, P. Bourdauducq, J.-L. Couturier, J. Kervennal, I. Suisse, J. A. Mortreux, *J. Mol. Catal. A:* **1995**, *97*, 29.
28 E. Monflier, P. Bourdauducq, J.-L. Couturier, FR 9.403.897 (**1994**).
29 A. Behr, T. Fischer, M. Grote, D. Schnitzmeier, *Chem. Ing. Tech.* **2002**, *74*, 1586.
30 A. Behr, M. Urschey, *J. Mol. Catal A:* **2003**, *197*, 101.
31 A. Behr, M. Urschey, V. Brehme, *Green Chem.* **2003**, *5*, 198.
32 J. Tsuji, M. Takahashi, *J. Mol. Catal.* **1981**, *10*, 107.
33 T. Prinz, W. Keim, B. Driessen-Hölscher, *Angew. Chem.* **1996**, *108*, 1835; *Angew. Chem., Int. Ed. Engl.* **1996**, *35*, 1708.
34 W. Keim, A. Durocher, P. Voncken, Erdöl, Kohle, Erdgas, *Petrochem.* **1976**, *29*, 31.
35 Z. Xia, D. Fell, *J. Prakt. Chem.* **1997**, *339*, 140.
36 J. Haggin, *Chem. Eng. News*, April 17, **1995**, 25.
37 Union Carbide Chem. (A. G. Abatjoglou, D. R. Bryant, R. R. Peterson), EP 3.509.222 (**1990**).

2.3.3
Membrane Techniques

José Sanchez Marcano

2.3.3.1
Membrane Separation Processes

Today, membrane separation processes are finding an extensive and ever-increasing use in the pharmaceutical, food, and petrochemical industries, as well as in environmental applications. Membrane separation processes present some advantages over more classical separation processes since they are less energy-consuming and more compact, and generally require less initial capital investment.

A membrane is a permeable or semi-permeable thin film made of a variety of different materials, from inorganic porous solids to different types of polymers. The main function of the membrane is to control the exchange of materials between two adjacent fluids (the phases can be similar or different). To do so, the thin film acts as a barrier, which separates different species either by sieving or by controlling their relative rate of transport through itself. Transport processes through the membrane are the result of a driving force typically associated with a gradient of concentration, pressure, temperature, electric potential, etc. The ability of a membrane to separate mixtures is determined by two parameters: permeability and selectivity. Permeability is defined as the flux (molar or volumetric flow per unit area) through a membrane scaled with respect to the membrane thickness and driving force. Thus, when the gradient involved is the partial pressure, the units of permeability are mol (or m^3) m m^{-2} Pa^{-1} s^{-1}. Selectivity is the second important parameter which characterizes the capability of a membrane to separate two species. "Ideal selectivity" is then defined as the ratio between the individual permeabilities of the two species.

Membranes can be classified according to the type of material their film is made of (organic or inorganic materials) or to the porosity of the thin layer (porous or dense). Choosing between a porous or a dense film as well as choosing the type of

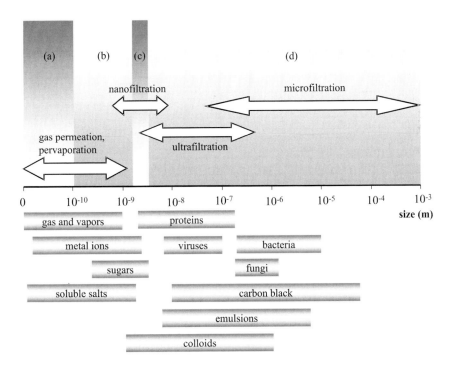

Figure 1 Membrane processes, different types of membranes and the sizes of species: (a) dense and ultramicroporous, (b) microporous, (c) mesoporous, (d) macroporous.

material depends on the separation process, the operating temperature, and the driving force. In addition, for applications concerning membrane reactors (MRs), where the reaction is coupled with the separation process, the thin film has also to be stable under reaction conditions.

Porous membranes can be made of polymers (polysulfones, polyacrylonitrile, polypropylene, silicones, perfluoropolymers, polyimides, polyamides, etc.), ceramics (alumina, silica, titania, zirconia, zeolites, etc.) or microporous carbons. Dense organic membranes are commonly used for molecular-scale separations involving gas and vapor mixtures, whereas the mean pore sizes of porous membranes is chosen considering the size of the species to be separated. Current membrane processes include microfiltration (MF), ultrafiltration (UF), nanofiltration (NF), gas and vapor separation (GS), and pervaporation (PV). Figure 1 indicates the types and sizes of species typically separated by these different separation processes.

Membranes are manufactured in a wide range of geometries; they include flat, tubular, and multi-tubular, hollow-fiber, and spiral-wound. The type of geometry of the membrane depends on the material the membrane is made of. Ceramic membranes generally come in tubular, multi-tubular and flat geometries, whereas most spiral-wound and hollow-fiber membranes are made of polymers.

2.3.3.2
The Coupling of a Membrane Separation Process with a Catalytic Reaction: the MR Concept

In MR processes, separation and reaction are combined into a single unit. The coupling of both processes therefore results in a continuous separation/feed of reactants/products, which often results in enhanced selectivity and/or yield. The MR concept was first applied in reactions in which the continuous extraction of products would enhance yield by shifting equilibrium. The catalytic reactions of this type that have been investigated include dehydrogenation and esterification. However, reactive separations also appear to be interesting for other reactions, such as hydrogenation and partial oxidation. Since the 1980s, many studies have been published on the application of membrane separations in catalytic processes. Reaction yield and selectivity have been reported to be strongly dependent not only on operating conditions but also on membrane permeability and selectivity. Detailed descriptions of the state-of-the-art of catalytic membrane reactors have already been published [1–4].

In the early stages of MRs, the two functions were coupled by connecting two distinct units – the reactor and the membrane separator – in series (Figure 2a). The membrane reactor design, shown in Figure 2 (b), which combines both process units into a single one, was the result of the development of the first process design into the second.

The design configuration of Figure 2 (b) results in obvious advantages over the first design: it is more compact, while capital and operating savings are realized by the elimination of intermediate steps. Other beneficial effects can result from the synergy between separation and reaction. Indeed, for reactions limited by thermo-

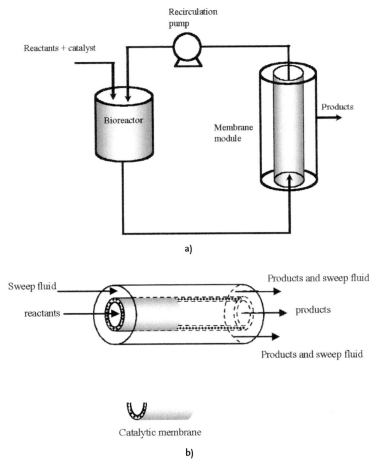

Figure 2 (a) A conventional system (reactor and membrane separator) and (b) a catalytic membrane reactor.

dynamic equilibrium, as is frequently the case with catalytic hydrocarbon dehydrogenation, it is obvious that the continuous and selective separation of hydrogen from the reaction vessel will result in better yields. Despite the advantages of the catalytic membrane reactor (CMR) of Figure 2 (b), the more conventional design of Figure 2 (a) is still frequently used, because of its simplicity and the easy control of its different operating parameters. In both MR designs shown in Figure 2, the membrane (in any configuration, whether tubular, hollow-fiber, or plate) defines two different chambers. These are the retentate chamber, where the reactants are fed and the reaction often takes place, and the permeate chamber, where the products are separated.

The impressive progress in the area of MRs since the 1980s has been associated with the development of membrane manufacturing and processes. In quite a short time, the field has progressed far beyond such initial applications as dehydrogenation reactions. A multitude of different configurations have been proposed in the literature in order to combine the membrane separation module and the reactor into a single unit. Sanchez and Tsotsis [1] have classified the catalytic MR into six basic configurations:

- the *packed-bed membrane reactor* (PBMR), in which the membrane provides only a separation function and where the reaction is provided by a packed bed of catalyst placed in the permeate or retentate sides;
- the *catalytic membrane reactor* (CMR), where the membrane simultaneously provides separation and reaction functions, to achieve which one could use either an intrinsically catalytic membrane (e.g., a zeolite or metallic membrane) or a membrane that has been made catalytic through impregnation or by ion exchange;
- the *packed-bed catalytic membrane reactor* (PBCMR), where a PBMR contains a catalytically active membrane in order to provide an additional reactive function;
- the *fluidized-bed membrane reactor* (FBMR) in which the packed bed has been replaced by a fluidized bed when good control of the process temperature is necessary;
- the *fluidized-bed catalytic membrane reactor* (FBCMR), with the same functional advantage as the FBMR;
- the *catalytic nonpermselective membrane reactor* (CNMR), in which the membrane is not selective, but is only used to provide a well-defined reactive interface [1].

Membrane reactors can also be classified according the role the membrane plays with respect to the removal/addition of species [1–5]. Indeed, MRs could be classified as reactive *membrane extractors* when the membrane's function is to remove one or more products; such action could result in an increased equilibrium yield, as in the catalytic dehydrogenation reaction applications previously described. In *membrane distributors,* the membrane is used as the distributor of one of the reactants. Such MRs find application for consecutive and parallel reactions (in the partial oxidation of hydrocarbons, for example). Controlling the addition of the oxidant through the membrane lowers its partial pressure and generally results in a better yield of the intermediate oxidation products. Another advantage obtained with this type of MR is the potential to avoid the thermal runaway phenomena that are typically associated with highly exothermic reactions. In some other reactions, the membrane is not even required to be selective; it only acts as a *membrane contactor* providing a controlled reactive (or extractive) interface between reactants (or products) flowing on the opposite sides of the membrane [6, 7]. In this configuration, the membrane is used to improve the contact between the different reactive phases that are fed separately into either side of the membrane; reactions requiring a precise stoichiometry and phase-transfer catalytic reactions are examples of such an application. Intimate contact between the reactants can also be carried out by forcing them to flow together through a membrane, which is endowed with catalytic sites for the reaction. This "flow-through" configuration can also be used, to decrease the internal diffusion limitations of the reaction rate. Indeed, the process enhancement is attributed to the intensification of the intraporous mass transfer, which is the result of the reactants being forced to flow through the membrane's pores. Multiphase reactions involving a catalyst with liquid and gaseous reactants can also be studied in reactive membrane contactors. As in the previous example, the primary advantage of the use of a membrane reactor consists in decreasing the mass-transfer limitations frequently encountered with such reactions in slurry- or trickle-bed reactors.

2.3.3.3
Application of the MR Concept to Homogeneous Catalytic Reactions (HCRs)

Since 1990 the MR concept has not been applied so intensively in HCR as in heterogeneous catalysis. Nevertheless, the application of the MR concept in this area seems to be a thrilling new topic, which opens enormous possibilities of research and development. Indeed, one of the major problems existing in HCR (in many forms, including transition metal catalyzed reactions and phase-transfer catalysis) is the separation of reaction products from catalysts (see Chapter 1). Therefore, the application of UF or NF in HCR for the separation and recycling of a soluble catalyst from the reaction mixture and/or for the continuous separation of the reaction products, as well as for the use of hollow-fiber membrane contactors in multiphase reactions (aqueous/solvent, aqueous/supercritical, solvent/gas, etc.), are some of the most exciting ideas in this area. The coupling of membrane techniques with multiphase HCR has been made possible with the recent development of membranes which are resistant to organic solvents. Indeed, this type of coupled process has already been performed in enzymatic reactions at laboratory and industrial scale [1]. The principle of this coupling is illustrated in Figure 3.

In general, homogeneous catalysts, such as soluble metal complexes, are expensive and their presence in the final product is often undesirable. One of the most frequently adopted solutions is the separation of the homogeneous catalysts from the reaction mixture by immobilization on a support. Nevertheless, immobilization is an arduous process, which often results in decreased catalytic activity and increasing leaching. Among the different membrane processes, NF, which presents a very low cut-off, is a very suitable technique to separate homogeneous catalysts from reaction products or to recuperate solvents.

Nair and co-workers [8] have coupled of semi-continuous NF with the Heck reaction. The objective was the synthesis of *trans*-stilbene from styrene and iodobenzene using $Pd(OAc)_2(PPh_3)_2$ as catalyst and $P(o\text{-tolyl})_3$ as stabilizing agent. They used solvent-resistant membranes and different aqueous/solvent systems (ethyl acetate and acetone/H_2O; methyl *tert*-butyl ether and acetone/H_2O; tetrahydrofuran/H_2O). The best conversion was obtained with the first-mentioned solvent mixture. A selectivity of 100% of *trans*-stilbene with a cumulative turnover number of 1200 was reported, where the rejection of the catalyst turned out to be as high as 97%. Therefore, the authors concluded that NF was a convenient technique to run catalytic reactions with catalyst recycling, since this method saves the catalyst, prevents the metal contamination of the products, and increases reactor productivity.

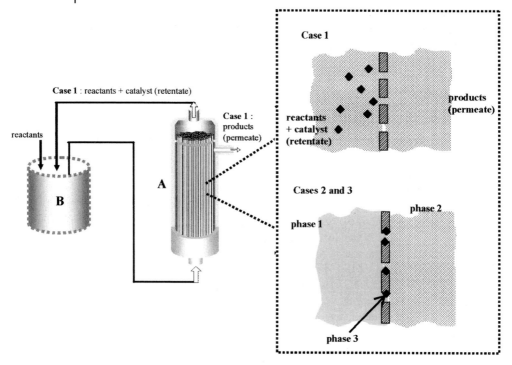

Figure 3 Some examples of the principle of coupling a membrane technique with an HCR. (a) Hollow-fiber membrane module; (b) reactor. Case 1, separation of a soluble catalyst by NF; case 2, MR contactor in two-phase reaction (only phase 1 and 2); case 3, MR contactor in three-phase reaction (phases 1, 2, and 3).

Some authors have grafted homogeneous catalysts to macromolecules in order to increase the kinetic size and thus avoid losses during UF or NF. Dwars et al. [9] used a micellar enlarged Rh-(2S,4S)-N-*tert*-butoxycarbonyl-4-diphenylphosphino-2-diphenylphosphinomethylpyrrolidine catalyst in a membrane reactor, equipped with a UF membrane for the enantioselective hydrogenation of α-amino acids. The chiral α-amino acid derivatives were thus obtained with good enantioselectivity and yield. The catalyst, embedded in micelles, obtained from triblock copolymers as surfactants, was retained and reused several times without any loss of activity and enantioselectivity. The authors observed a minimal leaching of the catalyst components when they used a regenerated cellulose acetate UF membrane from Amicon. Giffels et al. [10] used oxazaborolidines enlarged with soluble polymers as catalysts for the enantioselective reduction of ketones in a continuously operated MR equipped with an NF membrane. The amount of chiral product per mole of catalyst was enhanced from 10 (batch) up to 560 moles (MR). The chiral alcohols were obtained in good to excellent enantiomeric excess and space–time yield. Smet et al. [11] combined the hydrogenation of dimethyl itaconate with Ru-BINAP as a homogeneous chiral catalyst with an NF process. Two different configurations have been utilized, one in which the membrane was inserted into the reactor itself, and another in which the membrane was extraneous to the reactor. The polymeric membranes tested were shown to separate the Ru-BINAP catalyst completely. The hybrid reactor system was tested for about a month, and showed very stable activity and enantioselectivity. Laue et al. [12] attached a homogeneous hydrogenation catalyst onto a polymer. They used a membrane reactor with a UF membrane, which was capable of retaining the soluble, polymer-bound catalyst. Their membrane reactor achieved high space–time yields of up to 578 g L^{-1} day^{-1}, and enantioselectivities of up to 94%.

Ilinitch et al. [13, 14] have studied the catalytic oxidation of sulfides. For this reaction, macroporous polymeric membranes, impregnated with a sodium salt of tetra(sulfophthalocyanine)cobalt(II) (Co-TSPC), were used. The MR utilized included macroporous membranes and operated in a "flow-through" configuration.

A recent application reported by Vankelecom et al. [15] concerns the use of polydimethylsiloxane membranes in which an active catalyst ([5,10,15,20-tetrakis-(2,6-dichlorophenyl)porphyrinato]manganese(III)chloride) was incorporated. The membranes are not permselective but only act as interfacial contactors to carry out the epoxidation (but also hydrogenation and hydroformylation) of various organic compounds. In this case, the thin film also gives additional advantages such as allowing the reaction to take place without any co-solvent, which prevents loss of the catalyst. The catalytic membrane was then tested for 3-penten-2-ol epoxidation using hydrogen peroxide. The results showed that selectivity dramatically increased (from approximately 20 to 100%) when the catalytic membrane was used instead of the free catalyst. This result was explained by the reduced mobility of the ligand in the membrane, when compared with its free form in a solution. Langhendries et al. [16] used the same concept for the oxidation of cyclohexane, cyclodecane, and *n*-dodecane using *t*-butyl hydroperoxide as oxidant, and zeolite-encaged iron phthalocyanine as catalyst (embedded in a polymeric membrane). During he *n*-dodecane oxidation in the CMR, the alcohol and ketone products are exclusively recovered with the organic *n*-dodecane phase, which is another advantage of the CMR configuration.

Frusteri and co-workers [17] have studied the activation of alkanes (CH$_4$, C$_2$H$_6$, C$_3$H$_8$) under mild conditions (80–120 °C, 140 kPa) in a three-phase CMR containing superacid catalytic membranes. In this reactor, they studied the functionalization of the C$_1$–C$_3$ alkanes to oxygenate on superacid catalytic films mediated by the Fe^{2+}/H$_2$O$_2$ Fenton oxidation system. The Nafion®-based membrane is sandwiched between two porous Teflon plates with the catalytic side toward the liquid phase, which consists of a solution of H$_2$O$_2$ and [Fe^{2+}] or [Fe^{3+}]. The gas phase consists of a mixture of hydrocarbon and N$_2$. The two half-cells of the three-phase MR operate as batch reactors with separate recirculation for the two phases. The authors present a reaction pathway which involves the activation of the C–H bond of the alkane molecule on the superacid membrane sites and the subsequent reaction of the activated alkane with primary reactive intermediates, generated from the Fe^{2+}/H$_2$O$_2$ system.

Catalysts supported on polymeric membranes have also been applied for hydroformylation reactions. Feldman et al. [18], for example, have reported the synthesis of a cellulose acetate membrane containing $HRh(CO)(PPH_3)_3$, which was incorporated in the thin film by introducing it into the cellulose acetate solution before the casting procedure. The membranes thus obtained were tested in the hydroformylation of ethylene and propylene [19], and were shown to be very active (turnover rate numbers with the film-supported catalyst were about five times higher than those observed with the corresponding homogeneous catalysts). The n/i ratio was reported to be higher as well. Hydroformylation of ethylene has also been studied in a membrane reactor by Kim and Datta [20]. These authors utilized a three-layer membrane, which consisted of a porous matrix supporting a liquid-phase catalyst dissolved in a high boiling point organic solvent, which was sandwiched between two different hydrophobic polymeric membranes. The reaction was studied using a flat-disk membrane. In this configuration, the reactants (C_2H_4, CO, H_2) were fed into one compartment of the membrane reactor, and a sweep gas was used in the second compartment in order to eliminate the product continuously. The experimental results were in good agreement with the theoretical model presented and showed that very high separation factors and good conversion were possible for this type of CMR. More recently, Naughton and Drago [21] reported the hydroformylation of 1-hexene by using a water-soluble rhodium complex incorporated in a PEG thin film. The catalytic activity of the supported catalyst was then better than that observed with the unsupported rhodium complex. The experimental results obtained with the supported catalysts indicate that the CMR technique is promising for this application.

The coupling of membrane techniques with the technology of supercritical fluids is an emerging topic which begins to be intensively explored for extraction and reaction purposes [22, 23]. This concept offers the advantages of using benign high-density gases, i.e., the possibility of achieving a high concentration of gaseous reactants in the same phase as the substrates and catalyst (see also Chapter 6). In addition, the membrane is also used to recycle the catalyst as explained above. Goetheer et al. [24] have recently reported the hydrogenation of 1-butene using a fluorous derivative of Wilkinson's catalyst $[RhCl\{P-(C_6H_4-p-SiMe_2CH_2CH_2C_8F_{17})_3\}_3]$. The membrane was made of porous silica/alumina, an inorganic material which is resistant to supercritical CO_2. The silica top layer had an average pore diameter of 0.6 nm whereas the fluorous derivative of Wilkinson's catalyst, obtained from $RhCl(cis,cis$-1,5-cyclooctadiene) and $P[C_6H_4-p-SiMe_2(CH_2)_2C_8F_{17}]_3$, had a mean molecular size between 2 to 4 nm. The difference in size resulted in a retention of the catalyst higher than 99.9%. The authors reported stable operation and continuous production of n-butane at 353 K and 20 MPa (see also Chapter 4).

2.3.3.4
Conclusion

To conclude this chapter, as the preceding discussion has – hopefully – indicated, there is considerable interest today and significant emerging potential for the application of membrane reactors in catalytic homogeneous reactions. The effectiveness of the MR concept in HRC at laboratory scale has been improved in many cases by considering the separation of the reaction products from the reaction mixture containing the soluble catalyst (grafted or not). Other possible applications which concern the membrane contactors in phase-transfer catalysis will also have to be taken in consideration in a near future. Mixed-matrix membranes incorporating the catalyst into the membrane matrix are promising for a variety of applications. The main remaining challenge concerns the development of membranes that have to be sufficiently resistant toward the solvents and high pressures encountered in supercritical conditions. Nevertheless, before commercial success is achieved, the key challenge that remains is to prove the long-term resistance of the membranes.

References

1 J. SANCHEZ MARCANO, T. T. TSOTSIS, *Catalytic Membranes and Membrane Reactors*, Wiley-VCH, Weinheim **2002**.
2 J. SANCHEZ, T. T. TSOTSIS, Current Developments and Future Research in Catalytic Membrane Reactors, in *Fundamentals of Inorganic Membrane Science and Technology*, A. J. Burggraaf, L. Cot (Eds.), Elsevier Science B. V., Amsterdam **1996**.
3 A. DIXON, Innovations in Catalytic Inorganic Membrane Reactors, in *Catalysis, Vol. 14*, The Royal Society of Chemistry, Cambridge **1999**.
4 J. SANCHEZ, T. T. TSOTSIS, Reactive Membrane Separation, in *Reactive Separation Processes*, S. Kulprathipanja, Ed., Taylor & Francis, USA **2002**.
5 A. JULBE, D. FARRUSSENG, C. GUIZARD, *J. Membr. Sci.* **2001**, *181*, 3.
6 H. J. SLOOT, G. F. VERSTEEG, W. P. M. VAN SWAAIJ, *Chem. Eng. Sci.* **1990**, *45*, 2415.
7 M. TORRES, J. SANCHEZ, J. A. DALMON, B. BERNAUER, J. LIETO, *Ind. Eng. Chem. Res.* **1994**, *33*, 24.
8 D. NAIR, J. T. SCARPELLO, L. S. WHITE, L. M. FREITAS DOS SANTOS, I. F. J. VANKELECOM, G. LIVINGSTON, *Desalination* **2002**, *147*, 301.
9 T. DWARS, J. HABERLAND, I. GRASSERT, G. OEHME, U. KRAGL, *J. Mol. Catal. A:* **2001**, *168*, 8.
10 G. GIFELS, J. BELICZEY, M. FELDER, U. KRAGL, *Tetrahedron: Asymmetry* **1998**, *9*, 691.
11 K. D. SMET, I. F. J. VANKELECOM, P. A. JACOBS, *Proc. International Congress on Catalysis with Membrane Reactors-2000*, Zaragoza, Spain July 3–5 **2000**, 119.
12 S. LAUE, L. GREINER, J. WOLTINGER, A. LIESE, *Adv. Synth. Catal.* **2001**, *342*, 71.
13 O. M. ILINITCH, F. P. CUPERUS, L. V. NOSOVA, E. N. GRIBOV, *Catal. Today* **2000**, *56*, 137.
14 O. M. ILINITCH, F. P. CUPERUS, R. W. VAN GEMERT, E. N. GRIBOV, L. V. NOSOVA, *Sepn. Pur. Technol.* **2000**, *21*, 55.
15 I. F. J. VANKELECOM, P. A. JACOBS, *Catal. Today* **2000**, *56*, 147.
16 G. LANGHENDRIES, G. V. BARON, I. F. J. VANKELCOM, R. F. PARTON, P. A. JACOBS, *Catal. Today* **2000**, *56*, 13.
17 F. FRUSTERI, C. ESPRO, F. ARENA, E. PASSALACQUA, A. PATTI, A. PARMALIANA, *Catal. Today* **2000**, *61*, 37.
18 J. FELDMAN, I. W. SHIN, M. ORCHIN, *J. Appl. Polym. Sci.* **1987**, *34*, 969.
19 J. Feldman, M. Orchin, *J. Mol. Catal.* **1990**, *63*, 213.
20 J. S. KIM, R. DATTA, *AIChE J.* **1991**, *37*, 1657.
21 M. NAUGHTON, R. DRAGO, *J. Catal.* **1995**, *155*, 383.
22 G. M. RIOS, J. SANCHEZ, M. P. BELLEVILLE, S. SARRADE, *Chem. Eng. Rev.* **2002**, *18*, 1, 49.
23 J. ROMERO, C. GIJIU, J. SANCHEZ, G. RIOS, *Chem. Eng. Sci.* **2004**, 59, 7, 1569.
24 E. L. V. GOETHEER, A. W. VERKERK, L. J. P. VAN DEN BROEKE, E. DE WOLF, B. J. DEELMAN, G. VAN KOTEN, J. T. F. KEURENTJES, *J. Catal.* **2003**, *219*, 126.

2.3.4
Micellar Systems

Günther Oehme

2.3.4.1
Introduction

Micelles are spherical supramolecular assemblies of amphiphilic compounds with surface-active properties (surfactants, tensides) in a colloidal dimension [1] (see also Section 2.3.1). Typical micelle-forming molecules carry a hydrophilic headgroup and a hydrophobic tail. This is shown schematically in Figure 1.

Figure 1 Formation of aqueous micelles and reverse micelles.

Examples of such surfactants are cetyltrimethylammonium bromide (CTAB, $9.2 \cdot 10^{-4}$ M), polyoxyethylene(22)hexadecanol (Brij 58, $7.7 \cdot 10^{-5}$ M), and sodium dodecyl sulfate (SDS, $8.1 \cdot 10^{-3}$ M). The actual formation of micelles begins above a certain temperature (Krafft's point) and above a characteristic concentration (critical micelle concentration, CMC) which is given in parentheses. Driving forces for the formation of aqueous micelles are the solvation of the headgroup and the desolvation of the alkyl chain ("hydrophobic effect") [2]. The association–dissociation process is very fast and occurs within milliseconds, faster than the rate of the most thermally initiated reactions. Because of a very high polarity gradient between surface and core, micelles can enclose different organic species from the surrounding aqueous phase [3]. This incorporation can be described by an equilibrium constant as a basis for kinetic treatments and creates a micro-heterogeneous multiphase system [4]. Encapsulation of reactants in micelles often enhances or inhibits the reaction rates. Any reaction promotion has been called "micellar catalysis". The rate enhancement of organic reactions in micelles can be a combination of the following effects [5]:

- a medium effect caused by a lower dielectric constant in the interior;
- a stabilization of the transition state owing to an interaction with the head group;
- concentration of the reactants by incorporation into the micelle.

Because of the relatively high regularity in the micellar core some asymmetric reactions occur with an enhanced stereoselectivity. Other typical effects of organized surfactants are the alteration of chemical and photophysical pathways, of quantum efficiency and ionization potentials, of oxidation and reduction properties, and finally of products and charges [6]. The morphology of micelles can change to rod-like structures at increasing concentrations of surfactants. The suitability of micellar solutions as reaction media, depending on the structure of the surfactants and environmental influences, is discussed in a review by Taşcioğlu [7].

Depending on the structure of the amphiphiles and the composition of the solvent mixture, other types of aggregates are possible, such as vesicles, reverse micelles, and microemulsions [8]. Because of the presence and proximity of polar and apolar regions in the interior of surfactant assemblies, there are some similarities to enzymes and natural membranes. Supramolecular aggregates of surfactants have a membrane-mimetic chemistry [9].

2.3.4.2
Selected Reactions in Aqueous Micelles

Important topics include solvolytic reactions, oxidations, reductions, and C–C coupling reactions. The saponification of activated esters in aqueous micelles is a model for an enzyme-mimetic reaction. The influence of the micellar medium on the reaction rate and on the stereoselectivity has been investigated. Models of metalloenzymes were developed with the ligands 1–3 and metal ions like Ni^{II}, Cu^{II}, Zn^{II}, and Co^{II} [10]. The amphiphilic ligands which were embedded in micelles enhanced the reaction rate in comparison with a dispersion in surfactant-free water. The use of optically active ligands (e.g., 2) led to a moderate kinetic resolution of α-aminoesters.

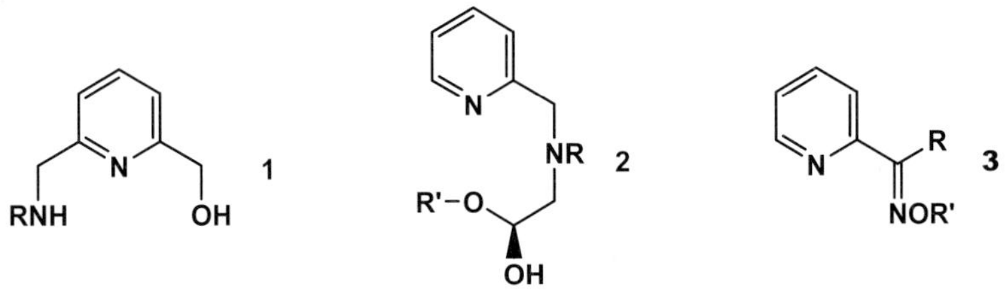

$R = CH_3, C_nH_{2n+1}$ $n = 8, 12, 13, 16$

Pioneering work in micellar catalysis was described by Menger et al. [11] in 1975: the oxidation of piperonal to the corresponding carboxylic acid by means of $KMnO_4$ in presence of CTAB.

An example of micellar oxidation is a biomimetic system of cytochrome P450 investigated by Monti et al. [12], who found in the epoxidation of styrene with NaOCl a significant influence of the type of surfactant: cetylpyridinium chloride promoted the reaction more than cetyltrimethylammonium bromide. One explanation could be a specific noncovalent interaction between catalyst, substrate and surfactant.

The asymmetric hydrogenation of dehydro-α-amino acid derivatives catalyzed by chiral rhodium–phosphine complexes is a successful example of an aqueous micellar system [Eq. (1)]. Normally, activity and selectivity are decreased by changing from an organic solvent to water but the addition of a small amount of a micelle-forming amphiphile leads to a significant increase in the reaction rate and in the enantioselectivity [13, 14].

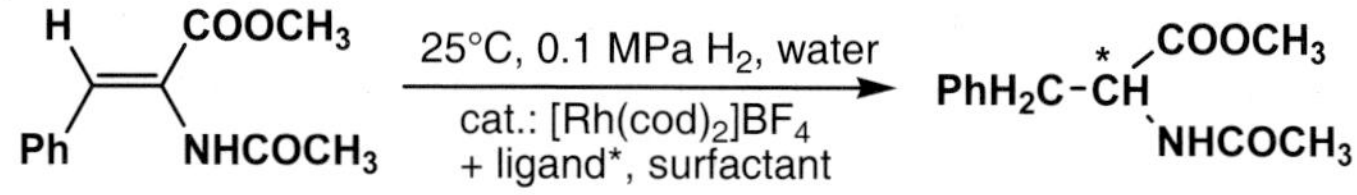

$$\text{(structure)} \xrightarrow[\substack{\text{cat.: } [Rh(cod)_2]BF_4 \\ + \text{ ligand*, surfactant}}]{25°C, 0.1 \text{ MPa } H_2, \text{ water}} PhH_2C\overset{*}{-}CH\overset{\diagup COOCH_3}{\underset{\diagdown NHCOCH_3}{}} \quad (1)$$

Yonehara et al. [15] recently extended the scope of surfactant-promoted enantio-selective hydrogenation to simple enamides.

The formation of C–C bonds is one of the most important reaction types in organic chemistry. This field has advanced rapidly due to the introduction of transition metal catalysts. Most C–C coupling reactions utilizing carbon monoxide are of industrial significance. In hydroformylation of long-chain olefins especially, a phase-transfer reagent or a micellar system is required when aqueous conditions are employed [16]. Fell and Papadogianakis [17] synthesized surface-active phosphines starting from tris(2-pyridyl)phosphine and different long-chain β-sultones 4.

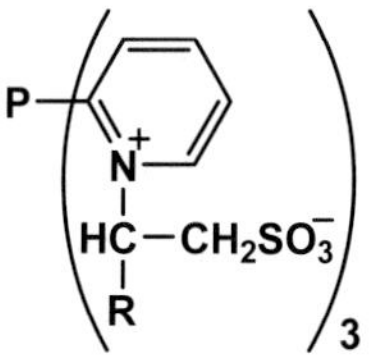

R = C_3H_7 to $C_{11}H_{23}$

New types of amphiphilic phosphines used by Hanson [18] and van Leeuwen as ligands resulted in high activities and satisfying regioselectivities (up to $n/i \sim 50$ for aldehydes with a xanthen-derived amphiphilic diphosphine [19]). A systematic investigation of surfactants during the hydroformylation of higher olefins by water-soluble Rh-TPPTS complexes was conducted by Chen et al. [20]. They found that only cationic amphiphiles gave a significant effect on yield and regioselectivity.

Suzuki-type C–C coupling reactions with palladium–phosphine complexes as catalysts can also be promoted by surfactants [21] in a toluene/water biphasic system.

Water-stable Lewis acids as catalysts for aldol reactions were developed by Kobayashi and co-workers [22]. A high promotion could be observed by combination of Lewis acid and surfactant (LASCs = Lewis acid–surfactant combined catalysts as shown in Eq. (2)). The surfactant is here the anion of dodecanesulfonic acid.

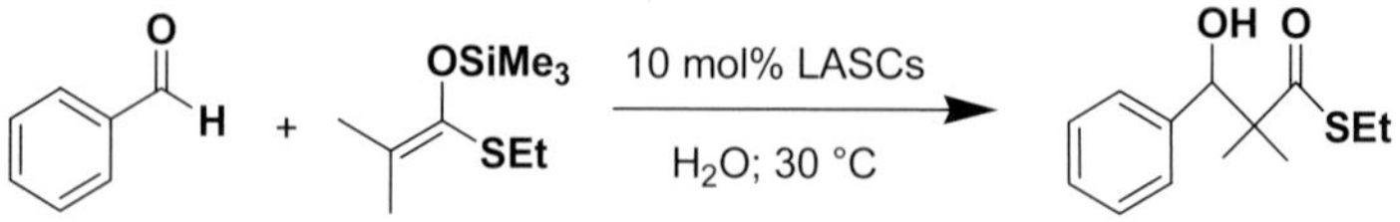

$$\text{(structure)} \xrightarrow[\substack{H_2O; 30 \,°C}]{10 \text{ mol}\% \text{ LASCs}} \text{(structure)} \quad (2)$$

In the same manner Engberts and co-workers [23] used the copper(II) salt of SDS as the catalyst in a Diels–Alder reaction of 3-(p-substituted phenyl)-1-(2-pyridyl)-2-propen-1-ones with cyclopentadiene in water as medium, and observed a spectacular enhancement of the reaction rate.

The palladium-catalyzed asymmetric alkylation of 1,3-diphenyl-2-propenyl acetate with dimethyl malonate could be influenced in activity and enantioselectivity by the surfactant cetyltrimethylammonium hydrogen sulfate [24].

2.3.4.3
Reactions in Reverse Micelles

Reverse micelles are formed by association of surfactants with colloidal drops of water in an organic medium (see Figure 1). A favored surfactant seems to be AOT (sodium bis(2-ethylhexyl)sulfosuccinate) but SDS and tetraalkylammonium salts have also proved to be useful. Like aqueous micelles, reverse micelles exist in highly diluted systems [25].

The hydrogen transfer reaction from 1,2-cyclohexanedimethanol to E-4-phenyl-3-buten-2-one is catalyzed by [RuCl(S-binap)(benzene)]Cl and accelerated by SDS [26]. The surfactant is essential for the rate enhancement and the authors proposed the existence of reverse micelles. Interesting new experiments on CH activation in *n*-heptane or methylcyclohexane were communicated by Elsevier and co-workers [27]. In a reverse micellar system containing NaPtCl$_4$ in D$_2$O, AOT, and the substrate, CH/CD exchange occurs under mild conditions (80 °C) regioselective in the methyl group. One of the main fields of application of reverse micelles is the entrapment of enzymes in the water cavity [28].

2.3.4.4
Perspectives

The solubilization of reactants in aqueous micelles and the achievement of rate and selectivity enhancement are sometimes unexpectedly high, but the main problem has been the separation of products, amphiphile, and catalyst after the reaction. One solution should be the immobilization of micelle-like structures on polymers [29] (cf. also Chapter 7). An early proposal to use immobilized amphiphiles like micelles was made by Brown and Jenkins [30]. Regen [31] used amphiphilic polymers for the first time in phase-transfer catalysis. A variety of different amphiphilized and amphiphilic polymers were successfully applied in the asymmetric hydrogenation of amino acid precursors [32]. The simplest type of immobilized surfactant is an admicelle [33]; this means that the surfactant (SDS) is adsorbed on alumina, thus forming a bilayer on its surface, which is stable against washing with water or other polar solvents. In analogy to admicelles, Milstein and co-workers [34] adsorbed the amphiphilic complex [Rh(4,4'-diheptadecyl-2,2'-bipyridine)-(1,5-hexadiene)]$^+$PF$_6^-$ on hydrophobic or hydrophilic porous glass particles and observed under the conditions of a Langmuir–Blodgett film, a high catalytic activity and selectivity for the hydrogenation of acetone in an aqueous medium. The complex was almost inactive in solution.

It should be concluded that catalysis with surfactant assemblies is an active and successful area of research. In the case of emulsion polymerization, phase-transfer processes, and analytical applications, micellar methods are of practical importance.

References

1 J. H. CLINT, *Surfactant Aggregation*, Blackie, Glasgow **1992**.
2 (a) C. TANFORD, *The Hydrophobic Effect: Formation of Micelles and Biological Membranes*, Wiley, New York **1980**; (b) W. BLOKZIJL, J. B. F. N. ENGBERTS, *Angew. Chem.* **1993**, *105*, 1610; *Angew. Chem., Int. Ed. Engl.* **1993**, *32*, 1545.
3 S. D. CHRISTIAN, J. F. SCAMEHORN (Eds.), *Solubilization in Surfactant Aggregates*, Marcel Dekker, New York **1995**.
4 C. A. BUNTON, in *Kinetics and Catalysis in Microheterogeneous Systems* (Eds.: M. GRÄTZEL, K. KALYANASUNDARAM), Marcel Dekker, New York **1991**, p.33.
5 J. M. BROWN, S. K. BAKER, A. COLENS, J. R. DARWENT, in *Enzymic and Non-enzymic Catalysis* (Eds.: P. DUNNILL, A. WISEMAN, N. BLAKEBROUGH), Horwood, Chichester **1980**, p. 111.
6 J. TEXTER (Ed.), *Reactions and Synthesis in Surfactant Systems*, Marcel Dekker, New York **2001**.
7 S. TAŞCIOĞLU, *Tetrahedron* **1996**, *52*, 11113.
8 U. PFÜLLER, *Mizellen, Vesikel, Mikroemulsionen*, Volk und Gesundheit, Berlin **1986**.
9 J. H. FENDLER, *Membrane Mimetic Chemistry*, Wiley, New York **1982**.
10 P. SCRIMIN, U. TONELLATO, in *Surfactants in Solution* (Eds.: K. L. MITTAL, D. D. SHAH), Plenum, New York **1991**, Vol. 11, p. 349.
11 F. M. MENGER, J. V. RHEE, H. K. RHEE, *J. Org. Chem.* **1975**, *40*, 3803.
12 D. MONTI, A. PASTORINI, G. MANCINI, S. BOROCCI, P. TAGLIATESTA, *J. Mol. Catal. A:* **2002**, *179*, 125.
13 G. OEHME, E. PAETZOLD, R. SELKE, *J. Mol. Catal.* **1992**, *71*, L1.
14 T. DWARS, U. SCHMIDT, C. FISCHER, I. GRASSERT, R. KEMPE, R. FRÖHLICH, K. DRAUZ, G. OEHME, *Angew. Chem.* **1998**, *37*, 2853; *Angew. Chem., Int. Ed.* **1998**, *37*, 2853.
15 K. YONEHARA, K. OHE, S. UEMURA, *J. Org. Chem.* **1999**, *64*, 9381.
16 H. BAHRMANN, S. BOGDANOVIC, in *Aqueous-Phase Organometallic Catalysis* (Eds.: B. CORNILS, W. A. HERRMANN), Wiley-VCH, Weinheim **1998**, p. 306.
17 B. FELL, G. PAPADOGIANAKIS, *J. Mol. Catal.* **1991**, *66*, 143.
18 B. HANSON, *Coord. Chem. Rev.* **1999**, *185*, 795.
19 J. N. H. REEK, P. C. J. KAMER, P. W. N. M. VAN LEEUWEN, in *Rhodium Catalyzed Hydroformylation* (Eds.: P. W. N. M. VAN LEEUWEN, C. CLAVER), Kluwer, Dordrecht **2000**, p. 253, 259.
20 H. CHEN, Y. LI, J. CHEN, P. CHENG, Y. HE, X. LI, *J. Mol. Catal. A:* **1999**, *149*, 1.
21 E. PAETZOLD, G. OEHME, *J. Mol. Catal. A:* **2000**, *152*, 69.
22 S. KOBAYASHI, K. MANABE, *Pure Appl. Chem.* **2000**, *72*, 1373.
23 S. OTTO, J. B. F. N. ENGBERTS, J. C. T. KWAK, *J. Am. Chem. Soc.* **1998**, *120*, 9517.
24 C. RABEYRIN, C. NGUEFACK, D. SINOU, *Tetrahedron Lett.* **2000**, *41*, 7461.
25 P. L. LUISI, in *Kinetics and Catalysis in Microheterogeneous Systems* (Eds.: M. GRÄTZEL, K. KALYANASUNDARAM), Dekker, New York **1991**, p. 115.
26 K. NOZAKI, M. YOSHIDA, H. TAKAYA, *J. Organomet. Chem.* **1994**, *473*, 253.
27 S. GAEMERS, K. KEUNE, A. M. KLUWER, C. J. ELSEVIER, *Eur. J. Inorg. Chem.* **2000**, 1139.
28 A. SANCHEZ-FERRER, F. GARCIA-CARMONA, *Enzyme Microb. Technol.* **1994**, *16*, 409.
29 (a) A. LASCHEWSKY, *Adv. Polym. Sci.* **1995**, *124*, 1; (b) G. OEHME, I. GRASSERT, E. PAETZOLD, R. MEISEL, K. DREXLER, H. FUHRMANN, *Coord. Chem. Rev.* **1999**, *185*, 585.
30 J. M. BROWN, J. A. JENKINS, *J. Chem. Soc., Chem. Commun.* **1976**, 458.
31 S. L. REGEN, *Angew. Chem.* **1979**, *91*, 464; *Angew. Chem., Int. Ed. Engl.* **1979**, *18*, 421.
32 T. DWARS, G. OEHME, *Adv. Synth. Catal.* **2002**, *344*, 239.
33 R. SHARMA (Ed.), *Surfactant Adsorption and Surface Solubilization*, ACS Symp. Ser. No. 615, ACS, Washington DC **1995**.
34 K. TÖLLNER, R. POPOVITZ-BIRO, M. LAHAV, D. MILSTEIN, *Science* **1997**, *278*, 2100.

2.3.5
Thermoregulated Phase-Transfer Catalysis

Zilin Jin and Yanhua Wang

2.3.5.1
Introduction

The most severe drawback in homogeneous catalysis is the separation of the catalyst from the reaction mixture. The industrial success of the aqueous two-phase hydroformylation of propene to *n*-butanal [1] in Ruhrchemie AG in 1984 represents the considerable progress in this field. However, aqueous/organic biphasic catalysis has its limitations when the water solubility of the starting materials proves too low, as in hydroformylation of higher olefins (see Chapter 1). To solve this issue, a variety of approaches have been attempted. Additions of co-solvents [2] or surfactants [3, 4] to the system or application of tenside ligands [5, 6] and amphiphilic phosphines [7, 8] are ways to increase the reaction rates. Other approaches such as fluorous biphase system (FBS; see Chapter 4) [9], supported aqueous phase catalysis (SAPC; see Section 2.6) [10], supercritical CO_2 (cf. Chapter 6) [11] and ionic liquids (cf. Chapter 5) [12] have also been introduced to deal with this problem.

   Recently, a new aqueous biphasic catalytic system based on the cloud point of nonionic tensioactive phosphine, termed "thermoregulated phase-transfer catalysis" (TRPTC) has been developed [13]. The concept of TRPTC as a "missing link" could not only provide a meaningful solution to the problem of catalyst/product separation, but also extricate itself from the limitation of low reaction rates of water-immiscible substrates.

2.3.5.2
Thermoregulated Ligands and Thermoregulated Phase-Transfer Catalysis

It is well known that the water solubility of nonionic surfactants with polyoxyethylene moieties as the hydrophilic group is based on the hydrogen bonds formed between poly(ethylene glycol) (PEG) chains and water molecules (Figure 1). The solubility of this type of surfactant decreases with a rise in temperature, and their aqueous solutions will undergo an interesting phase separation process (a miscibility gap) on heating to a low critical solution temperature – the "cloud point". A reasonable explanation attributes this phenomenon to the cleavage of hydrogen bonds. In addition, it is worth mentioning that such a process is a reversible one since the water solubility could be restored on cooling to a temperature lower than the cloud point [14]. A thermoregulated ligand (TRL) [15] is generally defined as a kind of nonionic surface-active phosphine ligand containing PEG chains as the hydrophilic group in the molecular structure, which demonstrates a special property of inverse temperature-dependent solubility in water. Bergbreiter et al. [16] firstly reported a "smart ligand" with a phosphorus-bonded block copolymer of ethylene oxide and

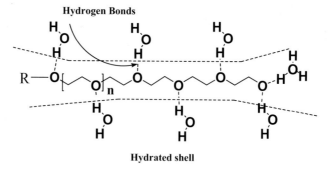

Figure 1 Hydrogen bonds formed between a polyether chain and water.

propylene oxide, which exhibits a property of inverse temperature-dependent solubility. Jin et al. synthesized a series of TRLs (**1–5**) by introducing PEG chains to phosphines [17–21] and investigated their water solubility and cloud point (C_p) [18].

$Ph_{3-m}P$ —⟨◯⟩— $[-O(CH_2CH_2O)_nH]_m$ **PETPP**   **1**

m = 1–3

$[RO-(CH_2CH_2O)_n]_mPPh_{3-m}$ m = 1, 2   **AEOPP**   **2**

P– $O(CH_2CH_2O)_nR$ **OPGPP**   **3**

Ph_2P —⟨◯⟩— SO_2N ⟨ $(CH_2CH_2O)_nH$ / $(CH_2CH_2O)_nH$ **PEO-DPPSA**   **4**

N ⟨ $(CH_2CH_2O)_nH$ / $(CH_2CH_2O)_nH$ **PEO-DPPPA**   **5**

PPh_2

TRPTC system based on the C_p of TRLs has been proposed [13]. The general principle of TRPTC is depicted in Figure 2. The TRPTC process can be described as follows: at a temperature lower than the cloud point, the catalyst is soluble in the aqueous phase. On heating to a temperature higher than the cloud point, however, the catalyst transfers from the aqueous phase into the organic phase. Thus, the catalyst and the substrate are in the same phase and the reaction proceeds in the

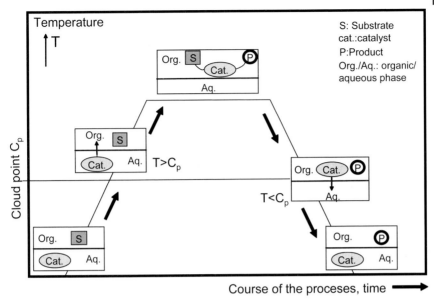

Figure 2 The general principle of TRPTC.

organic phase. As soon as the reaction is completed and the system is cooled to a temperature lower than the cloud point, the catalyst returns to the aqueous phase. The introduction of TRPTC is free from the shortcomings of classical biphasic catalysis, in which the scope of application is restrained by the water solubility of substrate. Moreover, the thermoregulated phase-transfer catalyst can be separated and recycled by simple phase separation.

The thermoregulated phase-transfer function of TRLs has been proven by means of the aqueous-phase hydrogenation of sodium cinnamate in the presence of an Rh/AEOPP complex as the catalyst [18]. As outlined in Figure 3, an unusual inversely temperature-dependent catalytic behavior has been observed.

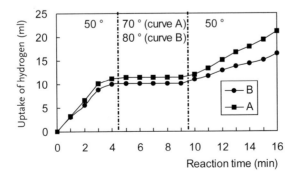

Figure 3 Hydrogenation of sodium cinnamate in water.
The cloud point of A is 64 °C and the cloud point of B is 72 °C.

2.3.5.3

Applications of Thermoregulated Phase-Transfer Catalysis

TRPTC has already been applied successfully in the hydroformylation of higher olefins and the CO selective reduction of nitroarenes, both on a laboratory scale.

2.3.5.3.1

Hydroformylation of Higher Olefins

Approaches in which surfactants [22, 23] and surface-active water-soluble phosphines [5, 6] are used to accelerate the rate of hydroformylation of high olefins in the aqueous two-phase system have been reported. Surface-active materials tend to make it possible for hydrophobic higher olefins to "enter" the aqueous phase through micellar solubilization (see Sections 2.3.1 and 2.3.4).

Thermoregulated phase-transfer catalysis has been used successfully for the aqueous biphasic hydroformylation of higher olefins [13, 18]. A reasonable explanation for the satisfactory catalytic reactivity is that it results from the thermoregulated properties of Rh/TRL complexes. As shown in Table 1, average turnover frequencies (TOFs) of 250 h^{-1} for 1-dodecene and 470 h^{-1} for styrene have been achieved. Even the hydroformylation of oleyl alcohol, an extremely hydrophobic internal olefin, exhibits a yield of 72% [24].

The recycling effect of the catalyst was also examined. Aqueous phase containing the Rh/PEO-DPPPA catalyst after phase separation was re-used 20 times in the hydroformylation of 1-decene [21]. It should be pointed out that leaching of Rh into the organic phase might be diminished with difficulty to less than the ppm level by means of a single-phase separation.

Breuzard et al. [25] prepared chiral polyether ligands derived from (*S*)-binaphthol and combined with the [Rh(cod)$_2$]BF$_4$ complex. This system has been used in the catalytic enantioselective hydroformylation of styrene in thermoregulated phase-transfer conditions, but the *ee* value is less than 25%.

Table 1 Two-phase hydroformylation of olefins catalyzed by nonionic phosphine-modified rhodium complexes.

Olefin	Ligand (PETPP)	N$^{a)}$	P/Rh	P [MPa]	T [°C]	Yield [%]	n/i	TOF [h^{-1}]
1-Hexene	*m* = 3	18	5	5.0	100	85	2.0	–
1-Octene	*m* = 3	18	5	5.0	100	88	2.8	–
1-Dodecene	*m* = 3	18	5	5.0	100	84	1.8	250
Styrene	*m* = 1	25	12	3.0	80	94	0.5	470
p-Chlorostyrene	*m* = 1	25	12	3.0	80	92	0.6	460
p-Methoxystyrene	*m* = 1	25	12	3.0	80	95	0.5	480
Cyclohexene	*m* = 1	25	5	6.0	120	82	–	140
Oleyl alcohol	*m* = 1	16	5	6.0	120	72	–	70

a) *N* – average number of ethylene oxide units per molecule of ligand.

2.3.5.3.2
Selective Reduction of Nitroarenes

Selective reduction of nitroarene to aromatic amine by the basic aqueous route using CO as the reducing agent leads to fine chemicals, due to the selective reduction of the nitro group while the other groups, like halogen, –C=C, –C=O, and –CN, could be kept free from hydrogenation or dehalogenation [Eq. (1)].

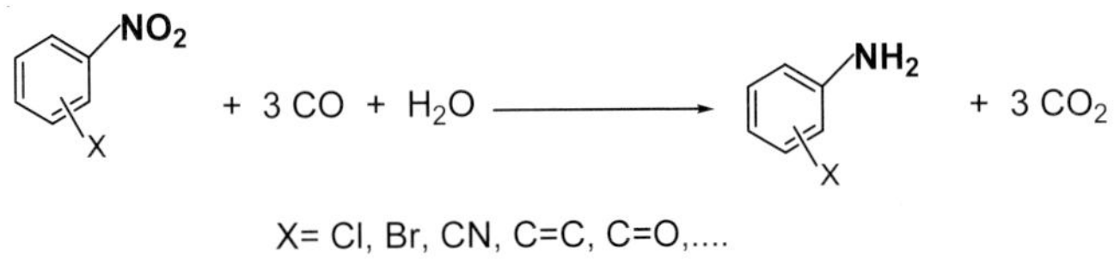

$$+ \ 3 \ CO + H_2O \longrightarrow \qquad\qquad + \ 3 \ CO_2 \tag{1}$$

X= Cl, Br, CN, C=C, C=O,....

Tafesh et al. [26] first began using water-soluble Pd complexes with TPPTS or BINAS as ligands in the CO selective reduction of nitroarenes, obtaining a yield less than 65%.

Recently, the CO selective reduction of nitroarenes catalyzed by $Ru_3(CO)_9(PETPP)_3$ [27] and $Ru_3(CO)_9(PEO\text{-}DPPSA)$ [28] has been reported by Jin et al. The results are listed in Table 2. After being used four times, the activity of the catalyst is not evidently decreased.

Table 2 CO selective reduction of several aromatic nitro compounds catalyzed by $Ru_3(CO)_9(PETPP)_3$.[a]

Reactant	Product	Conversion [%]	Yield [%]
$o\text{-}ClC_6H_4NO_2$	$o\text{-}ClC_6H_4NH_2$	99.2	99.0
$m\text{-}ClC_6H_4NO_2$	$m\text{-}ClC_6H_4NH_2$	99.6	99.4
$p\text{-}NO_2C_6H_4CN$	$p\text{-}NH_2C_6H_4CN$	97.4	97.0
$p\text{-}NO_2C_6H_4COCH_3$	$p\text{-}NH_2C_6H_4COCH_3$	95.8	95.4

a) Reaction conditions: $T = 140 \ °C$, $P_{CO} = 4.0 \ MPa$, $V_{toluene}/V_{water} = 4 : 4$, $Ru_3(CO)_9(PETPP)_3 = 0.01 \ mmol$.

2.3.5.4
Conclusion

Although the development of aqueous/organic two-phase catalysis has only occurred more than 20 years since its emergence, crucial advantages of this methodology have been proven as it overcomes the immanent problem of catalyst/product separation associated with homogeneous catalysis. Compared with classical aqueous/organic two-phase catalysis, the process of TRPTC is more "homogeneous" to some extent because the catalyst and substrate remain in the same organic phase at the reaction temperature. TRPTC is also quite different from FBS. While the initial fluorous/organic biphase of FBS becomes a single phase at an appropriate higher temperature, TRPTC maintains the aqueous/organic two-phase system throughout the reaction. It is the mobile catalyst that "transfers" between the two phases without any additive in response to temperature changes; the possibility of changing the status of the phase is one of the properties of the ligand.

The introduction of TRPTC to aqueous/organic two-phase is free from the shortcomings of classical aqueous two-phase catalysis, in which the scope of application is more or less restrained by the water solubility of the organic reactants. Obviously, the core of TRPTC is to use TRL with the cloud point in water. Therefore, to design and prepare ligands with higher catalytic reactivity at lower cost will be a main topic of scientific research and industrial exploitation of this strategy in the future. Just as Cornils remarks in Ref. [29]: "since the agent responsible for the merger and subsequent separation of the phases is the appropriately custom-designed ligand itself, there is no call for investing extra effort in the removal and recycling of a foreign additive, and this must therefore be regarded as a promising avenue for further exploration on a commercially realistic scale."

References

1 (a) B. Cornils, J. Falbe, Proc. *4th Int. Symp. on Homogeneous Catalysis*, Leningrad, Sept. **1984**, p. 487; (b) H. Bach, W. Gick, E. Wiebus, B. Cornils, *Prepr. Int. Symp. High-Pressure Chem. Eng.*, Erlangen/Germany, Sept. **1984**, p. 129; (c) H. Bach, W. Gick, E. Wiebus, B. Cornils, *Prepr. 8th ICC*, Berlin **1984**, Vol. V, p. 417; *Chem. Abstr.* **1987**, *106*, 198 051; cited also in A. Behr, M. Roeper, *Erdgas und Kohle* **1984**, *37* (11), 485; (d) H. Bach, W. Gick, E. Wiebus, B. Cornils, *Abstr. 1st IUPAC Symp. Org. Chemistry*, Jerusalem **1986**, p. 295; (e) E. Wiebus, B. Cornils, *Chem. Ing. Tech.* **1994**, *66*, 916; (f) B. Cornils, E. Wiebus, *CHEMTECH* **1995**, *25*, 33; (g) E. G. Kuntz, *CHEMTECH* **1987**, *17*, 570.
2 P. Purwanto, H. Delmas, *Catal. Today* **1995**, *24*, 135.
3 R. G. Nuzzo, S. L. Haynie, M. E. Wilson, G. M. Whitesides, *J. Org. Chem.* **1981**, *46*, 2861.
4 Ruhrchemie AG (H. Bahrmann, B. Cornils, W. Konkol, W. Lipps), DE 3 412 335 (**1985**).
5 B. Fell, G. Papadogianakis, *J. Mol. Catal.* **1991**, *66*, 143.
6 H. Ding, B. E. Hanson, T. Bartik, B. Bartik, *Organometallics* **1994**, *13*, 3761.
7 A. Buhling, P. C. J. Kamer, P. W. N. M. Van Leeuwen, *J. Mol. Catal. A:* **1995**, *98*, 69.
8 A. Buhling, P. C. J Kamer, P. W. N. M. Van Leeuwen, J. W. Elgersma, K. Goubitz, J. Fraanje, *Organometallics* **1997**, *16*, 3027.
9 I. T. Horvath, J. Rabai, *Science* **1994**, *266*, 72.
10 J. P. Arhancet, M. E. Davis, J. S. Merola, B. E. Hanson, *Nature* **1989**, *339*, 454.
11 D. Koch, W. Leitner, *J. Am. Chem. Soc.* **1998**, *120*, 13 398.
12 Y. Chauvin, H. Olivier, *CHEMTECH* **1995**, *25*, 26.
13 Z. L. Jin, X. L. Zheng, B. Fell, *J. Mol. Catal. A:* **1997**, *116*, 55.
14 N. Schonfeldt, *Surface-Active Ethylene Oxide Adducts*, Wissenschaftliche Verlagsges. mbH, Stuttgart **1976**.
15 Z. L. Jin, in *Catalysis from A to Z* (Eds.: B. Cornils, W. A. Herrmann, R. Schloegl, C. H. Wong), VCH, Weinheim **2000**, p. 581.
16 D. E. Bergbereiter, L. Zhang, V. M. Mariagnanam, *J. Am. Chem. Soc.* **1993**, *115*, 9295.
17 Z. L. Jin, Y. Y. Yan, H. P. Zuo, B. Fell, *J. Prakt. Chem.* **1996**, *338*, 124.
18 J. Y. Jiang, Y. H. Wang, C. Liu, F. S. Han, Z. L. Jin, *J. Mol. Catal. A:* **1999**, *147*, 131.
19 R. F. Chen, X. Z. Liu, Z. L. Jin, *J. Organomet. Chem.* **1998**, *571*, 201.
20 J. Y. Jiang, Y. H. Wang, C. Liu, Q. M. Xiao, Z. L. Jin, *J. Mol. Catal. A:* **2001**, *171*, 85.
21 C. Liu, J. Y. Jiang, Y. H. Wang, F. Wen, Z. L. Jin, *J. Mol. Catal. A:* **2003**, *198*, 23.
22 M. J. H. Russel, *Platinum Met. Rev.* **1998**, *32*, 179.
23 B. Fell, D. Leckel, Ch. Schobbeu, *Fat. Sci. Technol.* **1995**, *97*, 219.
24 F. Z. Kong, X. L. Zheng, Z. L. Jin, *Chin. Chem. Lett.* **2003**, *14* (9), 917.

25 J. A. J. Breuzard, M. L. Tommasino, M. C. Bonnet, M. Lemaire, *J. Organomet. Chem.* **2000**, *616*, 37.

26 A. Tafesh, M. Beller, *Tetrahedron Lett.* **1995**, *36*, 9305.

27 J. T. Mei, J. Y. Jiang, Q. M. Xiao, Y. M. Li, Z. L. Jin, *Chem. J. Chin. Univ.* **2000**, *21* (6), 894.

28 J. Y. Jiang, J. T. Mei, Y. H. Wang, F. Wen, Z. L. Jin, *Appl. Catal. A:* **2002**, *224*, 21.

29 B. Cornils, *Angew. Chem.* **1995**, *107*, 1709; *Angew. Chem., Int. Ed. Engl.* **1995**, *34*, 1575.

2.3.6
Environmental and Safety Aspects

Boy Cornils and Ernst Wiebus

Although the use of water as a liquid support in homogeneous processes has raised controversy in academia [1], twenty years of experience prove the decisive advantages of this aqueous technology. None of the predictions from "false prophets" arrived – there have been no problems with "detection in case of leakage" or with "incineration of bleed streams", with the tremendous heat of evaporation of water, or with the decomposition of water-sensitive compounds. Spills are hard to collect – but what liquid is easy to handle in that case? (see also p. 30) [2]. On the contrary, water does not ignite, does not burn, is odorless and colorless, and is ubiquitous: important prerequisites for the solvent of choice in homogeneous catalytic processes. Additionly, the favorable thermal properties make water doubly exploited as a mobile support and as a heat-transfer fluid. The only restriction regarding the general use of water in homogeneously catalyzed processes results from the fact that to attain adequate chemical reaction, there has to be a minimum solubility of the organic substrates in the aqueous catalyst phase – but this is a chemical engineering limitation, not one which has to be addressed to water as a catalyst support! Twenty years of experience with the most important aqueous-phase process, the Ruhr-chemie/Rhône-Poulenc (RCH/RP) hydroformylation process and its production of some five millions of tons of butyraldehydes provide unequivocal results: the process is extremely favorable to the environment and there have been no accidents as a consequence of using water.

The details and the background of RCH/RP's developments have been described elsewhere. The work of Kuntz (then at RP; see Section 2.4.1.1.1) on trisulfonated triphenylphosphine (TPPTS) and the industrial implementation and improvements developed by Ruhrchemie eventually laid the foundation for the subsequent successful commercialization [3]. TPPTS is the ideal ligand modifier for the oxo-active $HRh(CO)_4$. Without any expensive preformation steps, three of the four CO ligands can be substituted by the readily soluble (1100 g L^{-1} water) and nontoxic (LD_{50}, oral: > 5000 mg kg^{-1}) TPPTS, which yields the hydrophilic oxo catalyst $HRh(CO)[P(3-sulfophenyl-Na)_3]_3$. The trisulfonation, in particular, permits the fine adjustment of TPPTS and thus of the hydrophilic versus the hydrophobic properties; hydrophilicity is ranked in the order TPP*TS* > TPP*DS* > TPP*MS*, from the trisulfonated through to the monosulfonated species.

Because of the solubility of the Rh(I) complex in water and its insolubility in the oxo products, the oxo unit is essentially reduced to a continous stirred tank reactor followed by a phase separator (decanter) and a stripping column. This simplification (see Figure 3, Section 2.4.1.1.1) is a "green" improvement – speaking in the terms of the sustainability supporters – which contributes considerably to the favorable reduction in manufacturing costs of the new process. Propene and syngas are added to the noncorrosive catalyst solution, they react, and the reaction products and the catalyst phase pass into the decanter. While being degassed, they are thus separated into the aqueous catalyst solution and the organic butyraldehyde phase. The catalyst solution exchanges heat and produces steam in the heat exchanger, is replaced by the same amount of water (not catalyst solution!) dissolved in the crude aldehyde, and is returned to the oxo reactor. The crude aldehyde passes through a stripping column in which it is treated with fresh syngas in countercurrent and, if necessary, it is freed from unreacted alkene. No side reactions occur which decrease the selectivity and yield of the aldehyde, since stripping of the aldehydes is carried out in the absence of the oxo catalyst, a distinctive feature of RCH/RP's process. The reboiler of the subsequent "n/i" column, separating n- from isobutytaldehyde, is designed as a heat-absorbing falling-film evaporator incorporated into the oxo reactor, thus providing a neat, efficient method of recovering heat by transferring the heat of reaction from the reactor to cold n-butanal, which subsequently heats the n/i column. Whereas other oxo processes are steam consumers, the RCH/RP process including the distillation unit exports steam [4] – in energy terms a drastic improvement and even "a greener replacement" [5].

The catalyst is not sensitive to sulfur or other oxo poisons, which is another environmental advantage. Together with the simple but effective decantation, which allows the withdrawal of organic and other byproducts at the very moment of separation, accumulation of activity-decreasing poisons in the catalyst solution is prevented. Therefore no special treatment or purification step is necessary. This reduces the environmental burden still further. For a considerable time the oxo units at Ruhrchemie's plant site were even supplied with syngas derived from coal produced by the TCGP (Ruhrchemie/Ruhrkohle's version of the Texaco coal gasification process). In various cases this can be an important factor as far as local resources are concerned.

The oxo catalyst [HRh(CO)(TPPTS)$_3$], its formation from suitable precursors, and its handling and operation are described under "Lower Alkenes" in Section 2.4.1.1.3. The reaction system is self-adjusting – an important consideration for safety reasons – and thus control analyses are needed only at prolonged intervals. Owing to the high degree of automation, only two employees per shift supervise two or more oxo units with a total capacity of over 600 000 tons y^{-1}. The design obviates the need for certain equipment (e.g., feed and cooling pumps); the on-stream factor of the whole system exceeds 98%. Typical reaction conditions, crude product compositions of the RCH/RP process averaged out over a 20-year period, and a discussion of selectivities and activities are given in Ref. [6].

The high selectivity of the C$_3$ conversion toward C$_4$ products makes fractional distillation after aldehyde distillation unnecessary, reduces expenditure, and thus also minimizes the monetary and environmental load. The manufacture of the byproducts advantageously becomes part of the 2-EH (2-ethylhexanol) process since the heavy ends consist mainly of 2-ethyl-3-hydroxyhexanal, which, during downstream processing, is also converted to 2-EH. This (and the avoidance of the formation of butyl formates) is the reason for the considerable simplification of the process flow diagram compared with other process variants (cf. Figure 2 in Section 2.1).

The RCH/RP process exhibits considerable improvements in respect of environmental conservation, conservation of resources, and minimization of pollution. This can be demonstrated by various criteria and proved very convincingly by means of Sheldon's environmental factor E [7], which is far more suitable and constructive than Trost's "atom effiency" [8]. Sheldon defined the E factor as the amount of waste produced per kilogram of product, and specified the E factor for every segment of industry (Table 1).

Table 1 Environmental acceptability: the E factor.

Segment of chemical industry	Product tonnage	Byproducts/product [w/w]
Oil refining	10^8–10^9	< 0.1
Bulk chemicals	10^4–10^6	< 1–5
Fine chemicals	10^2–10^4	5–50
Pharmaceuticals	10^1–10^3	25–100+

As expected and as shown in Table 2, this environmental quotient for conventional oxo processes (using cobalt catalysts) and the production of the bulk chemical n-butyraldehyde is actually about 0.6–0.9, depending on the definition of the term "target product". This range indicates that the main byproduct of propene hydroformylation, isobutyraldehyde, is processed by a number of producers (e.g., to isobutanol, isobutyric acid, or neopentyl glycol) so that isobutyraldehyde is not always and everywhere a byproduct; thus when it becomes a target product the E factor falls from 0.9 to 0.6.

Table 2 E factors for oxo processes.

Process	Byproducts/products [w/w]	
	Isobutyraldehyde as product	Isobutyraldehyde as byproduct
Cobalt-high pressure	> 0.6	< 0.9
Rhodium RCH/RP process	< 0.04	< 0.1

Strictly speaking, this observation is included in Sheldon's wider statement, according to which the E factor is refined and becomes the "environmental quotient" EQ, depending on the nature of the waste. Since such quotients "are debatable and will vary from one company to another and even from one production to another" [7], they will not be discussed here. The crucial point is that, on the same basis (taking into account all the byproducts, including those used in ligand manufacture, etc. – i.e., as for life cycle assessment) conventional oxo processes have an *E* factor of 0.6–0.9, which falls to below 0.1 in the RCH/RP process: an important pointer to the enviromental friendliness of this technology. The low *E* factor of < 0.04 indicates that the utilization of material resources is improved more than tenfold: according to Sheldon's assessment, production of the bulk chemical "butyraldehyde" is classified alongside the highly efficient products of oil refining processes.

Whereas this important quotient is calculated solely from the product spectrum, process simplifications are a consequence of combining the Rh catalyst with the special biphasic process: as indicated in Figure 2 in Section 2.1, the reaction sequence and the procedure of separation and work-up of the products is considerably simplified. As stated in Ref. [9], the capital expenditure for the old Co technology is at least 1.9 times higher than for the RCH/RP process. The conservation of energy resources with the RCH/RP process is dramatic: the steam consumption figures for the Co process are very much higher than those for the Rh process and power consumption is twice as high as that of the RCH/RP process. For example, the compression costs alone for the required syngas are in a ratio (Co process/Rh process) of 1.7 : 1, and the volume of waste water from the RCH/RP process is 70 times lower. The fact that the water-soluble catalyst does not leave the oxo reactor and its immediate surroundings is an important reason why the Rh losses are in the ppb range. The reduced probability of leaks also increases effectiveness and safety.

The "solvent" water reliably averts the risk of fire inherent in the old Co processes as a result of leaking highly flammable metal carbonyls. The reaction system with its "built-in extinguishing system" reliably prevents such fires, and the painstaking measuring and monitoring procedure necessitated by the valuable catalyst metal rhodium, accompanied by constant simultaneous balancing of the RCH/RP process, permits any leaks from the aqueous system to be detected much earlier than was ever possible previously. This also applies to the cooling system, in which any leak from the falling-film evaporator would be noticed after a loss of only a few ppm of rhodium. The solvent water is available instantly everywhere; it is not odor-free, since because of its residual solubility for butyraldehydes it is as certain to be detected by the sensory/olfactory organs as are the product streams of other oxo processes.

Taking all the criteria into consideration, the RCH/RP biphasic oxo process is probably the soundest and "best-natured" variant in terms of the environment: it is a "green" and "sustainable" process which, in addition to its enviromental compatibility, has the advantage of being extremely cost-effective and reduces the manufacturing costs by a sensational 10%. During the manufacture of more than 5 MM tons of *n*-butyraldehyde the plants experienced no serious accident which could be attributed to the use of water as the "mobile support".

On the other hand, there are the advantages of a successful waste management system which ultimately lead to higher cost-effectiveness: the aqueous-biphase process is a typical example of "cleaner technology" or "soft chemistry" – and both expressions are involved here in the non-ideological meaning of the word [10].

Thus it can be taken for granted that among the homogeneously catalyzed processes the aqueous-biphase variants will be in the forefront of further development.

References

1 Z. JIANG, A. SEN, *Macromolecules* **1994**, *27*, 7215; NATO Advanced Research Workshop, *Aqueous Organometllic Chemistry and Catalysis*, Debrecen, Hungary, Aug./Sept. 1994, Kluwer, Dordrecht **1995**, p. 1.

2 See the discussion in Section 2.1.

3 See references in B. CORNILS, W. A. HERRMANN, *Aqueous-Phase Organometallic Catalysis*, 2nd ed., Wiley-VCH, Weinheim **2004**.

4 See Section 2.4.1.1.3, under "Lower Alkenes".

5 P. M. JENKINS, S. C. TSANG, *Green Chem.* **2004**, *6*, 69.

6 See Ref. [3], p. 384.

7 R. A. SHELDON, *CHEMTECH* **1994**, *24*, 38.

8 B. M. TROST, *Science* **1991**, *254*, 1471; *Angew. Chem.* **1995**, *107*, 285.

9 See Ref. [3], p. 343.

10 R. T. BAKER, W. TUMAS, *Science* **1999**, *284*, 1477; S. K. Ritter, *Chem. Eng. News* **2002**, May 20, 38.

2.4
Typical Reactions

2.4.1
Syntheses with Carbon Monoxide

2.4.1.1
Hydroformylation

2.4.1.1.1
Historical Development

Boy Cornils and Emile G. Kuntz

The history of biphasic homogeneous catalysis starts with Manassen's statement [1]:

> "... the use of two immiscible liquid phases, one containing the catalyst and the other containing the substrate, must be considered. The two phases can be separated by conventional means and high degrees of dispersion can be obtained through emulisification."

Roughly at the same time (in contradiction to a misleading statement by Papadogianakis [2]) and parallel to work done by Shell [3] (concerning nonaqueous systems) and by Joó [4] and others [5], one of us (EK, then at Rhône-Poulenc) devoted time and effort to starting practical work on aqueous biphasic catalysis with organometallic catalysts (mainly for hydroformylations), developing the basic principle and the currently well-known standard ligand triphenylphosphine trisulfonate, TPPTS (cf. Section 2.2.3.2).

At that time the general basis for industrial hydroformylation was the monophasic operation with cobalt or with modified Co catalysts [6]. The Wilkinson school of thought opened up a new field of rhodium chemistry by applying the remarkable catalytic properties of ligand-modified (especially triphenylphosphine-substituted) Rh carbonyls [7]. The high activity, productivity, and selectivity of Rh catalysts promised economically operated plants with only some (or some dozens of) kilograms of precious metals needed on the inventory. The Monsanto acetic acid process proved that the amount of precious metals involved represents only a minor part of the investment and the manufacturing costs – provided the "handling" of commercial amounts of precious metal catalysts would be feasible, especially with a "biphasic" approach for the separation of catalyst and substrate/products by decantation (cf. Section 2.3).

This was the starting point of Kuntz's work at Rhône-Poulenc, trying to transfer rhodium catalysts from the organic monophase to the aqueous biphase not by triphenylphosphine *mono*sulfonate (which had been tried by Joó or Wilkinson [4, 8]) but by TPPTS. From 1974 onward, the scope of different reactions using biphasic catalyst systems was tested in laboratory-scale experiments. Among these were

Multiphase Homogeneous Catalysis
Edited by Boy Cornils and Wolfgang A. Herrmann et al.
Copyright © 2005 Wiley-VCH Verlag GmbH & Co. KGaA, Weinheim
ISBN: 3-527-30721-4

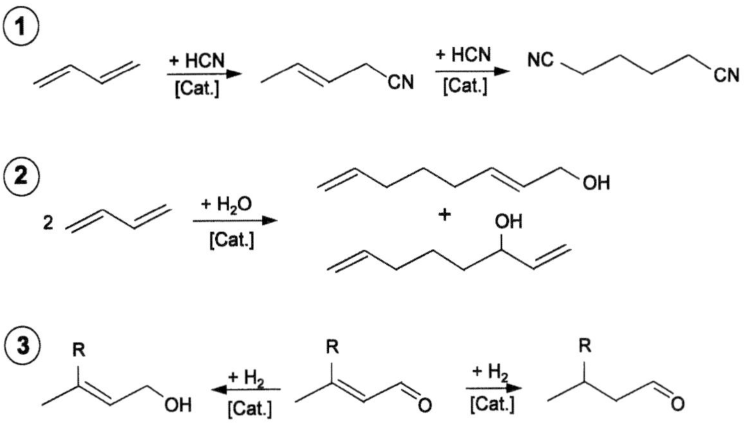

Figure 1 Industrially relevant reactions filed in Rhône-Poulenc's patents ([Cat.] = catalyst + TPPTS) [10]. ① Hydrocyanation; ② Telomerization; ③ Partial hydrogenation of unsaturated aldehydes.

butadiene hydrodimerization, hydrogenation, and hydroformylation (Figure 1 [9]). Even during this initial stage of experimental work it was shown that only a small fraction of the precious metal (less than 0.1 ppm Rh) is leached by the organic (product) phase. Various patent applications were filed to protect Rhône-Poulenc's main fields of interest. One of these was hydroformylation although Rhône-Poulenc was not an oxo producer.

Following earlier contacts, in 1982 Ruhrchemie (RCH) and Rhône-Poulenc (RP) joined forces to develop a continuously operated hydroformylation process for the manufacture of *n*-butyraldehyde from propene. On the basis of the ideas documented in RP's patent applications, RCH used its own long-standing expertise with its own biphasic catalytic cracking process and its long experience in converting laboratory-scale syntheses to commercial processes – as Ost observed [11] as long ago as 1907 in Z. Angew. Chem., the forerunner of Angewandte:

> "... For it is one thing to invent a process that is right in principle but a very different thing to introduce it on the industrial scale."

In many cases, RCH was very successful at transferring new laboratory-scale processes to the industrial dimension. Between 1982 and 1984, in a period of less than 24 months, RCH developed and tested a completely new process for which no prototype was available and for which various new apparatuses had to be invented. Using an scale-up factor of 1 : 24 000 the first production unit employing the "Ruhrchemie/Rhône-Poulenc oxo process" (RCH/RP) went on stream in July 1984 with an initial capacity of 100 000 tons per year (Figure 2).

Because of Ruhrchemie's commitment and status as an important oxo producer, development was primarily driven by product and commercial considerations. It was not until the 1990s that further development became science-driven, including especially all the scientific research work currently being conducted at universities as a result of the successful implementation of the RCH/RP process.

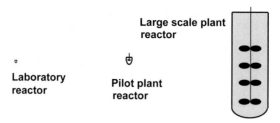

	Laboratory	OT-pilot plant	Plant
Temp. [°C]	125	125	122
Press. [bar]	30	50	50
Volume	5 L	50 L	120 m^3
Scale-up factor	1	1 : 10	1 : 24.000
TON	3,1 x 10^6	3,6 x 10^6	>> 5 x 10^6
n/iso-ratio	94 : 6	95 : 5	96 : 4

Figure 2 Development work at Ruhrchemie: the different stages of reactor design [12].

Focal points of RCH's work are described using Figure 3 as a schematic flowsheet.
Some of the conditions of the new process were based on RP's patents but all had to be adjusted to the rougher operation conditions of a commercial plant, to different qualities and purities of commercially available feedstocks (alkenes of different origin, syngas [from oil or coal], catalyst precursors, etc.), to a normal operation of 8760 h y^{-1}, and to the finely tuned relationships between, e.g., temperatures, pressures/partial pressures, concentrations of various organic and inorganic components in different phases, mass and heat transfer, and flow conditions of a continuous process. The economics of cooling and heat recovery

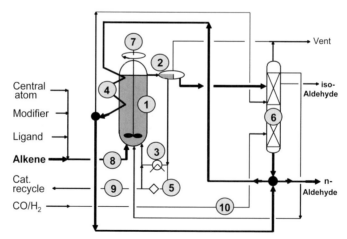

Figure 3 General Ruhrchemie design of the aqueous-biphasic hydroformylation process as the realization of Kuntz's inventions; focal points of R&D work as indicated.

and the utilization of unreacted substrates in off-gases or vents are of special importance.

Extensive chemical engineering work (including many details) comprised the following items: reactor design (**1** in Figure 3) together with the details of mass transport, behavior of phases/decanter (**2**), heat recovery/steam generation (**3**, **4**); rhodium and ligand recycle (**5**), fractionation (**6**), control units (**7**), feed (**8**, **10**), and catalyst work-up (**9**) [13]. Preceding work revealed the basic dependencies as far as pH values, addition of CO_2, salt effects, ionic strengths of solutions, the influence of catalyst modifiers, spectator effects, or microwave irradation and sonochemical treatment are concerned.

All the development work included the full work-up of the reactants, sidestreams, and products, reflecting the characteristics of the biphasic system, the testing of reaction parameters (even under random conditions), the dynamic responses of both the reactor and the catalytic system, product qualities, etc. Apart from the handling of the new aqueous catalyst system, a virtually new design of the hydroformylation process had to be prepared, involving much detailed work. Most of the engineering solutions to new questions arising from the Figure 3 system and numerous important variants have been patented. This included the preparation of TPPTS in commercial amounts and superior quality, which was also solely RCH's responsibility (see, for example, Ref. [14]).

All's well that ends well: the first commercial plant of RCH/RP's process went on stream in summer 1984, and today they produce approximately 890 000 metric tons per year of aldehydes, mainly C_4s from propene besides some C_5 (from butene) at two different locations. Typical reaction conditions, experiences, and performances over a 20-year average are given in Sections 2.3.6 and 2.4.1.1.3 (under "Lower Alkenes") and in Ref. [15].

References

1 J. MANASSEN, in *Catalysis Progress in Research* (Eds.: F. BASOLO, R. L. BURWELL), Plenum Press, London **1973**, p. 183.

2 G. PAPADOGIANAKIS, R. A. SHELDON, *New J. Chem.* **1996**, *20*, 175.

3 Among others: Shell Dev. Corp. (Eds.: W. KEIM, S. R. BAUR, H. CHUNG, P. W. GLOCKNER, H. VAN ZWET, D. CAMEL, R. F. MASON), US 3.635.937, 3.637.636, 3.644.563, 3.644.564, 3.647.914, 3.647.915, 3.661.803 (1972); W. KEIM, in *Fundamental Research in Homogeneous Catalysis* (Eds.: M. GRAZIANI AND M. GIONGO), Plenum Press, New York **1984**, Vol. 4, p. 131.

4 F. JOÓ, M. T. BECK, *React. Kin. Catal. Lett.* **1975**, *2*, 257; F. JOÓ, Z. TÓTH, M. T. BECK, *Inorg. Chim. Acta* **1977**, *25*, L61; F. JOÓ, *Aqueous Organometallic Catalysis*, Kluwer, Dordrecht **2001**, especially p. 1 ff.

5 See references in B. CORNILS, W. A. HERRMANN (Eds.), *Aqueous-Phase Organometallic Catalysis*, 2nd ed., Wiley-VCH, Weinheim **2004**, p. 351.

6 B. CORNILS, in *New Syntheses with Carbon Monoxide* (Ed.: J. FALBE), Springer, Berlin **1980**.

7 J. A. OSBORNE, G. WILKINSON, J. F. YOUNG, *Chem. Commun.* **1965**, 17; J. A. OSBORNE, F. H. Jardine, J. F. Young, G. Wilkinson, *J. Chem. Soc. A:* **1966**, 1711; T. A. Stephenson. G. Wilkinson, *J. Inorg. Nucl. Chem.* **1966**, *28*, 945.

8 A. F. BOROWSKI, D. J. COLE-HAMILTON, G. WILKINSON, *New J. Chem.* **1978**, *2*, 137.

9 Rhône-Poulenc (E. G. Kuntz), FR 2.314.910 (**1975**), FR 2.349.562 (**1976**), FR 2.338.253 (**1976**), FR 2.366.237 (**1976**).

10 B. Cornils, E. G. Kuntz, *J. Organomet. Chem.* **1995**, *502*, 177.

11 H. Ost, *Z. Angew. Chem.* **1907**, *20*, 212.

12 *The Scale-Up of Chemical Processes* (25th–28th Sept. 2000, St. Helier, Jersey); *Conference Proceedings*, pp. 27–37.

13 B. Cornils, *Org. Proc. Res. Dev.* **1998**, *2*, 121.

14 Ruhrchemie AG (R. Gärtner, B. Cornils, H. Springer, P. Lappe), EP 0.107.006 (**1983**).

15 E. Wiebus, B. Cornils, *Chem. Eng. Tech.* **1994**, *66*, 916; B. Cornils, E. Wiebus, *CHEMTECH* **1995**, *25* (1), 33; H.-W. Bohnen, B. Cornils, *Adv. Catal.* **2002**, *47*, 1.

2.4.1.1.2
Kinetics

Raghunath V. Chaudhari

Introduction

Homogeneously catalyzed processes often involve multiphase contacting due to the presence of reactants or products in gaseous, liquid, and solid phases, though the principal catalytic reaction occurs in a single phase in which the catalyst exists in a dissolved state. The most important aspects concerning the commercial viability of such processes are an understanding of the overall rate behavior, selectivity engineering, and catalyst recycle. In spite of several important developments in homogeneous catalysis, their applications in industry have been limited due to difficulties in identifying an economic and industrially feasible separation of products and catalysts. In order to solve this problem, numerous approaches to the immobilization of homogeneous catalysts have been explored. It is well known that several attempts to heterogenize homogeneous catalysts, which include polymer anchoring, encapsulation/tethering [1, 2] supported liquid-phase catalysis [3], and use of organometallic catalysts on mineral supports [4, 5], have not led to industrially viable alternatives. In this context, biphasic catalysis using water-soluble metal complexes has been the most significant development in recent years, in which the aqueous-phase catalyst is contacted with the immiscible organic phase containing reactants and products with or without gaseous reactants in a multiphase system (gas–liquid–liquid), thus allowing a simple catalyst-product separation. It was after the work of Kuntz [6] on the synthesis of the triphenylphosphine trisulfonate (TPPTS) ligand and its application in the hydroformylation of olefins that the research on water-soluble catalysis gained momentum. The concept has been proven on a commercial scale in the Ruhrchemie/Rhône-Poulenc process for the hydroformylation of propene to butyraldehyde [7]. The roles of different water-soluble ligands, their synthesis and stability, as well as other means of intensifying these gas–liquid–liquid catalytic reactions, have been extensively studied and the subject has been reviewed by Kalck and Monteil [8], Herrmann and Kohlpaintner [9], Cornils [10], and Beller et al. [11] among others. Herrmann and Kohlpaintner

[9] have shown that Rh complexes with other water-soluble ligands, such as BISBIS (sulfonated (2,2'-bis(diphenylphosphinomethyl)-1,1'-biphenyl) and NORBOS, give exceptionally high activities and n/i ratios. Uses of cationic [12], nonionic [13], and surface-active [14, 15] phosphines as water-soluble ligands have also been proposed. Of all the catalyst systems studied, Rh-TPPTS is the most suitable and commercially proven catalyst system for biphasic hydroformylation. Several modifications of the water-soluble catalysts using co-solvents [15], surfactants and micelle-forming reagents [16], a supercritical CO_2–water biphasic system [17], supported aqueous-phase catalysis [18], and catalyst-binding ligands (interfacial catalysis) [19] have been proposed to overcome the lower rates observed in biphasic catalysis due to poor solubilities of reactants in water (see Sections 2.2.3.2 and 2.3.3.3). So far, endeavors have been centered on innovating novel catalyst systems from the viewpoint of efficient catalyst recycle and rate enhancement, but limited information is available on the kinetics of biphasic hydroformylation.

The understanding of the overall rate of biphasic catalytic reactions is an equally important aspect in the evolution of an economical process. In the case of aqueous biphasic hydroformylation reactions, the rate will be governed by several factors, which include dissolution of CO, H_2, and olefins in organic and aqueous (catalyst) phases, their solubility, their partition coefficients, and the intrinsic kinetics of the reactions occurring in the aqueous phase [20]. The most important of these is knowledge of the kinetics, which is also essential to understand the reaction mechanism and elucidation of the rate-controlling step. The aim of this contribution is to present a review of the current status on the kinetics of hydroformylation of olefins using water-soluble catalysis and catalyst recycle studies (see also Section 2.3.1). The role of ligands, pH, co-solvents, and surfactants on the rate behavior will be discussed.

Overall Rate of Reaction

Hydroformylation of olefins using aqueous biphasic catalysts is an example of a gas–liquid–liquid catalytic system in which a reaction of two gaseous reactants (carbon monoxide and hydrogen) and gaseous or liquid-phase olefin occurs in the presence of a water-soluble catalyst in a liquid–liquid dispersion. The reaction of dissolved gases and olefins occurs in the aqueous (**bulk**) phase or organic aqueous interface (Figure 1). The overall rate of reaction in such multiphase catalytic systems depends on gas–liquid and liquid–liquid mass transfer, the solubility of gas-phase reactants in the organic and aqueous phases, the liquid–liquid equilibrium properties, and the intrinsic kinetics of the reaction in the aqueous phase. In addition, the dispersion characteristics of the droplets, the droplet size, and the bubble size can also influence the rate of reaction. Depending on the fractional hold-up of the aqueous phase, it will be either a continuous aqueous phase with dispersed organic droplets or a dispersed aqueous phase in a continuous organic medium. The coupled influence of mass transfer with chemical reaction is expected to be quite different in these two situations. To determine of intrinsic kinetics of such reactions, it is necessary to ensure that the experimental data are obtained

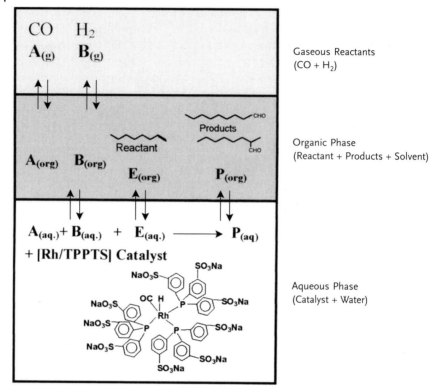

Figure 1 Schematic representation of biphasic hydroformylation.

under a kinetically controlled regime and the contributions of the interphase mass transfer and hydrodynamics of the liquid–liquid dispersion are not significant. The criteria for significance of gas–liquid and liquid–liquid mass-transfer steps for such systems have been described by Mills and Chaudhari [21]. In this section, examples of hydroformylation reactions using aqueous biphasic catalysts will be reviewed with a focus on kinetic studies, the effect of additives, co-solvents and surfactants, and catalyst recycle studies.

Kinetic Studies Without Additives

Hydroformylation of olefins using water-soluble catalysts in two-phase systems has been extensively studied [22, 23], the role of different types of water-soluble ligands and metal complexes in the activity, selectivity, and stability has been discussed for a variety of olefinic substrates. A few case studies in which kinetics and rate behavior have been addressed are reviewed here.

The hydroformylation of 1-octene using a water-soluble Rh-TPPTS catalyst in a biphasic medium was studied in the absence of any additive or co-solvent by Bhanage [24]. The experiments were carried out under conditions such that the aqueous phase containing the Rh-TPPTS catalyst was dispersed in the continuous organic phase consisting of 1-octene and toluene. The results indicated absence of hydrogenation products and the selectivity to hydroformylation products was greater than 98%, with the n/i ratio in the range 1–2 : 1. The aqueous Rh complex catalyst was found to retain its activity even after ten recycles, indicating negligible deactivation. The effects of the concentration of catalyst precursor, TPPTS, and of 1-octene, and the partial pressure of CO and hydrogen, were studied in a temperature range of 353–373 K. The rate of reaction was found to be first order with respect to catalyst concentration and hydrogen partial pressure. This was explained as a consequence of oxidative addition of hydrogen to acyl-carbonyl rhodium species as the rate-determining step [25]. The reaction rate was found to be 0.7th order with respect to CO partial pressure, in contrast to the CO inhibition observed for homogeneously catalyzed hydroformylation [26]. For a water-soluble catalyst, the concentration of dissolved carbon monoxide in the aqueous phase is very low compared to that in the organic phase; hence, formation of a dicarbonyl Rh species $[(RCO)Rh(CO)_2(TPPTS)_2]$, which is believed to be responsible for a negative-order dependence [25], is not very likely. Therefore, this difference in the trends is not truly due to any change in the reaction mechanism. In the homogeneous catalytic reaction, the rate varies linearly with carbon monoxide pressure in the lower region and only beyond a certain pressure of CO is rate inhibition observed. The 1-octene concentration dependence on the rate of hydroformylation showed an apparent reaction order of 1.7 (Figure 2), but this was due to an inappropriate account of the solubility variation with changes in 1-octene concentrations. Bhanage [24] has shown that if the variation in solubility is accounted for, the rate of reaction shows first-order dependence on 1-octene concentrations. A rate model [Eq. (1)], derived from

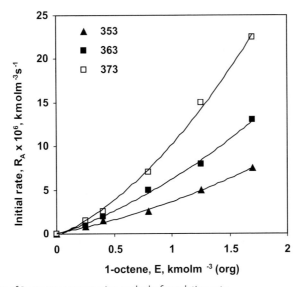

Figure 2 Effect of 1-octene concentration on hydroformylation rate.

the well-known mechanism of hydroformylation [25] assuming oxidative addition of hydrogen to acyl-carbonyl rhodium species as a rate-determining step, was found to represent the rate data satisfactorily:

$$R = \frac{k\,K_1\,K_2\,K_3\,A\,B\,C\,D}{1 + \alpha\,B} \tag{1}$$

where A = partial pressure of hydrogen, MPa; B = partial pressure of carbon monoxide, MPa; C = concentration of catalyst, kmol m^{-3}; D = concentration of olefin, kmol m^{-3}; k = reaction rate constant, s^{-1}; and K_1, K_2, K_3, K_4 and α are constants.

The effect of aqueous phase hold-up on the rate of hydroformylation for 900 and 1500 rpm is shown in Figure 3. At 1500 rpm, the rate versus ε_{1a} (aqueous phase hold-up) plot shows a maximum. For kinetic control, the rate is expected to vary linearly with catalyst loading. However, in a case where the reaction occurs essentially at the liquid–liquid interphase, it would depend on the liquid–liquid interfacial area even though liquid–liquid mass transfer is not rate limiting. For $\varepsilon_{1a} > 0.4$, phase inversion occurs and the interfacial area would be determined by the dispersed phase, which would be the organic phase. Since, for $\varepsilon_{1a} > 0.5$, ε_1 will decrease with an increase in ε_{1a}, a reduction in liquid–liquid interfacial area is expected. Hence, the observed results of a decrease in the rates with an increase in ε_{1a} indicate a possibility of interfacial reaction rather than a bulk aqueous-phase reaction [7]. For $d_1 < 0.3$ mm, a very large interfacial area ($a_1 = 6\,\varepsilon_1\,d\,/\,d_1$) in the range (4–5) $\cdot$ 10^4, 1/m is likely to exist compared to the volume of the aqueous phase.

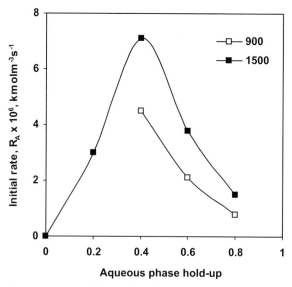

Figure 3 Effect of aqueous phase hold-up on hydroformylation rate.

Hydroformylation of ethylene using a water-soluble Rh-TPPTS catalyst has been investigated [27] using a toluene–water solvent system at 353 K. The effect of TPPTS concentration on rate shows a maximum at a P/Rh ratio of 8 : 1. The rate of reaction first increases with catalyst concentration, and above a certain value it remains constant. The effect of aqueous-phase hold-up shows a maximum in the rate at $\varepsilon_{1a} = 0.4$. The apparent reaction orders for the partial pressures of hydrogen and ethylene were found to be one and zero respectively. A strong inhibition in the rate with an increase in P_{CO} was observed. An interesting example of tandem synthesis of methacrolein in an aqueous biphasic system has been reported by Deshpande et al. [28], in which hydroformylation of ethylene and aldol condensation reactions occur in two immiscible liquid phases with a high yield of the product. Use of a two-phase system prevents contact of the hydroformylation and aldol catalysts, the interaction of which leads to deactivation.

Herrmann et al. [29] have reported the hydroformylation of propene using a Rh-BISBIS catalyst system in a continuous stirred reactor. The activity of this catalyst (45.5 mol g^{-1} min^{-1}) was found to be three times higher than that of a Rh-TPPTS catalyst (15 mol g^{-1} min^{-1}). The n/i ratio also improved from 94 : 6 (TPPTS) to 97 : 3 (BISBIS). They have also studied the hydroformylation of 1-hexene at 5 MPa CO/H$_2$ pressure in the temperature range 395–428 K, and observed that the activity increased from 0.73 to 10.73 (mol g^{-1} min^{-1}) when the temperature was raised from 395 to 428 K whereas the n/i ratio decreases from 97 : 3 to 94 : 6. The catalyst was found to be stable even after 16 h of a continuous run. Wachsen et al. [30] have shown that hydroformylation of propene using a Rh-TPPTS catalyst operates under conditions of significant gas–liquid mass-transfer resistance. A detailed analysis of the engineering aspects supported with experimental validation is necessary to understand the critical scale-up issues. The kinetics of hydroformylation of styrene using a biphasic HRh(CO)(TPPTS)$_3$ catalyst [31] showed that the rate was first order with respect to catalyst loading, H$_2$, CO, and styrene concentrations. The rate was found to be limited by transport of styrene from the organic to the aqueous phase under certain conditions.

The development of supported aqueous-phase catalysis (SAPC; see Section 2.6) opened the way to hydroformylating hydrophobic alkenes such as oleyl alcohol and octene [14]. SAPC involves dissolving an aqueous-phase HRh(CO)(TPPTS)$_3$ complex in a thin layer of water adhering to a silica surface. Such a catalyst shows a significantly high activity for hydroformylation. For classical liquid–liquid systems, the rate of hydroformylation decreases in the order 1-hexene > 1-octene > 1-decene; however, with SAP catalysts, these alkenes react at virtually the same rate and the solubility of the alkene in the aqueous phase is no longer the rate-determining factor [32]. The loss of catalytic activity due to depletion of water in the aqueous film is a major drawback of the SAPC catalysts.

Co-Solvent Effect

It has been shown that by using co-solvents such as ethanol, acetonitrile, methanol, ethylene glycol, and acetone, the rate of hydroformylation can be enhanced by

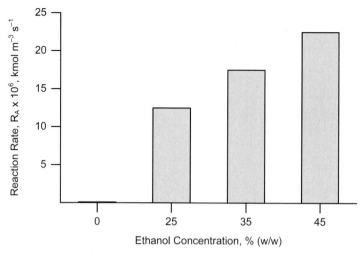

Figure 4 Effect of co-solvent on 1-octene hydroformylation.

severalfold [33, 34]. However, in some cases, a lower selectivity is obtained due to interaction of the co-solvent with products (e.g., formation of acetals by the reaction of ethanol and aldehyde). The hydroformylation of 1-octene with dinuclear $[Rh_2(\mu\text{-}SR)_2(CO)_2(TPPTS)_2]$ and $HRh(CO)(TPPTS)_3$ complex catalysts using various co-solvents has been investigated by Monteil et al. [33], who showed that ethanol was the best co-solvent. Purwanto and Delmas [34] have shown that the hydroformylation rate increases eight-fold in the presence of ethanol as a co-solvent (Figure 4) and studied the kinetics of hydroformylation of 1-octene using $[Rh(COD)Cl]_2$-TPPTS catalyst in the temperature range 333–343 K. First-order dependence was observed with respect to the concentration of catalyst and 1-octene. The effect of the partial pressure of hydrogen indicates a fractional order (0.6–0.7) and substrate inhibition was observed with a partial pressure of carbon monoxide. Equation (2) was proposed as the rate equation

$$R = \frac{k\,A\,B\,C\,D}{(1 + K_1\,A)\,(1 + K_2 + B)^2} \qquad (2)$$

The kinetics of hydroformylation of 1-octene using $[Rh(COD)Cl]_2$ as a catalyst precursor with TPPTS as a water-soluble ligand and ethanol as a co-solvent was further studied by Deshpande et al. [15] at different pH values. The rate increased by two- to fivefold when the pH increased from 7 to 10, while the dependence of the rate was found to be linear with olefin and hydrogen concentrations at both pH values. The rate of hydroformylation was found to be inhibited at higher catalyst concentrations at pH 7, in contrast to linear dependence at pH 10 (Figure 5). The effect of the concentration of carbon monoxide was linear at pH 7, in contrast to the usual negative-order dependence. At pH 10, substrate-inhibited kinetics was observed with respect to CO (Figure 6).

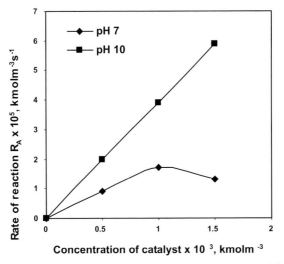

Figure 5 Effect of catalyst concentration on 1-octene hydroformylation rate at different pH values.

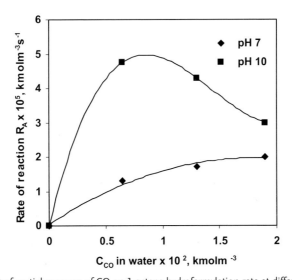

Figure 6 Effect of partial pressure of CO on 1-octene hydroformylation rate at different pH values.

Smith et al. [35] have reported a drop in the activity when the pH of the reaction medium was reduced from 6.8 to 5 for a Rh nitrate/AMPHOS catalyst system (AMPHOS = 1-N,N,N-trimethylamino-2-diphenylphosphinoethane iodide). Hydroformylation of 1-tetradecene with a water-soluble Rh-NABSDPP (NABSDPP: Na-butylsulfonated diphenylphosphine) catalyst gave poor rates in acidic (pH 2.5 to 6) medium. A seven- to eight-fold increase in the rates was observed when the pH was increased from 6 to 10 [36].

The kinetics of hydroformylation of styrene using $HRhCO(TPPTS)_3$ catalyst in a biphasic system with various co-solvents was investigated by Nair [31]. N-Methyl-pyrrolidone (NMP) and ethanol were found to enhance the rate by 7–9-fold compared to that in the absence of any co-solvent. However, with ethanol as a co-solvent, the aldehyde selectivity observed was 80% with acetals as side products. With NMP co-solvent the aldehyde selectivity was > 99%. A kinetic study at 373 K revealed that the rate was first-order dependent on catalyst concentration, fractional-order on CO, and first-order tending to zero-order on styrene concentration.

Surfactant Effect

The addition of various surfactants and micelle-forming agents in the biphasic hydroformylation of olefins was also considered as a tool for enhancement of the reaction rates (see Section 2.3.4). Whereas the presence of a surfactant leads to a lower droplet size in the dispersed phase, thus increasing the liquid–liquid interfacial area and hence the mass-transfer rate, the formation of emulsions is considered as a major drawback of this system. Mass-transfer effects in biphasic hydroformylation of 1-octene in the presence of cetyltrimethylammonium bromide (CTAB) was studied by Lekhal et al. [37]. A mass-transfer model based on Higbie's penetration theory was proposed to predict the rate of hydroformylation in a gas–liquid–liquid system.

Hydroformylation of 1-dodecene in the presence of various micelle-forming agents with aqueous $[RhCl(CO)(TPPTS)_2]$ catalyst was studied by Li et al. [38]. The solubility of 1-dodecene was enhanced in the presence of mixed micelles due to a decrease in the critical micelle concentration (CMC), and higher conversion and selectivity were observed in the mixed micellar solution compared to that for a single micelle type. For biphasic hydroformylation of 1-dodecene in the presence of CTAB, Yang et al. [39] reported several agitator configurations for improving the mixing, dispersion, and interphase mass transfer. An empirical form of macro-kinetic equation [Eq. (3)] was found to represent the initial rate data as well as the n/i ratio of aldehydes.

$$R = k\, N^m\, C_{\mathrm{CTAB}}{}^n\, D^p \left(\frac{V_\mathrm{O}}{V_\mathrm{W}}\right)^q \tag{3}$$

In another report a semi-empirical rate equation has been proposed combining mechanisms of homogeneous catalysis and interfacial parameters for biphasic hydroformylation of 1-dodecene with a water-soluble Rh complex catalyst [40].

Promoting Interfacial Catalysis

Chaudhari et al. [19] have shown that the rate of biphasic hydroformylation can be enhanced severalfold by using a catalyst-binding ligand to promote interfacial catalytic reaction. This approach involves the use of a ligand that is insoluble in the aqueous catalyst phase but has a strong affinity for the metal complex catalyst. The

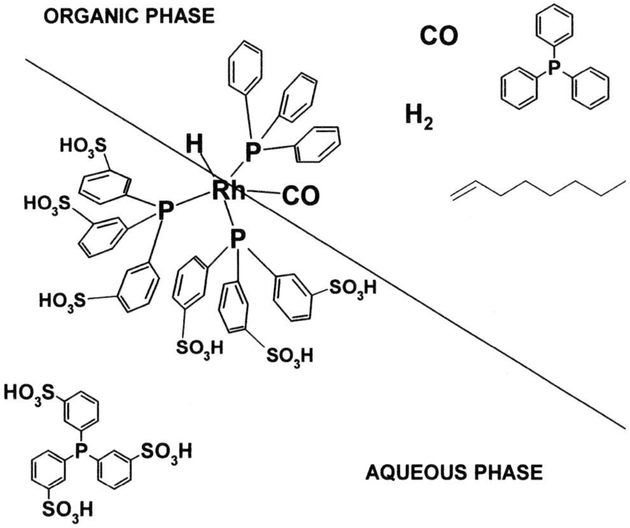

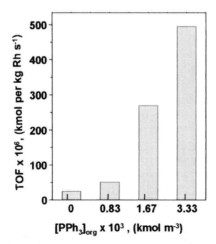

ORGANIC PHASE

CO

H₂

AQUEOUS PHASE

Figure 7 Schematic representation of interfacial catalysis.

interaction of the ligand and the catalyst takes place essentially at the liquid–liquid interface (Figure 7); thus the concentration of the catalytic species will be enriched at the interface, where it can access the reactants present in the organic phase in significantly higher concentrations with respect to the aqueous phase. This results in a dramatic increase in the rate of such a biphasic catalytic reaction, as indicated by experimental data on the hydroformylation of 1-octene using a Rh-TPPTS complex catalyst with triphenylphosphine as a catalyst-binding ligand [19]. The rate of the reaction was enhanced by 10–50-fold in the presence of catalyst-binding ligands when compared with the biphasic hydroformylation reaction (Figure 8).

$TOF \times 10^6$, (kmol per kg Rh s^{-1})

$[PPh_3]_{org} \times 10^3$, (kmol m^{-3})

Figure 8 Enhancement of hydroformylation rate by promoting interfacial catalysis.

This concept has also been demonstrated in reverse for the hydroformylation of a water-soluble olefin (allyl alcohol) with the organic phase containing a catalyst, $HRh(CO)(PPh_3)_3$. In this case, the catalyst is present in the organic phase, whereas the catalyst-binding ligand (TPPTS) is added to the aqueous phase.

The effect of reaction parameters, such as the concentrations of catalyst and olefin and the partial pressures of CO and hydrogen, on the rate of reaction has been studied at 373 K [24]. The rate varies linearly with catalyst concentration, olefin concentration, and partial pressure of hydrogen. A typical substrate-inhibited kinetics was observed with the partial pressure of carbon monoxide. Further, a rate equation [Eq. (4)] to predict the observed rate data has been proposed.

$$R = \frac{k\,A\,B\,D}{(1 + K_1\,B^2)\,(1 + K_2\,D)} \tag{4}$$

It is important to note that the kinetic trends for carbon monoxide were completely opposite to those without a catalyst-binding ligand. Since, under conditions of interfacial catalysis, a higher CO concentration is accessible to the catalytic species, substrate inhibition is observed. The aqueous-phase catalyst could be recycled several times without loss of activity.

Conclusions

The kinetics of hydroformylation of olefins using aqueous biphasic catalysts demonstrates that the rate behavior varies significantly for biphasic catalytic reactions depending on the ligands, additives, and co-solvents. In particular, the kinetics with respect to CO shows a variation for different systems. A major limitation to the rate of biphasic hydroformylation is the solubility of the olefin in the aqueous catalyst phase. Using co-solvent, catalyst-binding ligands, micelle-forming agents, supercritical CO_2–water biphasic systems, and SAPC, the rates are enhanced significantly. Although sufficient information on the intrinsic kinetics is now available, further studies on understanding of the role of gas–liquid and liquid–liquid mass transfer, the influence of dispersed-phase hold-up, drop size, and phase equilibrium properties are necessary. Interface mass-transfer parameters under gas–liquid–liquid hydroformylation conditions also need to be investigated.

References

1 K. Mukhopadhyay, A. B. Mandale, R. V. Chaudhari, *Chem. Mater.* **2003**, *15*, 1766.
2 K. Mukhopadhyay, R. V. Chaudhari, *J. Catal.* **2003**, *213*, 73.
3 Monsanto (P. R. Rony, J. F. Roth), US 3.855.307 (**1974**).
4 M. E. Davis, C. Saldarria, J. A. Rossin, *J. Catal.* **1987**, *103*, 520.
5 E. J. Rode, M. E. Davies, B. E. Hanson, *J. Catal.* **1987**, *96*, 574.
6 Rhône-Poulenc (E. Kuntz), FR 2.314.910 (**1975**), US 4.248.802 (**1981**).

7 (a) B. CORNILS, J. FALBE, *Proc. 4th Int. Symp. on Homogeneous Catalysis*, Leningrad, Sept. **1984**, p. 487; (b) H. BACH, W. GICK, E. WIEBUS, B. CORNILS, ABSTR. *1st IUPAC Symp. Org. Chemistry*, Jerusalem **1986**, p. 295; (c) E. WIEBUS, B. CORNILS, *Chem. Ing. Tech.* **1994**, *66*, 916; (d) B. CORNILS, E. WIEBUS, *CHEMTECH* **1995**, *25*, 33.

8 P. KALCK, F. MONTEIL, *Adv. Organomet. Chem.* **1992**, *34*, 219.

9 W. A. HERRMANN, C. W. KOHLPAINTNER, *Angew. Chem., int. Ed. Engl.* **1993**, *32*, 1524.

10 B. CORNILS, *Angew. Chem. Int. Ed. Engl.* **1995**, *34*, 1575.

11 M. BELLER, B. CORNILS, C. FROHNING, C. W. KOHLPAINTNER, *J. Mol. Catal.* **1995**, *104*, 17.

12 A. HESSLER, S. KUCKEN, O. STELZER, J. BLOTEVOGEL-BALTRONAT, W. S. SHELDRICK, *J. Organomet. Chem.* **1995**, *501*, 293.

13 D. J. DARENSBOURG, N. W. STAFFORD, F. JOÓ, J. H. REIBENSPIES, *J. Organomet. Chem.* **1995**, *488*, 99.

14 H. DING, B. E. HANSON, T. BARTIK, B. BARTIK, *Organometallics* **1994**, *13*, 3761.

15 R. M. DESHPANDE, PURWANTO, H. DELMAS, R. V. CHAUDHARI, *I EC Res.* **1996**, *35*, 3927.

16 HUA CHEN, YAOZHONG LI, JUNRU CHEN, PUMING CHENG, YU-E HE, XIANJUN LI, *J. Mol. Catal. A:* **1999**, *149*, 1.

17 B. M. BHANAGE, Y. IKUSHIMA, M. SHIRAI, M. ARAI, *Chem. Commun.* **1999**, 1277.

18 J. P. ARHANCET, M. E. DAVIES, J. S. MEROLA, B. E. HANSON, *Nature (London)* **1988**, *339*, 454.

19 R. V. CHAUDHARI, B. M. BHANAGE, R. M. DESHPANDE, H. DELMAS, *Nature (London)* **1995**, *373*, 501.

20 R. V. CHAUDHARI, A. BHATTACHARYA, B. M. BHANAGE, *Catal. Today* **1995**, *24*, 123.

21 P. L. MILLS, R. V. CHAUDHARI, *Catal. Today* **1997**, *37*, 367.

22 B. CORNILS, W. A. HERRMANN (Eds.), *Aqueous Phase Organometallic Catalysis: Concepts and Applications*, Wiley-VCH **1998**.

23 B. CORNILS, *Org. Proc. Res. Dev.* **1998**, *2*, 121.

24 B. M. Bhanage, Studies in hydroformylation of olefins using transition metal complex catalysts, Ph. D. Thesis, University of Pune **1995**.

25 D. EVANS, J. A. OSBORN, G. WILKINSON, *J. Chem. Soc. A* **1968**, 3133.

26 R. M. DESHPANDE, R. V. CHAUDHARI, *Ind. Eng. Chem. Res.* **1988**, *27*, 1996.

27 S. S. DIVEKAR, *Kinetic modeling of hydroformylation of olefins using homogeneous and biphasic catalysis*, Ph. D. Thesis, University of Pune **1995**.

28 R. M. DESHPANDE, M. M. DIWAKAR, A. N. MAHAJAN, R. V. CHAUDHARI, *J. Mol. Catal. A:* **2004**, *211*, 49.

29 W. A. HERRMANN, C. KOHLPAINTNER, H. BAHRMANN, W. KONKOL, *J. Mol. Catal.* **1992**, *73*, 191.

30 O. WACHSEN, K. HIMMLER, B. CORNILS, *Catal. Today* **1998**, *42*, 373.

31 V. S. Nair, Hydroformylation of olefins using homogeneous and biphasic catalysts, Ph. D. Thesis, University of Pune **1999**.

32 I. T. HORVATH, R. V. KASTRUP, A. A. OSWAID, E. J. MOZELESKI, *Catal. Lett.* **1989**, *2*, 85.

33 F. MONTEOL, R. QUEAU, P. KALCK, *J. Organomet. Chem.* **1994**, *480*, 177.

34 PURWANTO, H. DELMAS, *Catal. Today* **1995**, *24*, 134.

35 R. T. SMITH, R. K. UNGAR, L. J. SANDERSON, M. C. BAIRD, *Organometallics* **1983**, *2*, 1138.

36 S. KANAGASABAPATHY, *Studies in oxidative carbonylation and hydroformylation reactions using transition metal catalysts*, Ph. D. Thesis, University of Pune **1996**.

37 A. LEKHAL, R. V. CHAUDHARI, A. M. WILHEIM, H. DELMAS, *Catal. Today* **1999**, *48*, 265.

38 M. LI, Y. LI, H. CHEN, Y. HE, X. LI, *J. Mol. Catal. A:* **2003**, *194*, 13.

39 C. YANG, X. BI, Z. MAO, *J. Mol. Catal. A:* **2002**, *187*, 35.

40 Y. ZHANG, Z. MAO, J. CHEN, *Catal. Today* **2002**, *74*, 23.

2.4.1.1.3
Conversion of Alkenes

Carl-Dieter Frohning and Christian W. Kohlpaintner

a) Lower Alkenes

Introduction

Although olefins with variable chain lengths have been successfully hydroformylated in aqueous two-phase reactions, a distinction between lower and higher olefins is reasonable. The solubility of ethylene (C_2), propene (C_3), and C_4 olefins, herein referred to as lower olefins, in the aqueous catalyst phase is high enough to assure chemical reaction without phase-transfer limitations. Olefins with chain lengths greater than C_4 have a significantly lower solubility, thus making special means necessary to overcome such limitations (see under "Mechanism", below).

We will describe the basics of aqueous two-phase hydroformylation with TPPTS and rhodium complexes thereof [1] as they apply to C_3 and C_4 olefins according to the Ruhrchemie/Rhône-Poulenc process. Emphasis will be put on the commercial applications and the basic description of the processes.

Mechanism

The mechanism of the oxo reaction has been extensively studied in the past. A comparative study of the two commercially applied oxo catalysts $HRh(CO)(TPP)_3$ (TPP = triphenylphosphine) and $HRh(CO)(TPPTS)_3$ was performed by Horváth [2]. The latter, which is water-soluble, is considered to react according to the dissociative mechanism. However, remarkable differences exist in the catalytic activity and the selectivity of the organic and the water-soluble catalyst. The latter shows much lower specific activity but an increased selectivity to linear products in the hydroformylation of propene. From an Arrhenius plot it is concluded that the dissociation energy of TPPTS from $HRh(CO)(TPPTS)_3$ is about 30 ± 1 kcal mol^{-1} (1 kcal mol^{-1} = 4.18 kJ mol^{-1}). Compared with the dissociation energy of TPP from $HRh(CO)(TPP)_3$ (19 ± 1 kcal mol^{-1} [3]) the difference is greater than 10 kcal mol^{-1}, thus explaining the lower catalytic activity at comparable reaction conditions. Additionally, it was shown that $HRh(CO)(TPPTS)_3$, in contrast to its organic-soluble derivative, does not form $HRh(CO)_2(P)_2$ (P = TPPTS; **1**) at syngas pressures up to 200 bar. By dissociation of either carbon monoxide or TPPTS the unsaturated species $HRh(CO)(TPPTS)_2$ (**2**) and $HRh(CO)_2(TPPTS)$ (**3**) are generated, which are responsible for the formation of linear or branched aldehydes (Scheme 1). As $HRh(CO)(TPPTS)_2$ is formed by dissociation of TPPTS from $HRh(CO)(TPPTS)_3$ and $HRh(CO)_2(TPPTS)$ is obtained through an equilibrium reaction from $HRh(CO)_2(TPPTS)_2$, the observed increased selectivity to linear products becomes explicable.

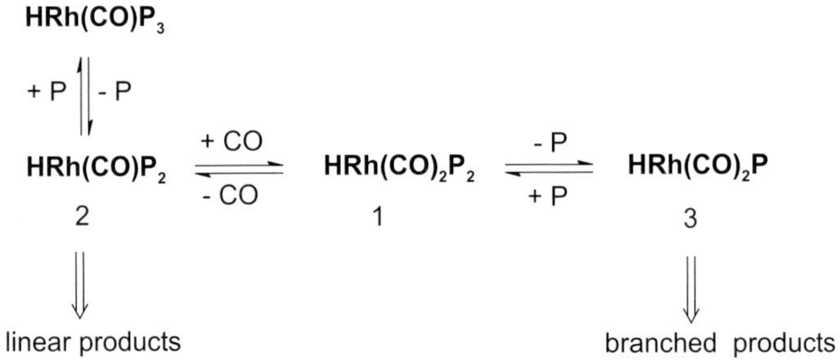

HRh(CO)P$_3$

$+$ P $\|$ $-$ P

HRh(CO)P$_2$ $\overset{+ \, CO}{\underset{- \, CO}{\rightleftharpoons}}$ **HRh(CO)$_2$P$_2$** $\overset{- \, P}{\underset{+ \, P}{\rightleftharpoons}}$ **HRh(CO)$_2$P**

2   1   3

$\Downarrow$ $\Downarrow$

linear products   branched products

Scheme 1 Alternative paths of the oxo reduction.

Kinetics

Limited data are available for the kinetics of the oxo synthesis with HRh(CO)-(TPPTS)$_3$. The hydroformylation of 1-octene was studied in a two-phase system in the presence of ethanol as a co-solvent to enhance the solubility of the olefin in the aqueous phase [4]. A rate expression was developed which was nearly identical to that of the homogeneous system, the exception being a slight correction for low hydrogen partial pressures [Eq. (1)]; see also Section 2.4.1.1.1.

$$R_0 = k \, \frac{[octene]_0 [cat][H_2][CO]}{(1 + K_{H_2}[H_2]) \, (1 + K_{CO}[CO])^2} \tag{1}$$

A rigid reaction rate model, established under idealized conditions, becomes complex and complicated when it is transferred to the hydroformylation of lower olefins under conditions relevant to industrial practice, as the mass-transfer phenomena involved in a triphasic system (gas/liquid/liquid) in large reactors have to be taken into account. The resulting algorithm in general is limited to a narrow bandwidth of operating conditions, thereby diminishing the applicability and reliability of such models. The substitution of rigid models by data-driven models (e.g., so-called artificial neural networks) has been under consideration for some years by one of the authors (CDF).

Recent Developments

In order to develop highly active and selective catalysts for propene hydroformylation, several ligands based on biphenyl or binaphthyl structures were synthesized and have been applied in the oxo synthesis [5]. A mixture of six-, seven- and eightfold sulfonated NAPHOS, called BINAS, together with rhodium is the most active water-soluble oxo catalyst known today. Even at very low phosphine to rhodium ratios (P/Rh), *n/i* selectivities of 98 : 2 are achieved.

A challenge in synthesizing new water-soluble ligands is the direct functionalization of new or previously known organic phosphines. The plethora of functionalized phosphines available today have been categorized and discussed in

[6]. In the particular case of sulfonated phosphines the introduction of the sulfonato group is difficult as the phosphines tend to oxidize during treatment with oleum (sulfur trioxide dissolved in concentrated sulfuric acid) and require specific synthesis methods [7]. Some information about special phosphines and their manufacturing conditions are given in [8].

Commercial Applications

The industrial hydroformylation of short-chained olefins such as propene and butenes is nowadays almost exclusively performed by so-called LPO (LPO = low-pressure oxo) processes, which are rhodium-based. In other words, the former high-pressure technology based on cobalt has been replaced by the low-pressure processes, which cover more than 80% of the total C_4 capacity due to their obvious advantages [9, 30].

Two basic variants of LPO processes exist: the homogeneous processes, e.g., the catalyst and the substrate are present in the same liquid phase; and the two-phase process (Ruhrchemie/Rhône-Poulenc process, RCH/RP) applying a water-soluble catalyst. The homogeneous processes dominate the field by far, a consequence as much of their early development as of the licensing policy for the two-phase process. Both types of process use rhodium as the catalyst metal in combination with a suitable phosphine as ligand. More precisely, the ligand is triphenylphosphine (TPP) in most cases, applied as such in the homogeneous case and in its water-soluble (sulfonated) variant (TPPTS) in the RCH/RP process. Some phosphite ligands have also gained commercial importance for LPO processes recently [10, 11], but no metal other than rhodium has been successfully applied commercially in this technology.

Economic and Ecological Aspects

The RCH/RP process has been in operation since the early 1980s, producing some 5 million tonnes of *n*-butanal [12, 13]. The water-soluble catalyst $HRh(CO)(TPPTS)_3$ combines the advantages of a homogeneous catalyst (high activity, high selectivity) with those of a heterogeneous one. These advantages, in addition to a highly efficient recovery of process heat, lead to a superior technology which also results in a cost advantage compared with the classical homogeneous processes. The RCH/RP process has its strengths in the efficiency of material usage (raw materials, energy and byproduct credits) along with smaller fixed costs due to the ease of operation. The overall cost advantage is estimated to be roughly 10% compared with the standard processes (see Section 2.5.1).

The ecological benefits of this modern process are obvious and can be summarized as follows:

- usage of water as a nontoxic, nonflammable solvent;
- efficient usage of C_3 raw material (propene);
- high selectivity toward the desired products;
- excellent atom economy;

- energy consumption minimized, e.g., net steam exporter;
- efficient recovery of catalyst (loss factor = $1 \cdot 10^{-9}$);
- ligand toxicity is not critical (LD_{50}, oral > 5 g kg^{-1});
- almost zero environmental emissions (cf. also Section 2.3.6).

Along with other water-based reactions the RCH/RP process has been critically reviewed with respect to its environmental attractiveness by Sheldon [14]. Overall the RCH/RP process, besides a technical success, is an outstanding example for the impact of modern technology on both economic and ecological aspects at the same time.

C_3 Process Description

Ruhrchemie AG was the first to seize upon the idea of applying a water-soluble rhodium catalyst and thus commercializing a process which had been elaborated on a laboratory scale by Rhône-Poulenc earlier [15, 16].

It took only two years of intensive research to develop the technical concept and to erect the first plant, which went on stream in 1984 [17]. By 1987 the second unit was already built, followed by a third unit in 1997. Today the total estimated capacity in n-butanal amounts to more than 400 000 t y^{-1} having been 350 000 t y^{-1} in 1995 [18, 19]. An additional plant for the production of n-pentanal from n-butene was brought on stream in 1995.

Basically, the requirements for a process using an aqueous catalyst phase are the same as for the homogeneous processes. The reaction of propene, hydrogen, and carbon monoxide takes place in the first liquid phase, the aqueous catalyst solution. The second, organic, layer is formed by the reaction product, e.g., butanals. Intimate contact between the catalyst solution and the gaseous reactants has to be provided by intensive gas dispersion at the bottom of the reactor together with sufficient stirring. The two liquid phases form an intimate admixture (emulsion) upon stirring which occupies most of the reactor volume, leaving only a small headspace as an internal reservoir for the gaseous reactants. A heating/cooling device is necessary in order to enable start-up of the reactor and to control the exothermic hydroformy-lation reaction (about 28 kcal mol^{-1} or 118 kJ mol^{-1}). Finally, the mixture of liquid and gaseous products has to be withdrawn from the reactor, products and catalyst have to be separated, and the latter has to be recycled.

A simplified scheme for the RCH/RP unit is presented in Figure 1 [1, 12, 13]. The reactor (1) is essentially a continuous stirred tank reactor equipped with a gas inlet, a stirrer, a heat exchanger and a catalyst recycle line. Catalyst and reactants are introduced at the bottom of the reactor. Vent gas is taken from the head of the reactor and from the phase separator. Control of the liquid volume inside the reactor is simple: the liquid mixture, composed of catalyst solution and aldehydes, leaves via an overflow and is transferred to a phase separator (2), where it is partially degassed. The separation of the aqueous catalyst solution (density of the catalyst solution ≈ 1100 g L^{-1}) and the aldehydes occurs rapidly and completely, favored by the difference in densities (density of aldehyde layer ≈600 g/L due to dissolved gases).

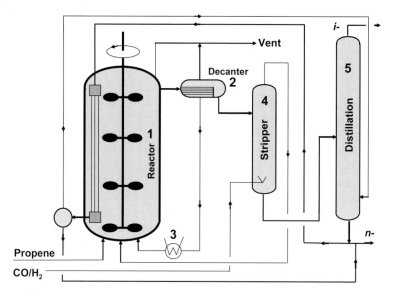

Figure 1 The RCH/RP unit.

The catalyst solution passes through a heat exchanger (3) and produces process steam that is consumed in downstream operations. Some water is extracted from the catalyst solution by its physical solubility in the aldehydes (about 1.3% w/w) which may be replaced before the catalyst solution reenters the reactor.

The subsequent stripping column (4) is important. From the raw organic phase coming from the phase separator and entering at the top of the stripping column, the dissolved reactants are removed by a fresh countercurrent stream of synthesis gas. The pressure inside the stripping column is kept slightly higher than the pressure in the reactor; e.g., no additional mechanical compression nor heating is necessary to recycle unconverted reactants. The resulting crude aldehydes are virtually free of propene as well as propane and contain only minimum amounts of dissolved synthesis gas. The head gas from the stripping column is fed back into the reactor.

The crude aldehydes are split into *n*- and isobutanals in the distillation column (5). The heat required is supplied by the hydroformylation itself: the reboiler of the distillation is a falling-film evaporator which is incorporated in the synthesis reactor using *n*-butanal as the heat carrier. This system has clear advantage over the classical hydroformylation processes, as the RCH/RP process not only uses the heat of reaction efficiently but is also a net steam exporter.

The favorable combination of stripping column, distillation, and heat recovery system is closely linked with the properties of the catalyst solution. The raw aldehydes are virtually free from ingredients of the catalyst solution, thus avoiding any of the well-known side reactions which take place in the presence of even traces of catalyst during thermal treatment. The fact that highly reactive *n*-butanal may be used as a heat-transfer medium well underlines this statement.

Table 1 RCH/RP process: typical data.

	Range	Typical value
Reaction conditions		
Temperature [°C]	110–130	120
Pressure [bar]	40–60	50
CO/H_2 ratio	0.98–1.03	1.01
Propene conversion [%]	85–99	95
Propene purity [%]	85–99.9	95
Product composition [%]		
Isobutanal	4–8	4.5
n-Butanal	95–91	94.5
Isobutanol	< 0.1	< 0.1
n-Butanol	0.5	0.5
Butyl formates	traces	traces
Heavy ends	0.2–0.8	0.4
Selectivity to C_4 products	> 99	> 99.5
Selectivity to C_4 aldehydes	99	99
n/i ratio	93 : 7–97 : 3	95 : 5

The active catalyst species $HRh(CO)(TPPTS)_3$ is generated during the start-up of the reactor. Rhodium is introduced in any suitable form, e.g., as an acetate or as the salt of another organic acid. The resulting solution, with a rhodium concentration in the range of 200–350 ppm, is brought to reaction temperature under synthesis gas pressure, leading to the formation of the active yellow complex. No induction period is observed when the reaction is started immediately by adding propene and synthesis gas after the reaction temperature of about 120 °C has been reached, making start-up and close-down operations extremely easy. Some typical data for the hydroformylation of propene are summarized in Table 1 [8].

C_4 Process Description

It is an intrinsic characteristic – not only in the presence of aqueous catalyst solutions – that the rate of the oxo reaction in comparable conditions declines with increasing chain length of the olefin. This fact is attributed, inter alia, to the decreasing solubility of higher olefins in the aqueous catalyst solution, which correspondingly leads to low olefin concentrations and thus reduced reaction rates [20]. There are not too many options of ways to overcome the problem: an increase in temperature (with a negative impact on long-term ligand stability and n/i-ratio), an increase in rhodium concentration (cost factor), or addition of substances improving the solubility of the olefins (complicating the simple basic process). For the hydroformylation of n-butene a slight increase in rhodium concentration is sufficient to ensure appreciable space–time yields in industrially relevant conditions.

The cheapest source for n-1-butene is "raffinate II", a C_4 cut from which butadiene (by extraction) and isobutene (by conversion into methyl t-butyl ether) have been removed. The remaining mixture of C_4 hydrocarbons contains about 50–65% of *n*-1-butene, the remainder consisting of *cis/trans* *n*-2-butene and saturated butanes. A high concentration of *n*-1-butene in the raffinate is desirable, for obvious reasons. On the other hand the price of "raffinate II" is directly proportional to its content in *n*-1-butene. Therefore it is an unconditional requirement that the process be compatible with different concentrations of *n*-1-butene in the feedstock.

The most valuable product of C_4 hydroformylation is *n*-pentanal, whereas the isomers 2-methylbutanal and 3-methylbutanal are less in demand and lower in value. A catalyst with high selectivity should not catalyze the hydroformylation of 2-butene and convert 1-butene predominantly to *n*-pentanal. Both requirements are fulfilled by the Rh/TPPTS system [21]. However, despite the high regioselectivity in hydroformylation, a side reaction occurs which diminishes the overall selectivity. Part of the *n*-1-butene isomerizes under reaction conditions parallel to the hydroformylation reaction to *n*-2-butene, which is not hydroformylated in the presence of Rh/TPPTS under regular reaction conditions.

The reaction temperature to be chosen depends on several aspects. High temperatures favor the activity of the catalyst system and increase the partial pressure of *n*-1-butene (b.p. –6.1 °C) but have a negative impact on the long-term stability of the ligand. As a compromise, a reaction temperature 5–10 °C higher than in the hydroformylation of propene is acceptable. Also, with respect to the partial pressure of *n*-1-butene the overall pressure is lower: about 40 bar has been proven suitable. The stripping column, as the central unit in the process, deserves special attention: in order to remove dissolved butenes and butane completely from the oxo crude product a balance between temperature and pressure conditions has to be established.

The process design corresponds to the RCH/RP process for the hydroformylation of propene; for example, by slight adjustment of the conditions propene as well as *n*-1-butene may be processed in the same unit [21].

Deactivation Phenomena

All commercially applied rhodium/phosphine catalysts deactivate with time, for different reasons and in several ways. The most obvious is a decline in activity, as it directly reduces the unit capacity and frequently also leads to increased consumption of materials as unconverted olefins have to be vented. The reasons are found in the rhodium inventory: there are always some losses which decrease the rhodium concentration in the reaction medium, either by carry-over with the products during thermal separation from the catalyst (homogeneous systems), or by being swept out with the products from the two-phase system. Although these losses normally are in the ppb range with respect to the rhodium concentration in the products, they may well accumulate into substantial losses if the lifetime of the catalyst charge exceeds months or even years. In most kinetic expressions for the reaction rate the rhodium concentration is first-order or close to, thus directly influencing that rate. On the other hand, the formation of inactive rhodium species may leave the rhodium inventory virtually unchanged, although the catalyst loses activity [22]. Finally, the formation of modified phosphines has been proven to occur under reaction conditions, e.g. TPPTS can be converted to propyldisulfophenylphosphine (PDSPP), which acts as a stronger electron donor than TPPTS and thus occupies coordination sites on the rhodium [6].

The deactivation mechanism for TPPTS has been elucidated in some detail. The primary idea of *ortho*-metallation of the phenyl ring has been abandoned in the meantime as it definitely plays no role. Instead the deactivation is initiated by the oxidative insertion of the rhodium metal into the P–C bond of the triphenyl-phosphine ligand. An analogous mechanism for the TPPTS degradation has been outlined for the homogeneous system with TPP as ligand [23].

In continuous operation phosphorus-containing consecutive products are formed which also influence the activity of the rhodium center and thus contribute to the catalyst deactivation. One of the main degradation products from TPPTS is the sodium salt of *m*-formylbenzene sulfonic acid, which indicates the insertion of the rhodium atom into the P–C bond.

The aryl–rhodium species in which rhodium has replaced one phosphorus atom presumably exists as a phosphido-bridged dimer which is inactive. This compound may subsequently be converted to a series of consecutive products, e.g., alkyldiaryl-phosphines, which act as catalyst poisons.

Besides the degradation reactions, all phosphines are oxidized by traces of oxygen, which are always present in the olefins introduced. Synthesis gas, generated mainly by partial oxidation of hydrocarbons, may well contain small amounts of oxygen which are removed by special gas purification systems. Nevertheless, oxidation also plays a role in the losses of ligand in long-term operation. A certain concentration of "active" ligand, e.g. phosphorus in the oxidation state +III, is necessary to ensure stability of the rhodium and a sufficiently high n/i ratio; maintenance of this concentration is achieved best by adding small portions of ligand at short intervals over the whole run rather than adding greater quantities of fresh ligand now and then.

At the very beginning it was noticed that some sulfur dioxide together with sulfur-containing components was swept out with the butanals formed, causing partial poisoning of the catalysts in the subsequent hydrogenation steps. The source of the sulfur dioxide is clearly identified today: during sulfonation of TPP with oleum, i.e., in the presence of a surplus of sulfur trioxide, some of the TPP is oxidized to TPPO, yielding sulfur dioxide as a reduced product, which in part is preserved in the neutralized catalyst solution. The problem was resolved by treating the acidic aqueous solution of TPPTS with some inert gas (nitrogen, carbon dioxide), thus leading to complete removal of the sulfur dioxide [24].

The TPPTS catalyst system itself is not sensitive toward sulfur and most of the other common poisons for hydroformylation catalysts. One reason is the continuous withdrawal of organic and other byproducts with the product phase and the vent stream from the decanter (see Figure 1), avoiding the accumulation of poisons in the catalyst solution.

Outlook and Future Developments

The hydroformylation of propene and butene in the presence of an aqueous catalyst phase has proven successful since the early 1980s. In particular, the hydroformylation of propene has acquired merit from the very beginning in 1984 and has encountered virtually no problems, even in large-scale units. Comparison with other hydroformylation processes based on the conventional homogeneous principle has shown some distinct technical and economic advantages:

- phase separation is an elegant and efficient method to recover catalyst and oxo crude;
- space–time yields in biphasic and homogeneous processes are on a comparable level;
- the n/i ratio is very high (95 : 5) and can be shifted if desired;
- losses in rhodium and TPPTS are negligible;
- the absence of thermal strain reduces the formation of high-boiling byproducts;
- the technical equipment is extremely simple and reliable;
- the energy usage system is process-integrated (net energy exporter);
- selectivity is high overall with respect to propene input;
- there are no environmental emissions.

However, despite the advantages cited for the hydroformylation of propene and butene it has to be admitted that the biphasic system also nears its limits when olefins with increasing chain lengths are considered. Due to the decreasing solubility of the olefins in the aqueous catalyst phase, the reaction rate slows down, leading to an unacceptably low space–time yield (please see Figure 3 in Section 2.1).

Several proposals have been published to solve this problem, e.g., by using poly-ether-substituted triphenylphosphines [25–27] (see also Section 2.2.3.3). These types of phosphines show inverse temperature-dependent solubility in water that enables them to act as thermoregulated phase-transfer ligands [25] (see Section 2.3.5). So far, little is known about their applicability in technical operation. On the other hand, the homogeneous systems are also facing problems with long-chain olefins, but at a different process stage: whereas the hydroformylation still proceeds acceptably, the recovery of the oxo crude by distillation from the catalyst residue generates byproducts and destroys the catalyst or the ligand, respectively. These facts explain the survival of "ancient" cobalt hydroformylation processes for the hydroformylation of C_6–C_{10} olefins until a convenient solution has been found for one or the other variant. It may be noted that Celanese operates a ligand-modified homogeneous rhodium catalyst for hydroformylation of C_6 and C_8 olefins very successfully at its Bay City (TX, USA) plant.

The extension of the biphasic principle to higher olefins may be accomplished by changing the ligand [29], e.g., from TPPTS to bisphosphines, some of which have already proven to be valuable tools to increase the specific activity combined with high n/i selectivity [4]. Increasing the specific (or better: intrinsic) activity of rhodium in the aqueous two-phase system may be coupled with understanding of the relevant mass-transport phenomena. In this case the role of the phase boundaries as a potential barrier for the chemical reaction will have to be carefully analyzed (see Sections 2.3.1, 2.3.4, and 2.4.1.1.2).

The potential of the biphasic system has not been fully elucidated so far, and there is still a broad field for research activity with this simple but highly efficient technique.

References

1 E. Wiebus, B. Cornils, *CHEMTECH* **1995**, *25* (1), 33.
2 I. T. Horvath, R. V. Kastrup, A. A. Oswald, E. J. Mozeleski, *Catal. Lett.* **1989**, *2*, 85.
3 R. V. Kastrup, J. S. Merola, E. J. Mozeleski, R. J. Kastrup, R. V. Reisch, in *ACS Symp. Ser.* (Eds.: E. C. Alyea, D. W. Meek) **1982**, *196*, 34.
4 P. Purwanto, H. Delmas, *Catal. Today* **1995**, *24*, 135.
5 (a) W. A. Herrmann, C. W. Kohlpaintner, H. Bahrmann, W. Konkol, *J. Mol. Catal.* **1992**, 73, 191; (b) W. A. Herrmann, C. W. Kohlpaintner, R. B. Manetsberger, H. Bahrmann, H. Kottmann, *J. Mol. Catal. A:* **1995**, 97, 65; (c) H. Bahrmann, H. Bach, C. D. Frohning, H. J. Kleiner, P. Lappe, D. Peters, D. Regnat, W. A. Herrmann, *J. Mol. Catal. A:* **1997**, 116, 49.
6 W. A. Herrmann, C. W. Kohlpaintner, *Angew. Chem., Int. Ed. Engl.* **1993**, *32*, 1524.
7 W. A Herrmann, C. W. Kohlpaintner, *Inorg. Synth.* **1998**, *32*, 8.
8 W. A. Herrmann, G. P. Albanese, R. B. Manetsberger, P. Lappe, H. Bahrmann, *Angew. Chem., Int. Ed. Engl.* **1995**, *34*, 811.
9 (a) C. D. Frohning, C. W. Kohlpaintner, H. W. Bohnen, in *Applied Homogeneous Catalysis with Organometallic Compounds* (Eds.: B. Cornils, W. A. Herrmann), Wiley-VCH, Weinheim **2002**, Vol. 1, p. 31; (b) H. W. Bohnen, B. Cornils, *Adv. Catal.* **2002**, *47*, 1.
10 C. W. Kohlpaintner, in *Encyclopedia of Catalysis*, keyword: *Hydroformylation Industrial*, on-line edition **2002**.
11 B. Cornils, W. A. Herrmann, *J. Catal.* **2003**, *216*, 23.
12 B. Cornils, E. Wiebus, *Chem. Ing. Tech.* **1994**, *7*, 916.
13 B. Cornils, E. Wiebus, *Recl. Trav. Chim. Pays-Bas* **1996**, *115*, 211 ; (b) B. Cornils, E. Wiebus, *Hydrocarb. Process.* **1996**, *3*, 63.
14 G. Papadogianakis, R. A. Sheldon, *New J. Chem.* **1996**, *20*, 175.
15 Rhône-Poulenc Ind. (E. G. Kuntz et al.), FR 2.230.654 (**1983**), FR 2.314.910 (**1975**), FR 2.338.253 (**1976**), FR 2.349.562 (**1976**), FR 2.366.237 (**1976**), FR 2.473.504 (**1979**), FR 2.478.078 (**1980**), FR 2.550.202 (**1983**), FR 2.561.650 (**1984**).
16 E. G. Kuntz, *CHEMTECH* **1987**, *17*, 570.
17 B. Cornils, E. G. Kuntz, *J. Organomet. Chem.* **1995**, *502*, 177.
18 *ECN* **1995**, *Jan. 15–22*, 29.
19 *Europa Chemie* **1995**, *1*, 10.
20 O. Wachsen, K. Himmler, B. Cornils, *Catal. Today* **1998**, *42*, 373.
21 H. Bahrmann, C. D. Frohning, P. Heymanns, H. Kalbfell, P. Lappe, D. Peters, E. Wiebus, *J. Mol. Catal. A:* **1997**, *116*, 35.
22 E. B. Walczuk, P. C. J. Kamer, P. W. N. M. van Leeuwen, *Angew. Chem.* **2003**, *115*, 4813.
23 (a) J. A. Kulpe, *Dissertation* **1989**, Technische Universität München;
    (b) C. W. Kohlpaintner, *Dissertation* **1992**, Technische Universität München;
    (c) R. A. Dubois, P. E. Garrou, K. Lavin, H. R. Allock, *Organometallics* **1984**, *3*, 649;
    (d) R. M. Deshpande, S. S. Divekar, R. V. Gholap, R. V. Chaudhari, *J. Mol. Catal.* **1991**, *67*, 333.
24 Celanese Chemicals Europe GmbH (H. W. Bohnen, R. Fischer, W. Zgorzelski, K. Schalapski, E. Wiebus, W. Greb, J. Herwig), DE 100.07.341 (**2001**).

25 Z. Yin, Y. Yan, H. Zuo, B. Fell, *J. Prakt. Chem.* **1996**, *338*, 124 and *J. Mol. Catal.* **2001**, *171*, 85.

26 B. Fell, D. Leckel, C. Schobben, *Fat. Sci. Technol.* **1995**, *97*, 219.

27 P. Wentworth, A. M. Vandersteen, K. D. Janda, *J. Chem. Soc., Chem. Commun.* **1997**, 759.

28 E. Monflier, G. Fremy, Y. Castanet, A. Mortreux, *Angew. Chem.* **1995**, *107*, 2450.

29 Most recent reviews: (a) F. Jóo, À. Kathó, *J. Mol. Catal. A:* **1997**, *116*, 3; (b) B. Cornils, W. A. Herrmann, R. W. Eckl, *J. Mol. Catal. A:* **1997**, *116*, 27.

30 ECN **2003**, Nov. 17–23, 16.

b) Higher Alkenes

Helmut Bahrmann, Sandra Bogdanovic, and Piet W. N. M. van Leeuwen

Introduction

The Ruhrchemie/Rhône-Poulenc [1] process for the hydroformylation of short-chain alkenes such as propene and butene (cf. Section 2.4.1.1.3) combines a facile catalyst recycling with high selectivity and sufficiently high conversion rates to provide a commercially viable large-scale manufacturing process for butanal [2] and valeraldehyde [3]. Higher alkenes (> C_8) are not suitable for the RCH/RP process as run in Oberhausen.

Here, we summarize briefly investigations into the two-phase hydroformylation of higher alkenes with aqueous rhodium–triphenylphosphine bisulfonate (Rh–TPPTS) catalyst systems.

Two-Phase Hydroformylation of Higher Alkenes with Rh–TPPTS as Catalyst System

The Unmodified Ruhrchemie/Rhône-Poulenc Process

Few data are available in the academic literature on the Rh–TPPTS catalyst system. This section will provide some information on the effect of various reaction parameters (pressure, P/Rh ratio, rhodium concentration, alkene chain length, etc.) in the two-phase hydroformylation of higher alkenes with the aqueous catalyst system Rh–TPPTS. The alkene *n*-1-hexene was most thoroughly investigated.

Preparation and Effects of the Basic Catalyst HRh(CO)(TPPTS)$_3$ [4]

The conversion versus time diagram (Figure 1) illustrates the dependence of the reaction rate on the chain length of the alkene. The reaction proceeds according to first-order kinetics, i.e., the consumption rate of the substrate alkene is proportional to the concentration of the substrate.

Severe effects have also been determined for the reaction pressure, but especially for the P/Rh ratio: *n/iso* selectivities from 38 : 62 to 98 : 2 were observed. A review was recently published [5].

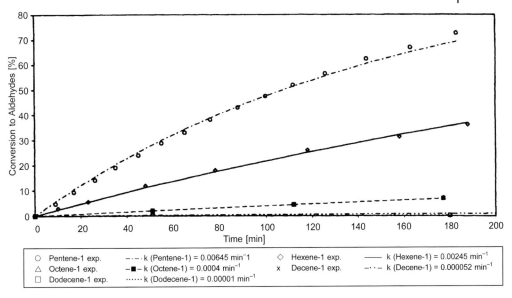

Figure legend:

- ○ Pentene-1 exp.   —·—· k (Pentene-1) = 0.00645 min⁻1   ◇ Hexene-1 exp.   —— k (Hexene-1) = 0.00245 min⁻¹
- △ Octene-1 exp.   —■— k (Octene-1) = 0.0004 min⁻¹   x Decene-1 exp.   —··— k (Decene-1) = 0.000052 min⁻¹
- ☐ Dodecene-1 exp.   ······ k (Dodecene-1) = 0.00001 min⁻¹

Figure 1 Dependence of reaction rate on chain length of alkenes.

Other Water-Soluble Phosphines

BINAS and BISBIS

Special chelating ligands, such as BINAS (sulfonated 2,2′-bis(diphenylphosphino-methyl)-1,1′-binaphthylene) [6, 7] and BISBIS (sulfonated 2,2′-bis(diphenyl-phosphinomethyl)-1,1′-biphenyl) [8], were found to be very useful ligands for the twophase hydroformylation of higher alkenes. Under standard conditions with BINAS a relatively high conversion of 36% is achieved (Table 1). In the presence of only 0.24 wt.% of tetradecyltrimethylammonium BINAS in the catalyst phase, the conversion rate rises to 77% (3 h) or 84% (6 h). In addition it should be mentioned that no decrease of the excellent *n/iso* ratio of 99 : 1 is observed with BINAS, as opposed to TPPTS.

A high selectivity toward linear aldehydes at a low P/Rh ratio could also be achieved with sulfonated fluorophosphine [tris(*p*-fluorophenyl)phosphine] [9].

Dinuclear rhodium(I) complexes with TPPTS, containing thiolato bridging ligands in aqueous phase, were found to be transformed into the monomer HRh(CO)(TPPTS)₃ under reaction conditions [10].

Recyclable catalysts based on Rh complexes and xanthene diphosphine ligands were used in the hydroformylation of 1-octene. Amphiphilic diphosphines spontaneously form aggregates that are remarkably stable at elevated temperatures and are specially suitable for the aqueous two-phase catalytic process. The observed TOF in the hydroformylation of 1-octene using ligands that form vesicles was up to 14 times higher compared to ligands that do not form aggregates [11]. Electron microscopy experiments showed that these ligands and their complexes form

Table 1 Comparison of TPPTS and BINAS.

	No additive		$C_{14}Me_3N$ action		
	BINAS[a]	TPPTS	BINAS[a]		TPPTS
	3 h	*3 h*	*3 h*	*6 h*	*3 h*
Additive [%]	–	–	0.24	–	0.86
Rate of conversion [%]	36	22	77	84	74
n/i rate[b]	99 : 1	98 : 2	99 : 1	99 : 1	91 : 9

P/Rh ratio 15 : 1.

a) Relative to the whole catalyst solution.
b) Average values from 3 to 30 single experiments (standard conditions; $T = 125$ °C,
   $p = 2.5$ MPa, Ru concentration = 300–400 ppm, P/Rh ratio = 80–100 : 1, reaction time = 3 h).

There are several methods for the preparation of the active hydroformylation catalyst $HRh(CO)(TPPTS)_3$. In analogy to its homogeneous counterpart , it is possible to synthesize the catalyst directly from $RhCl_3$ and TPPTS under syngas pressure.

vesicles in H_2O if the hydrophobic part of the ligand is large enough. The formed aggregates are stable at elevated temperatures (90 °C), and their presence leads to a significant enhancement of the solubility of 1-octene in aqueous solution. Furthermore, recycling experiments show that the TOF and the high selectivity toward the more valuable linear aldehyde remains the same in four consecutive runs. The aggregates stay intact during the recycling and the active Rh complex is retained in the H_2O-phase quantitatively [12].

Two novel phosphines, tris[*p*-(10-phenyldecyl)phenyl]phosphine and 2,2'-bis{di}-[*p*-(10-phenyldecyl)phenylphosphinomethyl]-1,1'-biphenyl, were successfully synthesized and sulfonated in H_2SO_4. The resulting water-soluble surface active phosphines were applied to the rhodium-catalyzed hydroformylation of higher alkenes. It is found that these two ligands are not only excellent for octene hydroformylation, but catalyze tetradecene hydroformylation under biphasic conditions as well. Rates and selectivities are superior to TP PTS-modified rhodium catalysts under the same reaction conditions [13].

Nonionic tensioactive water-soluble phosphines that act as ligands for rhodium-catalyzed hydroformylation of higher alkenes under biphasic conditions have been tested, too. Phosphines discussed are $P[p\text{-}C_6H_4(OCH_2CH_2)_nOH]_3$ ($n = 18$, 25) and $Ph_2P[p\text{-}C_6H_4(OCH_2CH_2)_nOH]$ ($n = 16$, 25). The rhodium catalyst combined with these ligands gave an average turnover frequency of 182 h^{-1} for 1-hexene. More importantly, recovery and re-use of catalyst is possible because of the inverse temperature-dependent water solubility of the phosphines [14].

Phosphonated Ligands

New and efficient routes to modify phosphines with phosphonic acid groups have been developed. Phosphonate-phosphines showed high solubilities in water and were used to immobilize rhodium catalysts in the aqueous phase of biphasic systems. In the two-phase hydroformylation of propene, some of the novel catalysts showed activities and regioselectivities similar to those of Rh/TPPTS. Amphiphilic Rh/phosphonate-phosphine catalysts were found widely superior to Rh/TPPTS in the hydroformylation of 1-octene [15].

The Combination of Triphenylphosphine Monosulfonate (TPPMS) and Polar Solvents

Several processes for the hydroformylation of higher alkenes have been suggested on the basis of the water-soluble ligand TPPMS. In contrast to TPPTS, which is almost exclusively soluble in water, TPPMS can be used in both aqueous and polar organic media.

Abatjoglou et al. from the Union Carbide Corporation (UCC) presented a homogeneous process for the hydroformylation of higher alkenes, combined with an aqueous two-phase catalyst recovery [16]. The key discovery is that alkali-metal salts of monosulfonated triphenylphosphine form reverse micelles in organic media in the presence of certain solubilizing agents which are stable under the reaction conditions. These systems can easily be induced to separate into a nonpolar product phase and a polar catalyst phase, thereby providing the catalyst recovery typical of two-phase reactions. The separation of the micelles can be accomplished by either raising the temperature or cooling the reaction mixture. In the case of N-methylpyrrolidine (NMP)-solubilized systems the addition of water brings about a sharp separation into an organic phase and a catalyst phase. The product phase, however, has to be extracted with water to eliminate traces of catalyst components completely.

Fell et al. published a simplified version of the above-mentioned process design of UCC in which the hydroformylation reaction of 1-tetradecene is performed homogeneously with a Rh/Li–TPPMS catalyst system in the presence of methanol. When the methanol is distilled off after almost complete conversion, the catalyst complex precipitates and can be separated by filtration or extraction with water [17].

The idea of using monosulfonated or monocarboxylated triphenylphosphines in a biphasic reaction medium in the presence of amphiphilic reagents has already been patented, in 1981, by the Johnson–Matthey Corporation. However, the recycling of the catalyst complex published in this patent was not complete so that no technical process could be established in those early days of two-phase hydroformylation [18].

Outlook

What are the major challenges for the two-phase hydroformylation of higher alkenes?

The most important applications of higher oxo products are plasticizer alcohols in the C_8–C_{11} range and synthetic detergent alcohols in the C_{12}–C_{18} range, with a worldwide consumption of 1.5 million tons [19] and 1.2 million tons in 1995, respectively. Compared with cobalt, rhodium as catalyst metal is favorable with respect to the raw material economy and the energy balance in the hydroformylation of higher alkenes. A biphasic hydroformylation process would bear the advantage that the long-chain aldehydes can be separated from the catalyst simply by phase separation. For alkenes above C_{10} the crude aldehyde cannot be separated from the unreacted alkene by distillation. Therefore, such a process would be required to achieve complete conversion in continuous operation.

Since the major raw materials for higher plasticizer alcohols are internal alkenes from polygas units (e.g., diisobutene, tripropenes), this market requires the development of even more efficient biphasic catalyst systems for internal and branched alkenes.

Scientifically, another major challenge is the development of a biphasic hydro-formylation process for internal alkenes combining isomerization and hydroformy-lation of linear internal alkenes and affording predominantly terminal hydro-formylation products. Such a technology would be of primary interest for the fine chemical and the detergent alcohol markets.

References

1 (a) W. H. HERRMANN, C. W. KOHLPAINTNER, *Angew. Chem.* **1993**, *105*, 1588; *Angew. Chem., Int. Ed. Engl.* **1993**, *32*, 1524; (b) E. WIEBUS, B. CORNILS, *Chem. Ing. Tech.* **1994**, *66*, 916.

2 B. CORNILS, E. WIEBUS, *CHEMTECH* **1995**, 33.

3 H. BAHRMANN, C. D. FROHNING, P. HEYMANNS, H. KALBFELL, P. LAPPE, D. PETERS, E. WIEBUS, *J. Mol. Catal. A:* **1997**, *116*, 35.

4 W. A. HERRMANN, J. A. KULPE, W. KONKOL, H. BAHRMANN, *J. Organomet. Chem.* **1990**, *389*, 85.

5 H. BAHRMANN, S. BOGDANOVIC, P. W. N. M. VAN LEEUWEN, in *Aqueous Phase Organometallic Catalysis* (Eds.: B. CORNILS, W. A. HERRMANN), 2nd ed., p. 391, Wiley-VCH, Weinheim **2004**.

6 (a) Hoechst AG (W. A. HERRMANN, R. MANETSBERGER, H. BAHRMANN, C. W. KOHLPAINTNER, P. LAPPE), EP 571.819 (1992); (b) W. A. HERRMANN, C. W. KOHLPAINTNER, R. B. MANETSBERGER, H. BAHRMANN, H. KOTTMANN, *J. Mol. Catal.* **1995**, *97*, 65; (c) H. BAHRMANN, K. BERGRATH, H.-J. KLEINER, P. LAPPE, C. NAUMANN, D. PETERS, D. REGNAT, *J. Organomet. Chem.* **1996**, *520*, 97.

7 (a) H. BAHRMANN, H. BACH, C. D. FROHNING, H.-J. KLEINER, P. LAPPE, D. PETERS, D. REGNAT, W. A. HERRMANN, *J. Mol. Catal.* **1997**, *116*, 49.

8 W. A. HERRMANN, C. W. KOHLPAINTNER, H. BAHRMANN, W. KONKOL, *J. Organomet. Chem.* **1992**, 196.

9 B. FELL, G. H. PAPADOGIANAKIS, *J. Prakt. Chem.* **1994**, *336*, 591; (b) Hoechst AG (G. H. PAPADOGIANAKIS, B. FELL, H. BAHRMANN), EP 489.330 (1990).

10 F. MONTEIL, L. MIQUEL, R. QUEAU, P. KALCK, in *Aqueous Organometallic Chemistry and Catalysis* (Eds.: I. T. HORVÁTH, F. JOÓ), NATO ASI Series, Kluwer, Dorecht **1995**, p. 131.

11 J. N. H. RECK, A. J. SANDEC, M. SCHREUDER-GOEDHEJT, P. C. J. KAMER, P. W. N. M. VAN LEEUWEN, *Erdöl, Erdgas, Kohle* **2001**, *117*, 134.

12 M. SCHREUDER-GOEDHEJT, B. E. HANSON, J. N. H. RECK, P. C. J. KAMER, P. W. N. M. VAN LEEUWEN, *J. Amer. Chem. Soc.* **2000**, *122*, 1650.

13 B. E. HANSON, H. DING, C. W. KOHLPAINTNER, *Catal. Today* **1998**, *42*, 421.

14 Y. WANG, J. JIANG, Q. MIAO, X. WU, Z. JIN, *Catal. Today* **2002**, *74*, 85.

15 (a) S. BISCHOFF, M. KANT, *Catal. Today* **2001**, *66*, 183; (b) S. BISCHOFF, M. KANT, *Ind. Eng. Chem. Res.* **2000**, *39*, 4908.

16 (a) *Chem. Eng. News* **1995**, *73*, 25: Review on the 209th ACS National Meeting in Anaheim, **1995**; (b) Union Carbide Corporation (Eds.: A. G. ABATJOGLOU, D. R. BRYANT), US 4.731.486 (1988).

17 Z. XIA, B. FELL, *J. Prakt. Chem.* **1997**, *339*, 140.

18 Johnson–Matthey Public Ltd. Co. (M. J. H. RUSSEL, B. A. MURRER), US 4.399.312 (1983).

19 This figure does not include 2-ethylhexanol. *n*-Butyraldehyde-derived 2-ethylhexanol is by far the most important plasticizer alcohol, with an estimated worldwide consumption of 2.4 million t.

c) Functionalized Alkenes

Eric Monflier

Functionalized olefins can be classified in two groups: the δ-functionalized olefins in which the functional group is not directly branched on the double bond but on an alkyl chain of the olefin as in the case of oct-7-en-1-al or linoleic alcohol, and the α-functionalized olefins in which the functional group is directly branched on the double bond as in the case of methyl acrylate or phenyl vinyl ether. The results described for these two groups will be discussed separately. Hydroformylation of water-soluble olefins in two-phase system with water-insoluble catalysts is far beyond the scope of this chapter and will not be discussed here [1, 2].

Biphasic Hydroformylation of *d*-Functionalized Olefins

Hydroformylation of ω-alkene carboxylic acid methyl esters catalyzed by a Rh/TPPTS system was initially investigated by Fell et al. [Eq. (1)]. As expected in a biphasic medium, low molecular ω-alkene carboxylic acid methyl esters such as methyl 4-pentenoate can be hydroformylated efficiently without any additives whereas methyl esters of higher ω-alkene carboxylic acids such as methyl 13-tetradecenoate require the presence of mass-transfer promoters such as surfactants [3] or chemically modified β-cyclodextrins [4].

P(CO/H$_2$): 10 MPa ; 120 °C

2-10 hours ; Phosphine / Rh : 60

[Rh / P(C$_6$H$_4$SO$_3$Na)$_3$]

n = 0, 1, 5, 6, 9

Conversion: up to 100 %
Aldehydes selectivity: up to 100 %

(1)

Unsaturated fat chemicals can also be hydroformylated in the presence of surfactants. For instance, the Johnson-Matthey Co. has reported that oleic acid methyl ester or linoleic acid methyl ester can be hydroformylated in micellar media using a water-soluble rhodium complex of monocarboxylated triphenylphosphine as catalyst [5]. Interestingly, linolenic acid methyl ester can be hydroformylated to the triformyl derivative with a selectivity of 55% with a Rh/TPPTS catalytic system in the presence of hexadecyltrimethylammonium bromide [Eq. (2)] [6].

P(CO/H$_2$): 10 MPa ; 120 °C

6 hours ; Phosphine / Rh : 20

[Rh / P(C$_6$H$_4$SO$_3$Na)$_3$]

$H_3C\text{-}(CH_2CH=CH)_3(CH_2)_7CO_2CH_3$

$H_3C\text{-}(CH_2CH \ CH)_3(CH_2)_7CO_2CH_3$

Conversion: up to 100 %
Triformyl selectivity: 55 %

(2)

Hydroformylation of oleyl alcohol into formylstearyl alcohol has been successfully achieved with a 96.6% yield by using a Rh/TPPTS complex dissolved in an aqueous film supported on a high surface area silica gel [7]. This supported catalyst also made it possible to hydroformylate allyl 9-decenyl ether and 3-methyl-2-(2-pentenyl)-2-cyclopenten-1-one (*cis*-jasmone). However, with the latter substrate, the aldehyde yields did not exceed 38% [8].

The DSM Co. has described an attractive approach for the synthesis of adipic acid or 6-aminocaproic acid precursors by claiming the hydroformylation of 3-pentenoic acid into 5-formylvaleric acid in biphasic medium [Eq. (3)].

$$\text{H}_3\text{C}-\text{CH}=\text{CH}-\text{CH}_2\text{CO}_2\text{H} \xrightarrow[\begin{array}{c}\text{Pt}\,/\,\boxed{\quad\text{PAr}_2\quad}\\ \text{Ar: C}_6\text{H}_4\text{SO}_3\text{Na}\end{array}]{P(\text{CO/H}_2): 5\text{ MPa}\,;\,100\ °\text{C}\,;\,4\text{ hours}} \quad \begin{array}{c}\text{O}\diagdown\text{C}\diagup\text{H}\\ \vdots\\ \text{H}_2\text{C}=\text{C}=\text{C}-\text{CH}_2\text{CO}_2\text{H}\\ \quad\ \text{H}\quad\text{H}\end{array} \quad (3)$$

Conversion: 79 %
Formylvaleric acid selectivity: 80.3 %

With a water-soluble platinum complex of tetrasulfonated *trans*-1,2-bis(diphenyl-phosphinomethylene)cyclobutane as catalyst, the 5-formylvaleric acid selectivity reached 62% [9]. The same catalytic system allows also the hydroformylation of *trans*-3-pentenenitrile with 91.4% selectivity.

Rhodium/poly(enolate-*co*-vinyl alcohol-*co*-vinyl acetate) catalysts have been developed for the selective biphasic hydroformylation of functionalized olefins [10]. Although the conversions were low (< 25%), excellent selectivities for the hydro-formylation of methyl 3,3-dimethylpenten-4-onate can be achieved with such water-soluble polymer-anchored rhodium catalysts (see Chapter 7). Indeed, only the linear aldehyde was obtained during the hydroformylation of methyl 3,3-dimethylpenten-4-onate.

To our knowledge, the only industrial application of the water-soluble catalyst for the hydroformylation of δ-functionalized olefins has been developed by Kuraray [11]. In this process, oct-7-en-1-al is hydroformylated into 1,9-nonanedial by using a rhodium catalyst and the monosulfonated triphenylphosphine (cf. also Section 2.4.4.2).

Hydroformylation of various unsarurated alcohols with different catalytic systems has been investigated by two research groups. Ziolkowski et al. have reported the hydroformylation of 1-buten-3-ol, 2-methyl-2-propen-1-ol and 2-buten-1-ol with a catalytic system containing Rh(acac)(CO)$_2$ and the water-soluble phosphine Ph$_2$PCH$_2$CH$_2$CONHC(CH$_3$)$_2$CH$_2$SO$_3$Li (PNS). In all cases, the main products of the hydroformylation were 2-hydroxytetrahydrofuran derivatives formed via a hydroxyaldehyde cyclization [Eq. (4)] [12].

$$\xrightarrow[\text{[Rh / PNS]}]{\begin{array}{c}P(\text{CO/H}_2): 1\text{ MPa}\,;\,80\ °\text{C}\\ 2\text{ hours ;Phosphine / Rh : 3}\end{array}} \quad (4)$$

Conversion: 93 %
Yield: 91 %

As described by Paganelli et al., hydroformylation of 1,1-diarylallyl alcohols such as 1,1-bis(p-fluorophenyl)-2-propenol can be carried out conveniently in a biphasic system using a water-soluble complex formed by a mixture of [Rh(COD)Cl]₂ and TPPTS [13, 14] or sulfonated Xantphos [14]. Interestingly, the retroaldolization reaction leading to byproducts seems to take place with more difficulty in a biphasic medium.

A beneficial effect of water on the reaction rate and aldehyde selectivity was observed during the Rh/TPPTS-catalyzed hydroformylation of the water-soluble N-allylacetamide. Thus, the reaction rates and aldehyde selectivities in water are much higher than those observed in organic solvents with a Rh/PPh₃ catalyst. Unfortunately, the linear/branched aldehydes ratios were rather low and the water-soluble catalyst cannot be recovered [2].

Surprisingly, 2,5-dimethoxy-2,5-dihydrofuran did not form the expected aldehydes under CO/H₂ pressure (3 MPa) in the presence of the Rh/TPPTS system. Instead of hydroformylation, hydrogenation was the main reaction path in water, where 2,5-dimethoxytetrahydrofuran and its hydrolysis product were obtained. However, hydroformylation occurred when the surface-active phosphine $Ph_2P(CH_2)_{10}PO_3Na_2$ was used [15].

Biphasic Hydroformylation of α-Functionalized Olefins

The first work on α-functionalized olefins was focused on the hydroformylation of acrylic esters [Eq. (5)] [16–19].

P(CO/H₂): 5 MPa ; 50 °C
2 hours ; Phosphine / Rh : 10
[Rh / P(C₆H₄SO₃Na)₃]

R: methyl, ethyl,
2-ethoxyethyl
butyl 2 ethylhexyl

Conversion: up to 100 %
Aldehydes selectivity: up to 99 %
iso / n aldehydes ratio: 130

(5)

As in the case of N-allylacetamide, immobilization of the catalyst in the aqueous phase results in an enhancement of the catalytic activity [17]. Indeed, hydroformylation rates of acrylic esters that are soluble in water were much higher in a biphasic system than those observed under comparable homogeneous conditions. Except for 2-ethylhexyl acrylate, the initial rate was increased by a factor of 2.4, 12, 2.8, and 14 for methyl, ethyl, butyl, and 2-ethoxyethyl acrylate, respectively [18]. The peculiar enhancement of the catalytic activity for the hydroformylation of water-soluble acrylates was attributed to the formation of hydrogen bonding between water and the carboxyl group of the acrylate [19]. The decrease in the activity with 2-ethylhexyl acrylate was due to mass-transfer limitation. As a matter of fact, the hydroformylation of this substrate can be achieved by using an aqueous phase supported rhodium catalyst [18] or chemically modified cyclodextrins [20]. More surprising is the unprecedented observation that immobilizing of the Rh/TPPTS catalyst on a wet silica gel yields an extremely active supported aqueous-phase catalyst for the hydroformylation of a series of acrylic esters [20]. Such an increase in the reaction rate was not observed when the Rh/TPPTS complex was immobilized on poly-electrolyte-coated latex particles [21].

Rhodium-catalyzed hydroformylation of aryl vinyl ethers can be achieved efficiently in the presence of TPPTS [13], water-soluble polymers [10, 22], or human serum albumin as ligand [23]. The conversions were comparable with those reached in a homogeneous medium but the chemoselectivities can be lower. Indeed, at 100 °C, aryl vinyl ethers can be cleaved into the corresponding phenols (up to 11%).

Finally, water-soluble phosphorylated BINAPs were tested as ligands in aqueous biphasic rhodium-catalyzed hydroformylation of vinyl acetate. Compared with catalysts prepared with the parent ligand in a homogeneous medium, the chemo-, regio- and enantioselectivities were markedly lower [24].

Conclusion

Although the scope of the aqueous biphasic hydroformylation of functionalized olefins still needs to be deeply investigated, these few studies demonstrate clearly that functionalized olefins can be hydroformylated efficiently in an aqueous biphasic medium. However, it should be kept in mind that water is not only an inert mobile phase. Water can also act as a reactant or a coordinating solvent that modifies catalytic species. So, in some cases, unexpected increases or decreases in the activity or selectivity can be observed.

References

1 See, for example: (a) M. McCarthy, H. Stemmer, W. Leitner, *Green Chem.* **2002**, *4*, 501; (b) R. M. Deshpande, S. S. Divekar, B. M. Bhanage, R. V. Chaudhari, *J. Mol. Catal.* **1992**, *75*, L19.
2 (a) G. Verspui, G. Elbertse, G. Papadogianakis, R. A. Sheldon, *J. Organomet. Chem.* **2001**, *621*, 337; (b) G. Verspui, G. Elbertse, F. A. Sheldon, M. A. Hacking, R. A. Sheldon, *Chem. Commun.* **2000**, 1363.
3 B. Fell, G. Papadogianakis, C. Schobben, *J. Mol. Catal.* **1995**, *101*, 179.
4 Centre National de la Recherche Scientifique (E. Monflier, Y. Castanet, A. Mortreux), WO 96/22267 (**1996**).
5 Johnson-Matthey Co. (M. J. H. Russel, B. A. Murrer), US 4.399.312 (**1983**).
6 B. Fell, D. Leckel, C. Schobben, *Fat. Sci. Technol.* **1995**, *97*, 219.
7 J. P. Arhancet, M. E. Davies, J. S. Merola, B. E. Hanson, *J. Catal.* **1990**, *121*, 327.
8 Virginia Tech. Intellectual Properties (J. P. Arhancet, M. E. Davies, B. E. Hanson), US 4.947.003 (**1990**).
9 DSM Co. (O. J. Gelling, I. Toth), WO 95/18783 (**1995**).
10 J. Chen, H. Alper, *J. Am. Chem. Soc.* **1997**, *119*, 893.
11 M. Matsumoto, N. Yoshimura, M. Tamura, Kuraray Co., US4.510.332 (**1985**).
12 E. Mieczynska, A. M. Trzeciak, J. J. Ziolkowski, *J. Mol. Catal. A:* **1999**, *148*, 59.
13 S. Paganelli, M. Zanchet, M. Marchetti, G. Mangano, *J. Mol. Catal. A:* **2000**, *157*, 1.
14 (a) C. Botteghi, M. Marchetti, S. Paganelli, F. Persil-Paoli, *Tetrahedron* **2001**, *57*, 1631; (b) C. Botteghi, T. Corrias, M. Marchetti, S. Paganelli, O. Piccolo, *Org. Process Rev. Dev.* **2002**, *6*, 379.
15 S. Bischoff, M. Kant, *Catal. Today* **2000**, *58*, 241.

16 G. Frémy, E. Monflier, R. Grzybek, J. J. Ziolkowski, A. M. Trzeciak, J. F. Carpentier, Y. Castanet, A. Mortreux, *J. Organomet. Chem.* **1995**, *505*, 11.

17 G. Frémy, E. Monflier, J. F. Carpentier, Y. Castanet, A. Mortreux, *Angew. Chem., Int. Ed. Engl.* **1995**, *34*, 1474.

18 G. Frémy, E. Monflier, J. F. Carpentier, Y. Castanet, A. Mortreux, *J. Catal.* **1996**, *162*, 339.

19 G. Frémy, E. Monflier, J. F. Carpentier, Y. Castanet, A. Mortreux, *J. Mol. Catal. A:* **1998**, *129*, 35.

20 (a) E. Monflier, G. Frémy, Y. Castanet, A. Mortreux, *Angew. Chem., Int. Ed. Engl.* **1995**, *34*, 2269; (b) E. Monflier, G. Frémy, S. Tilloy, Y. Castanet, A. Mortreux, *Tetrahedron Lett.* **1995**, *36*, 9481.

21 S. Mecking, R. Thomann, *Adv. Mater.* **2000**, *12*, 953.

22 A. Nait, H. Alper, *J. Am. Chem. Soc.* **1998**, *120*, 1466.

23 (a) M. Marchetti, G. Mangano, S. Paganelli, C. Botteghi, *Tetrahedron Lett.* **2000**, *41*, 3717; (b) C. Bertucci, C. Botteghi, D. Giunta, M. Marchetti, S. Paganelli, *Adv. Synth. Catal.* **2002**, *344*, 556.

24 A. Köckritz, S. Bischoff, M. Kant, R. Siefken, *J. Mol. Catal. A:* **2001**, *174*, 119.

2.4.1.2
Carbonylations

Matthias Beller and Jürgen G. E. Krauter

Apart from hydroformylations, the potential advantages of two-phase catalysis for other carbonylation reactions have not been thoroughly evaluated. Only a few examples of carbonylation reactions under biphasic conditions have been described.

Metal-catalyzed *reductive* carbonylation of nitroaromatics using CO has been the subject of intensive investigation in recent years because of the commercial importance of amines, urethanes, and isocyanates [1]. Biphasic operation could offer interesting horizons regarding the ease of catalyst recycling. Thus, palladium catalysts have been applied in the presence of water-soluble ligands such as TPPTS or BINAS [2] for the carbonylation of substituted nitroaromatics (Scheme 1).

Interestingly, the nitro group can be selectively reduced to an amino group at TONs of some thousands, even in the presence of halide substituents or a vinyl group [3] although recycling of the catalyst was not possible due to decompsition. The reductive carbonylation of 5-hydroxymethylfurfural under aqueous biphasic conditions yields the 5-methyl derivative instead of dicarbonylation [4].

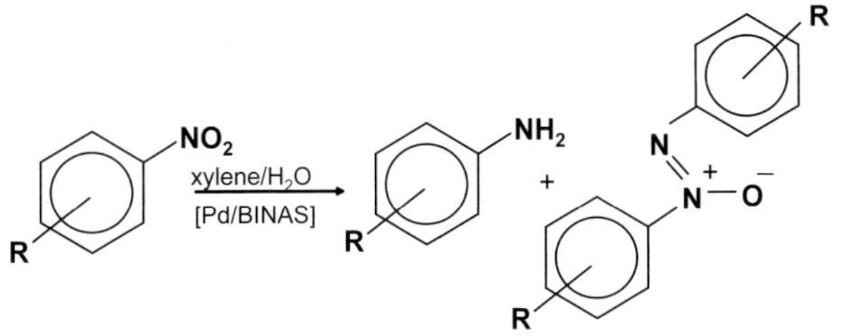

Scheme 1 Biphasic reaction of nitroaromatics.

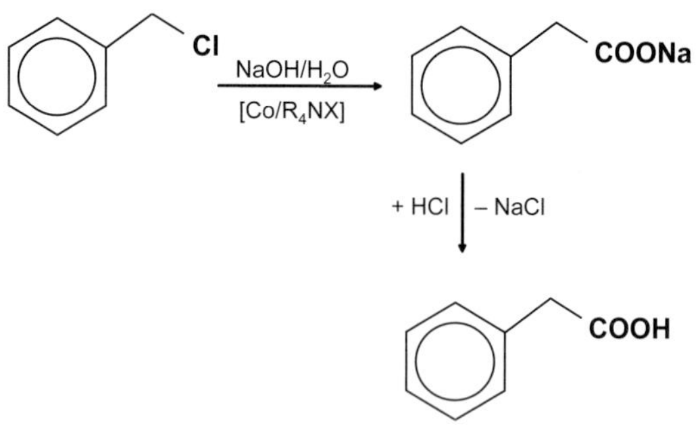

Scheme 2 Hydrocarbonylation of benzyl chloride.

Mechanistically closely related are hydroxycarbonylations (or carboxylations) of alkyl, ary, benzyl, or allyl halides [5]. Thus, aryl iodides have been carbonylated with various Pd salts in water – lacking phosphine ligands. With TONs of 100 000 the results are remarkable although the reaction is not truly biphasic. When ligands such as TPPTS are used the activity of the Pd catalyst was decreased considerably. The hydrocarbonylation of chloroarenes to the corresponding carboxylic acids, especially the commercially interesting phenylacetic acids, proceeds via the Na salts with $Co_2(CO)_8$ and a benzyltrialkylammonium surfactant in a biphasic medium employing diphenyl ether and aqueous 40% NaOH as a solvent (Scheme 2).

Another proposal from the industry used a Pd catalyst dissolved in aqueous TPPTS, yieding turnover frequencies (TOFs) of 135 h^{-1} [6]. Apart from aryl–X derivatives and, more interestingly from an industrial point of view, metal-catalyzed carbonylation of substituted benzyl halides to give the corresponding phenylacetic acids has been investigated [7]. Two-phase systems are applied with the catalyst and substrate being dissolved in the organic phase and the product formed is dissolved in an excess of alkaline aqueous solution. Despite significant disadvantages such as indispensable addition of phase-transfer agents and additional salt as a byproduct, the carbonylation of benzyl chloride to give phenaylacetc acid for use in perfume constitutents and pesticides has been reported to be practiced on a commercial scale by Montedison [8]. Other metal catalysts besides Co carbonyl which have been utilized for biphasic carbonylation of benzylic halides to carboxylic acids under phase-transfer conditions include Pd(0)– and Ni–carbonyl complexes [9]. It is assumed that catalysis takes place in the organic phase in all these reactions. However, by the use of the water-soluble ligand TPPMS (see Section 2.2.3.2) a Pd catalyst which is active in the water phase is formed [10]. Nevertheless, the addition of surfactants is effective in accelerating this reaction. This effect is attributed not to a simple surface activation but to a counterphase-transfer catalysis.

On the bases of the Pd-catalyzed carbonylation of benzylic halides Sheldon et al. investigated the functionalization of 5-hydroxymethylfurfural to 5-formylfuran 2-acetic acid in an aqueous medium in the presence of a water-soluble Pd/TPPTS catalyst (Scheme 3) [4]. They found that the hydroxy group displays similar reactivity under acidic conditions as the benzylic halides.

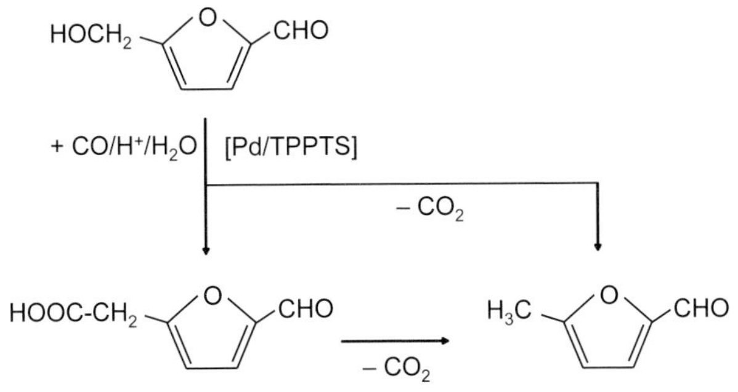

Scheme 3 Carbonylation of 5-hydroxymethylfurfural.

In general, for carbonylations, Pd is preferable to Ni as the catalyst metal with respect of catalyst efficiency. Thus Okano et al. [10] described some other efficient Pd-catalyzed carbonylations of allyl chloride and substituted allylic halides (e.g., Scheme 4).

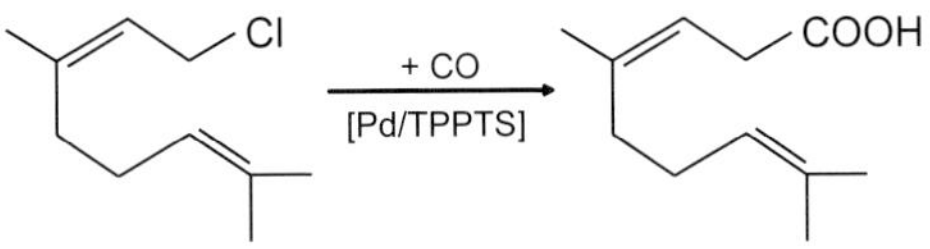

Scheme 4 Carbonylation of allylic chlorides.

In greater detail, the water-soluble Pd complex [PdCl$_2$TPPMS] has been used in a two-phase system consisting of aqueous NaOH/benzene. Clearly, the isomerization depends on the concentration of the base and was therefore suppressed by a method of continuous addition to the aqueous medium.

Biphasic hydrocarboxylations of alkenes yield carboxylic acids with a typical linear to branched (*n/i*) ratio which ranges from 1 to 1.4 (Scheme 5).

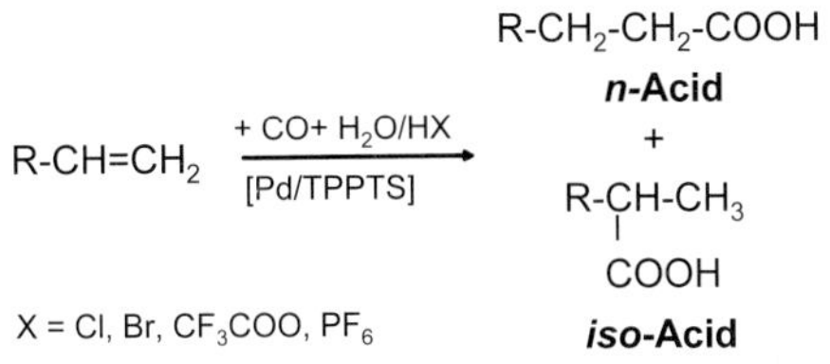

Scheme 5 Hydrocarboxylations of alkenes.

A Brønsted acid HX may be used as a co-catalyst. Long-chain alkenes give only insufficient conversion due to low solubility and isomerization side reactions. In order to overcome these problems the addition of co-solvents (such as β-cyclodextrins; cf. Section 2.2.3.3) was recommended. Their advantageous effect was rationalized by a host–guest complex of the cyclic carbohydrate and the alkene feed which prevents isomerization of the double bond [5].

The biphasic carbonylation of isopropylallylamines leads to *N*-isopropylbutyro-lactam, whereas with syngas the competitive hydroformylation yields the *N*-iso-propylpyrrolidine (Scheme 6) [11]. In this case, water under the conditions of the water-gas shift reaction serves as a hydrogen source.

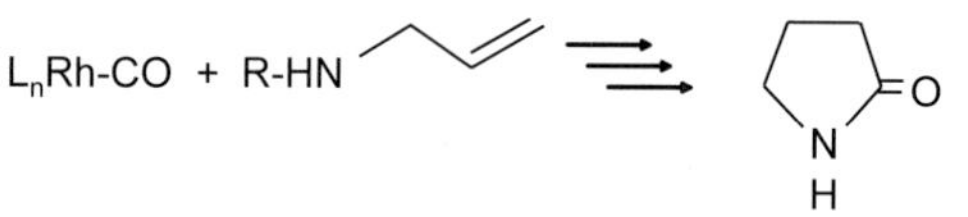

Scheme 6 Carbonylation of isopropylallyamine.

So far, nearly all the reported reactions in two-phase systems suffer similarly to their homogeneous counterparts from low catalyst efficiency. Thus it was predicted for the future that careful design of water-soluble catalyst sytems will make it possible to overcome these problems, and indeed some recent protocols describe the formation of phenylacetic acid or ibuprofen on a semi-technological scale [5, 12].

References

1 M. Beller, J. G. E. Krauter, in B. Cornils, W. A. Herrmann (Eds.), *Aqueous-Phase Organometallic Catalysis*, 2nd ed., Wiley-VCH, Weinheim **2004**, p. 501.

2 B. Cornils, W. A. Herrmann, R. Schlögl, C.-H. Wong, *Catalysis from A to Z*, 2nd ed., Wiley-VCH, Weinheim **2003**.

3 A. M. Tafesh, M. Beller, *Tetrahedron Lett.* **1995**, *36*, 9305; F. Ragaini, S. Cenini, *J. Mol. Catal. A:* **1996**, *105*, 145.

4 G. Papadogianakis, L. Maat, R. A. Sheldon, *J. Chem. Soc.,Chem. Commun.* **1994**, 2659 and *J. Mol. Catal. A:* **1997**, *116*, 179.

5 F. Bertoux, E. Monflier, Y. Castanet, A. Mortreux, *J. Mol. Catal. A:* **1999**, *143*, 11.

6 C. W. Kohlpaintner, M. Beller, *J. Mol. Catal. A:* **1997**, *116*, 259.

7 M. Beller, B. Cornils, C. D. Frohning, C. W. Kohlpaintner, *J. Mol. Catal. A:* **1995**, *104*, 17; P. Iayasree, S. P. Gupte, R. V. Chaudhari, *Stud. Surf. Sci. Catal.* **1998**, *113*, 875.

8 L. Cassar, *Chim. Ind.* **1985**, *67*, 256.

9 L. Cassar, M. Foa, *J. Organomet. Chem.* **1977**, *134*, C15; H. Alper, H. des Abbayes, *J. Organomet. Chem.* **1977**, *134*, C11; H. Alper, K. Hashem, J. Heveling, *Organometallics* **1982**, *1*, 775; I. Amer, H. Alper, *J. Am. Chem. Soc.* **1989**, *111*, 927.

10 T. Okano, I. Uchida, T. Nakagaki, H. Konishi, *J. Mol. Catal.* **1989**, *54*, 2589; T. Okano, T. Hayashi, J. Kiji, *Bull. Chem. Soc. Jpn.* **1994**, *67*, 2339.

11 R. G. de Rosa, J. D. Ribeiro de Campos, R. Buffon, *J. Mol. Catal. A:* **2000**, *153*, 19.

12 (a) F. Bertoux, S. Tilloy, E. Monflier, Y. Castanet, A. Motreux, *J. Mol. Catal. A:* **1999**, *138*, 53; (b) G. Verspui, G. Papadogianakis, R. A. Sheldon, *Catal. Today* **1998**, *42*, 449; (c) F.-W. Li, L.-W. Xu, C.-G. Xia, *Appl. Catal. A:* **2003**, *253*, 509.

2.4.2
Hydrogenation and Hydrogenolysis

2.4.2.1
Hydrogenation

Ferenc Joó and Ágnes Kathó

2.4.2.1.1
Introduction

Hydrogenation has been *the* prototype reaction of homogeneous catalysis by transition metal complexes [1]. The number of papers and patents on homogeneous hydrogenation is enormous and this reaction is treated in detail in numerous reviews and monographs [2–4].

Being a highly polar solvent, water is *not* a good medium in which to dissolve molecular hydrogen and the usual substrates of catalytic hydrogenations, mostly apolar organics. Conversely, its immiscibility with many of the common organic solvents makes possible the realization of hydrogenation processes in biphasic solvent systems. A specific value of water as solvent lies in the fact that there are catalysts (such as $K_3[Co(CN)_5]$) and substrates (e.g., carbohydrates) which do not dissolve in common nonpolar organic solvents.

Recent research concerning the application of supercritical (sc) fluids and ionic liquids (IL) as solvents in homogeneous catalysis, [5, 6] opened the way to the development of biphasic water/$scCO_2$ [7] and water/IL [8] systems for the hydrogenation of various substrates with water-soluble catalysts (see also Chapters 5 and 6).

2.4.2.1.2
Catalysts and Mechanisms of Hydrogenation in Aqueous Solution

Solubility of the catalysts in water can be due either to their *overall* charge or to their water-soluble ligands. Most frequently, derivatives of well-known tertiary phosphines (modified by sulfonation, carboxylation, phosphonation, etc.) serve as such ligands (structures **1–16**). Among these, *sulfonated* arylphosphines are easily available, stable and well soluble in water over a wide pH range.

Obviously, there is a great deal of analogy between the mechanisms of hydrogen activation and hydrogenation in aqueous and nonaqueous systems. The most common made of activation of dihydrogen is its homolytic splitting by oxidative addition [Eq. (1)].

$$[RhCl(TPPMS)_3] + H_2 = [H_2RhCl(TPPMS)_3] \tag{1}$$

Aqueous organometallic catalysis, however, is not a mere duplication of what had already been observed in organic solvents, and indeed, special effects of the

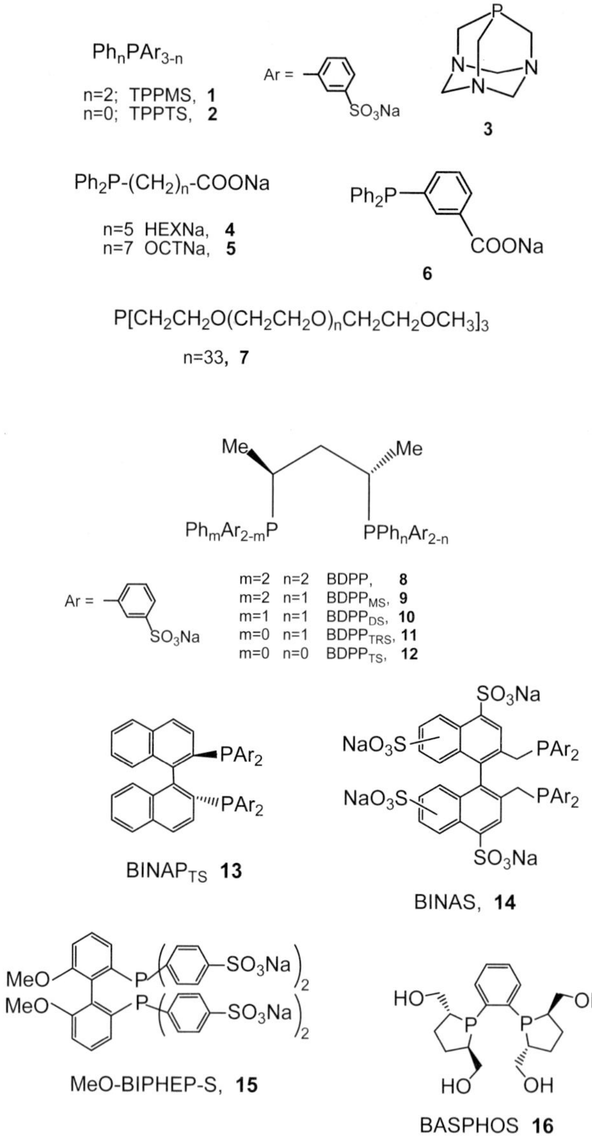

Ph_nPAr_{3-n}

n=2; TPPMS, **1**
n=0; TPPTS, **2**

Ar = (structure with SO_3Na)

3

$Ph_2P-(CH_2)_n-COONa$

n=5 HEXNa, **4**
n=7 OCTNa, **5**

Ph_2P- (structure with COONa)

6

$P[CH_2CH_2O(CH_2CH_2O)_nCH_2CH_2OCH_3]_3$

n=33, **7**

(structure with Me and Me)

$Ph_mAr_{2-m}P$ PPh_nAr_{2-n}

Ar = (structure with SO_3Na)

m=2	n=2	BDPP,	**8**
m=2	n=1	BDPP$_{MS}$,	**9**
m=1	n=1	BDPP$_{DS}$,	**10**
m=0	n=1	BDPP$_{TRS}$,	**11**
m=0	n=0	BDPP$_{TS}$,	**12**

(binaphthyl structure with PAr_2 and PAr_2)

BINAP$_{TS}$ **13**

(binaphthyl structure with SO_3Na, NaO_3S, PAr_2, NaO_3S, PAr_2, SO_3Na)

BINAS, **14**

MeO / MeO (structure with P, SO_3Na, SO_3Na)

MeO-BIPHEP-S, **15**

HO (structure with P, P, OH) HO OH

BASPHOS **16**

aqueous solvent can be encountered. For example, in the oxidative addition of H_2 to *trans*-[IrCl(CO)(TPPMS)$_2$] yielding *trans*-[H$_2$IrCl(CO)(TPPMS)$_2$] a change of the solvent from toluene to water brought about a 50-fold increase in the reaction rate [9]. Such a great increase in rate should be a consequence of water favoring a polar transition state.

Many transition metal hydrides are sufficiently acidic to undergo proton dissociation in the presence of bases or in solvents of suitable solvation power [10]. Whereas in benzene solutions formation of [HRuCl(PPh$_3$)$_3$] takes place only in the presence of an added base such as triethylamine [Eq. (2)], reactions of $[RuCl_2(TPPMS)_2]_2$ with H_2 in water are spontaneous [Eqs. (3)–(5)]. In addition to the temperature and phosphine excess, positions of equilibria (3)–(5) depend critically on the pH of the solution [11].

$$[RuCl_2(PPh_3)_3] + H_2 + Et_3N \rightarrow [HRuCl(PPh_3)_3] + Et_3NH^+ + Cl^- \qquad (2)$$

$$[RuCl_2(TPPMS)_2]_2 + 2\ H_2 \rightleftharpoons [HRuCl(TPPMS)_2]_2 + 2\ H^+ + 2\ Cl^- \qquad (3)$$

$$[RuCl_2(TPPMS)_2]_2 + 2\ H_2 + 2\ TPPMS \rightleftharpoons 2\ [HRuCl(TPPMS)_3] + 2\ H^+ + 2\ Cl^- \quad (4)$$

$$[RuCl_2(TPPMS)_2]_2 + 4\ H_2 + 4\ TPPMS \rightleftharpoons 2\ [H_2Ru(TPPMS)_4] + 4\ H^+ + 4\ Cl^- \quad (5)$$

$[RuCl_2(TPPMS)_2]_2$ is an active catalyst (precursor) for the hydrogenation of water-soluble olefins, such as maleic, fumaric, and crotonic acids. For these hydrogenations detailed kinetic studies [12] revealed the same reaction mechanism as had been suggested earlier for the hydrogenation of maleic acid in DMF solutions catalyzed by $[RuCl_2(PPh_3)_3]$ [4] (Scheme 1). It could be concluded that neither the sulfonation of the phosphine ligand nor the replacement of an organic solvent by water had any effect on the reaction mechanism of alkene hydrogenation by the Ru(II)–phosphine catalysts.

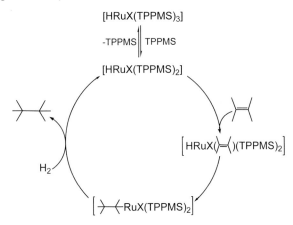

Scheme 1 Mechanism for $[HRuX(TPPMS)_2]_2$-catalyzed hydrogenation of water-soluble olefins.

2.4.2.1.2
Hydrogenation of Alkenes, Alkynes, and Arenes

Several copper, silver, ruthenium, rhodium, and cobalt compounds (such as $RuCl_3 \cdot aq$, $[RuCl_4(bipy)]^{2-}$ (bipy = 2,2′-bipyridine), $RhCl_3 \cdot aq$, bis(dimethylglyoximato)cobalt derivatives (cobaloximes), etc.) have been found to catalyze hydrogenations in aqueous solutions [4]. Although important for the early research into homogeneous catalysis, these catalysts did not gain synthetic significance.

$[HCo(CN)_5]^{3-}$ is readily formed under mild conditions from $Co(CN)_2$, KCN, and H_2 [Eqs. (6) and (7)]. It is an active catalyst for the hydrogenation of a variety of unsaturated substrates but the catalysis suffers from several drawbacks such as the rapid "aging" of the catalyst with a loss of activity, and the necessity of using highly basic aqueous solutions.

$$Co(CN)_2 + 3 \; KCN \rightleftharpoons K_3[Co(CN)_5] \tag{6}$$

$$2 \; K_3[Co(CN)_5] + H_2 = 2 \; K_3[HCo(CN)_5] \tag{7}$$

Conjugated dienes (such as 2,4-hexadienoic acid or sorbic acid) and polyenes can be selectively hydrogenated to monoenes; unactivated alkenes are totally unreactive [13]. Unfortunately, the possibilities for modification of the catalyst by ligand alteration are very restricted [14].

Selective hydrogenation of sorbic acid to *trans*-3-hexenoic acid was also achieved with a $[Ru(CO)(Cp^*)\{P(CH_2CH_2CH_2OH)_3\}][CF_3SO_3]$ ($Cp^* = \eta^5\text{-}C_5Me_5$) catalyst [15].

Hydrogenation of polymers results in improved thermal and oxidative stability. In the hydrogenation of polybutadiene using the $[RhCl(4)_2]_2$ and $[RhCl(5)_2]_2$ catalysts, the pendant (terminal) vinyl units were hydrogenated preferentially over the internal double bonds [Eq. (8)] [16].

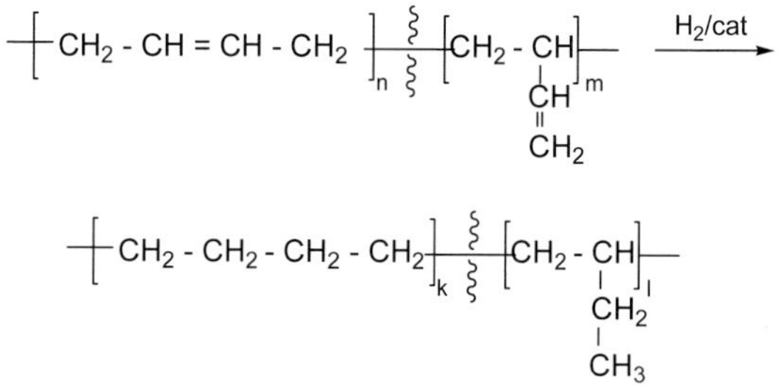

$$\tag{8}$$

The catalytic modification of lipid membranes, either in model systems (liposomes) or in living cells [17], is a special application of homogeneous hydrogenation of olefins in aqueous media. An ideal catalyst efficiently reduces the unsaturated fatty acid units in the polar lipids at low temperatures (0–40 °C) in an aqueous environment, does not effect transformations other than hydrogenation, can be totally removed from the cell after the reaction is completed, and has no "self-effect", such as toxicity. So far the most investigated homogeneous catalyst for biomembrane hydrogenation is $[Pd(QS)_2]$, QS = 1,2-dioxy-9,10-anthraquinone-3-sulfonic acid (Alizarin Red) [18]. Such biological conversions are treated in more detail in Section 2.4.7.

In aqueous solution *mer*-$[Ir(H)(H)Cl(PMe_3)_3]$ catalyzed the hydrogenation of various alkynes to alkanes [19].

Hydridoarene clusters of Rh and Ru are moderately active catalysts of hydrogenation of simple olefins [20]. Conversely, benzene and monosubstituted benzenes can be efficiently hydrogenated in aqueous biphasic systems with hydridoareneruthenium cluster catalysts, such as $[Ru_3(\mu_2\text{-}I\,I)_2(\mu_2\text{-}OH)(\mu_3\text{-}O)(\eta^6\text{-}C_6H_6)(\eta^6\text{-}C_6Me_6)_2]^+$ [21].

Lignin phenols were hydrogenated to the corresponding cyclohexanols with Ru(II)/ TPPMS or TPPTS catalysts, resulting in the inhibition of the light-induced yellowing of lignin and lignin-rich wood pulps [22]. A recent study [23] called attention to the possible formation of metal colloids in aqueous arene hydrogenations.

The reaction rates and selectivities of enantioselective hydrogenations in water are often much inferior to those of analogous systems using organic solvents [24]. Among the tertiary phosphine ligands for enantioselective hydrogenation in aqueous solutions are structures **8–16** (above).

The water-soluble variants of the highly successful Ru(II)–BINAP catalyst have been prepared using the 5,5'-disulfonato-BINAP [25], and the tetrasulfonate BINAP$_{TS}$ (**13**) ligands [26]. In water, [Ru(R-BINAP$_{TS}$)Cl$_2$] catalyzed the hydrogenation of (Z)-α-acetamidocinnamic and (Z)-α-acetamidoacrylic acid with 87.7% (R) and 68.5% (R) *ee*, respectively. The water-soluble catalyst showed the same dependence on hydrogen pressure as the parent [Ru(BINAP)Cl$_2$] (a sharp decline of *ee* with increasing p$_{H2}$ [26]). The cationic [Rh(BASPHOS, **16**)(COD)][BF$_4$] catalyst showed 99.6% *ee* in the hydrogenation of (Z)-α-acetamidoacrylic acid [27].

Water-soluble atropisomeric diphosphines, such as the (S)-(+)- and (R)-(–)-MeO-BIPHEP tetrasulfonate (**15**) were used as components of Ru(II)- and Rh(I)-based hydrogenation catalysts [28]. In some cases high substrate/catalyst ratios could be used (up to 10 000 : 1), a strong requirement for practical applications [29].

The solubility of hydrogen in water is $8 \cdot 10^{-4}$ M, which is about 20% of the solubility in MeOH, $3.86 \cdot 10^{-3}$ M (both at 20 °C, 0.1 MPa total pressure [30]). The limited solubility of H$_2$ may influence both the chemoselectivity (hydrogenation versus isomerization of olefins) [31] and the enantioselectivity [26, 32] of a given reaction.

2.4.2.1.3
Hydrogenation of Compounds with C=O and C=N Bonds

Several water-soluble ruthenium complexes, with P = TPPMS, TPPTS, or PTA ligands (cf. Section 2.2.3.2), catalyze the selective reduction of crotonaldehyde, 3-methyl-2-butenal (prenal), and *trans*-cinnamaldehyde to the corresponding unsaturated alcohols (Scheme 2) [33–36]. Chemical yields are often close to quantitative in reasonable times and the selectivity toward the allylic alcohol is very high (> 95%). The selectivity of the reactions is critically influenced by the pH of the aqueous phase [11] as well as by the H$_2$ pressure [37]. The hydrogenation of propionaldehyde, catalyzed by Ru(II)/TPPTS complexes, was dramatically accelerated by the addition of inorganic salts [38], too. In sharp contrast to the Ru(II)-based catalysts, in hydrogenation of unsaturated aldehydes rhodium(I) complexes preferentially promote the reaction of the C=C double bond, although with incomplete selectivity [33, 39].

Both aliphatic and aromatic unsaturated aldehydes were reduced exclusively to unsaturated alcohols by hydrogen transfer from aqueous sodium formate in a two-phase system with or without an organic solvent. The reactions proceeded smoothly with either [RuCl$_2$(TPPMS)$_2$] [40] or [RuCl$_2$(PTA)$_4$] catalyst [35].

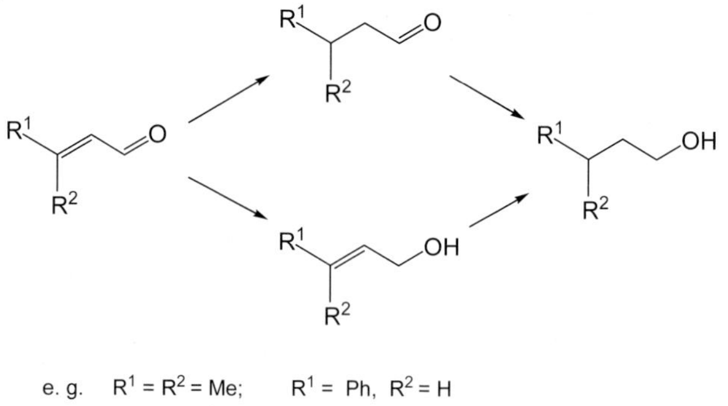

e. g. $R^1 = R^2 = Me$; $R^1 = Ph$, $R^2 = H$

Scheme 2 Selective reduction of unsaturated aldehydes to alcohols, catalyzed by water-soluble Ru complexes.

Hydrogenation of ketones is less facile and less selective than the reduction of aldehydes. [$H_2Ru(TPPTS)_4$] proved to be an active catalyst in hydrogenation of 2-butanone, cyclohexanone, and benzylacetone (80 °C, 3.5 MPa H_2) [34]. The same catalyst was also rather selective toward the formation of the saturated ketone in the hydrogenation of *trans*-4-hexen-3-one [Eq. (9)], yielding only 2–7% of 3-hexanol. Similar C=C/C=O selectivity was found with the [$(C_5R_5)RuCl(PTA)_2$] (R = H or CH_3) catalysts, too [41].

Importantly, ethyl and methyl acetoacetate were hydrogenated with Ru(II)/ 5,5'-disulfonato-BINAP and Ru(II)/MeOBIPHEP-S (**15**) with 91% *ee* and 93% *ee*, respectively [28].

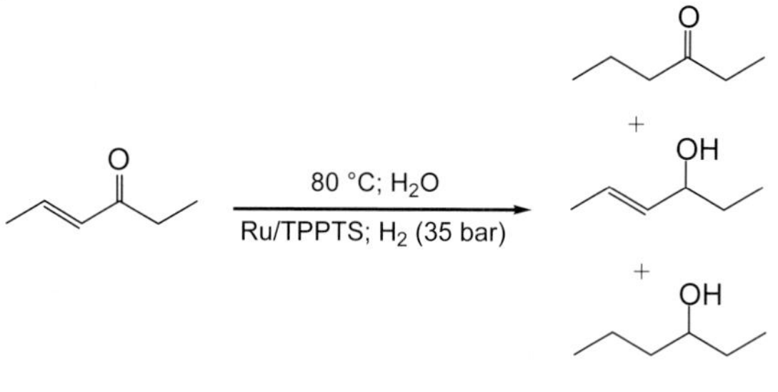

$$\text{80 °C; } H_2O \qquad \text{Ru/TPPTS; } H_2 \text{ (35 bar)} \tag{9}$$

The organometallic aqua complex [$Ir(Cp^*)(H_2O)_3$]$^{2+}$ served as a suitable catalyst precursor for the hydrogenation of water-soluble aldehydes, ketones, and olefins [42]. [$Ru(bipy)(\eta^6\text{-}C_6Me_6)(H_2O)$]$^{2+}$ [43] and [$Ir(bipy)(Cp^*)(H_2O)$] [44] were active in the hydrogenation of various ketones by hydrogen transfer from aqueous sodium formate. In hydrogen-transfer reductions of acetophenone derivatives in aqueous or biphasic systems 96% *ee* was obtained with [$Ru(p\text{-cymene})Cl$]$_2$ + (S)-proline amides [45], 94% *ee* with [$MCl_2(Cp^*)$]$_2$ (M = Rh, Ir) + water-soluble aminosulfon-amides [46], and 84% *ee* with phosphonated N,N'-dimethyl-1,2-diphenylethane-1,2-diamine ligands [47].

Efficient hydrogenations of carbohydrates such as fructose, D-glucose, and D-mannose was achieved with [$HRuCl(TPPTS)_3$] and with Ru(II)/TPPMS and Ru(II)/TPPTS catalysts prepared in situ [12, 48, 49].

Imines, such as *N*-benzylacetophenoneimine, are relatively stable to hydrolysis and were hydrogenated in water/ethyl acetate two-phase solvent mixtures [50, 51]. In a benzene – AOT – water reverse micellar solution (AOT = bis(2-ethylhexyl)-sulfosuccinate), [Rh(NBD)(BDPP)]$^+$ catalyzed the reaction with 82% *ee*, compared to the 68% *ee* in neat benzene [52].

Ketoximes and oximes of 2-oxo acids are hydrogenated to amines [53]. 2-Amino acids can be prepared in high yields by reductive amination of 2-oxo acids in an aqueous NH_3 solution [Eq. (10)] [54]. Aromatic aldehydes were converted with high selectivity to benzylamines by reductive amination with aqueous ammonia catalyzed by a Rh(I)/TPPTS catalyst prepared in situ [55]. In a related, ingenious one-pot process, alkenes were directly converted to primary amines by sequential hydroformylation, condensation with aqueous ammonia, and hydrogenation of the resulting imines [56].

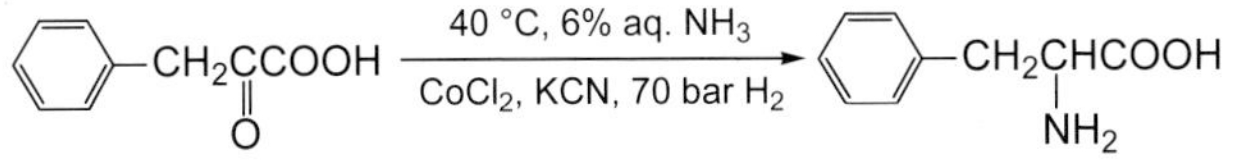

$$\text{Ph-CH}_2\text{CCOOH} \xrightarrow[\text{CoCl}_2,\text{ KCN, 70 bar H}_2]{40 \text{ °C, 6\% aq. NH}_3} \text{Ph-CH}_2\text{CHCOOH} \qquad (10)$$

2.4.2.1.4
Hydrogenolysis of C–O, C–N, C–S, and C–Halogen Bonds

Hydrogenolysis of allyl acetate with [PdCl$_2$L$_2$] (L = **1, 6, 7**) catalyst proceeded smoothly in heptane/aqueous sodium formate mixtures at 80 °C [57]. The catalyst, formed in situ from [Pd(OAc)$_2$] and TPPTS, could be used for the selective removal of allylic protecting groups; such closely related protecting groups as dimethylallylcarbamates and allyloxycarbonates could also be distinguished [58, 59].

Asymmetric hydrogenolysis of sodium *cis*-epoxysuccinates leads to malic acid derivatives which are useful building blocks in natural product synthesis [Eq. (11)] [60].

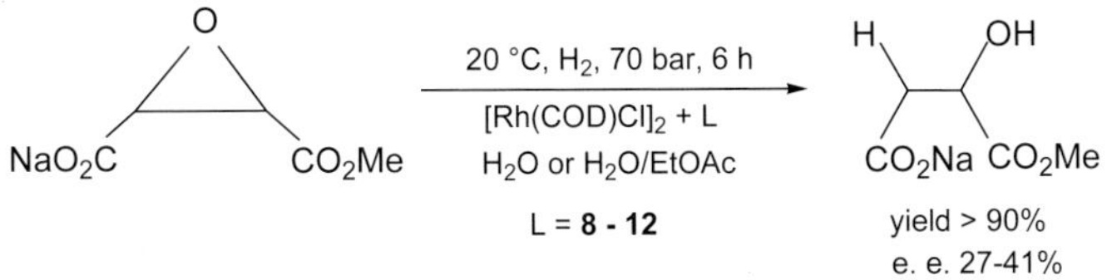

$$\xrightarrow[\substack{[\text{Rh(COD)Cl}]_2 + \text{L} \\ \text{H}_2\text{O or H}_2\text{O/EtOAc} \\ \text{L} = \textbf{8 - 12}}]{20 \text{ °C, H}_2, \text{ 70 bar, 6 h}} \qquad (11)$$

yield > 90%
e. e. 27-41%

The removal of sulfur from petroleum is commonly achieved by hydrogenation on heterogeneous catalysts (in a hydrodesulfurization, or HDS, process) but a few homogeneously catalyzed reactions are also known [61, 62]. For biphasic processes, see Section 2.4.2.2.

Hydrogenolysis of the carbon–halogen bond is an important reaction, both from synthetic and from environmental points of view. The results of the early experiments with [HCo(CN)$_5$]$^{3-}$ as catalyst are summarized in Ref. [4]. A variety of organic halides could be effectively dehalogenated with aqueous sodium formate with [PdCl$_2$L$_2$] (L = various sulfonated phosphines) [57], [RuCl$_2$(TPPMS)$_2$]$_2$ and [Ru(H$_2$O)$_3$(PTA)$_3$]$^{2+}$ [63], and [Ir(bipy)(Cp*)(H$_2$O)] [64] catalysts.

2.4.2.1.5
Miscellaneous Hydrogenations

Hydrogenation of nitro compounds to amines can be achieved with $[HCo(CN)_5]^{3-}$ as catalyst; the reaction is often accompanied by reductive dimerization [4]. The complexes prepared from $PdCl_2$ and TPPTS or BINAS (14) catalyzed the selective hydrogenation of nitroarenes to anilines at 100 °C under 2 MPa CO [65]. Catalytic hydrogenation of chloronitroaromatics often leads to dehalogenation. Importantly, in DMSO-containing water, the $[Pd(OAc)_2]$ + TPPTS catalyst hydrogenated 5-chloro-2-nitrophenol to 5-chloro-2-aminophenol with outstanding selectivity [66]. Anilines can be obtained from nitroarenes under water-gas shift (WGS) conditions, too [Eq. (12)] [67]. $[Rh_6(CO)_{16}]$ [68] and $[Ru_3(CO)_{12}]$ (with added aliphatic amines [69]) are among the most active catalysts.

$$PhNO_2 + 3\ CO + H_2O \rightarrow PhNH_2 + 3\ CO_2 \tag{12}$$

Hydrogenation of carbon dioxide, bicarbonates, and carbonates can also be achieved in aqueous solutions [70–73]. Up till now, however, only formic acid and/or formate salts were observed as the sole products of these hydrogenations. An application as a CO_2 sink is not in sight.

References

1 R. NOYORI, S. HASHIGUCHI, in *Applied Homogeneous Catalysis with Organometallic Compounds* (Eds.: B. CORNILS, W. A. HERRMANN), VCH, Weinheim **1996**, Section 2.9.
2 F. JOÓ, Á. KATHÓ, in *Aqueous Phase Organometallic Catalysis* (Eds.: B. CORNILS, W. A. HERRMANN), 1st ed., Wiley-VCH, Weinheim **1998**, Section 6.2.
3 F. JOÓ, *Aqueous Organometallic Catalysis*, Kluwer, Dordrecht **2001**.
4 B. R. JAMES, *Homogeneous Hydrogenation*, Wiley, New York **1973**.
5 D. J. ADAMS, P. J. DYSON, S. J. TAVENER, *Chemistry in Alternative Reaction Media*, Wiley, Chichester **2004**.
6 F. JOÓ, in *Encyclopedia of Catalysis* (Ed.: I. T. HORVÁTH), Wiley, New York **2002**, Vol. 1, p. 737.
7 G. B. JACOBSON, C. TED LEE JR., K. P. JOHNSTON, W. TUMAS, *J. Am. Chem. Soc.* **1999**, *121*, 11902.
8 P. J. DYSON, D. J. ELLIS, T. WELTON, *Can. J. Chem.* **2001**, *79*, 705.
9 D. P. PATERNITI, P. J. ROMAN JR., J. D. ATWOOD, *J. Chem. Soc., Chem. Commun.* **1996**, 2659.
10 S. S. KRISTJÁNSDÓTTIR, J. R. NORTON, in *Transition Metal Hydrides* (Ed.: A. DEDIEU), VCH, New York **1992**, Chapter 9.
11 F. JOÓ, J. KOVÁCS, A. C. BÉNYEI, Á. KATHÓ, *Angew. Chem.* **1998**, *110*, 1024; *Angew. Chem., Int. Ed.* **1998**, *37*, 969.
12 Z. TÓTH, F. JOÓ, M. T. BECK, *Inorg. Chim. A* **1980**, *42*, 153.
13 J. KWIATEK, *Catal. Rev.* **1967**, *1*, 37.
14 J. T. LEE, H. ALPER, *J. Org. Chem.* **1990**, 55, 1854.
15 B. DRIESSEN-HÖLSCHER, J. HEINEN, *J. Organomet. Chem.* **1998**, *570*, 141.
16 D. C. MUDALIGE, G. L. REMPEL, *J. Mol. Catal. A:* **1997**, *116*, 309.
17 L. VÍGH, F. JOÓ, in *Applied Homogeneous Catalysis with Organometallic Compounds* (Eds.: B. CORNILS, W. A. HERRMANN) VCH, Weinheim **1996**, Section 3.3.10.2.
18 F. JOÓ, N. BALOGH, L. I. HORVÁTH, G. FILEP, I. HORVÁTH, L. VÍGH, *Anal. Biochem.* **1991**, *194*, 34.

19 T. X. Le, J. S. Merola, *Organometallics* **1993**, *12*, 3798.

20 G. Süss-Fink, A. Meister, G. Meister, *Coord. Chem. Rev.* **1995**, *143*, 97.

21 M. Faure, A. Tesouro Vallina, H. Stoeckli-Evans, G. Süss-Fink, *J. Organomet. Chem.* **2001**, *621*, 103.

22 T. Q. Hu, M. Ezhova, T. Y. H. Wong, A. Z. Lu, B. R. James, *Abstr. Int. Symposium on Homogeneous Catalysis, ISHC-13* (Tarragona) **2002**, p. 66.

23 C. Daguenet, P. J. Dyson, *Catal. Commun.* **2003**, *4*, 153.

24 A. Wolfson, I. F. J. Vankelecom, S. Geresh, P. A. Jacobs, *J. Mol. Catal. A:* **2003**, *198*, 39.

25 Takasago Inc. (T. Ishizaki, H. Kumobayashi), EP Appl. 544.455 (**1993**).

26 K. Wan, M. E. Davis, *Tetrahedron: Asymm.* **1993**, *4*, 2461.

27 J. Holz, D. Heller, R. Stürmer, A. Börner, *Tetrahedron Lett.* **1999**, *40*, 7059.

28 R. Schmid, E. A. Broger, M. Cereghetti, Y. Crameri, J. Foricher, M. Lalonde, R. K. Müller, M. Scalone, G. Schoettel, U. Zutter, *Pure Appl. Chem.* **1996**, *68*, 131.

29 Hoffmann-La Roche AG (M. Lalonde, R. Schmid), EP Appl. 667.530 (**1995**).

30 W. F. Linke, A. Seidell, *Solubilities of Inorganic and Metal-Organic Compounds*, American Chemical Society, Washington, DC **1958**.

31 Y. Sun, C. LeBlond, J. Wang, D. G. Blackmond, *J. Am. Chem. Soc.* **1995**, *117*, 12 647.

32 I. Tóth, B. E. Hanson, M. E. Davis, *Tetrahedron: Asymm.* **1990**, *1*, 913.

33 J. M. Grosselin, C. Mercier, G. Allmang, F. Grass, *Organometallics* **1991**, *10*, 2126.

34 M. Hernandez, P. Kalck, *J. Mol. Catal. A:* **1997**, *116*, 131.

35 D. J. Darensbourg, F. Joó, M. Kannisto, Á. Kathó, J. H. Reibenspies, D. J. Daigle, *Inorg. Chem.* **1994**, *33*, 200.

36 R. A. Sánchez-Delgado, M. Medina, F. López-Linares, A. Fuentes, *J. Mol. Catal. A:* **1997**, *116*, 167.

37 G. Papp, J. Elek, L. Nádasdi, G. Laurenczy, F. Joó, *Adv. Synth. Catal.* **2003**, *345*, 172.

38 E. Fache, C. Santini, F. Senocq, J. M. Basset, *J. Mol. Catal.* **1992**, *72*, 337.

39 D. J. Darensbourg, N. W. Stafford, F. Joó, J. H. Reibenspies, *J. Organomet. Chem.* **1995**, *488*, 99.

40 A. Bényei, F. Joó, *J. Mol. Catal.* **1990**, *58*, 151.

41 D. N. Akbayeva, L. Gonsalvi, W. Oberhauser, M. Peruzzini, F. Vizza, P. Brüggeller, A. Romerosa, G. Sava, A. Bergamo, *Chem. Commun.* **2003**, 264.

42 N. Makihara, S. Ogo, Y. Watanabe, *Organometallics* **2001**, *20*, 497.

43 S. Ogo, T. Abura, Y. Watanabe, *Organometallics* **2002**, *21*, 2964.

44 T. Abura, S. Ogo, Y. Watanabe, S. Fukuzumi, *J. Am. Chem. Soc.* **2003**, *125*, 4149.

45 H. Y. Rhyoo, H.-J. Park, W. H. Suh, Y. K. Chung, *Tetrahedron Lett.* **2002**, *43*, 269.

46 T. Thorpe, J. Blacker, S. M. Brown, C. Bubert, J. Crosby, S. Fitzjohn, J. P. Muxworthy, J. M. J. Williams, *Tetrahedron Lett.* **2001**, *42*, 4041.

47 C. Maillet, T. Praveen, P. Janvier, S. Minguet, M. Evain, C. Saluzzo, M. Lorraine Tommasino, B. Bujoli, *J. Org. Chem.* **2002**, *67*, 8191.

48 S. Kolarić, V. Šunjić, *J. Mol. Catal. A:* **1996**, *110*, 189.

49 A. W. Heinen, G. Papadogianakis, R. A. Sheldon, J. A. Peters, H. van Bekkum, *J. Mol. Catal. A:* **1999**, *142*, 17.

50 C. Lensink, E. Rijnberg, J. G. de Vries, *J. Mol. Catal. A:* **1997**, *116*, 199.

51 J. Bakos, Á. Orosz, B. Heil, M. Laghmari, P. Lhoste, D. Sinou, *J. Chem. Soc., Chem. Commun.* **1991**, 1684.

52 J. M. Buriak, J. A. Osborn, *Organometallics* **1996**, *15*, 3161.

53 J. Kwiatek, I. L. Mador, J. K. Seyler, *Adv. Chem. Ser.* **1963**, *37*, 201.

54 A. J. Birch, D. H. Williamson, *Org. React.* **1977**, *24*, 1.

55 T. Gross, A. M. Seayad, M. Ahmad, M. Beller, *Org. Lett.* **2002**, *4*, 2055.

56 B. Zimmermann, J. Herwig, M. Beller, *Angew. Chem., Int. Ed.* **1999**, *38*, 2372.

57 T. Okano, I. Moriyama, H. Konishi, J. Kiji, *Chem. Lett.* **1986**, 1463.

58 S. Lemaire-Audoire, M. Savignac, G. Pourcelot, J.-P. Genêt, J.-M. Bernard, *J. Mol. Catal. A:* **1997**, *116*, 247.

59 R. Widehem, T. Lacroix, H. Bricout, E. Monflier, *Synlett* **2000**, 722.

60 J. Bakos, Á. Orosz, S. Cserépi, I. Tóth, D. Sinou, *J. Mol. Catal. A:* 1997, *116*, 85.

61 R. A. Sánchez-Delgado, *Organometallic Modeling of the Hydrodesulfurization and Hydrodenitrogenation Reactions*, Kluwer, Dordrecht 2002.

62 M. A. Busolo, F. Lopez-Linares, A. Andriollo, D. E. Páez, *J. Mol. Catal. A:* 2002, *189*, 211.

63 A. Cs. Bényei, Sz. Lehel, F. Joó, *J. Mol. Catal. A:* 1997, *116*, 349.

64 S. Ogo, N. Makihara, Y. Kaneko, Y. Watanabe, *Organometallics* 2001, *20*, 4903.

65 A. M. Tafesh, M. Beller, *Tetrahedron Lett.* 1995, *36*, 9305.

66 Hoechst AG (H. Bahrmann, B. Cornils, A. Dierdorf, S. Haber), DE Appl. 19.619.359 (1996).

67 K. Nomura, *J. Mol. Catal. A:* 1995, *95*, 203.

68 F. Joó, H. Alper, *Can. J. Chem.* 1985, *63*, 1157.

69 K. Kaneda, M. Yasumura, T. Imanaka, S. Teranishi, *J. Chem. Soc., Chem. Commun.* 1982, 935.

70 W. Leitner, *Coord. Chem. Rev.* 1996, *153*, 257.

71 H. Hayashi, S. Ogo, T. Abura, S. Fukuzumi, *J. Am. Chem. Soc.* 2003, *125*, 14 266.

72 J. Elek, L. Nádasdi, G. Papp, G. Laurenczy, F. Joó, *Applied Catal. A:* 2003, *255*, 59.

73 P. G. Jessop, F. Joó, C.-C. Tai, *Coord. Chem. Rev.* 2004, *248*, 2425.

2.4.2.2
Hydrogenation, Hydrogenolysis, and Hydrodesulfurization of Thiophenes

Claudio Bianchini and Andrea Meli

2.4.2.2.1
Introduction

Hydrodesulfurization [HDS, Eq. (1)] is the process by which sulfur is removed from fossil materials upon treatment with a high pressure of H_2 (3.5–17 MPa) at high temperature (300–425 °C) in the presence of *heterogeneous* catalysts, generally transition metal sulfides (Mo, W, Co, Ni) supported on alumina [1]. About 90% of the sulfur in fossil materials is contained in thiophenic molecules, which comprise an enormous variety of substituted thiophenes, and benzo[*b*]thiophenes, dibenzo[*b,d*]thiophenes as well as other fused-ring thiophenes, most of which are generally less easily desulfurized over heterogeneous catalysts than any other sulfur compound in petroleum feedstocks (e.g., thiols, sulfide, and disulfides).

$$C_aH_bS + c\,H_2 \rightarrow H_2S + C_aH_d \tag{1}$$

The current HDS technologies are capable of reducing the sulfur contents in gasoline and diesel fuel to the marketing limits (< 50 ppm in 2005) [2]. This achievement requires a high consumption of energy, with consequent emission of greenhouse gases, and large quantities of hydrogen.

Since all the problems affecting HDS are destined to get worse, due to the increase in sulfur levels in the global supplies of crude over the next decade [3], intensive research efforts are being devoted to both improve traditional hydrotreating catalysis and develop alternative strategies for achicving low and ultra low sulfur in fuels.

Most alternative approaches to desulfurization of sulfur compounds contained in raw oils involve multiphase systems and reactors for processes such as: oxidation to water-soluble sulfones catalyzed by either metal compounds or protic acids [4]; ultrasonic degradation, using shock waves to break carbon–sulfur bonds [5]; biodesulfurization using a series of enzyme-catalyzed reactions [3]; liquid–liquid extraction using substrates capable of forming hydrocarbon-insoluble adducts with thiophenes [6]; degradation by means of supercritical water [7]; or hydrogenation in multiphase systems using homogeneous or heterogenized single-metal sites [8].

Reference to important work is provided in this section, which, however, is exclusively concerned with hydrogenation reactions of thiophenes either in aqueous biphasic systems, wherein the metal catalyst resides in the aqueous phase and can be recycled by phase separation, or in liquid–solid systems wherein the solid phase is constituted of a heterogenized metal catalyst that can be recycled by filtration.

2.4.2.2.2
Hydrogenation Reactions

The principal mechanisms proposed for the heterogeneous HDS of a prototypical thiophenic molecule, namely benzo[b]thiophene (BT), are illustrated in Scheme 1. The hydrogenation of C–C double bonds occurs in step **a**, involving the regioselective reduction of BT to dihydrobenzo[b]thiophene (DHBT), as well as in step **e**, where styrene is reduced to ethylbenzene [1].

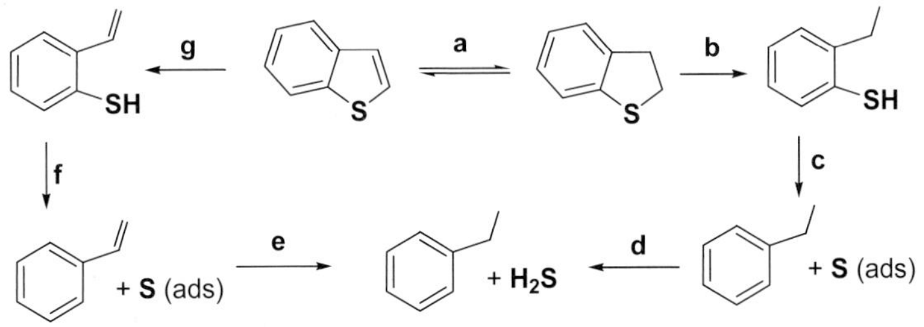

Scheme 1 Heterogeneous HDS of benzo[b]thiophene.

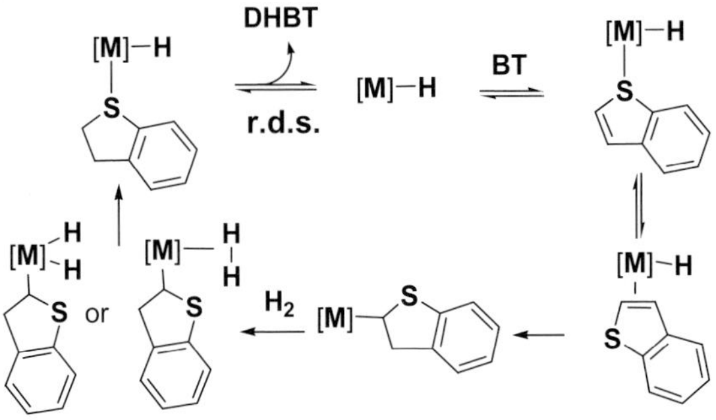

Scheme 2 Hycrogenation of benzo[b]-thiophene catalyzed by metal complexes (r.d.s. = rate determining step).

Irrespective of the phase system (homogeneous, aqueous biphasic, heterogeneous single-site), the hydrogenation mechanisms of thiophene (T) or BT catalyzed by metal complexes comprise the usual steps of H_2 oxidative addition, η^2-C,C coordination of the substrate, hydride transfer to form dihydrobenzothienyl, and elimination of DHBT by hydride/dihydrobenzothienyl reductive coupling (Scheme 2) [8].

Hydrogenation in Aqueous Biphasic Systems with Water-Soluble Metal Catalysts

The aqueous biphasic hydrogenation of T or BT to the corresponding cyclic thioether was primarily achieved with the use of water-soluble ruthenium(II) catalysts supported by triphenylphosphine trisulfonate (TPPTS) or triphenylphosphine monosulfonate (TPPMS) ligands [9] (see Section 2.2.3.2). The reactions were performed in water/decalin under relatively harsh experimental conditions (130–170 °C, 7–11 MPa H_2) leading to the selective reduction of the heterocyclic ring. It was observed that nitrogen compounds did not inhibit the hydrogenation of either T or BT. In some cases, in fact, a promoting effect was observed.

Efficient and robust rhodium and ruthenium catalysts for the hydrogenation of BT to DHBT have been obtained using the backbone-sulfonated polyphosphines $NaO_3S(C_6H_4)CH_2)_2C(CH_2PPh_2)_2$ ($Na_2DPPPDS$) [10] and $NaO_3S(C_6H_4)CH_2C$-$(CH_2PPh_2)_3$ (NaSULPHOS) [11]. The ruthenium(II) binuclear complex Na[{Ru-(SULPHOS)}$_2$(μ-Cl)$_3$] [12] and the mononuclear complex [Ru(MeCN)$_3$(SULPHOS)]-(SO_3CF_3) [13] showed comparable activity in water/decalin or water/n-heptane, suggesting the formation of the same catalytically active species.

Hydrogenation in Liquid–Solid Systems with Heterogenized Metal Catalysts

The first attempts to hydrogenate sulfur heterocycles with a supported metal catalyst date back to 1985 when Fish reported that Rh(PPh$_3$)$_3$Cl tethered to 2% crosslinked phosphinated polystyrene-divinylbenzene was able to selectively hydrogenate various heteroaromatics, including BT to DHBT (benzene, 85 °C, ca. 2 MPa H_2) with rates three times faster than those observed in homogeneous phase with Rh(PPh$_3$)$_3$Cl [14]. This rate enhancement was attributed to steric requirements surrounding the active metal center in the tethered complex, which would favor the coordination of the heterocycles by disfavoring that of PPh$_3$.

The most active and fully recyclable single-site catalyst for the hydrogenation of thiophenes in naphthas or model hydrocarbon solvents is still the silica-supported single-site complex [Ru(NCMe)$_3$(SULPHOS)](OSO$_2$CF$_3$)/SiO$_2$ [15]. Upon hydrogenation (3 MPa H_2), this Ru(II) complex forms a very active catalyst for the selective hydrogenation of BT to DHBT with a TOF as high as 2000 mol of BT converted (mol of catalyst · h)$^{-1}$ (Scheme 3). Remarkably, the TOF did not practically change even when a new feed containing 2000 equiv of BT in 2 ml of n-octane was injected into the reactor after 1 h reaction, showing that DHBT does not compete with BT for coordination to ruthenium. Unlike BT and T, DBT was not hydrogenated by any of the catalysts investigated irrespective of the metal oxidation state or the phase system.

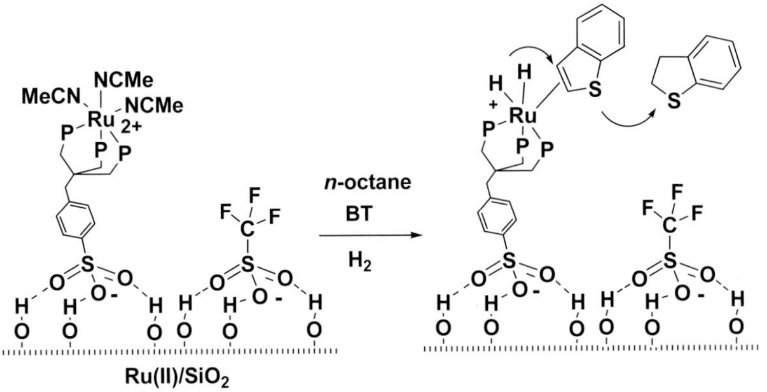

Scheme 3 Activation of the silica-supported Ru(II) complex catalyst and hydrogenation of BT.

2.4.2.2.3
Hydrogenolysis/Desulfurization Reactions

The reaction which transforms a thiophenic substrate into the corresponding thiol is referred to as hydrogenolysis (Scheme 4). Highly energetic metal fragments with filled orbitals of appropriate symmetry are necessary to lower the barrier to C–S insertion which occurs via $d\pi$(metal) $\rightarrow \pi^*$(C–S) transfer [8, 16]. The steric crowding at the metal center must be great enough to disfavor the η^2-C,C bonding mode of the substrate, but not so great as to impede the coordination of the substrate via the sulfur atom.

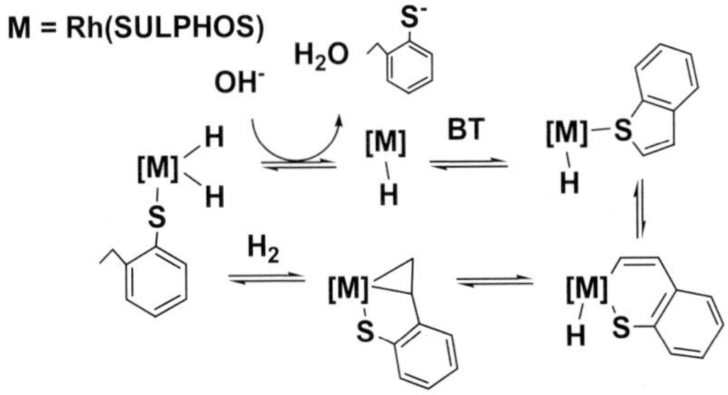

Scheme 4 Catalytic mechanism in hydrogenolysis of thiophenes.

Hydrogenolysis in Aqueous Biphasic Systems with Water-Soluble Metal Catalysts

The aqueous biphasic hydrogenolysis of BT has been accomplished in either water/ *n*-decalin or water/naphtha mixtures with the rhodium(I) precursor [Rh(COD)-(SULPHOS)] (COD = cycloocta-1,5-diene) [17]. Rather harsh reaction conditions (160 °C, 3 MPa H_2) and an equivalent amount of NaOH were required for high conversions of BT to 2-ethylthiophenolate. In these conditions, the thiolate product was totally recovered in the aqueous phase, leaving the hydrocarbon phase formally "desulfurized" (Scheme 5).

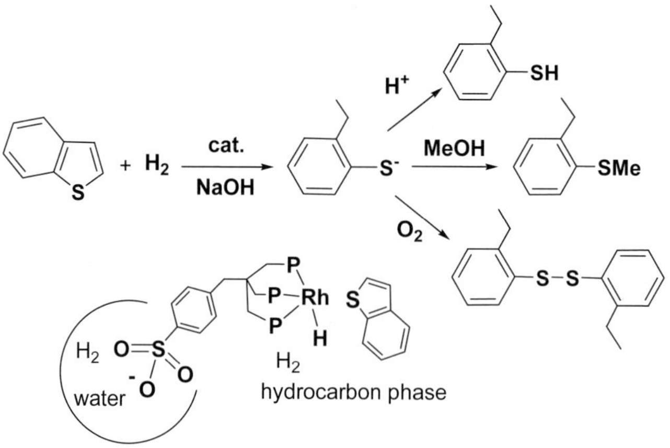

Scheme 5 Aqueous biphasic hydrogenolysis of benzo[*b*]thiophene.

Hydrogenolysis in Liquid–Solid Systems with Heterogenized Metal Catalysts

The tripodal triphosphine moiety –C(CH$_2$PPh$_2$)$_3$ was anchored to a crosslinked styrene-divinylbenzene polymer yielding a polymeric material, POLYTRIPHOS, which reacted with a CH$_2$Cl$_2$ solution of [RhCl(COD)]$_2$ in the presence of AgPF$_6$ to give the polystyrene-supported complex [Rh(COD)(POLYTRIPHOS)]PF$_6$ (Rh 0.94 wt.%) (Scheme 6) [18].

Upon hydrogenation with 3 MPa H$_2$, this supported rhodium complex generated an effective catalyst for the hydrogenolysis of BT to 2-ethylthiophenol (TOF 48) and ethylbenzene (TOF 2), which still represents the first evidence of a successful single-site catalyst in the desulfurization of a thiophenic substrate. Indeed, under rather harsh experimental conditions (THF, KOBu^t, 160 °C, 3 MPa H$_2$), BT was mainly converted to a mixture of 2-ethylthiophenol and ethylbenzene (ET). No trace of the hydrogenation product DHBT was observed. As well, no rhodium leaching was observed, while the catalyst was recycled several times with no loss of catalytic activity. The surprising desulfurization of BT to ET has been interpreted in terms

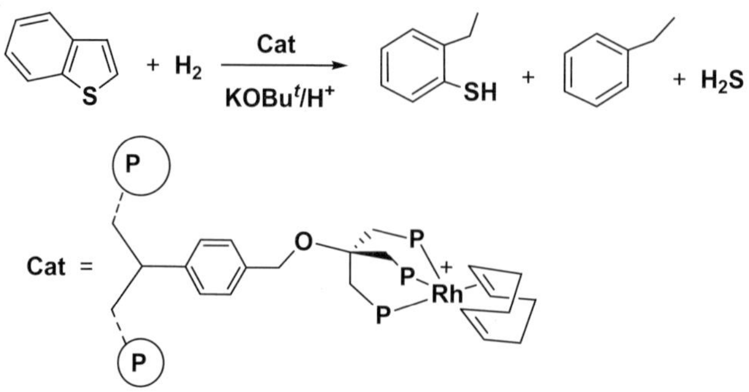

Scheme 6 Hydrogenolysis and desulfurization of benzo[*b*]thiophene with a supported Rh complex.

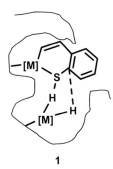

1

of the concomitant action of two metal sites kept in close proximity by the flexible polystyrene matrix (**1**). Evidence showing the need for at least two metal sites to promote the desulfurization of a thiophenic substrate has several precedents in the literature [19].

2.4.2.2.4
Perspectives

Nowadays, deep desulfurization of gasoline and naphtha can be achieved by traditional hydrotreating, yet with technologies that require an exceedingly high energy consumption and cause a strong environmental impact. The future in the production of clean fuels from fossil materials is represented by multiphase catalysis as well as other multiphase absorption/extraction techniques [1–8]. Multiphase catalysis will involve both reduction and oxidation processes whereby sulfur removal can be achieved by well-defined metal sites capable of lowering the energy barriers to both C–S bond cleavage and disruption of the thiophene-ring aromaticity.

References

1 (a) T. Kabe, A. Ishihara, W. Qian, *Hydrodesulfurization and Hydrodenitrogenation*, Wiley-VCH, Tokyo **1999**; (b) H. Topsøe, B. S. Clausen, F. E. Massoth, *Hydrotreating Catalysis*, Springer-Verlag, Berlin **1996**.

2 See: http://www.epa.gov/otaq/gasoline.htm.

3 D. J. Monticello, *CHEMTEC* **1998**, *28*, 38.

4 (a) R. Angelici, *Abstracts ISHHC 11*, Northwestern University, USA, July 20–25, **2003**; (b) K. Yazu, Y. Yamamoto, T. Furuya, K. Miki, K. Ukegawa, *Energy & Fuels* **2001**, *15*, 1535; (c) N. d'Alessandro, L. Tonucci, M. Bonetti, M. Di Deo, M. Bressan, A. Morvillo, *New J. Chem.* **2003**, *27*, 989; (d) Y. Wang, G. Lente, J. H. Espenson, *Inorg. Chem.* **2002**, *41*, 1272; (e) Unipure Corporation (A. S. Rappas, V. P. Nero, S. J. DeCanio), US 6.406.616 (**2002**); (f) Unipure Corporation (A. S. Rappas), US 6.402.940 (**2002**).

5 (a) Sulphco Inc. (R. W. Gunnerman), US 6.500.219 (**2002**); (b) Sulphco Inc. (T. F. Yen, H. Mei, S. H.-M. Lu), US 6.402.939 (**2002**).

6 (a) R. T. Yang, A. J. Hernandez-Maldonado, F. H. Yang, *Science* **2003**, *301*, 79; (b) Y. Shiraishi, T. Naito, T. Hirai, I. Komosawa, *Chem. Commun.* **2001**, 1256; (c) V. Meille, E. Schulz, M. Vrinat, M. Lemaire, *Chem. Commun.* **1998**, 305; (d) A. Bosmann, L. Datsevich, A. Jess, A. Lauter, C. Schmitz, P. Wasserscheid, *Chem. Commun.* **2001**, 2494.

7 Exxon Research and Engineering Company (M. Siskin, D. T. Ferrughelli, A. R. Katritzky, W. N. Olmstead), US 5.611.915 (**1997**).

8 (a) C. Bianchini, A. Meli, in *Aqueous-Phase Organometallic Catalysis – Concepts and Applications* (Eds.: B. Cornils, W. A. Herrmann), VCH, Weinheim **1998**, p. 477; (b) C. Bianchini, A. Meli, F. Vizza, in *Applied Homogeneous Catalysis with Organometallic Compounds* (Eds.: B. Cornils, W. A. Herrmann), Wiley-VCH, Weinheim **2002**,Vol. 3, p. 1099.

9 (a) INTEVEP S.A. (D. E. Páez, A. Andriollo, R. A Sánchez-Delgado, N. Valencia, F. López-Linares, R. Galiasso), US 08/657.960 (**1996**); (b) INTEVEP S. A. (D. E. Páez, A. Andriollo, R. A Sánchez-Delgado, N. Valencia, F. López-Linares, R. Galiasso), Sol. Patente Venezolana 96-1630 (**1996**); (c) M. A. Busolo, F. Lopez-Linares, A. Andriollo, D. E. Paez, *J. Mol. Catal. A:* **2002**, *189*, 211.

10 (a) CNR (C. Bianchini, A. Meli, F. Vizza), PCT 97/06493 (**1999**); (b) CNR (C. Bianchini, A. Meli, F. Vizza), IT 96A000272 (**1996**).

11 C. Bianchini, P. Frediani, V. Sernau, *Organometallics* **1995**, *14*, 5458.

12 I. Rojas, F. Lopez Linares, N. Valencia, C. Bianchini, *J. Mol. Catal. A:* **1999**, *144*, 1.

13 (a) C. Bianchini, A. Meli, S. Moneti, W. Oberhauser, F. Vizza, V. Herrera, A. Fuentes, R. A. Sánchez-Delgado, *J. Am. Chem. Soc.* **1999**, *121*, 7071; (b) C. Bianchini, A. Meli, S. Moneti, F. Vizza, *Organometallics* **1998**, *17*, 2636.

14 R. H. Fish, A. D. Thormondsen, H. Heinemann, *J. Mol. Catal.* **1985**, *31*, 191.

15 (a) C. Bianchini, V. Dal Santo, A. Meli, S. Moneti, M. Moreno, W. Oberhauser, R. Psaro, L. Sordelli, F. Vizza, *J. Catal.* **2003**, *213*, 47; (b) C. Bianchini, V. Dal Santo, A. Meli, W. Oberhauser, R. Psaro, F. Vizza, *Organometallics* **2000**, *19*, 2433.

16 (a) A. W. Myers, W. D. Jones, *Organometallics* **1996**, *15*, 2905; (b) A. W. Myers, W. D. Jones, S. M. McClements, *J. Am. Chem. Soc.* **1995**, *117*, 11704; (c) L. Dong, S. B. Duckett, K. F. Ohman, W. D. Jones, *J. Am. Chem. Soc.* **1992**, *114*, 151; (d) W. D. Jones, L. Dong, *J. Am. Chem. Soc.* **1991**, *113*, 559.

17 (a) C. Bianchini, A. Meli, V. Patinec, V. Sernau, F. Vizza, *J. Am. Chem. Soc.* **1997**, *119*, 4945; (b) C. Bianchini, J. Casares, A. Meli, V. Sernau, F. Vizza, R. A. Sánchez-Delgado, *Polyhedron* **1997**, *16*, 3099; (c) C. Bianchini, V. Herrera, M. V. Jiménez, A. Meli, R. A. Sánchez-Delgado, F. Vizza, *J. Am. Chem. Soc.* **1995**, *117*, 8567.

18 C. Bianchini, M. Frediani, F. Vizza, *Chem. Commun.* **2001**, 479.

19 (a) C. Bianchini, A. Meli, S. Moneti, F. Vizza, *Organometallics* **1997**, 16, 5696; (b) D. A. Vicic, W. D. Jones, *J. Am. Chem. Soc.* **1999**, *121*, 7606.

2.4.3
Oxidations

2.4.3.1
Partial Oxidations

Roger A. Sheldon

2.4.3.1.1
Introduction

Metal-catalyzed hydrogenations, carbonylations, hydroformylations, etc. involve transition metals in low oxidation states coordinated to soft ligands, e.g., phosphines, as the catalytically active species and organometallic compounds as reactive intermediates. Performing such reactions in aqueous/organic biphasic media generally involves the use of water-soluble variants of these ligands, e.g., sulfonated triarylphosphines such as **1** (see Section 2.2.3.2). In contrast, catalytic oxidations involve transition metals in high oxidation states as the active species, generally coordinated to relatively simple hard ligands, e.g. carboxylate. Reactive intermediates tend to be coordination complexes rather than organometallic species.

Water often has an inhibiting effect on catalytic oxidations owing to strong coordination to the hard metal center hampering coordination of a less polar substrate, such as a hydrocarbon (cf. Sections 2.2.1 and 2.2.2). Coordination of complex nitrogen- and/or oxygen-containing ligands can lead to the generation of more active oxidants by promoting the formation of high oxidation states. For example, in heme-dependent oxygenases and peroxidases the formation of active high-valent oxo-iron complexes is favored by coordination to a macrocyclic porphyrin ligand. Hence, we have limited our frame of reference to systems in which the reaction takes place in the aqueous phase using transition metal complexes of water-soluble ligands as catalysts.

2.4.3.1.2
Water-Soluble Ligands

Much of this research falls into the category of biomimetic oxidations. Hence, water-soluble porphyrins and the structurally related phthalocyanines have been widely used (see structures **1–4** for examples).

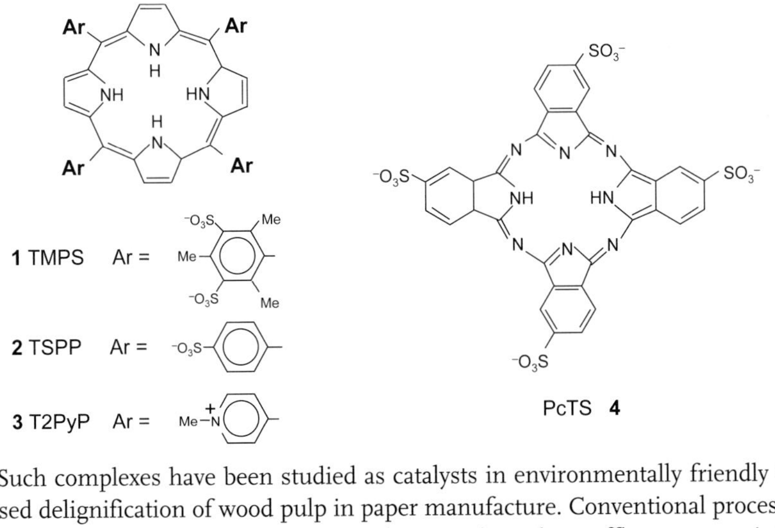

1 TMPS   Ar =

2 TSPP   Ar =

3 T2PyP Ar =

PcTS **4**

Such complexes have been studied as catalysts in environmentally friendly O_2-based delignification of wood pulp in paper manufacture. Conventional processes involve the use of Cl_2 or ClO_2 as oxidants and produce effluents containing chlorinated phenols. For example, Wright and co-workers [1] oxidized lignin model compounds such as **5** with O_2 [Eq. (1)] in the presence of $Na_3Fe(III)$ (PcTS), $Na_3Co(III)$ (TSPP), and $Na_3Rh(III)$ (TSPP) catalysts. The latter gave the highest rates and selectivities.

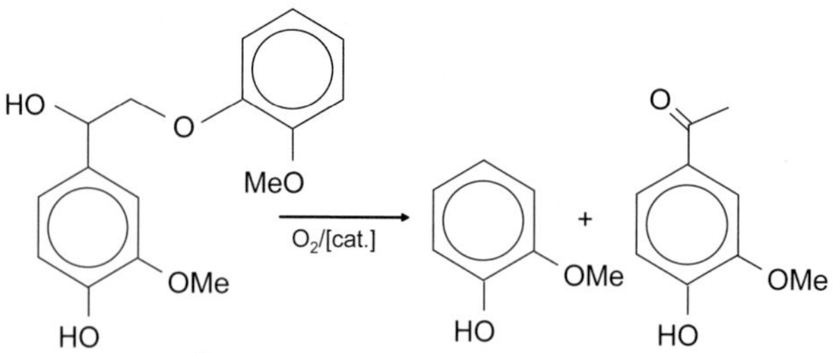

$$(1)$$

Similarly, Hampton and Ford [2] studied the Fe(PcTS)-catalyzed autoxidation of 3,4-dimethoxybenzyl alcohol as a model for delignification. They concluded, however, that the catalyst degrades too fast to be useful for delignification. In this context it is worth mentioning that water-soluble polyoxometallates such as $Pv_2Mo_{10}O_{40}{}^{5-}$ have also been used as catalysts for delignification with O_2 [3].

Fe(PcTS) [4] and Fe(TMPS) [5] have also been examined as catalysts for the oxidative destruction of chlorinated phenols in waste water, using H_2O_2 or $KHSO_5$ as the oxidant. For example, 2,4,6-trichlorophenol underwent facile oxidation to 2,6-dichloro-1,4-benzoquinone with Fe(PcTS)/H_2O_2 [4] or Fe(TMPS)/$KHSO_5$ [5]. Similarly, Fe(III) and Mn(III) complexes of T2MPyP catalyzed the oxidation of phenols with $KHSO_5$ [6].

Water-soluble manganese complexes of 1,4,7-trimethyl-1,4,7-triazacyclononane (**6**) and related ligands are highly effective catalysts for low-temperature bleaching of stains [7]. Polyphenolic compounds were used as appropriate models for tea stains. The same complexes were shown to catalyze the selective epoxidation of styrene and 4-vinylbenzoic acid with aqueous H_2O_2 [Eq. (2)] in aqueous MeOH or water, respectively. However, large amounts of H_2O_2 (10 equiv.) were required, indicating that considerable nonproductive decomposition occurs. Subsequently, it was shown that nonproductive decomposition of the hydrogen peroxide could be largely suppressed by the addition of oxalate [8] or ascorbic acid [9] as co-catalysts, or by anchoring the ligand to a solid support [10]. More recently, the use of glyoxylic acid methyl ester hemiacetal as a co-catalyst was shown to afford an even more effective epoxidation catalyst, enabling high conversions with only a 30% excess of

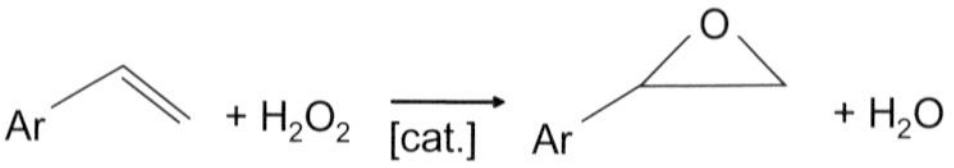

$$(2)$$

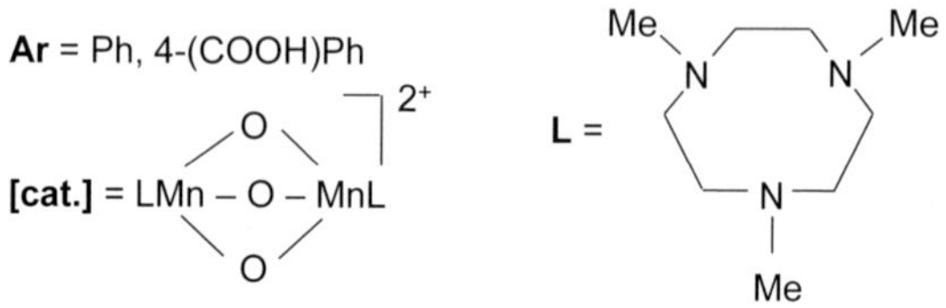

hydrogen peroxide [11]. Interestingly, the corresponding *cis*-diols were observed as byproducts in many cases and a concerted mechanism via a manganese(III)–*cis*-diol complex was proposed to explain their formation. The binuclear manganese complex of **6** also catalyzed the oxidation of benzylic alcohols, to the corresponding benzaldehydes, with aqueous hydrogen peroxide [12].

Collins and co-workers [13] have developed a series of iron(III) complexes of macrocyclic tetradentate nitrogen ligands, so-called TAML oxidant activators, with greatly enhanced stability toward oxidative and hydrolytic degradation (**7**). They are efficient, water-soluble activators of hydrogen peroxide, over a broad pH range, with a wide variety of potential applications, e.g., to replace chlorine bleaching in the pulp and paper industry and for use in water effluent treatment in the textiles industry. Applications in organic synthesis have not, as yet, been explored.

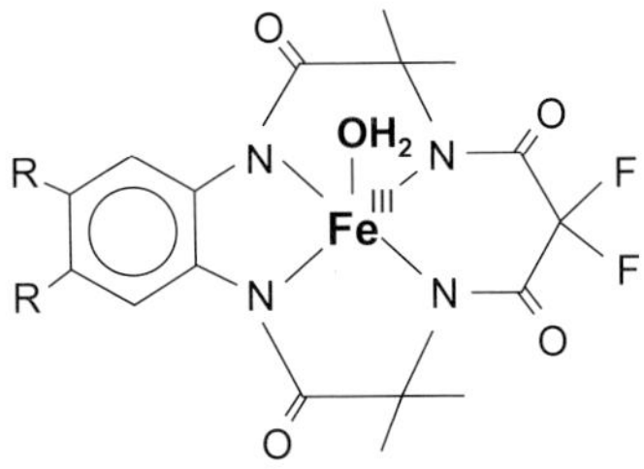

7

Recently, we described [14–16] the use of a water-soluble palladium(II) complex of sulfonated bathophenanthroline (**8**) as a stable, recyclable catalyst for the aerobic oxidation of alcohols in a two-phase aqueous–organic medium [e.g. Eq. (3)]. Reactions were generally complete in 5 h at 100 °C/3 Mpa air with as little as 0.25 mol% catalyst. No organic solvent is required (unless the substrate is a solid) and the product is easily recovered by phase separation. The catalyst is stable and remains in the aqueous phase, facilitating recycling to the next batch.

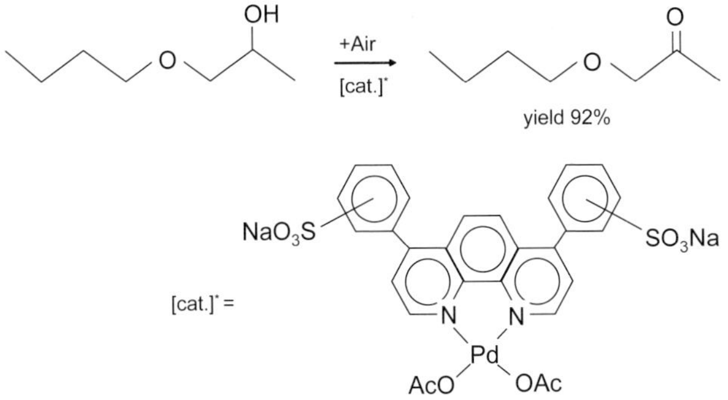

$$(3)$$

yield 92%

A wide range of primary and secondary alcohols were oxidized with TOFs ranging from 10 to 100 h^{-1}, depending on the structure and the solubility of the alcohol in water (since the reaction occurs in the water phase the alcohol must be at least sparingly soluble in water). Secondary alcohols afforded the corresponding ketones in > 99% selectivity in virtually all cases studied. Primary alcohols afforded the corresponding carboxylic acids via further oxidation of the initially formed aldehyde, e.g., 1-hexanol afforded 1-hexanoic acid in 95% yield. It is important to note that this was achieved without the necessity to neutralize the carboxylic acid product with 1 equiv. of base. When the reaction was performed in the presence of 1 mol% of the stable free radical, TEMPO (2,2,6,6-tetramethylpiperidinoxyl), over-oxidation was suppressed and the aldehyde was obtained in high yield, e.g. 1-hexanol afforded hexanal in 97% yield.

Compared to most existing systems for the aerobic oxidation of alcohols, the Pd–bathophenanthroline system is an order of magnitude more reactive, requires no organic solvent, involves simple product isolation and catalyst recycling, and has a broad scope in organic synthesis.

A catalytic cycle was proposed [15] for the reaction in which, consistent with the observed half-order in palladium, the active catalyst is formed by initial dissociation of a hydroxyl-bridged palladium(II) dimer. This is followed by coordination of the alcohol and β-hydrogen elimination affording the carbonyl product and palladium(0). The latter is reoxidized to palladium(II) by dioxygen. More recently, electronic [17] and steric [18] effects of substituents in the phenanthroline ligands on the rates and substrate scope of these reactions were studied. Results were in accordance with the proposed mechanism and afforded an optimized catalyst which was highly active (turnover frequencies > 1500 h^{-1}) and tolerated a wide variety of functional groups in the alcohol substrate.

2.4.3.1.3
Concluding Remarks

Important advances have been made in the last few years in the design of water-soluble, oxidatively stable ligands. In combination with appropriate metal ions they afford stable, water-soluble catalysts for oxidations with dioxygen or hydrogen peroxide in aqueous–organic biphasic systems. Up till now synthetic applications have generally been limited to olefin epoxidations with hydrogen peroxide (Mn) and aerobic oxidation of alcohols (Pd). These methodologies constitute green alternatives – clean oxidants, no need for organic solvents, facile product separation and catalyst recycling – for traditional oxidations. In the future we expect that these methodologies will be further applied in organic synthesis. In particular, iron complexes offer significant advantages from both an economic and an environmental viewpoint.

References

1 P. A. WATSON, L. J. WRIGHT, T. J. FULLERTON, *J. Wood Chem. Technol.* **1993**, *13*, 371; 391; 411.
2 K. W. HAMPTON, W. T. FORD, *J. Mol. Catal. A:* **1996**, *113*, 167.
3 I. A. WEINSTOCK, R. H. ATILLA, R. S. REINER, M. A. MOEN, K. E. HAMMEL, C. J. HOUTMAN, C. L. HILL, M. K. HARRUP, *J. Mol. Catal. A:* **1997**, *116*, 59.

4 A. SOROKIN, B. MEUNIER, *J. Chem. Soc., Chem. Commun.* **1994**, 1799; A. SOROKIN, J.-L. SÉRIS, B. MEUNIER, *Science* **1995**, *268*, 1163.

5 R. S. SHAKLA, A. ROBERT, B. MEUNIER, *J. Mol. Catal. A:* **1996**, *113*, 45.

6 N. W. J. KAMP, J. R. LINDSAY SMITH, *J. Mol. Catal. A:* **1996**, *113*, 131.

7 R. HAGE, J. E. IBURG, J. KERSCHNER, J. H. KOEK, E. L. M. LEMPERS, R. J. MARTENS, U. S. RACHERIA, S. W. RUSSELL, T. SWARTHOFF, M. R. P. VAN VLIET, J. B. WARNAAR, L. VAN DER WOLF, B. KRIJNEN, *Nature* **1994**, *369*, 637; R. Hage, *Recl. Trav. Chim. Pays-Bas* **1996**, *115*, 385.

8 D. E. DE VOS, B. F. SELS, M. REYNAERS, Y. V. SUBBA RAO, P. A. JACOBS, *Tetrahedron Lett.* **1998**, *39*, 3221.

9 A. BERKESSEL, C. A. SKLORZ, *Tetrahedron Lett.* **1999**, *40*, 7965.

10 D. E. DE VOS, S. DE WILDEMAN, B. F. SELS, P. J. GROBET, P. A. JACOBS, *Angew. Chem., Int. Ed.* **1999**, *38*, 980.

11 J. BRINKSMA, L. SCHMIEDER, G. VAN VLIET, R. BOARON, R. HAGE, D. E. DE VOS, P. L. ALSTERS, B. L. FERINGA, *Tetrahedron Lett.* **2002**, *43*, 2619.

12 C. ZONDERVAN, R. HAGE, B. L. FERINGA, *Chem. Commun.* **1997**, 419.

13 T. J. COLLINS, *Acc. Chem. Res.* **2002**, *35*, 782 and references cited therein.

14 G. J. TEN BRINK, I. W. C. E. ARENDS, R. A. SHELDON, *Science* **2000**, *287*, 1636.

15 G. J. TEN BRINK, I. W. C. E. ARENDS, R. A. SHELDON, *Adv. Synth. Catal.* **2002**, *344*, 355.

16 R. A. SHELDON, I. W. C. E. ARENDS, G. J. TEN BRINK, *Acc. Chem. Res.* **2002**, *35*, 774.

17 G. J. TEN BRINK, I. W. C. E. ARENDS, M. HOOGENRAAD, G. VERSPUI, R. A. SHELDON, *Adv. Synth. Catal.* **2003**, *345*, 497.

18 G. J. TEN BRINK, I. W. C. E. ARENDS, M. HOOGENRAAD, G. VERSPUI, R. A. SHELDON, *Adv.Synth. Catal.* **2003**, *345*, 1341.

2.4.3.2
Wacker-Type Oxidations

Eric Monflier

The Wacker-type oxidation of olefins is one of the oldest homogeneous transition metal-catalyzed reactions [1]. The most prominent example of this type of reaction is the oxidation of ethylene to acetaldehyde by a $PdCl_2/CuCl_2/O_2$ system (Wacker–Hoechst process). In this industrial process, oxidation of ethylene by Pd(II) leads to Pd(0), which is reoxidized to Pd(II) via reduction of Cu(II) to Cu(I). To complete the oxidation–reduction catalytic cycle, Cu(I) is classically reoxidized to Cu(II) by O_2 [2, 3]. The use of bidentate ligands [4], bicomponent systems constituted of benzoquinone and iron(II) phthalocyanine [5] or chlorine-free oxidants such as ferric sulfate [6], heteropoly acid [7], and benzoquinone [8], make it possible to increase the selectivity reaction by avoiding the formation of chlorinated products.

The Wacker reaction has been applied to numerous simple olefins such as α-olefins and cycloalkenes, or to functionalized olefins such nitroethylene, acrylonitrile, styrene, allyl alcohol, or maleic acid [3]. The carbonyl group is formed at the carbon atom of the double bond where the nucleophile would add in a Markovnikov addition. Reversal of the regioselectivity has only been observed with particular substrates such as 1,5-dienes [9]. Conversion and selectivity for the oxidation of these olefins were found to be very dependent on the water solubility of the olefin. Indeed, high molecular weight olefins do not react under the standard biphasic Wacker conditions due to their low solubility in water. Furthermore, the products obtained are often highly contaminated by chlorinated products and isomerized olefins [8, 10, 11]. In order to overcome these problems, numerous studies have been undertaken. Only the most significant approaches for the oxidation of higher α-olefins (1-hexene and above) will be developed here.

2.4.3.2.1
Co-Solvents or Solvents

The use of a co-solvent or solvent is the simplest solution to overcome mass-transfer limitations [8, 11, 12]. Among the different solvents described in the literature (DMSO, acetone, THF, dioxane, acetonitrile, ethanol), dimethylformamide appears to be the most suitable. Indeed, oxidation of 1-dodecene into 2-dodecanone can be achieved with yields greater than 80% in water–DMF mixtures.

2.4.3.2.2
Phase Transfer Catalysis

Phase-transfer catalysis constitutes another alternative to increase oxidation rates of water-insoluble terminal alkenes [13]. Although some results in the literature are contradictory [14], it seems that only quaternary ammonium salts containing at least one long-chain alkyl group are suitable as phase-transfer catalysts. Interestingly, rhodium and ruthenium complexes can also be used instead of palladium for the oxidation of terminal olefins [15]. With these catalysts, symmetrical quaternary ammonium salts such as tetrabutylammonium hydrogen sulfate are effective. The rate of palladium-catalyzed oxidation of terminal olefins can also be improved by using polyethylene glycol (PEG) instead of quaternary ammonium salts [16]. Thus, the rate of PEG-400-induced oxidation of 1-decene is three times faster than cetyltrimethylammonium bromide-catalyzed oxidation under the same conditions (see also Section 2.3.5).

2.4.3.2.3
Microemulsion Systems

These systems have also been proposed for the oxidation of sparingly water-soluble olefins [17]. The microemulsion system consisted of formamide, 1-hexene, 2-propanol as co-surfactant and $C_9H_{19}-C_6H_4-(OCH_2CH_2)_8OH$ as surfactant. In such a medium, the oxidation rates of 1-hexene to 2-hexanone were three times faster than those observed in the water–DMF mixture under similar condition.

2.4.3.2.4
Immobilized Catalysts

Immobilized catalysts have also been described for the oxidation of water-insoluble olefins. Most work has been done with polymer-anchored palladium catalysts [18].

For instance, palladium supported on a highly rigid cyanomethylated polybenz-imidazole produces a highly effective catalyst for the oxidation of 1-decene, with activity higher than homogeneous systems in some cases [19]. The use of palladium and copper salts dissolved in an aqueous film supported on a high surface-area silica gel has also been proposed by Davies et al. to perform the oxidation of higher olefins (the SAPC concept; see Section 2.6). Unfortunately, the ketone yields are rather low (< 25%) and significant isomerization of the olefin occurred [20].

2.4.3.2.5
Inverse Phase-Transfer Catalysis

The principle of inverse phase-transfer catalysis (cf. Section 2.2.3.3) has been successfully applied to the oxidation of higher olefins [21]. The success of this oxidation is mainly due to the use of β-cyclodextrin functionalized with hydrophilic or lipophilic groups. The best results have been obtained with a multicomponent catalytic system constituted of $PdSO_4$, $H_9PV_6Mo_6O_{40}$, $CuSO_4$, and randomly O-methylated β-cyclodextrin [22]. Interestingly, β-cyclodextrin modified by nitrile-containing groups was found to be more efficient than native β-cyclodextrin to perform 1-octene oxidation [23]. The rate increase was attributed to the formation of a ternary inclusion complex between the olefin, the palladium salt, and the modified cyclodextrin. Recently, the effect of other inverse phase-transfer catalysts such as sulfonated calixarenes was also investigated [24]. Interestingly, the sulfonated calixarenes displayed substrate selectivity, confirming that internal cavity of the calixarenes recognizes the substrates.

2.4.3.2.6
Summary

Although the Wacker-type oxidation of olefins has been applied since the early 1980s, processes involving higher olefins are still the subject of investigations due to their poor solubility in water. Particularly interesting in this context is the inverse phase-transfer catalysis using water-soluble host molecules. Indeed, upon a careful choice of the substituent, these receptor molecules avoid the isomerization into internal olefins or make it possible to perform substrate selective oxidations that cannot be achieved a biphasic medium with conventional transition metal catalysts.

References

1 G. W. PARSHALL, S. D. ITTEL, in *Homogeneous Catalysis. The Applications and Chemistry of Catalysis by Soluble Transition Metal Complexes*, 2nd ed., Wiley-Interscience, New York 1992, pp. 138–142.
2 R. JIRA, W. BLAU, D. GRIMM, *Hydrocarbon Proc.* 1976, 97.
3 R. JIRA, in *Applied Homogeneous Catalysis with Organometallic Compounds* (Eds.: B. CORNILS, W. A. HERRMANN), VCH, Weinheim 1996, pp. 374–394.

4 G. J. ten Brink, I. W. Arends, G. Papadogianakis, R. A. Sheldon, *Chem. Commun.* **1998**, 2359.

5 (a) J. E. Bäckvall, R. B. Hopkins, *Tetrahedron Lett.* **1988**, *29*, 2885; (b) J. E. Bäckvall, R. B. Hopkins, H. Greenberg, M. M. Mader, A. K. Awasthi, *J. Am. Chem. Soc.* **1990**, *112*, 5160; (c) J. E. Bäckvall, A. K. Awasthi, Z. D. Renko, *J. Am. Chem. Soc.* **1987**, *109*, 4750.

6 H. Hasegawa, M. Triuchijima, Maruzen Oil Co., GB 1.240.889 **(1971)**.

7 (a) K. I. Matveev, *Kinet. Catal. (Engl. Transl.)* **1977**, *18*, 716; (b) A. Lambert, E. G. Derouane, I. V. Kozhevnikov, *J. Catal.* **2002**, *211*, 445.

8 W. A. Clement, C. M. Selwitz, *J. Org. Chem.* **1964**, *29*, 241.

9 T. L. Ho, M. H. Chang, C. Chen, *Tetrahedron Lett.* **2003**, *44*, 6955.

10 P. M. Henry, *J. Am. Chem. Soc.* **1966**, *88*, 1595.

11 (a) J. Tsuji, H. Nagashima, H. Nemoto, *Org. Synth.* **1984**, *62*, 9; (b) D. R. Fahey, E. A. Zuech, *J. Org. Chem.* **1974**, *39*, 3276.

12 T. Yokota, A. Sakakura, M. Tani, S. Sakaguchi, Y. Ishii, *Tetrahedron Lett.* **2002**, *43*, 8887.

13 (a) K. Januszkiewicz, H. Alper, *Tetrahedron Lett.* **1983**, *47*, 5159; (b) Phillips Petroleum Co. (T. P. Murtha, T. K. Shioyama), US 4.434.082 **(1984)**.

14 (a) C. Lapinte, H. Rivière, *Tetrahedron Lett.* **1977**, *43*, 3817; (b) Phillips Petroleum Co. (P. R. Stapp), US 4.237.071 **(1980)**.

15 K. Januszkiewicz, H. Alper, *Tetrahedron Lett.* **1983**, *24*, 5163.

16 H. Alper, K. Januszkiewicz, D. J. H. Smith, *Tetrahedron Lett.* **1985**, *26*, 2263.

17 I. Rico, F. Couderc, E. Perez, J. P. Laval, A. Lattes, *J. Chem. Soc., Chem. Commun.* **1987**, 1205.

18 F. R. Hartley, in *Supported Metal Complexes, Catalysis by Metal Complexes* (Eds.: R. Ugo, B. R. James), D. Reidel **1985**, pp. 293–294.

19 (a) D. C. Sherrington, H. G. Tang, *J. Catal.* **1993**, *142*, 540; (b) D. C. Sherrington, H. G. Tang. *J. Mol. Catal.* **1994**, *94*, 7.

20 J. P. Arhancet, M. E. Davis, B. E. Hanson, *Catal. Lett.* **1991**, *11*, 129.

21 E. Monflier, E. Blouet, Y. Barbaux, A. Mortreux, *Angew. Chem., Int. Ed. Engl.* **1994**, *33*, 2100.

22 E. Monflier, S. Tilloy, E. Blouet, Y. Barbaux, A. Mortreux, *J. Mol. Catal. A:* **1996**, *109*, 27.

23 E. Karakhanov, A. Maximov, A. Kirillov, *J. Mol. Catal. A:* **2000**, *157*, 25.

24 E. Karakhanov, T. Buchneva, A. Maximov, M. Zavertyaeva, *J. Mol. Catal. A:* **2002**, *184*, 11.

2.4.3.3
Methyltrioxorhenium(VII) in Oxidation Catalysis

Fritz E. Kühn

2.4.3.3.1
Introduction

Since 1989 organorhenium(VII) oxides, especially the very stable methyltrioxorhenium (MTO), have proven to be excellent catalyst precursors for a surprisingly broad variety of processes, among them various oxidation reactions [1–3]. This section gives a brief summary of the behavior of MTO and its peroxo derivatives in aqueous multiphase systems. Particular emphasis is given to the well-examined olefin epoxidation reactions.

2.4.3.3.2
Synthesis of Methyltrioxorhenium(VII)

MTO was first synthesized in 1979 in a quite time-consuming (weeks) and low-scale (milligrams) synthesis [4]. The breakthrough toward possible applications came only about ten years later [5]. An additional improvement was the prevention of any significant Re loss as unwanted byproducts [6, 7]. A further modification of the synthesis avoids the moisture-sensitive and expensive dirhenium heptaoxide as starting material, using instead Re powder or perrhenates [8, 9]. This method is of particular additional interest since it allows recyclization the catalyst decomposition products from reaction solutions (Scheme 1).

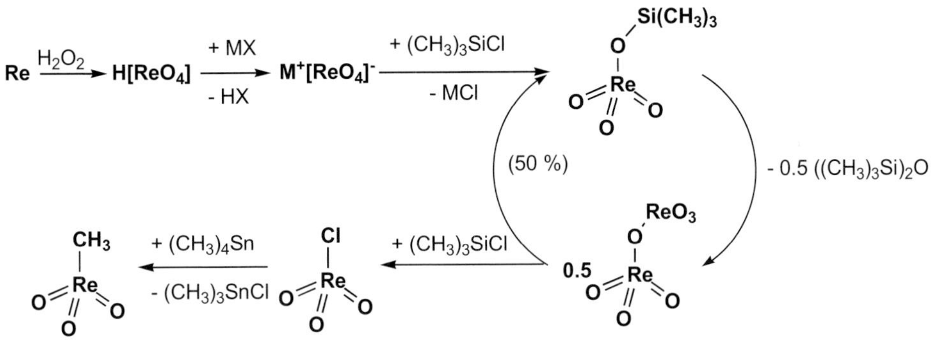

$$Re \xrightarrow{H_2O_2} H[ReO_4] \xrightarrow[-HX]{+ MX} M^+[ReO_4]^- \xrightarrow[-MCl]{+ (CH_3)_3SiCl}$$

(50 %)

$- 0.5\ ((CH_3)_3Si)_2O$

$+ (CH_3)_4Sn$ / $- (CH_3)_3SnCl$

$+ (CH_3)_3SiCl$

0.5

Scheme 1 Synthesis of methyltrioxorhenium(VII).

2.4.3.3.3
Behavior of Methyltrioxorhenium in Aqueous Solutions

MTO hydrolyzes rapidly in basic aqueous solutions and much more slowly in acidic media. At low concentrations (σ_{MTO} (concentration) < 0.008 M) the formation of methane gas and perrhenate was detected. At higher concentrations a second reaction, a faster reversible polymerization–precipitation, takes place to yield a golden-colored solid of the empirical composition $\{H_{0.5}[(CH_3)_{0.92}ReO_3]\}$ (poly-MTO) in about 70% yield [Eq. (1)] [10–15]. The reaction follows first-order reversible kinetics. The rate of polymerization–precipitation is independent of the concentration of H^+, and the reaction does not occur in the presence of oxidants [10]. The structure of the crystalline domains of poly-MTO can be described as double layers of corner-sharing CH_3ReO_5 octahedra (AA') with intercalated water molecules (B) in an ...AA'BAA'... layer sequence. It adopts the three-dimensional extended ReO_3 motif in two dimensions as a $\{ReO_2\}_n$ network. The oxo groups of two adjacent layers are face to face with the intercalated water layer. Hydrogen bridges are formed between the oxo groups and the water molecules, enhancing the structure of the polymer. The double layers are interconnected by van der Waals attractions generated by the nonpolar methyl groups, which are orientated inside the double layer (Figure 1). These structural features explain the observed lubricity of poly-MTO.

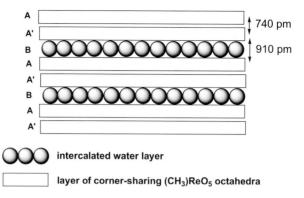

OOO intercalated water layer

☐ layer of corner-sharing (CH₃)ReO₅ octahedra

Figure 1 Structure of the crystalline domains of poly-MTO.

Substoichiometry with respect to the CH_3/Re ratio of $4.6 : 5$ and partial reduction by extra hydrogen equivalents are responsible for a high electric conductivity of poly-MTO [13, 14]. Poly-MTO also contains amorphous areas with a defect stacking of double layers and with a smaller content of water. The conductivity of poly-MTO is attributed to occasional demethylation and to the inclusion of extra hydrogen in the lattice. A theoretical study of MTO dimers, trimers, and tetramers has been performed. The tetramer already provides a good model for the interpretation of the IR and Raman spectra of poly-MTO and assigns the bridging oxygen atoms as the best candidates to bind the excess protons present in poly-MTO [16]. The terminal oxygens of monomeric MTO seem to be easily available for the formation of bridging oxygens [17].

$$O=\overset{\overset{\displaystyle CH_3}{|}}{\underset{\underset{\displaystyle O}{\|}}{Re}}\diagdown O + H_2O \longrightarrow \text{"}\{H_{0.5}[(CH_3)_{0.92}ReO_3]\}_\circ + O_2 + (HReO_4 + CH_4)\text{"} \qquad (1)$$

MTO

poly-MTO
(ca 70 %)

(ca. 30 %)

2.4.3.3.4
Catalyst Formation and Catalytic Applications

An important step in the understanding of the role of MTO in epoxidation and oxidation catalysis was the isolation and characterization of the reaction product of MTO with excess H_2O_2, i.e., a bisperoxo complex of stoichiometry $(CH_3)Re(O_2)_2O \cdot H_2O$ [18, 19]. This reaction takes place in any organic solvent or water. The structures of $(CH_3)Re(O_2)_2O \cdot H_2O$ and $(CH_3)Re(O_2)_2O \cdot (O=P(N(CH_3)_2)_3)$ (X-ray diffraction) were determined; the structure of ligand-free complex $(CH_3)Re(O_2)_2O$ is known from the gas phase [19, 20].

Experiments with the isolated bis(peroxo) complex $(CH_3)Re(O_2)_2O \cdot H_2O$ have shown that it is an active species in olefin epoxidation catalysis and several other catalytic reactions [19, 20]. In-situ experiments show that the reaction of MTO with

1 equiv. of H_2O_2 leads to a monoperoxo complex of the composition $(CH_3)Re(O_2)O_2$ [21, 22]. $(CH_3)Re(O_2)O_2$ has not been isolated and exists solely in equilibrium with MTO and $(CH_3)Re(O_2)_2O \cdot H_2O$. The monoperoxo complex is also catalytically active in oxidation processes. Kinetic experiments indicate that the rate constants for the transformation of most substrates into their oxidation products by catalysis with the mono- and bisperoxo complexes are of a comparable order of magnitude [21–23]. These results are supported by density functional calculations [24–26]. The transition states in the olefin epoxidation process starting from $(CH_3)Re(O_2)O_2$ and $(CH_3)Re(O_2)_2O \cdot H_2O$ are not different enough in energy to exclude one of these two catalysts totally from the catalytic process. The activation parameters for the coordination of H_2O_2 to MTO have also been determined. They indicate a mechanism involving nucleophilic attack. The protons lost in converting H_2O_2 to a coordinated O_2^{2-} ligand are transferred to one of the terminal oxygen atoms, which remains on the Re as the aqua ligand L. The rate of this reaction is not pH-dependent [27]. Two catalytic pathways for the olefin epoxidation may be described, depending on the concentration of the hydrogen peroxide used. With 85% hydrogen peroxide, only $(CH_3)Re(O_2)_2O \cdot H_2O$ appears to be responsible for the epoxidation activity (Scheme 2, cycle A). When a solution of 30 wt.% or less H_2O_2 is used, the monoperoxo complex, $(CH_3)Re(O_2)O_2$, is also taking part in the epoxidation process and a second catalytic cycle is involved as shown in Scheme 2, cycle B. For both cycles, a concerted mechanism is suggested in which the electron-rich double bond of the alkene attacks a peroxidic oxygen of $(CH_3)Re(O_2)_2O \cdot H_2O$. It has been inferred from experimental data that the system may involve a spiro arrangement [24–29].

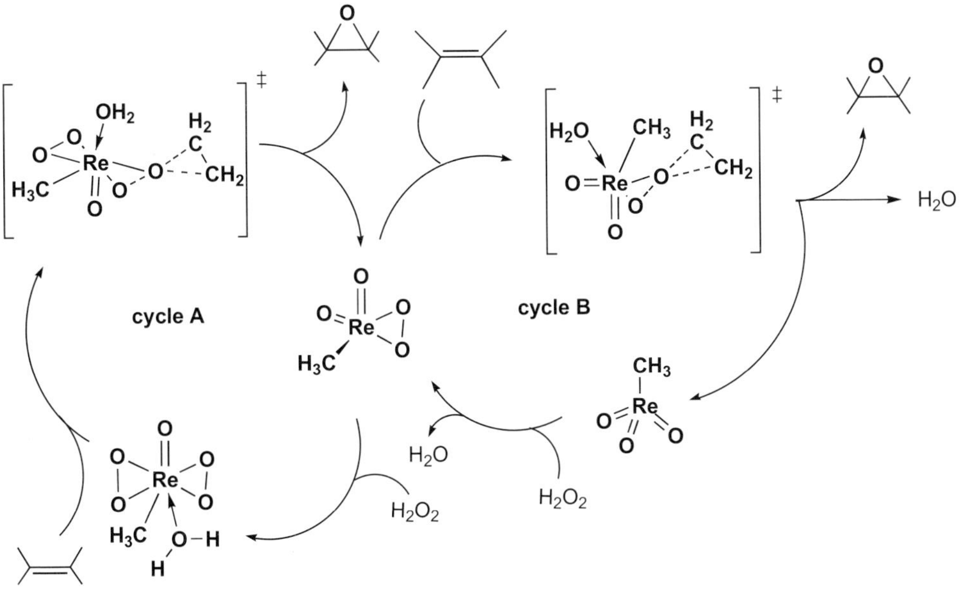

Scheme 2 MTO-catalyzed oxidation of olefins.

The most important drawback of the MTO-catalyzed process is the concomitant formation of diols instead of the desired epoxides, especially in the case of more sensitive substrates [30]. It was quickly detected that the use of Lewis base adducts of MTO significantly decreases the formation of diols due to the reduced Lewis acidity of the catalyst system. However, while the selectivity increases, the conversion decreases [30–32]. It was found that biphasic systems (water phase/organic phase) and addition of a significant excess of pyridine as Lewis base not only hamper the formation of diols but also increase the reaction velocity in comparison to MTO as catalyst precursor [33, 34].

Additionally it was shown that 3-cyanopyridine and especially pyrazole as Lewis bases are even more effective and less problematic than pyridine itself, while pyridine N-oxides are less efficient [35–37]. The Brønsted basicity of pyridines lowers the activity of hydronium ions, thus reducing the rate of opening the epoxide ring [38]. MTO forms trigonal-bipyramidal adducts with monodentate N-bases and (distorted) octahedral adducts with bidentate Lewis bases [39–41]. The monodentate Lewis-base adducts of MTO react with H_2O_2 to form mono- and bisperoxo complexes analogous to that of MTO, but coordinated by one Lewis-base molecule instead of H_2O. From the Lewis-base–MTO complexes to the bisperoxo complexes a clear increase in electron deficiency at the Re center can be observed by spectroscopic methods. The activity of the bisperoxo complexes in olefin epoxidation depends on the Lewis bases, the redox stability of the ligands, and the excess of Lewis base used. The peroxo complexes of the MTO–Lewis bases are, in general, more sensitive to water than MTO itself [40].

Furthermore, in the presence of olefins which are not readily transformed to their epoxides, 2,2'-bipyridine can be oxidized to its N-oxide by the MTO/H_2O_2 system [42]. Low to moderate stereo induction values (up to about 40% *ee* with conversions of around 10% at −5 °C reaction temperature) can be achieved when prochiral olefins, e.g., *cis*-β-methylstyrene or α-pinene are epoxidized with chiral amine adducts of MTO [43].

MTO has also been successfully applied as an olefin epoxidation catalyst in ionic liquids ([44, 45]; see also Chapter 5).

Alternative strategies to improve MTO-catalyzed oxidations have made use of host–guest inclusion chemistry [46–51].

A particularly important role of water and pH values becomes evident when looking at the catalyst deactivation processes. While MTO and its peroxo complexes are quite stable in acidic media, basic conditions lead to significantly reduced catalyst stabilities. In spite of the extraordinarily strong Re–C bond [52], characteristic of MTO, the cleavage of this bond plays a prominent role in the decomposition processes of these complexes [10, 53–55]. Concerning MTO, the full kinetic pH profile for the base-promoted decomposition to CH_4 and ReO_4^- was examined. Spectroscopic and kinetic data give evidence for mono- and dihydroxo complexes of formulae $CH_3ReO_3(OH^-)$ and $CH_3ReO_3(OH^-)_2$ prior to and responsible for the decomposition process. In the presence of hydrogen peroxide, $(CH_3)Re(O_2)O_2$ and $(CH_3)Re(O_2)_2O \cdot H_2O$ decompose to methanol and perrhenate with a rate that is dependent on $[H_2O_2]$ and $[H_3O]^+$. The complex peroxide and pH dependencies are explained by two possible pathways: attack of either hydroxide on $(CH_3)Re(O_2)O_2$ or HO_2^- on MTO. The bisperoxo complex decomposes much more slowly to yield O_2 and MTO [55]. Thus critical concentrations of strong nucleophiles have to be avoided; a great excess of hydrogen peroxide stabilizes the catalyst. It turned out to be advantageous to keep the steady-state concentration of water during the oxidation reaction as low as possible to depress catalyst deactivation.

Quite recently the possibility of MTO-catalyzed reactions utilizing dioxygen from the air instead of H_2O_2 as the oxidizing agent has been reported for some special cases [56, 57].

2.4.3.3.5
Summary

In summary, epoxidations with the MTO/H_2O_2 display several advantages. MTO is easily available, active in low concentrations of both MTO (0.05 mol%) and H_2O_2 (< 5 wt.%), it works over a broad temperature range (−40 to +90 °C) and is stable in water under acidic conditions as well as in basic media in special cases. Furthermore, the MTO/H_2O_2 system works in a broad variety of solvents, ranging from highly polar ones (e.g., nitromethane, water) to ones with low polarity (e.g., toluene). However, the reactions between MTO/H_2O_2 and alkenes are ca. one order of magnitude faster in semi-aqueous solvents (e.g., 85% H_2O_2) than in methanol. The rate constants for the reaction of MTO/H_2O_2 with aliphatic alkenes correlate closely with the number of alkyl groups on the alkene carbons. The reactions become significantly slower when electron-withdrawing groups such as −OH, −CO, −Cl, and −CN are present in the substrates.

A major advantage of MTO is that it does not decompose hydrogen peroxide. This is in striking contrast to many other oxidation catalysts. Turnover numbers of up to 2500 mol product per mole catalyst (reaction conditions: 0.1 mol% MTO, 5 mol% pyrazole) and turnover frequencies of up to 14 000 mole product per mole catalyst per hour have been reported, with typical MTO concentrations of 0.1–1.0 mol%. However, these impressive results have only been reached in perfluorinated alcohols as solvents with cyclohexene as substrate [58].

High selectivity (epoxide versus diol) can be adjusted by temperature control, trapping of water, or the use of certain additives, such as aromatic Lewis-base ligands, which additionally accelerate the epoxidation reactions. Selectivities of > 95% can be reached.

References

1 K. A. JØRGENSEN, Chem. Rev. 1989, 89, 447.
2 W. A. HERRMANN, F. E. KÜHN, Acc. Chem. Res. 1997, 30, 169.
3 C. C. ROMÃO, F. E. KÜHN, W. A. HERRMANN, Chem. Rev. 1997, 97, 3197.
4 I. R. BEATTIE, P. J. JONES, Inorg. Chem. 1979, 18, 2318.
5 W. A. HERRMANN, J. G. KUCHLER, J. K. FELIXBERGER, E. HERDTWECK, W. WAGNER, Angew. Chem., Int. Ed. Engl. 1988, 27, 394.

6 W. A. Herrmann, F. E. Kühn, R. W. Fischer, W. R. Thiel, C. C. Romão, *Inorg. Chem.* **1992**, *31*, 4431.

7 W. A. Herrmann, in *Applied Homogeneous Catalysis with Organometallic Compounds* (Eds.: B. Cornils, W. A. Herrmann), Vol. 3, 2nd ed., p. 1319, Wiley-VCH, Weinheim **2002**.

8 W. A. Herrmann, R. M. Kratzer, R. W. Fischer, *Angew. Chem., Int. Ed. Engl.* **1997**, *36*, 2652.

9 W. A. Herrmann, R. M. Kratzer, *Inorg. Synth.* **2002**, *33*, 110.

10 G. Laurenczy, F. Lukács, R. Roulet, W. A. Herrmann, R. W. Fischer, *Organometallics* **1996**, *15*, 848.

11 W. A. Herrmann, R. W. Fischer, W. Scherer, *Adv. Mater.* **1992**, *4*, 653.

12 W. A. Herrmann, R. W. Fischer, *J. Am. Chem. Soc.* **1995**, *117*, 3223.

13 W. A. Herrmann, W. Scherer, R. W. Fischer, J. Blümel, M. Kleine, W. Mertin, R. Gruehn, J. Mink, C. C. Wilson, R. M. Ibberson, L. Bachmann, M. R. Mattner, *J. Am. Chem. Soc.* **1995**, *117*, 3231.

14 H. S. Genin, K. A. Lawler, R. Hoffmann, W. A. Herrmann, R. W. Fischer, W. Scherer, *J. Am. Chem. Soc.* **1995**, *117*, 3244.

15 M. R. Mattner, W. A. Herrmann, R. Berger, C. Gerber, J. K. Gimzewski, *Adv. Mater.* **1996**, *8*, 654.

16 S. Köstlmeier, G. Pacchioni, W. A. Herrmann, N. Rösch, *J. Organomet. Chem.* **1996**, *514*, 111.

17 P. Gisdakis, N. Rösch, É. Bencze, I. S. Gonçalves, F. E. Kühn, *Eur. J. Inorg. Chem.* **2001**, 989.

18 W. A. Herrmann, R. W. Fischer, D. W. Marz, *Angew. Chem., Int. Ed. Engl.* **1991**, *30*, 1638.

19 W. A. Herrmann, R. W. Fischer, W. Scherer, M. U. Rauch, *Angew. Chem., Int. Ed. Engl.* **1993**, *32*, 1157.

20 W. A. Herrmann, J. D. G. Correia, G. R. J. Artus, R. W. Fischer, C. C. Romão, *J. Organomet. Chem.* **1996**, *520*, 139.

21 A. Al-Ajlouni, H. Espenson, *J. Am. Chem. Soc.* **1995**, *117*, 9234.

22 J. H. Espenson, *J. Chem. Soc., Chem. Commun.* **1999**, 479.

23 W. Adam, C. R. Saha-Möller, O. Weichold, *J. Org. Chem.* **2000**, *65*, 5001.

24 P. Gisdakis, N. Rösch, *Eur. J. Org. Chem.* **2001**, *4*, 719.

25 P. Gisdakis, I. V. Yudanov, N. Rösch, *Inorg. Chem.* **2001**, *40*, 3755.

26 C. di Valentin, R. Gandolfi, P. Gisdakis, N. Rösch, *J. Am. Chem. Soc.* **2001**, *123*, 2365.

27 O. Pestovski, R. v. Eldik, P. Huston, J. H. Espenson, *J. Chem. Soc., Dalton Trans.* **1995**, 133.

28 P. Gisdakis, W. Antonczak, S. Köstlmeier, W. A. Herrmann, N. Rösch, *Angew. Chem., Int. Ed.* **1998**, *37*, 2211.

29 W. Adam, Mitchell, C. M., *Angew. Chem., Int. Ed. Engl.* **1996**, *35*, 533.

30 W. A. Herrmann, R. W. Fischer, M. U. Rauch, W. Scherer, *J. Mol. Catal.* **1994**, *86*, 243.

31 W. Adam, C. M. Mitchel, C. R. Saha-Möller, *J. Org. Chem.* **1999**, *64*, 3699.

32 G. S. Owens, M. M. Abu-Omar, *J. Chem. Soc., Chem. Commun.* **2000**, 1165.

33 J. Rudolph, K. L. Reddy, J. P. Chiang, K. B. Sharpless, *J. Am. Chem. Soc.* **1997**, *119*, 6189.

34 H. Adolfsson, A. Converso, K. B. Sharpless, *Tetrahedron Lett.* **1999**, *40*, 3991.

35 C. Copéret, H. Adolfsson, K. B. Sharpless, *J. Chem. Soc., Chem. Commun.* **1997**, 1565.

36 W. A. Herrmann, H. Ding, R. M. Kratzer, F. E. Kühn, J. J. Haider, R. W. Fischer, *J. Organomet. Chem.* **1997**, *549*, 319.

37 W. A. Herrmann, F. E. Kühn, M. R. Mattner, G. R. J. Artus, M. Geisberger, J. D. G. Correia, *J. Organomet. Chem.* **1997**, *538*, 203.

38 W. D. Wang, J. H. Espenson, *J. Am. Chem. Soc.* **1998**, *120*, 11 335.

39 F. E. Kühn, A. M. Santos, P. W. Roesky, E. Herdtweck, W. Scherer, P. Gisdakis, I. V. Yudanov, C. Di Valentin, N. Rösch, *Chem. Eur. J.* **1999**, *5*, 3603.

40 P. Ferreira, W. M. Xue, É. Bencze, E. Herdtweck, F. E. Kühn, *Inorg. Chem.* **2001**, *40*, 5834.

41 S. M. Nabavizadeh, *Inorg. Chem.* **2003**, *42*, 4204.

42 M. Nakajima, Y. Sasaki, H. Iwamoto, S. I. Hashimoto, *Tetrahedron Lett.* **1998**, *39*, 87.

43 M. J. Sabatier, M. E. Domine, A. Corma, *J. Catal.* **2002**, *210*, 192.

44 G. S. Owens, M. M. Abu-Omar, *J. Mol. Catal. A:* **2002**, *187*, 215.

45 G. S. Owens, A. Durazo, M. M. Abu-Omar, *Chem. Eur. J.* **2002**, *8*, 3053.

46 W. Adam, C. M. Mitchell, C. R. Saha-Möller, O. Weichold, *J. Am. Chem. Soc.* **1999**, *121*, 2097.

47 D. Sica, D. Musumeci, F. Zollo, S. de Marino, *Eur. J. Org. Chem.* **2001**, *19*, 373.

48 R. Saladino, P. Carlucci, M. C. Danti, C. Crestini, E. Mincione, *Tetrahedron* **2000**, 10 031.

49 R. Buffon, U. Schuchardt, *J. Braz. Chem. Soc.* **2003**, *14*, 347.

50 A. O. Bouh, J. H. Espenson, *J. Mol. Catal. A:* **2003**, *200*, 43.

51 R. Bernini, E. Mincione, M. Cortese, R. Saladino, G. Gualandi, M. C. Belfiore, *Tetrahedron Lett.* **2003**, *44*, 4823.

52 C. Mealli, J. A. Lopez, M. J. Calhorda, C. C. Romão, W. A. Herrmann, *Inorg. Chem.* **1994** *33*, 1139.

53 K. A. Brittingham, J. H. Espenson, *Inorg. Chem.* **1999**, *38*, 744.

54 J. H. Espenson, H. Tan, S. Mollah, R. S. Houk, M. D. Eager. *Inorg. Chem.* **1998**, *37*, 4621.

55 M. M. Abu Omar, P. J. Hansen, J. H. Espenson, *J. Am. Chem. Soc.* **1996**, *118*, 4966.

56 V. B. Sharma, S. L. Jain, B. Sain, *Tetrahedron Lett.* **2003**, *44*, 2655.

57 V. B. Sharma, S. L. Jain, B. Sain, *Tetrahedron Lett.* **2003**, *44*, 3235.

58 M. C. A. van Vliet, I. W. C. E. Arends, R. A. Sheldon, *J. Chem. Soc., Chem. Commun.* **1999**, 821.

2.4.4
Addition Reactions

2.4.4.1
Hydrocyanation

Henry E. Bryndza and John A. Harrelson Jr.

Hydrogen cyanide is a remarkably versatile C_1 building block. Its use, however, has been limited by its relatively difficult synthesis/purification as well as by its high flammability, its tendency toward base-catalyzed explosive polymerization, and its toxicity. Most of the recent hydrocyanation literature comes from industry rather than academic laboratories.

If toxicity and handling difficulties are the drawbacks of HCN as a feedstock, the versatility of the nitrile functional group as a synthon is a significant advantage. Moreover, the full miscibility of HCN with water at 25 °C offers a potential advantage to pursuing hydrocyanation catalysis in aqueous media. Therefore, hydrocyanation reactions offer the potential for generating new nitrogenous products which are useful intermediates, including the addition of HCN to C=C, C=O, and C=N double bonds to generate new alkyl nitriles, cyanohydrins, and aminonitriles, respectively. All of these reactions have commercial impact today [1], although aqueous hydrocyanation catalysis remains an emerging area of technology.

The addition of HCN to activated alkenes, as in Michael additions, has been known and commercially practiced for many years. Addition of HCN to isophorone [Eq. (1)] is an example [2].

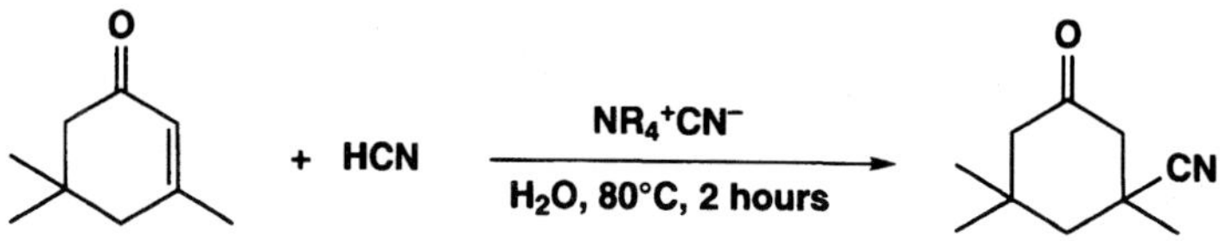

$$\text{(structure)} + \text{HCN} \xrightarrow[\text{H}_2\text{O, 80°C, 2 hours}]{\text{NR}_4{}^+\text{CN}^-} \text{(structure with CN)} \quad (1)$$

More common is the general acid- or base-catalyzed addition of HCN to ketones and aldehydes to give cyanohydrins [Eq. (2)] [1]. Because of the propensity of HCN to spontaneously and exothermically polymerize under basic conditions, general acid catalysis is sometimes favored over basic media, as was the case in the recent Sumitomo [3] and Upjohn work [4, 5]. Applications of aqueous media have been reported to lead to asymmetric hydrocyanation catalysis [Eq. (3)] [6, 7].

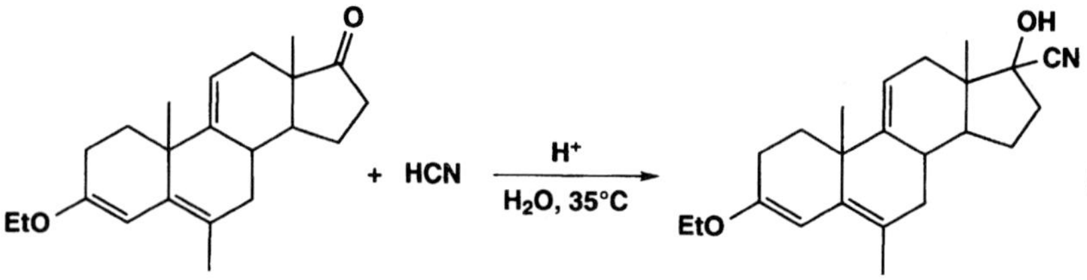

$$\text{(steroid structure)} + \text{HCN} \xrightarrow[\text{H}_2\text{O, 35°C}]{\text{H}^+} \text{(steroid structure with OH, CN)} \quad (2)$$

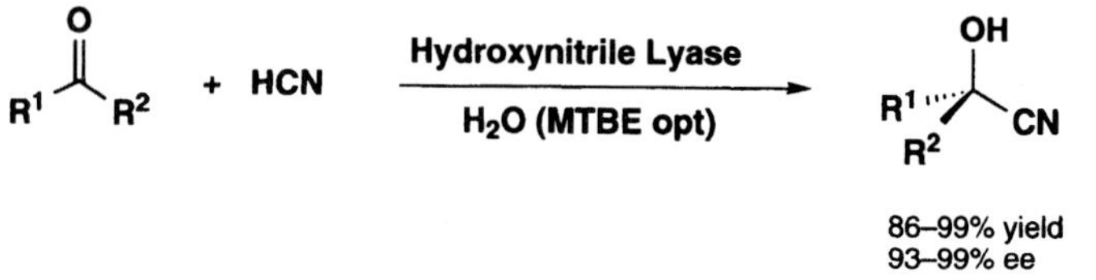

$$R^1 \overset{O}{\underset{}{\|}} R^2 + \text{HCN} \xrightarrow[\text{H}_2\text{O (MTBE opt)}]{\text{Hydroxynitrile Lyase}} R^1 \overset{OH}{\underset{R^2}{\|}} CN \quad (3)$$

86–99% yield
93–99% ee

Another variant of this HCN addition to polar C=X bonds is the Strecker synthesis of aminonitriles from ketones or aldehydes, HCN, and ammonia [Eq. (4)] [8]. A number of patents to Distler and co-workers at BASF do teach the use of aqueous media to facilitate isolation of pure products [9], and researchers at Grace [10], Stauffer [11], Mitsui [12], and Hoechst [13] have reportedly used aqueous media to control the rates of formaldehyde aminohydrocyanation to isolate intermediate addition products in good yields. The use of aqueous media has also proven advantageous when amides, rather than nitriles, are the desired products.

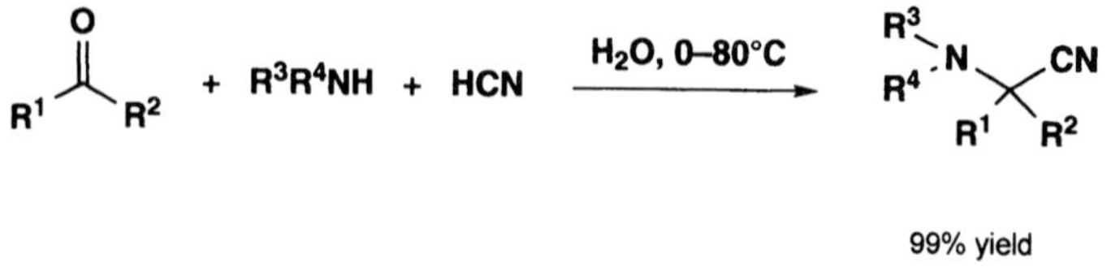

$$R^1 \overset{O}{\underset{R^2}{\|}} + R^3R^4NH + \text{HCN} \xrightarrow[]{\text{H}_2\text{O, 0–80°C}} \overset{R^3}{\underset{R^4}{}}N \overset{CN}{\underset{R^1 \quad R^2}{}} \quad (4)$$

99% yield

The addition of HCN to C=C double bonds can be effected in low yields to produce Markovnikov addition products. However, through the use of transition metal catalysts, the selective anti-Markovnikov addition of HCN to alkenes can take place. The most prominent example of the use of aqueous media for transition metal-catalyzed alkene hydrocyanation chemistry is the three-step synthesis of adiponitrile from butadiene and HCN [Eqs. (5–7)]. First discovered by Drinkard at DuPont [14], this nickel-catalyzed chemistry can use a wide variety of phosphorus ligands [15] and is practiced commercially in nonaqueous media by both DuPont and Butachimie, a DuPont/Rhône-Poulenc joint venture. Since the initial reports of Drinkard, first Kuntz [16] and, more recently, Huser and Perron [17, 18] from Rhône-Poulenc have explored the use of water-soluble ligands for this process to facilitate catalyst recovery and recycle from these high-boiling organic products.

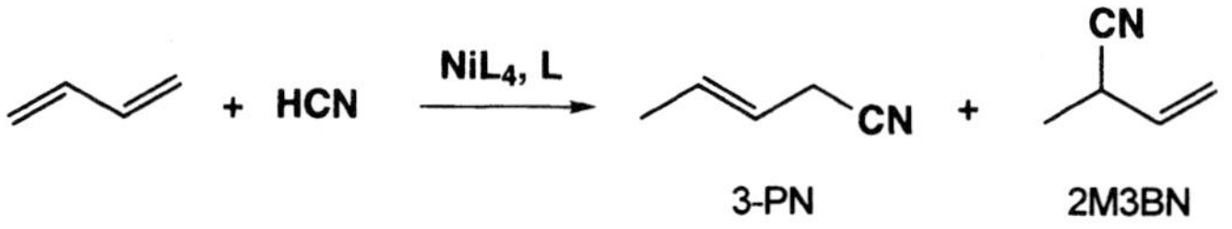

$$\text{+ HCN} \xrightarrow{\text{NiL}_4,\ \text{L}} \underset{\text{3-PN}}{\diagup\!\!\diagdown\!\!\diagup^{\text{CN}}} + \underset{\text{2M3BN}}{\overset{\text{CN}}{\diagup\!\!\diagdown}} \tag{5}$$

$$\underset{\text{2M3BN}}{\overset{\text{CN}}{\diagup\!\!\diagdown}} \xrightarrow{\text{NiL}_4,\ \text{L}} \underset{\text{3-PN}}{\diagup\!\!\diagdown\!\!\diagup^{\text{CN}}} \tag{6}$$

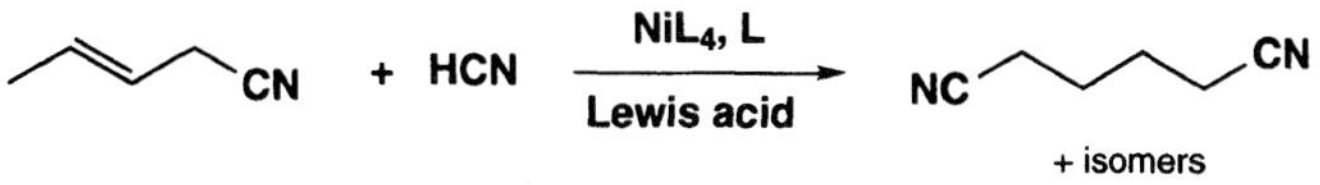

$$\diagup\!\!\diagdown\!\!\diagup^{\text{CN}} + \text{HCN} \xrightarrow[\text{Lewis acid}]{\text{NiL}_4,\ \text{L}} \text{NC}\diagup\!\!\diagdown\!\!\diagup\!\!\diagdown^{\text{CN}} \tag{7}$$
$$\text{+ isomers}$$

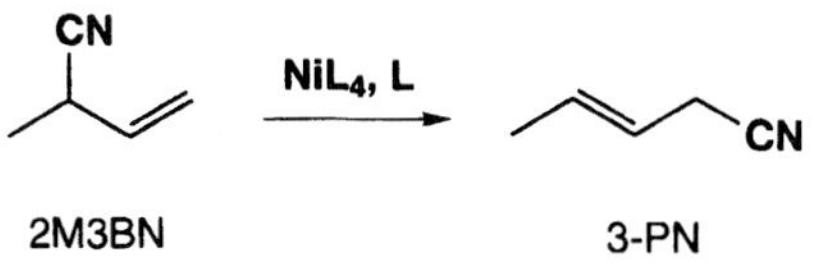

Huser et al. have extended Kuntz's work to the isomerization of 2-methyl-3-butenenitrile (2M3BN) to 3-PN (isomerization step; Eq. (6) 92% yield) [17–19].

A final example of aqueous media used in the hydrocyanation of butadiene is provided by Waddan at ICI [20]. In this chemistry, copper nitrate salts in aqueous media (among many others) are used for the oxidative dihydrocyanation of butadiene to dicyanobutenes [Eq. (8)].

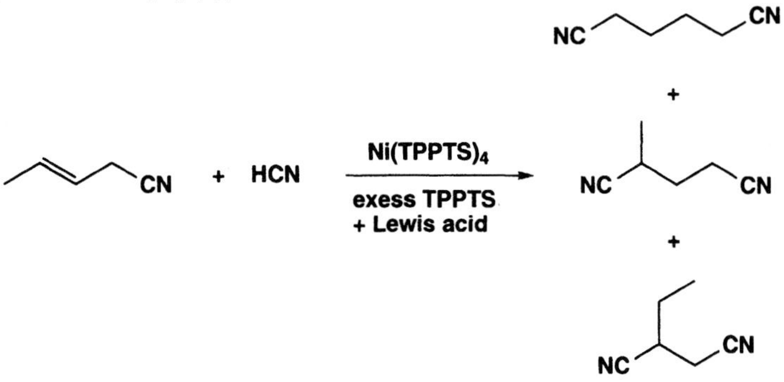

$$\diagup\!\!\diagdown\!\!\diagup^{\text{CN}} + \text{HCN} \xrightarrow[\substack{\text{exess TPPTS}\\ \text{+ Lewis acid}}]{\text{Ni(TPPTS)}_4} \begin{array}{c} \text{NC}\diagup\!\!\diagdown\!\!\diagup\!\!\diagdown^{\text{CN}} \\ + \\ \text{NC}\diagup\!\!\diagdown\!\!\diagup^{\text{CN}} \\ + \\ \text{NC}\diagup\!\!\diagup^{\text{CN}} \end{array} \tag{8}$$

Another type of aqueous cyanide chemistry is the oxidative coupling of cyanide to produce oxamide [Eq. (9)]. Both batch and continuous reaction have been demonstrated at Hoechst by Riemenschneider and Wegener [21], who report advantages of aqueous media not only in concurrently hydrolyzing the coupled products but also in facilitating product isolation from the reaction medium. This clever combination of reactive solvent is reportedly the basis for the commercial production of oxamide [2].

$$\diagup\!\!\!\!\diagdown \quad + \text{ HCN } + \text{ 0.5 O}_2 \xrightarrow[\substack{\textbf{solvent or}\\\textbf{biphasic media}}]{\textbf{Cu(II)salts}} \quad \text{NC} \diagup\!\!\!\!\diagdown\!\!\!\!\diagup \text{CN} \quad (9)$$

95% yield

Even from the limited examples of hydrocyanation reactions, it is clear that HCN is a versatile reagent and that its chemistry is generally compatible with aqueous media. Some traditional advantages which recommend the use of aqueous or biphasic media include the high solubility of HCN in water, the facility in removing products from catalysts, the ability to control rates of reactions more finely and the concomitant hydrolysis of nitrile functional groups to amides along with C–C bond formation. Commercial applications of aqueous hydrocyanations have been emerging, and improved catalysts and separation technologies coming into play in the chemical industry.

References

1 K. WEISSERMEHL, H.-J. ARPE, *Industrial Organic Chemistry*, 2nd ed., VCH, Weinheim **1993**, p. 43 ff., 280.

2 Nippon Chemical Co., Ltd. (H. TAKAHOSO, N. TAKAHASHI, K. MIDORIKAWA, T. SATO), US 5.179.221 (**1993**), JP 04.279.559 (**1992**), EP 502.707B (**1995**), DE 69.205.601E (**1995**).

3 Sumitomo, JP 04.198.173 (**1992**).

4 Upjohn (J. G. REID, T. DEBIAK-KROOK), WO 95/30.684, WO 95/30.684 (**1995**).

5 Upjohn (J. G. REID, T. DEBIAK-KROOK), AU 9.523.628 (**1995**), EP 759.928 (**1977**).

6 Allied (B. J. ARENA), US 4.959.467 (**1990**), WO 89/07.591 (**1989**), ES 2.010.137)**1989**), EP 402.399B (**1993**), JP 03.502.692W (**1991**), CA 1.321.785C (**1993**), DE 68.910.812 (**1993**).

7 Duphar International Research (H. W. GELUK, W. T. LOOS), US 5.350.871 (**1994**), EP 547.655 (**1993**), CA 2.084.855 (**1993**), HUT 064.105 (**1993**), JP 053.176.065 (**1993**), HU 209.736B (**1994**); Degussa (F. EFFENBERGER, J. EICHHORN, J. ROOS), DE 19.506.728 (**1976**).

8 Anonymous, Cyanides (Hydrogen Cyanides), in *Kirk-Othmer Encyclopedia of Chemical Technology*, 2nd ed., **1965**, Vol. 6, pp. 5832–5883.

9 BASF AG (H. DISTLER, K. L. HOCK), US 4.478.759 (**1984**), EP 45.386B (**1984**), DE 3.162.511G (**1984**); BASF AG (H. DISTLER, E. HARTERT, H. SCHLECHT), US 4.113.764 (**1978**), US 4.134.889 (**1979**), BE 854.313 (**1977**), DE 2.620.445B (**1979**), DE 2.621.450B (**1979**), DE 2.621.728B (**1979**), JP 52.136.124 (**1977**), DE 2.625.935B (**1979**), FR 2.350.332 (**1978**), DE 2.660.191 (**1978**), GB 1.577.662 (**1980**), US 4.164.511 (**1979**), US 4.022.815 (**1977**), BE 837.970 (**1976**), FR 2.299.315 (**1976**), GB 1.526.481 (**1978**), DE 2.503.582B (**1979**), CA 1.056.402 (**1979**), CH 619.926 (**1980**); BASF AG (H. DISTLER), EP 36.161B (**1983**), DE 3.010.511 (**1981**), UP 56.145.251 (**1981**), DE 3.160.741G (**1983**), JP 89.036.461B (**1989**).

10 W. R. GRACE, J. L. SU, M. B. SHERWIN, US 4.895.971 (**1990**), EP 367.364B (**1994**), DE 68.912.878E (**1994**), ES 2.061.961T3 (**1994**).

11 Stauffer (C. C. GRECO, W. STAMM), US 3.862.203 (**1976**), GB 1.244.329 (**1971**).

12 Mitsui Toatsu Chem. Inc., JP 02.009.874 (**1990**), JP 2.575.823B (**1997**).

13 Hoechst (K. WARNING, M. MITZLAFF, H. JENSEN), DE 2.634.048 (**1978**), DE 2.634.047 (**1978**).

14 Leading examples from an extensive patent literature include: DuPont (W. D. DRINKARD JR., R. V. LINDSEY JR.), US 3.536.748 (**1970**); DuPont (W. D. DRINKARD JR., B. W. TAYLOR), US 3.579.560 (**1971**); DuPont (Y. T. CHIA), US 3.676.481 (**1972**); DuPont (W. D. DRINKARD), US 3.739.011 (**1973**).

15 C. A. TOLMAN, R. J. MCKINNEY, J. D. DRULINER, W. R. STEVENS, *Adv. Catal.* **1985**, *33*, 1.

16 Rhône-Poulenc (E. KUNTZ), US 4.087.452 (**1978**), DE 2.700.904 (**1983**), BE 850.301 (**1977**), NL 77.00.262 (**1977**), BR 77.00171 (**1977**), FR 2.338.253 (**1977**), JP 52.116.418 (**1977**), GB

1.542.824 (**1979**), SU 677.650 (**1979**), CA 1.082.736 (**1980**), JP 82.061.270B (**1982**), IT
1.083.151B (**1985**), NL 188.158B (**1991**).
17 Rhône-Poulenc (M. HUSER, R. PERRON), US 5.486.643 (**1996**), EP 647.619 (**1995**),
FR 2.710.909 (**1995**), NO 9.403.748 (**1995**), CA 2.177.725 (**1995**), JP 07.188.143 (**1995**),
JP 2.533.073 (**1996**), EP 0.647.619A (**1994**).
18 Rhône-Poulenc (M. HUSER, R. PERRON), US 5.488.129 (**1996**), WO 96/33.969 (**1996**),
EP 650.959 (**1995**), FR 2.711.987 (**1995**), NO 9.404.153 (**1995**), CA 2.134.940 (**1995**),
JP 07.188.144 (**1995**), WO 97/12.857 (**1997**), FR 2.739.378 (**1977**).
19 Rhône-Poulenc (D. HORBEZ, M. HUSER, R. PERRON), WO 97/24.184 (**1997**), FR 2.743.011
(**1997**); Rhône-Poulenc (M. HUSER, R. PERRON), WO 97/24.183 (**1997**), FR 2.743.010 (**1997**),
and examples contained in Refs. [13–15].
20 ICI (D. Y. WADDAN), EP 032.299B (**1984**), JP 56.100.755 (**1981**), DE 3.066.441 (**1984**).
21 Hoechst (W. RIEMENSCHNEIDER, P. WEGENER), US 3.989.753 (**1976**), NL 7.402.197 (**1974**),
SE 7.402.352 (**1974**), FR 2.219.153 (**1974**), DD 110.854 (**1975**), JP 110.854 (**1975**),
JP 50.029.516 (**1975**), DE 2.403.120C (**1982**), AT 7.401.402 (**1976**), GB 1.458.871 (**1976**),
HUT 021.762 (**1977**), IL 44.209 (**1977**), CH 590.213 (**1977**), CS 7.401.122 (**1977**), CA
1.026.377 (**1978**), RO 70.980 (**1981**), JP 83.033.220B (**1983**), NL 177.400B (**1985**), SZ
7.401.011 (**1974**), BE 811.533 (**1974**), DE 2.308.941B (**1975**), DE 2.402.354C (**1982**),
DE 2.402.352 (**1975**), SU 631.069 (**1978**).

2.4.4.2
Hydrodimerization

Noriaki Yoshimura

The linear telomerization reaction of dienes (the taxogen) with nucleophiles such
as alcohols, amines, carboxylic acids, water, etc. (the telogen), catalyzed by ligand-
modified Pd or Ni complexes, provides an elegant method for the synthesis of
useful compounds. With water as telogen, the telomerization becomes a hydrodimerization. In the case of butadiene as taxogen, the reaction product is the versatile
2,7-octadien-1-ol (**1**) which may the basis for a series of various derivatives
(Scheme 1).

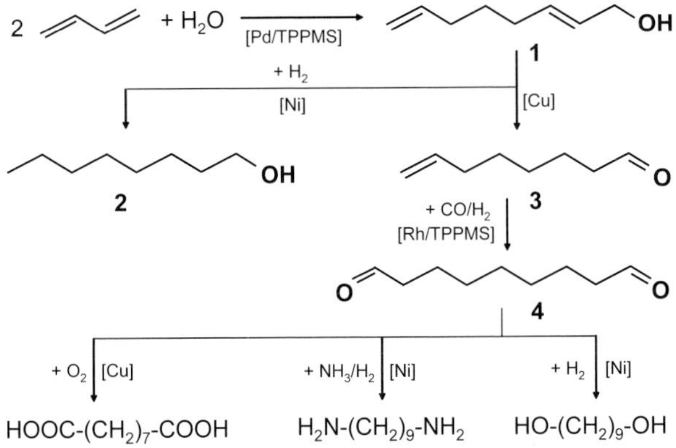

Scheme 1

2,7-Octadien-1-ol (**1** in Scheme 1) is a highly reactive compound which on the one hand can be hydrogenated over a fixed- bed Ni catalyst at 130–180 °C and a pressure of 3–8 MPa to 1-octanol (**2**). This octanol has a considerable market as a raw material for plasticizers for PVC and is produced with a capacity of approximately 5000 t y^{-1}. On the other hand, **1** can be hydrogenated/dehydrogenated over a copper chromite catalyst at 220 °C to yield 7-octenal (**3**). This aldehyde is hydroformylated to the dialdehyde **4** which is then hydrogenated to give 1,9-nonanediol. The dialdehyde can also give, on air oxidation in an acetic solvent with a copper catalyst, azelaic acid and, on reductive amination in ammonia in the presence of a nickel catalyst, 1,9-nonanediamine. The hydroformylation step mentioned is a variant of the RCH/RP process using TPP*M*S instead of TPP*T*S (cf. also Sections 2.2.3.2.1 and 2.4.1.1).

The hydrodimerization reaction to 1-octanol was developed commercially in 1991 by Kuraray using an aqueous homogeneous catalyst [1, 2], i.e., a ligand-modified palladium catalyst (Figure 1).

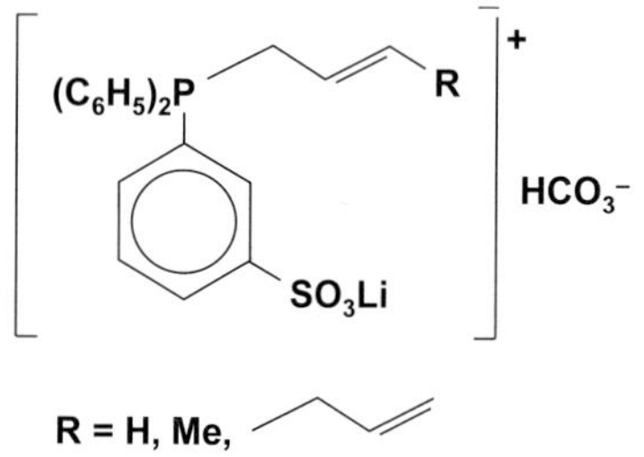

R = H, Me,

Figure 1 The TPPMS-modified Pd catalyst.

The catalyst is the phosphonium salt of the Li salt of TPPMS (cf. Section 2.2.3.2). This catalyst shows, as desired, sufficient activity at a high P/Pd ratio and no appreciable time-dependent deterioration of the activity after repeated catalyst cycles. The formation of phosphine oxides by traces of oxygen in the feed is also minimized. It has been demonstrated that the catalyst is stabilized at high levels if carbonate anions are present along with monoamines or tertiary ammonium carbonates.

To ensure a continuous operation and the separability of the reaction products from the catalyst solution, the use of sulfolane (tetrahydrothiophene-1,1-dioxide) is advantageous. Compared to other tested solvents, only sulfolane gives high yields (> 90%) and high selectivities (92%) with a ratio of 92 : 8 of the desired 2,7-octadien-1-ol versus the other isomer 1,7-octadien-3-ol.

For the telomerization of butadiene, distillation methods cannot be employed to separate the product from the reaction mixture containing the catalyst because the palladium complex has a lower thermal stability and high-boiling compounds would accumulate in the catalyst-containing solution that has been recycled. Therefore the extraction method with hexane as extraction agent (cf. Section 2.3.2) has been chosen (Figure 2).

In the present reaction, water acts as a nucleophile and the product hardly dissolves in it. Therefore, the process is capable of retaining the catalyst component

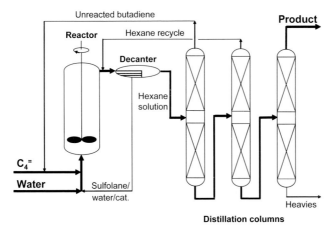

Figure 2 Extraction method for separating butadiene hydrodimerization products from the aqueous catalyst solution.

in the aqueous solution being used, and of extracting the product selectively. Thus, in order to solubilize the catalyst components in the aqueous sulfolane solution used, a hydrophilic group (e.g., a sulfonic acid salt) has been introduced into the phosphonium salt ligand and the procduct is extracted with an aliphatic saturated hydrocarbon such as hexane. This method has the advantages that (1) the catalyst and product can be separated without heating; (2) the extraction equilibrium is achieved for all compounds so that the accumulation of catalyst poisons and high-boiling byproducts is minimal. The catalyst lifetimes and the elution losses of catalyst components are within commercially acceptable ranges.

The total process consists thus of four steps: hydrodimerization, extraction, hydrogenation, and distillation [3]. The catalyst is separately prepared from a Pd salt, the phosphonium salt ligand, and a solution of triethylammonium hydrogen carbonate in aqueous sulfolane. Butadiene and water react at temperatures of 60–80 °C under a total pressure of CO_2 of 1–2 MPa, to achieve a 2,7-octadien-1-ol selectivity of 90–91%. In the following extraction step, 50–70% of the reaction products are extracted with hexane, and the aqueous sulfolane containing the catalyst, parts of the product, and the triethylammonium hydrogen carbonate returns to the reactor. The loss of catalyst is only in the range of a few ppm. After unreacted butadiene and hexane have been recovered from the extraction mixture, 2,7-octadien-1-ol is purified by distillation and subsequently hydrogenated.

According to Scheme 1, one of the products is 1-octanol, a valuable linear alcohol based on cheap butadiene instead of expensive 1-octene in combination with its hydroformylation (cf. Section 2.4.1.1). Alternatively, 2,7-octadien-1-ol can readily be converted to 7-octenal in a yield of at least 80%. This unsaturated aldehyde is then hydroformylated to a linar C_9 dialdehyde (**4** in Scheme 1). The dialdehyde may be converted by hydrogenation to 1,9-nonanediol, by hydrogenative amination to a nonanediamine, or by oxidation to azelaic acid: quite a whole, interesting product range.

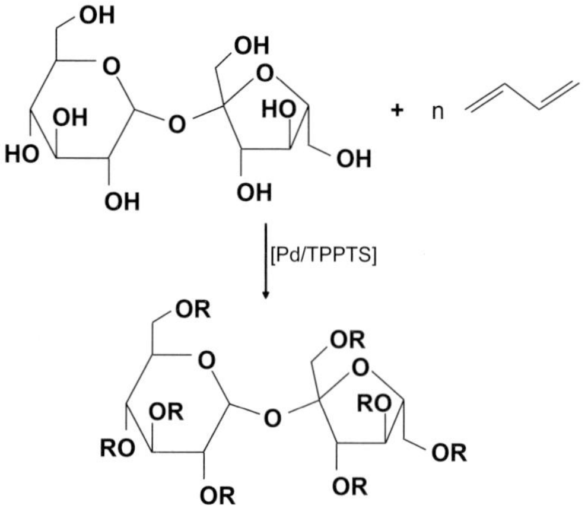

Scheme 2 Biphasic etherification of carbohydrates.

It may be noted that other homogeneous telomerization reactions of butadiene with ammonia lead to trioctadienylamines. When carried out in an aqueous two-phase operation with Pd/TPPTS, primary and secondary octadienylamines are obtained [4]. Telomerization of butadiene with formic acid or its salts using the above-mentioned process can produce 1,7-octadiene [5]. Substituting isoprene for butadiene leads to dimethyloctadiene. The biphasic Pd-catalyzed telomerization of butadiene with carbohydrates in aqueous operation is also an important reaction and yields the desired ethers (Scheme 2) [6].

As known from other publications [7], no other processes have been realized so far.

References

1 N. Yoshimura, in *Applied Homogeneous Catalysis with Organometallic Compounds*, 2nd ed. (Eds.: B. Cornils, W. A. Hermann), Wiley-VCH, Weinheim **2002**, Vol. 1, p. 361.
2 N. Yoshimura, in *Aqueous-Phase organometallic Catalysis*, 2nd ed. (Eds.: B. Cornils, W. A. Herrmann), Wiley-VCH, Weinheim **2004**, p. 540 ff.
3 Kuraray Ind. (Y. Tokitoh, N. Yoshimura et al.), US 5.057.631 (**1991**), US 5.118.885 (**1992**), US 4.417.079 (**1983**).
4 T. Prinz, W. Keim, B. Driessen-Hölscher, *Angew. Chem., Int. Ed. Engl.* **1996**, *35*, 1708.
5 Kuraray Ind. (T. Tsuda, N. Yoshimura), EP 0.704.417 (1996); Shell (K. Nozaki), US 4.180.694 (1979).
6 Eridania (A. Motreux, F. Petit et al.), FR 2.693.188 (1992).
7 For example: (a) Elf-Atochem (E. Monflier, P. Bourdauducq, J.-L. Coutourier), US 5.345.007 (**1993**); (b) A. Behr, M. Urschey, *J. Mol. Catal. A:* **2003**, *197*, 101; (c) Mitsubishi (K. Sato, Y. Seto, I. Nakajima), US 5.600.032 and 5.600.034 (1995); (d) Celanese (A. Tafesh, M. Beller, J. Krause), US 6.150.298 (1999); (e) Celanese (Y. Chauvin, L. Magna, G. P. Niccolai), US 6.525.228 (**2001**).

2.4.4.3
Olefin Metathesis

Wolfgang A. Herrmann

2.4.4.3.1
Introduction

Olefin metathesis is a most charming and curious, yet industrially useful C–C bond-forming reaction: alkylidene groups are exchanged at catalytic metal centers to form new olefins [1–3]. Thus, the most stable bonds in olefins undergo cleavage, initiated by metal-alkylidene species (Scheme 1).

While the mechanistic principles were settled some time ago [4], structurally defined, tailored catalysts came along only recently, especially for functionalized olefins. Olefin metathesis, previously a most parameter-sensitive reaction, is no longer restricted to simple olefins, and it can now also be conducted in polar solvents. Ring-opening metathesis (ROMP), acyclic α,ω-diene metathesis (ADMET), ring-closure metathesis (RCM), and just simple metathesis are the main applications.

Famous examples are the well-established Norsorex® process (using RuX_3 catalysts), the *Shell higher-olefins process* (SHOP), and the cyclooctene ROMP to form Vestenamer® rubber products [3, 5, 6].

Many industrial processes employ multiphase catalysis, predominantly gas-phase reactions at heterogeneous catalysts [1–6]. The nature of the active catalytic site is speculative in most of these cases. However, the generation of metal-attached carbenes is likely, even starting from simple olefins carrying active hydrogen; cf. Scheme 2. In other cases, activators containing carbene precursor groups, such as $Sn(CH_3)_4$, are used to generate the active sites at the surface, e.g. via the equilibrium

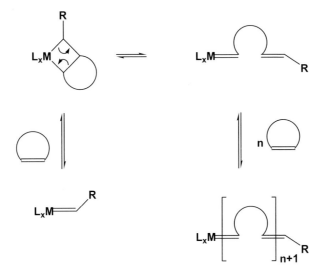

Scheme 1

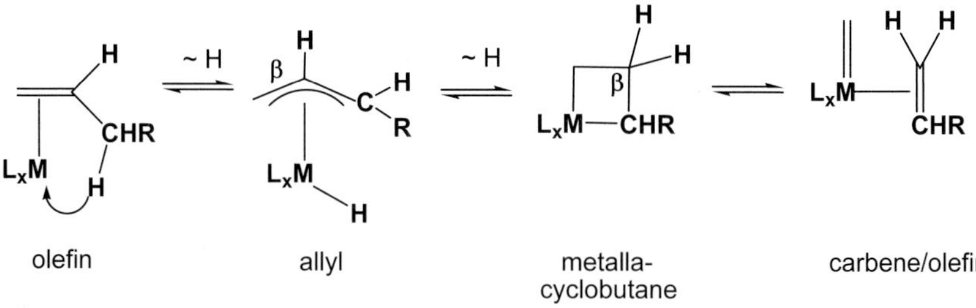

olefin   allyl   metalla-   carbene/olefin
                              cyclobutane

Scheme 2

$M-CH_3 \rightleftharpoons HM=CH_2$. Well-defined, single-component catalysts initiate living polymerization of olefins [7].

2.4.4.3.2
Ruthenium Catalysts

Water as a co-catalyst was early reported in the ROMP of (strained) cycloolefins, with the catalysts being the hydrates of $RuCl_3$, $OsCl_3$, and $CrCl_3$ in refluxing ethanol [Eq. (1)] [8]. Redox equilibria to generate Ru(II) active catalysts seem to be engaged but the mode of metal carbene formation finally remains unclear. These catalysts, however, initiate the aqueous-phase polymerization of substituted norbornenes in the presence of anionic emulsifiers and suitable reducing agents [9], even though the overall yield of polymers is still low.

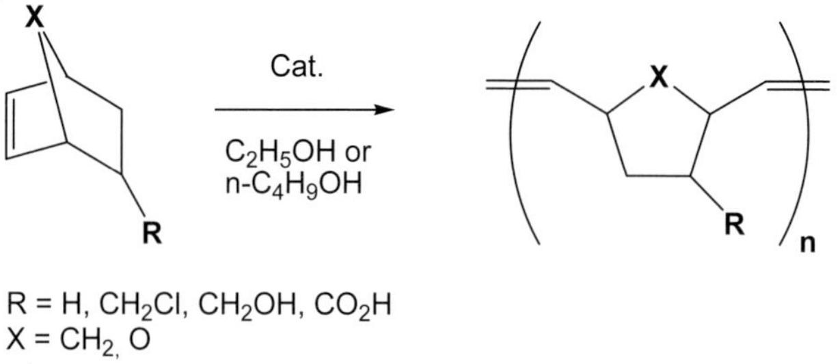

$$(1)$$

R = H, CH$_2$Cl, CH$_2$OH, CO$_2$H
X = CH$_2$, O

Water-soluble bis(allyl)ruthenium hydrate catalysts (**1** and **2**) initiate emulsion polymerization of norbornene, yielding surprisingly specific *cis* polymers (85–90% *cis*) [10].

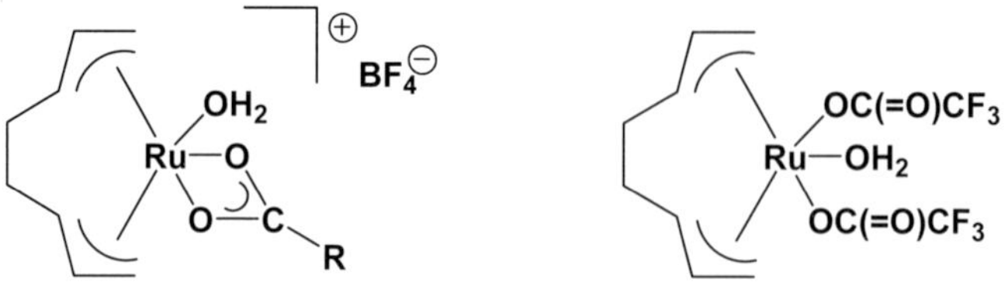

1   **2**

The polymers made by use of the other water-soluble catalysts normally have no defined stereochemistry, with the *cis/trans* ratio varying considerably from catalyst to catalyst. As a rule, the *trans* products are preferred.

This lack of selectivity changes with the structurally defined Ru(II)-carbene catalysts **3** introduced by Grubbs [11, 12]. The breakthrough, however, was achieved by the Herrmann strategy using N-heterocyclic carbenes (NHC instead of, or in addition to, phosphanes) in catalysts **4** and **5** [13–15]. Catalyst **5** first exemplified the principle of combining a strongly metal-attached NHC ligand with a phosphane ligand that undergoes dissociation to make a vacant coordination site available for the initial step. A wide array of now easily available NHCs is possible, including the C–C-saturated derivatives **6** and **7** that, in metathesis, are more active but of shorter catalyst lifetime. Larger-ring systems have also been employed, yielding different polymer properties than conventional poly(dicyclopentadiene) [16].

No initiation periods are observed with the new catalysts, in contrast to the "classical" RuX_3- and related catalysts. Living polymerization has been observed by NMR spectroscopy. The polydispersity of products formed from hydrophilic or hydrophobic monomers is typically very narrow (PD 1.10).

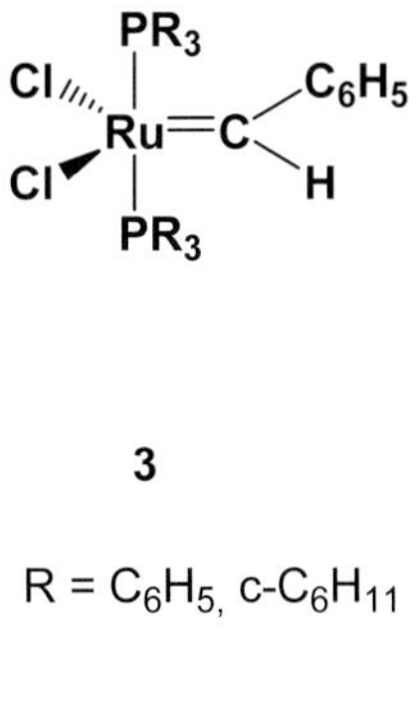

3

R = C_6H_5, c-C_6H_{11}

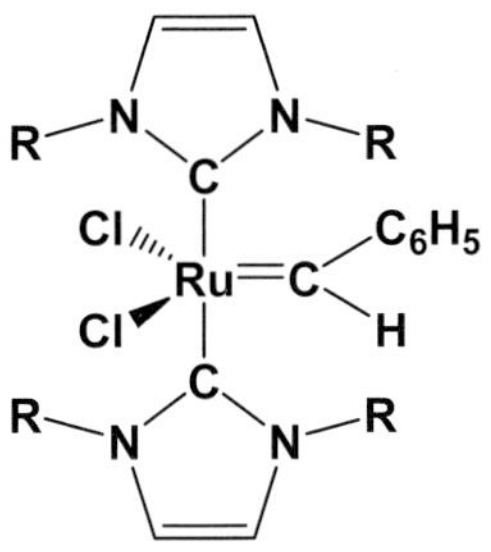

4

R = alkyl, aryl

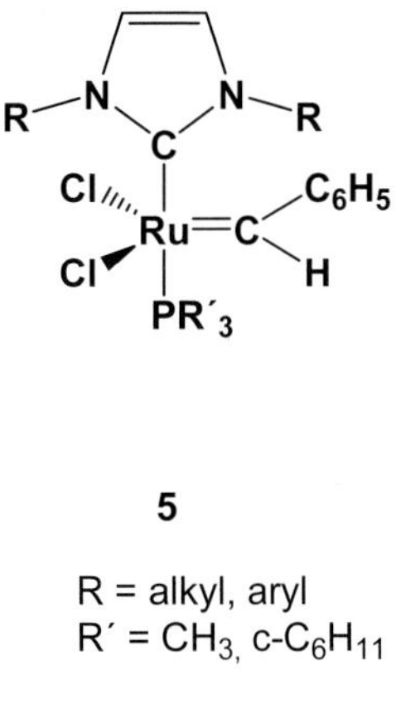

5

R = alkyl, aryl
R′ = CH_3, c-C_6H_{11}

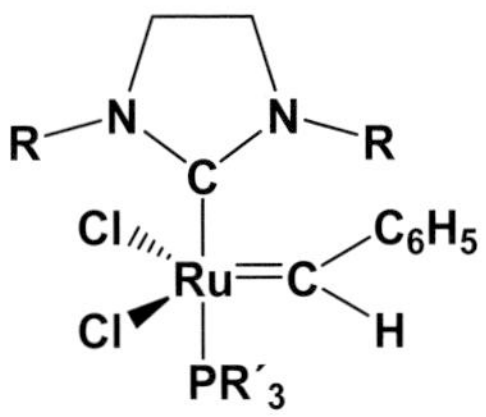

6

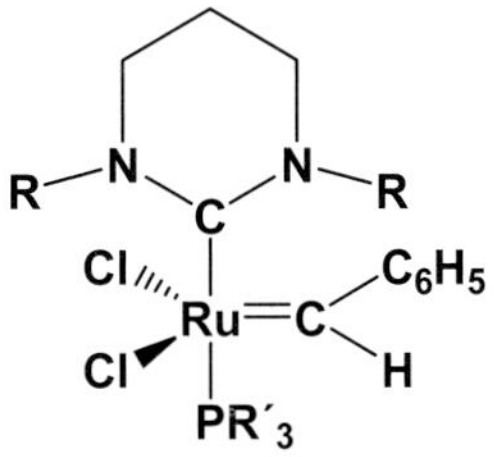

7

R = alkyl, aryl
R′ = CH_3, c-C_6H_{11}

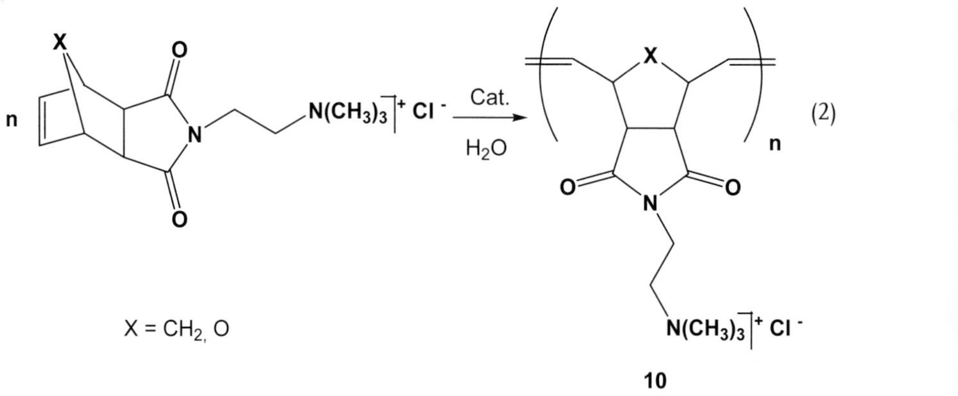

$$X = CH_2, O \qquad (2)$$

10

Well-defined block copolymers have been obtained via sequential monomer addition. Water-soluble, biologically active glycopolymers have also been made by this type of Ru(II) catalyst [18].

Replacement of PR_3 in compounds **3** by $P(C_6H_5)_2(C_6H_4\text{-}m\text{-}SO_3^-Na^+)$ ("TPPMS") and $[P(c\text{-}C_6H_{11})_2\{CH_2CH_2N(CH_3)_3\}]^+Cl$ gives water-soluble catalysts (e.g., **8**). While TPPMS is not electron-donating enough to act as an active catalyst [20], the alkyl-derived species **8** gives quick ROMP of the water-soluble monomers **9** to yield the polymer **10** at 45–80% conversion [Eq. (2)] [21].

2.4.4.3.3
Rhenium Catalysts

A novel class of metathesis catalysts derives from methyltrioxorhenium, CH_3ReO_3 (MTO) [22]. Combined with certain solid supports such as SiO_2/Al_2O_3 or Nb_2O_5, highly active catalysts are generated [23, 24]. They even tolerate functional groups such as ketones, esters, and carboxylic acids. Based upon NMR spectroscopic evidence [25] and in accord with theoretical studies, surface-attached $Re=CH_2$ species initiate metathesis via a novel type of tautomerism [Eq. (3)].

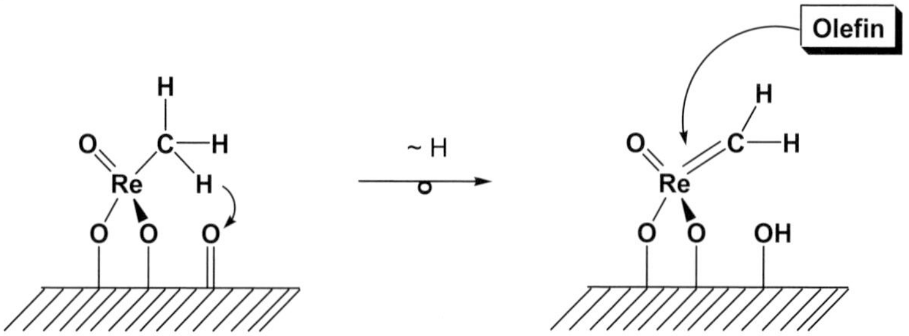

$$(3)$$

The catalyst precursor MTO was previously difficult to make, but it is now available from cheap starting materials on a multi-kilogram scale [26]. It is thus the ideal candidate for industrial uses, be it in metathesis or in oxidation [22].

2.4.4.3.4
Conclusion

In the meantime, olefin metathesis is applicable to all kinds of substrates, including those bearing functional and protic groups. The tailoring of structurally defined metal-carbene initiators allows for a defined stereochemistry of products. In many cases, metathesis occurs in two-phase processes: either as gas-phase reactions of simple olefins on heterogeneous catalysts, or as ring-opening polymerization with organic- or aqueous-phase catalysts.

References

1 K. J. Ivin, J. C. Mol, *Olefin Metathesis*, Academic Press, London **1997**.

2 V. Dragutan, A. T. Balaban, M. Dimonie, *Olefin Metathesis and Ring Opening Polymerization of Cyclo-Olefins*, 2nd ed., Wiley-Interscience, New York **1985**.

3 J. C. Mol, in *Applied Homogeneous Catalysis with Organometallic Compounds* (Eds.: B. Cornils, W. A. Herrmann), 2nd ed., p. 328, Wiley-VCH, Weinheim **2002**.

4 R. R. Schrock, *Acc. Chem. Res.* **1990**, *23*, 158.

5 W. A. Herrmann, *Kontakte (Darmstadt)* **1991**, *3*, 29.

6 W. A. Herrmann, B. Cornils, *Angew. Chem., Int. Ed. Engl.* **1997**, *36*, 1048.

7 R. H. Grubbs, W. Tumas, *Science* **1989**, *243*, 907.

8 F. W. Michelotti, W. P. Keaveney, *J. Polym. Sci. A* **1965**, *6* (1), 224.

9 R. E. Rinehart, H. P. Smith, *Polym. Lett.* **1965**, *3*, 1049.

10 S. Wache, *J. Organomet. Chem.* **1995**, *494*, 235.

11 (a) S. T. Nguyen, L. K. Johnson, R. H. Grubbs, *J. Am. Chem. Soc.* **1992**, *114*, 3974; (b) S. T. Nguyen, R. H. Grubbs, *J. Am. Chem. Soc.* **1993**, *115*, 9858; (c) P. Schwab, R. H. Grubbs, J. W. Ziller, *J. Am. Chem. Soc.* **1996**, *100*; (d) P. Schwab, M. B. France, J. W. Ziller, R. H. Grubbs, *Angew. Chem., Int. Ed. Engl.* **1995**, *34*, 2039.

12 (a) R. H. Grubbs, *Pure Appl. Chem.* **1994**, *A31*(11), 1829; (b) R. H. Grubbs, O. M. Lynn, in *Aqueous-Phase Organometallic Chemistry* (Eds.: B. Cornils, W. A. Herrmann), p. 466, Wiley-VCH, Weinheim **1998**.

13 (a) T. Weskamp, W. C. Schattenmann, M. Spiegler, W. A. Herrmann, *Angew. Chem., Int. Ed.* **1998**, *37*, 2490; (b) T. Weskamp. F. J. Kohl, W. Hieringer, D. Gleich, W. A. Herrmann, *Angew. Chem., Int. Ed.* **1999**, *38*, 2416.

14 W. A. Herrmann, *Angew. Chem., Int. Ed.* **2002**, *41*, 1290.

15 Degussa AG (W. A. Herrmann, W. Schattenmann, T. Weskamp), US 6.635.768 (**2003**).

16 (a) Nippon Zeon Comp. Ltd. (M. Sakamoto, S. Okada, Y. Tsunogae, S. Ikeda, W. A. Herrmann, K. Öfele), JP 2003.089.689 (**2003**); (b) W. A. Herrmann, K. Öfele, S. Okada, S. Ikeda, M. Sakamoto, Y. Tsunogae, WO 2003.027.079 (**2003**); (c) M. Sakamoto, S. Okada, Y. Tsunogae, S. Ikeda, W. A. Herrmann, K. Öfele, US Pat. pending (**2004**); (d) Compounds **7** have been reported recently in the literature; J. Yun, E. R. Marinez, R. H. Grubbs, *Organometallics* **2004**, *23*, 4172.

17 (a) D. M. Lynn, S. Kanaoka, R. H. Grubbs, *J. Am. Chem. Soc.* **1996**, *118*, 784; (b) C. Fraser, R. H. Grubbs, *Macromolecules* **1995**, *28*, 7248.

18 D. D. Manning, X. Hu, P. Beck, L. L. Kiessling, *J. Am. Chem. Soc.* **1997**, *119*, 3161.

19 B. Mohr, D. M. Lynn, R. H. Grubbs, *Organometallics* **1996**, *15*, 4317.

20 E. L. Dias, S. T. Nguyen, R. H. Grubbs, *J. Am. Chem. Soc.* **1997**, *119*, 3887.

21 D. M. Lynn, B. Mohr, R. H. Grubbs, *J. Am. Chem. Soc.* **1998**, *120*, 1627.

22 (a) W. Wang, W. A. Herrmann, R. Kratzer, J. Espenson, *Inorganic Syntheses* **2002**, *33*, 110; (b) W. A. Herrmann, F. R. Kühn, *Chemtracts – Organic Chemistry* **2001**, *14*, 59; (c) W. A. Herrmann, R. W. Fischer, F. E. Kühn, *Chem. Unserer Zeit* **1999**, *33*, 192;

(d) W. A. HERRMANN, F. E. KÜHN, G. M. LOBMAIER, *Aqueous-Phase Organometallic Catalysis* (Eds.: B. CORNILS, W. A. HERMANN), p. 529, Wiley-VCH, Weinheim **1998**; (e) W. A. HERRMANN, F. R. KÜHN, *Acc. Chem. Res.* **1997**, *30*, 169; (f) W. A. HERRMANN, C. C. ROMAO, F. E. KÜHN, *Chem. Rev.* **1997**, *97*, 3197; (g) F. E. KÜHN, M. GROARKE, in *Applied Homogeneous Catalysis with Organometallic Compounds* (Eds.: B. CORNILS, W. A. HERRMANN), 2nd ed., p. 1304.

23 W. A. HERRMANN, W. WAGNER, U. VOLKHARDT, DE 3.940.196 (**1989**), Hoechst AG.

24 (a) W. A. HERRMANN, W. WAGNER, U. N. FLESSNER, U. VOLKHARDT, H. KOMBER, *Angew. Chem., Int. Ed. Engl.* **1991**, *30*, 1636; (b) W. A. HERRMANN, R. W. FISCHER, D. W. MARZ, *Angew. Chem., Int. Ed. Engl.* **1991**, *30*, 1638; (c) W. A. HERRMANN, MEI WANG, *Angew. Chem., Int. Ed. Engl.* **1991**, *30*, 1641.

25 (a) R. BUFFON, A. CHOPLIN, M. LECONTE, J.-M. BASSET, R. TOUROUDE, W. A. HERRMANN, *J. Mol. Catal.* **1992**, *72*, L7–L10; (b) R. BUFFON, A. AUROUX, F. LEFEBVRE, M. LECONTE, A. CHOPLIN, J. M. BASSET, W. A. HERRMANN, *J. Mol. Catal.* **1992**, *76*, 287.

26 (a) Aventis (R. W. FISCHER, W. A. HERRMANN, R. KRATZER), US 6.180.807 (**2001**); (b) CataTech (W. A. HERRMANN, S. HIRNER, F. E. KÜHN), DE 000.000 (**2004**); (c) W. A. HERRMANN, in *Applied Homogeneous Catalysis with Organometallic Compounds* (Eds.: B. CORNILS, W. A. HERRMANN), 2nd ed., p. 1319; (d) W. A. HERRMANN, R. M. KRATZER, R. W. FISCHER, *Angew. Chem., Int. Ed. Engl.* **1997**, *36*, 2652.

2.4.4.4
Heck and Other C–C Coupling Reactions

Wolfgang A. Herrmann, Claus-Peter Reisinger, and Peter Härter

2.4.4.4.1
Introduction

Here, the term "Heck-type reaction" is taken to include palladium-catalyzed C–C coupling processes where vinyl or aryl derivatives are functionalized with olefins, alkynes, or organometallic reagents [see Eqs. (1) and (2)] [1]. Aryl and vinyl chlorides are most reluctant to undergo Pd-catalysed activation, as expected from the C–X bond dissociation energies [2].

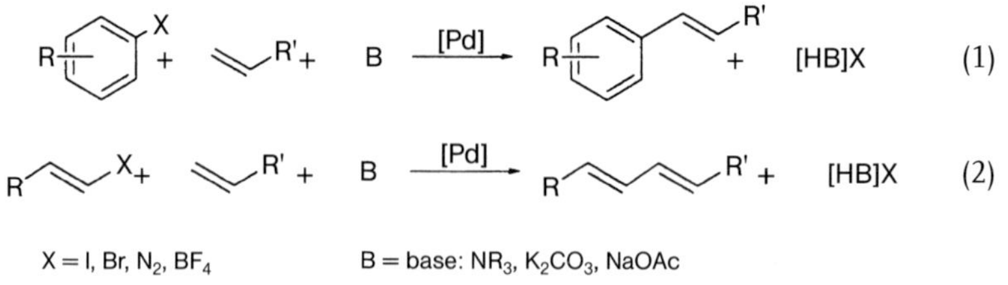

$$R \cdots \text{—} X + \text{—}R' + B \xrightarrow{[Pd]} R \cdots \text{—}\text{—}R' + [HB]X \qquad (1)$$

$$R \text{—}X + \text{—}R' + B \xrightarrow{[Pd]} R \text{—}\text{—}R' + [HB]X \qquad (2)$$

$$X = I, Br, N_2, BF_4 \qquad\qquad B = \text{base: } NR_3, K_2CO_3, NaOAc$$

Although most applications of Heck-type reactions are carried out in polar aprotic media, there are several successful approaches using partially or completely aqueous solution and aqueous–organic biphasic systems. Furthermore, the methodology was expanded to N–C and P–C bond-forming reactions; in addition, cross-coupling reactions will also be discussed here. In most cases, the recovery and re-use of the water-soluble catalyst is of minor importance, because the procedures are developed for the laboratory scale with respect to fine chemical synthesis. The main advantage of this approach may be the significant change in thermodynamics, resulting in milder reaction conditions and improvements in chemo- and regioselectivity. Recently, a novel type of biphasic system has been introduced using fluorous hydrocarbons/organic phases as solution media [43]. The main impetus for the employment of the fluorous biphase concept (FBS) is the easy recovery of the catalyst, a subject which is of crucial importance in industrial applications [44]. This methodology also provides a possibility of performing reactions under biphasic conditions with substrates which are labile toward water (see Chapter 4).

2.4.4.4.2
Catalysts and Reaction Conditions

Palladium is one of the most versatile and efficient catalyst metals in organic synthesis. Solubility in water is achieved by utilization of simple palladium(II) salts or water-soluble ligands, such as TPPTS and TPPMS. The active catalysts for heck-type reactions are zerovalent palladium(0) species [3], which are often generated in situ by thermal decomposition of a Pd(II) precursor or by the application of a reducing agent, e.g., 1–6 equiv. of a phosphine in the presence of base generates Pd(0) and the phosphine oxide [4].

The isolation of water-soluble palladium (0) complexes was achieved by gel-permeation chromatography for $Pd(TPPTS)_3$ [5]; an X-ray determination by Casalnuovo and Calabrese for $Pd(TPPMS)_3$ [6] is the first published structure of a transition metal complex containing a sulfonated phosphine.

Furthermore, the combination of palladium(II) salts with tetrabutylammonium halide additives, called "Jeffery conditions", is an efficient system for Heck-type reactions [7a], but the mechanistic implications are unknown. Also, nonionic phosphine ligands, such as triphenylphosphine which yields $Pd(PPh_3)_4$, are applied in water-miscible organic solvents, like DMF and acetonitrile. In these cases, the application of water is of crucial importance, but the role is often not well investigated.

2.4.4.4.3
Olefination

Since the pioneering work by Beletskaya and co-workers [8] the intra- and (more commonly) intermolecular arylation of olefins has been shown to proceed very smoothly in aqueous medium in the presence of palladium acetate. At the beginning, the methodology seemed to be limited to aryl iodides under a strong influence of the base: it was shown that the presence of potassium acetate instead of carbonate yielded lower reaction temperatures and higher rates [Eq. (3)].

Several years later, a similar approach succeeded even in the application of de-activated bromoanisole [9]. Further investigations by Jeffery indicated the rate- and selectivity-enhancing ability of tetraalkylammonium salts in Heck-type reactions [7].

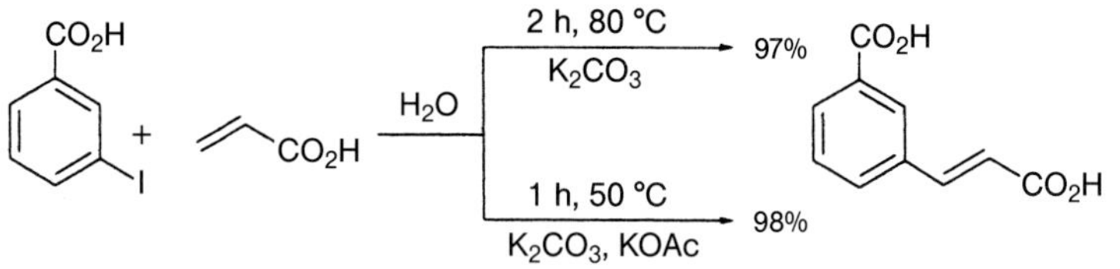

$$(3)$$

This approach was adopted by Daves for the coupling of iodo derivatives of nitrogen heterocycles with cyclic enol ethers and furanoid glycals in a water/ethanol mixture, using tetrabutylammonium chloride as a promoter [10]. Surprisingly, the use of absolute ethanol as reaction solvent was ineffective.

Furthermore, comparative studies with arylphosphine ligands in aqueous organic media demonstrated the superior activity of palladium tri(o-tolyl)phosphine complexes [11] with an unusual combination of 10 mol% tributylamine with 1.5 equiv of potassium carbonate in water [Eq. (4)] [9].

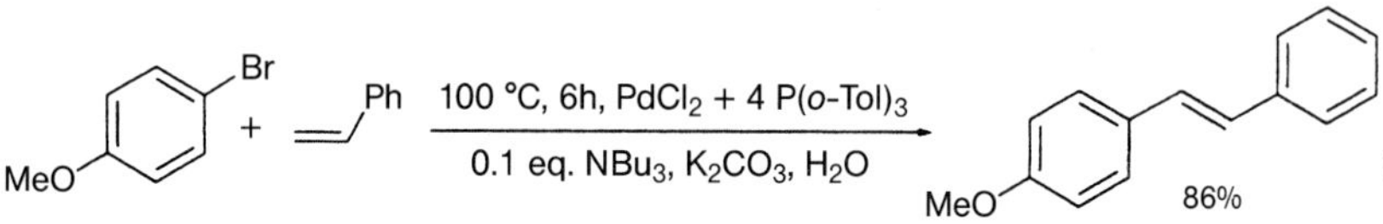

$$(4)$$

If water-soluble phosphine ligands are applied, extremely mild reaction conditions can be achieved. Especially, Pd(TPPMS)$_3$, which converts 4-iodotoluene, is tolerant of a broad range of functional groups, including those present in unprotected nucleotides and amino acids [6]. Interestingly, even the coupling of a donor-substituted iodoarenes and cyclic olefins can be conducted by palladium acetate with TPPTS at only 25 °C in aqueous acetonitrile. However, the low rates observed require a reaction time of up to 48 h for high conversions [12].

The application of ethylene in Heck reactions often shows different activities from other olefins, because of Wacker-type side reactions. It was found, however, that iodo- and acceptor-substituted bromoarenes are cleanly converted in aqueous media to the corresponding styrenes utilizing a palladiium–TPPMS complex [13]. Furthermore, high-purity o- and p-vinyltoluenes were prepared in a dimethyl-formamide/water mixture with palladium tri(o-tolyl)phosphine complexes [14]. Here, the role of water may be the dissolution of the inorganic base (potassium carbonate) in the organic media.

Even superheated (to 260 °C) or supercritical (to 400 °C; see Chapter 6) water was employed in the Heck reaction with several catalyst precursors and aryl halides with styrene. However, all conversions show large amounts of side products and the yields were in the 5–30% range, indicating radical intermediates and byproducts from decomposition of the arene starting material [15].

The progress of tandem Heck reactions in organic synthesis [16] led to their first application in the aqueous phase. Hence, a double Heck reaction on a substrate for which β-hydride elimination is possible results in three tricyclic products (Scheme 1) [17].

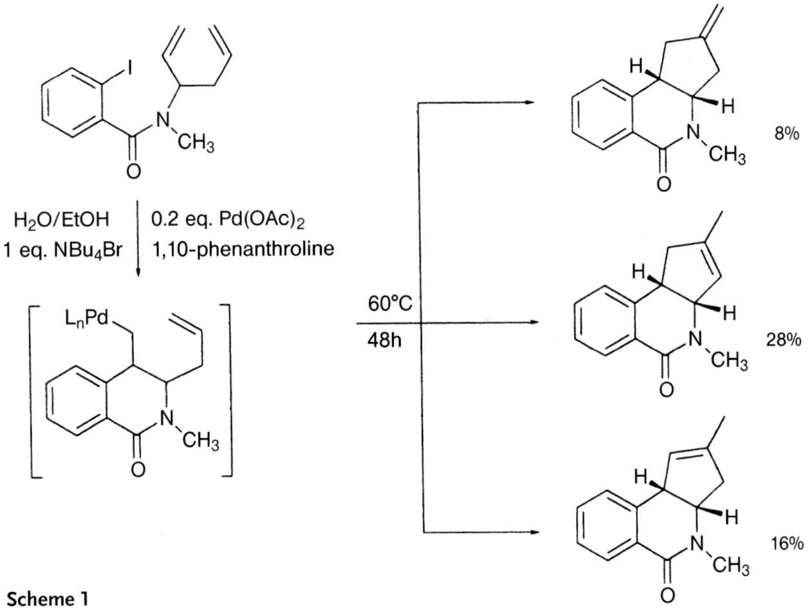

Scheme 1

Surprisingly, the application of 1,10-phenanthroline as a ligand suppresses β-hydride elimination completely and raises the total yield of double cyclization products to 52% [18]. In addition, an efficient one-pot procedure for Heck reactions starting with aniline derivatives, forming arenediazonium salts with sodium nitrite in 42% aqueous HBF$_4$, was reported [Eq. (5)] [19]. The process has several advantages: short reaction times, high catalytic turnover frequency, superior reactivity of the diazonium nucleofuge, and, most significantly, the use of aqueous reaction conditions. Therefore, this route toward ring-modified phenylalanine and tyrosine was used via a ring nitration and reduction sequence, thus expanding the field of artificial amino acids [20].

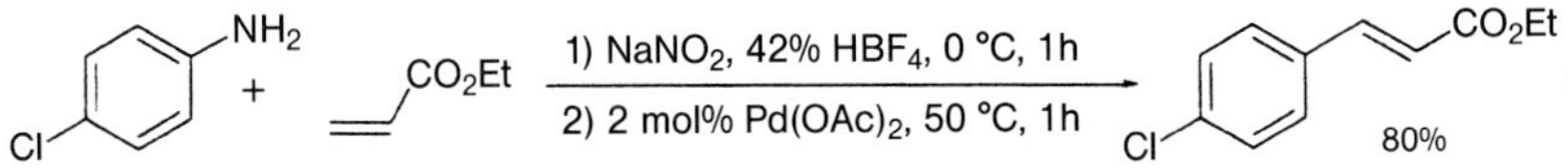

$$(5)$$

2.4.4.4.4
Alkyne Coupling

The palladium-catalyzed coupling of terminal acetylenes with aryl and vinyl halides is a widely used reaction in organic synthesis [21]. Hence, the application of water-soluble palladium complexes was first reported in aqueous acetonitrile with Pd(TPPMS)$_3$ and CuI as promoter, but was limited to aryl iodides [6]. The advantages of this catalyst already mentioned are low reaction temperatures and short reaction times with high yields [Eq. (6)]. Further ligand variations with TPPTS [12] and guanidino-functionalized phosphines [22] revealed that this methodology works also without any CuI promoter, when higher amounts of palladium (10 mol%) are used.

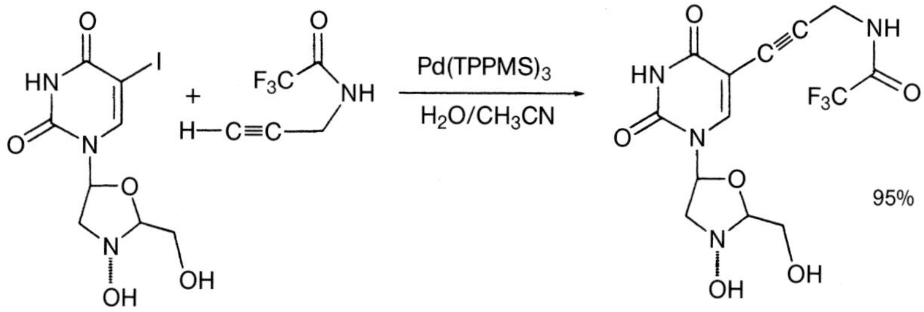

(6)

95%

Furthermore, Bumagin and Beletskaya reported the first coupling in neat water in the presence of a small amount tributylamine and potassium carbonate as base [23]. Surprisingly, the catalyst system consists of water-insoluble triphenylphosphine with PdCl$_2$ and CuI at room temperature, resulting in high yields with aryl iodides and phenylacetylene. The role of cuprous iodide was noted to be important to facilitate the reaction, which may be rationalized by two connected catalytic cycles.

In addition, the application of "Jeffery's conditions" by Sinou and co-workers, with extra triphenylphosphine and tetrabutylammonium hydrogen sulfate, confirmed that CuI is not essential to success in alkyne coupling reactions [24]. Moreover, they reported the most efficient coupling of bromoanisole with propargyl alcohol in 81% yield.

2.4.4.4.5
Cross-Coupling Reactions

Suzuki Coupling

The Suzuki coupling is defined by the presence of boron-containing coupling reactions. Thus, the palladium-catalyzed reaction of aryl or alkenyl halides with alkenylboronates or arylboronic acids is a regio- and stereoselective bond formation affording, in particular, unsymmetrical substituted biaryls [25]. Once again, the first application of this approach in the aqueous phase was reported by Casalnuovo and Calabrese, demonstrating the high efficiency of their Pd(TPPTS)$_3$-based catalyst system. Thus, 4-bromopyridine was coupled with *p*-tolylboronic acid in a water/methanol/benzene solvent mixture in 98% yield [Eq. (7)] [6].

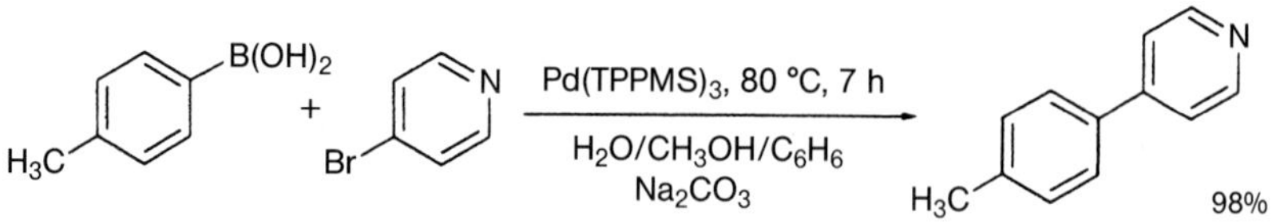

(7)

98%

Later, the same methodology was applied by Wallow and Novak for the synthesis of water-soluble poly(*p*-phenylene) derivatives via the "poly-Suzuki" reaction of 4,4′-biphenylylene bis(boronic acid) with 4,4′-dibromodiphenic acid in aqueous dimethylformamide [26]. These aromatic, rigid-chain polymers exhibit outstanding thermal stability (decomposition above 500 °C) and play an important role in high-

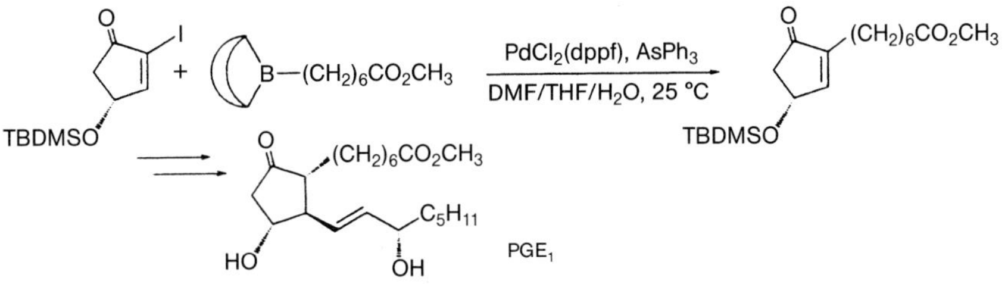

TBDMSO = t - butyldimethylsilyl
Scheme 2   dppf = 1,1 - bis (diphenylphosphinoferrocene)

performance engineering materials [27], conducting polymers [28], and nonlinear optical materials [29] (see also Section 7).

Furthermore, this regio- and stereoselective bond formation between unsaturated carbon atoms was applied to the synthesis of functionalized dienes under extremely mild conditions. Thus, even vinylic boronic esters containing an allylic acetal moiety and alkenylboronate having a chiral protected allylic alcohol were obtained successfully with vinylic iodides under aqueous conditions [30]. In addition, an exceptionally simple and efficient synthesis of a prostaglandin (PGE_1) precursor was reported by Johnson, by applying a DMF/THF/water solvent mixture with a 1,1-bis(diphenylphosphino)ferrocene palladium catalyst [31]. It is curious that the presence of water is an absolute necessity in order to succeed in this approach (Scheme 2).

It is noteworthy that 9-alkyl-9-BBN (9-BBN = 9-boracyclo[3.3.1]nonyl) reagents are easily prepared by hydroboration of the corresponding olefin, demonstrating the high variability of this approach in organic synthesis.

Stille Coupling

The Stille coupling depends on tin-containing reagents. Although the cross-coupling of organotin reagents with organic halides proceeds under extremely mild conditions, it seems to be the most unexplored field of palladium-catalyzed reactions [32], because of the high toxicity of the volatile tetraorganotin compounds. Thus, the first application in an aqueous medium was reported by Daves in 1993, describing the synthesis of a pyrimidine derivative formed by in-situ hydrolysis of the intermediate enol ether [33].

In 1995, Beletskaya [34] and Collum [35] reported independently the application of alkyltrichlorostannanes instead of tetraorganotin compounds, overcoming the disadvantage of three inert anchoring groups and technologically more important, because of their lower toxicity and availability via economic direct synthesis from tin(II) compounds [36]. Furthermore, the hydrolysis of the tin–halide bond in water results in higher water solubility, activation of the C–Sn bond toward electrophiles (e.g., in transmetallation) and less toxic byproducts. The reaction may be accomplished via intermediate anionic hydroxo complexes [37] produced in situ in aqueous alkaline solution, and proceeds in most cases in 3 h at 90–10 °C [Eq. (8)].

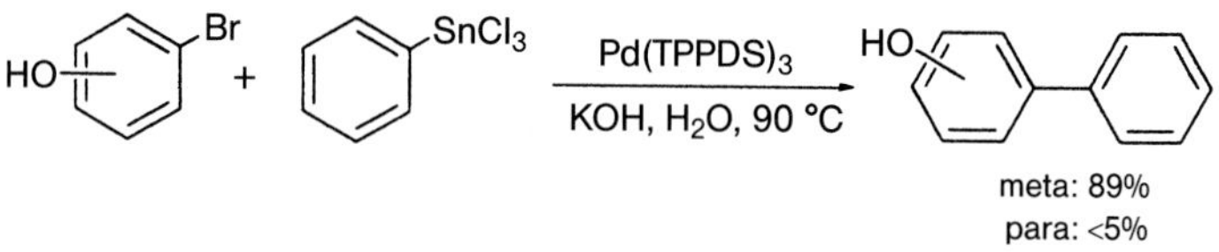

$$
\text{HO} \cdots \overset{\text{Br}}{\bigodot} + \overset{\text{SnCl}_3}{\bigodot} \xrightarrow[\text{KOH, H}_2\text{O, 90 °C}]{\text{Pd(TPPDS)}_3} \text{HO} \cdots \bigodot\!\!-\!\!\bigodot \tag{8}
$$

meta: 89%
para: <5%

For insoluble development in Heck-type reactions is P–C and N–C bond formation, which results from coupling of aryl halides with phosphorous compounds [38] and amines [39]. The first application in aqueous medium was achieved by coupling of a dialkyl phosphite with an aromatic iodide to give an arylphosphonate in 99% yield. In 1996, Stelzer and co-workers presented a P–C cross-coupling reaction between primary and secondary phosphines and functional aryl iodides to water-soluble phosphines [Eq. (9)], which are potentially applicable as ligands in aqueous-phase catalysis [40].

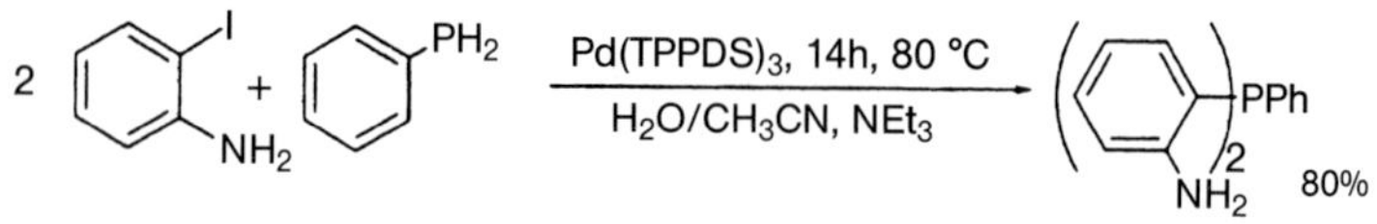

$$
2 \;\; \overset{\text{I}}{\underset{\text{NH}_2}{\bigodot}} + \overset{\text{PH}_2}{\bigodot} \xrightarrow[\text{H}_2\text{O/CH}_3\text{CN, NEt}_3]{\text{Pd(TPPDS)}_3,\ 14\text{h},\ 80\ °\text{C}} \left(\underset{\text{NH}_2}{\bigodot} \right)_{\!2}\!\!\text{PPh} \quad 80\% \tag{9}
$$

A rather unusual procedure has been published for the palladium- and copper-catalyzed synthesis of triarylamines, using an alkaline water–ethanol emulsion stabilized by cetyltrimethylammonium bromide [41]. Anyway, this method overcomes the problem in the synthesis of *N*-aryl carbazoles [Eq. (10)], which are not accessible by the method developed by Hartwig and Buchwald [42].

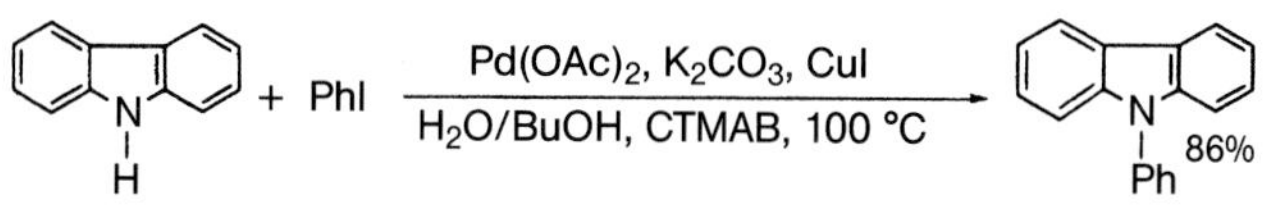

$$
\underset{\text{H}}{\overset{}{\bigodot\!\!\!\!\bigodot_{N}}} + \text{PhI} \xrightarrow[\text{H}_2\text{O/BuOH, CTMAB, 100 °C}]{\text{Pd(OAc)}_2,\ \text{K}_2\text{CO}_3,\ \text{CuI}} \underset{\text{Ph}}{\overset{}{\bigodot\!\!\!\!\bigodot_{N}}} \quad 86\% \tag{10}
$$

2.4.4.4.6
Conclusion

The advantages of Heck-type reactions in the aqueous phase are demonstrated by the large number of successful approaches presented here. The change in the thermodynamics caused by using water as the reaction medium results in milder reaction conditions, higher yields, and improvements in chemo- and regioselectivity. The trend in the last few years has moved toward systems which are designed to improve the catalyst recovery. The results reported in this field give great promise for powerful catalyst systems in the near future.

References

1 (a) H. A. DIECK, R. F. HECK, *J. Am. Chem. Soc.* **1974**, *96*, 1133; (b) R. F. HECK, *Org. React.* **1982**, *27*, 385.

2 (a) W. A. HERRMANN, *Applied Homogeneous Catalysis with Organometallic Compounds* (Eds.: B. CORNILS, W. A. HERRMANN), VCH, Weinheim **1996**, p. 712; (b) C.-P. REISINGER, Ph. D. Thesis, Technische Universität München, Germany **1997**.

3 C. AMATORE, E. BLART, J. P. GENET, A. JUTAND, S. LEMAIRE-AUDOIRE, M. SAVIGNAC,
J. Org. Chem. **1995**, *60*, 6829.

4 F. OZAWA, A. KUBO, T. HAYASHI, *Chem. Lett.* **1992**, 2177.

5 W. A. HERRMANN, J. KELLNER, H. RIEPL, *J. Organomet. Chem.* **1990**, *389*, 103.

6 A. L. CASALNUOVO, J. C. CALABRESE, *J. Am. Chem. Soc.* **1990**, *112*, 4324.

7 (a) T. JEFFERY, *Tetrahedron Lett.* **1994**, *35*, 3051; (b) T. JEFFERY, *Tetrahedron* **1996**, *52*, 10 113.

8 N. A. BUMAGIN, P. G. MORE, I. P. BELETSKAYA, *J. Organomet. Chem.* **1989**, *371*, 397.

9 N. A. BUMAGIN, V. V. BYKOV, L. I. SUKHOMLINOVA, T. P. TOLSTAYA, I. P. BELETSKAYA,
J. Organomet. Chem. **1995**, *486*, 259.

10 H.-C. ZHANG, G. D. DAVES JR., *Organometallics* **1993**, *12*, 1499.

11 W. A. HERRMANN, C. BROSSMER, K. ÖFELE, C.-P. REISINGER, T. PRIERMEIER, M. BELLER,
H. FISCHER, *Angew. Chem., Int. Ed. Engl.* **1995**, *34d*, 1844; *Angew. Chem.* **1995**, *107*, 1989.

12 J. P. GENET, E. BLART, M. SAVIGNAC, *Synlett* **1992**, 715.

13 J. KIJI, T. OKANO, T. HASEGAWA, *J. Mol. Catal.* **1995**, *97*, 73.

14 R. A. DEVRIES, A. MENDOZA, *Organometallics* **1994**, *13*, 2405.

15 (a) P. REARDON, S. METTS, C. CRITTENDON, P. DAUGHERITY, E. J. PARSONS, *Organometallics*
1995, *14*, 3810; (b) J. DIMINNIE, S. METTS, E. J. PARSONS, *Organometallics* **1995**, *14*, 4023.

16 A. DE MEIJERE, F. E. MEYER, *Angew. Chem., Int. Ed. Engl.* **1994**, *33*, 2379; *Angew. Chem.* **1994**,
106, 2473.

17 D. B. GROTJAHN, X. ZHANG, *J. Mol. Catal.* **1997**, *116*, 99.

18 R. BRESLOW, *Acc. Chem. Res.* **1991**, *24*, 159.

19 (a) S. SENGUPTA, S. BHATTACHARYA, *J. Chem. Soc., Perkin Trans.* **1993**, 1943;
(b) K. KIKUKAWA, K. NAGIRA, N. TERAO, F. WADA, T. MATSUDA, *Bull. Chem. Soc. Jpn.* **1979**, *52*,
2609.

20 S. SENGUPTA, S. BHATTACHARYA, *Tetrahedron Lett.* **1995**, *36*, 4475.

21 (a) J. TSUJI, *Palladium Reagents and Catalysis: Innovations in Organic Synthesis*, John Wiley,
Chichester **1995**; (b) W. A. HERRMANN, C.-P. REISINGER, C. BROSSMER, M. BELLER,
H. FISCHER, *J. Mol. Catal. A:* **1996**, *198*, 51.

22 H. DIBOWSKI, F. P. SCHMIDTCHEN, *Tetrahedron* **1995**, *51*, 2325.

23 N. A. BUMAGIN, L. I. SUKHOMLINOVA, E. V. LUZIKOVA, T. P. TOLSTAYA, I. P. BELETSKAYA,
Tetrahedron Lett. **1996**, *37*, 897.

24 J.-F. NGUEFACK, V. BOLITT, D. SINOU, *Tetrahedron Let.* **1996**, *37*, 5527.

25 A. SUZUKI, *Pure Appl. Chem.* **1991**, *63*, 419.

26 T. I. WALLOW, B. M. NOVAK, *J. Am. Chem. Soc.* **1991**, *113*, 7411.

27 A. E. ZACHARIADES, R. S. PORTER (Eds.), *The Strength and Stiffness of Polymers*, Marcel
Dekker, New York **1983**.

28 R. L. ELSENBAUMER, L. W. SHACKLETTE, *Handbook of Conducting Polymers* (Ed.: T. A. SKOTHEIM),
Marcel Dekker, New York **1986**.

29 D. J. WILLIAMS, *Angew. Chem., Int. Ed. Engl.* **1984**, *23*, 640; *Angew. Chem.* **1984**, *96*, 637.

30 J. P. GENET, A. LINQUIST, E. BLART, V. MOURIES, M. SAVIGNAC, M. VAULTIER, *Tetrahedron
Lett.* **1995**, *36*, 1443.

31 C. R. JOHNSON, M. P. BRAUN, *J. Am. Chem. Soc.* **1993**, *115*, 11 014.

32 (a) T. N. MITCHELL, *Synthesis* **1992**, 803; (b) J. K. STILLE, *Angew. Chem., Int. Ed. Engl.* **1986**,
25, 508; *Angew. Chem.* **1986**, *98*, 504.

33 H.-C. ZHANG, G. D. DAVES JR., *Organometallics* **1993**, *12*, 1499.

34 A. I. ROSHCHM, N. A. BUMAGIN, I. P. BELETSKAYA, *Tetrahedron Lett.* **1995**, *36*, 125.

35 R. RAI, K. B. AUBRECHT, D. B. COLLUM, *Tetrahedron Lett.* **1995**, *36*, 3111.

36 A. G. DAVIES, P. J. SMITH, *Comprehensive Organometallic Chemistry* (Eds.: G. WILKINSON,
F. G. A. STONE, E. W. ABEL), Pergamon, Oxford **1982**, Vol. 2, p. 519.

37 M. DEVAUD, *Rev. Chim. Miner.* **1967**, *4*, 921.

38 (a) O. HERD, A. HESSLER, M. HINGST, M. TEPPER, O. STELZER, *J. Organomet. Chem.* **1996**, *522*,
69; (b) A. L. CASALNUOVO, J. C. CALABRESE, *J. Am. Chem. Soc.* **1990**, *112*, 4324.

39 (a) A. S. GURAM, R. A. RENNELS, S. L. BUCHWALD, *Angew. Chem., Int. Ed. Engl.* **1995**, *34*,
1348; *Angew. Chem.* **1995**, *107*, 1456; (b) J. LOUIE, J. F. HARTWIG, *Tetrahedron Lett.* **1995**, *36*,

3609; (c) M. BELLER, T. H. RIERMEIER, C.-P. REISINGER, W. A. HERRMANN, *Tetrahedron Lett.* **1997**, *38*, 2073.

40 O. HERD, A. HESSLER, M. HINGST, M. TEPPER, O. STELZER, *J. Organomet. Chem.* **1996**, *522*, 69.

41 D. V. DAVYDOV, I. P. BELETSKAYA, *Russ. Chem. Bull.* **1995**, *44*, 1141 (Engl. Transl.).

42 C.-P. Reisinger, Diss. TU München, **1998**; cf. Ref. [2b].

43 I. HORVÁTH, *Acc. Chem. Res.* **1998**, *31*, 641.

44 C. C. TZSCHUCKE, C. MARKERT, W. BANNWARTH, S. ROLLER, A. HEBEL, R. HAAG, *Angew. Chem.* **2002**, *114*, 4136.

45 C. C. TZSCHUCKE, C. MARKERT, H. GLATZ, W. BANNWARTH, *Angew. Chem., Int. Ed.* **2002**, *41*, 4500.

46 J. MOINEAU, G. POZZI, S. QUICI, D. SINOU, *Tetrahedron Lett.* **1990**, *40*, 7683.

47 C. MARKERT, W. BANNWARTH, *Helv. Chim. Acta* **2002**, *85*, 1877.

48 S. SCHNEIDER, W. BANNWARTH, *Helv. Chim. Acta* **2001**, *84*, 735.

49 S. SCHNEIDER, W. BANNWARTH, *Angew. Chem., Int. Ed.* **2000**, *39*, 4142.

2.4.4.5
Aminations

Birgit Drießen-Hölscher (†)

2.4.4.5.1
Introduction

All kinds of amines are of great importance for synthetic chemists in basic research as well as for the chemical industry. But the synthesis of amines in terms of an optimum material balance, short sequences, and selectivities is still a challenge. In recent years, great efforts have been made worldwide to develop new amination reactions [1].

This section deals with catalyzed aminations in aqueous phases, so the developments discussed here must have some water in the reaction mixture. All reactions known to synthesize amines via catalysts in the presence of water can be divided into four categories and are discussed in the following order: Hartwig–Buchwald aminations, telomerization to yield amines, reductive aminations/hydroaminomethylations, and allylic aminations.

2.4.4.5.2
Hartwig–Buchwald Aminations

The palladium-catalyzed amination of aryl halides and sulfonates has emerged as a valuable method for the preparation of aromatic amines [2]. Numerous ligands and catalysts have been reported to effect this type of cross-coupling. This reaction, known as the Hartwig–Buchwald amination, is shown in Eq. (1).

$$\text{ArX} + \text{HNNR}^1 \xrightarrow[\text{toluene, 80-100 °C}]{\text{PdL, NaOBu}^t} \text{ArNRR}^1 \tag{1}$$

$$\text{X} = \text{Cl, Br, I, OSO}_2\text{CF}_3 \quad \text{R} = \text{H, alkyl, aryl} \quad \text{R}^1 = \text{aryl}$$

Under such conditions re-use of the palladium catalyst would be difficult. This was the motivation of Boche et al. to use a two-phase protocol by dissolving the six-fold sulfonated BINAS-6 (**1**) and palladium acetate in water [3].

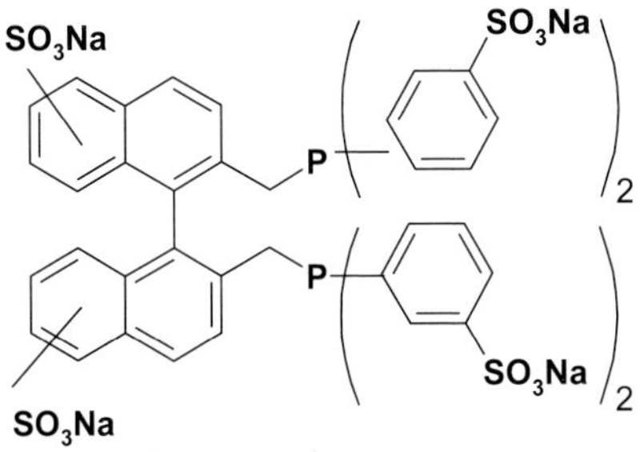

1

The reaction of 4-bromoacetophenone with *N*-methylaniline in water using NaOH as the base and the aforementioned catalyst afforded the tertiary amine in 36% yield. By adding co-solvents, such as methanol, the yield can be increased to 91%. In conclusion, the advantages are the facile catalyst/product separation, the reusability of the water-soluble palladium catalyst, and the use of NaOH instead of the expensive NaOBu^t as the base.

Aqueous aminations with palladium catalysts that are not soluble in water have recently been described by Buchwald et al. [4]. The use of **2**/Pd$_2$dba$_3$ and KOH in water without co-solvent gives excellent results in the amination of aryl chlorides containing nitro, trifluoromethyl, or pyridyl groups. Indoles and hindered aryl nonaflates can be aminated in good yields. In summary, co-solvents and phase-transfer catalysts are not necessary.

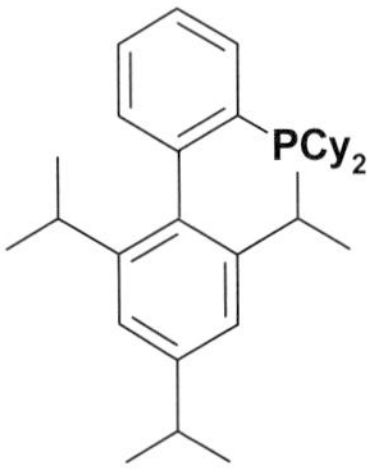

2

Aminations that occur in the presence of inexpensive and air-stable bases have become more important. Hartwig et al. demonstrated that low-cost alkali metal hydroxides can serve as stoichiometric base for the palladium-catalyzed cross-coupling of various amines with aryl chlorides and bromides [5].

Acceptable reaction rates and high yields are obtained with the combination of phase-transfer catalyst and 1 equiv of water and KOH, or with the combination of phase-transfer catalyst and concentrated aqueous NaOH or KOH. In summary, the new base system performed as well as NaOBu^t in certain reactions. In a recent paper, K$_3$PO$_4$ was delivered as an aqueous solution which was approximately 17% of the total volume of the reactions [6]. The coupling of a complex bromoarene and a fluorophore was tested with [CpPd (allyl)] as palladium source. Mixtures of dioxane and xylene gave much higher yields compared to NaOBu^t as the base in this medium.

Intramolecular palladium–catalyzed aryl amination chemistry is used to synthesize benzimidazoles. When NaOH in toluene is used as a base, the reaction rate is very slow. But aqueous reaction conditions (20% water/xylene) led to complete conversion of an amidine to benzimidazole **3** after two hours at reflux [7].

3

When this chemistry was attempted with microwave heating, a quantitative conversion to **3** occurred in just 20 min at 200 °C. But 10 mol% of a phase-transfer catalyst was necessary. The reaction conditions were optimized, leading to lower catalyst concentrations and shorter reaction times.

2.4.4.5.3
Telomerization to Yield Amines

Telomerizations have proven to be of great industrial value [8] and the products obtained play an important role as intermediates for the production of fine and bulk chemicals (see Section 2.4.4.2). The telomerization of ammonia and butadiene (Scheme 1) has been extensively studied using homogeneously catalyzed one-phase reactions [9]. In this case the main products obtained are the tertiary octadienylamines. The nucleophilicity increases in the order ammonia, primary octadienylamine, secondary octadienylamine so that the reaction cannot be stopped. A summary of all observed and possible products is given in Scheme 1.

These consecutive reactions can be avoided by using the biphasic method [10], which allows the selective production of the primary amines **4** and **5**. The reaction takes place in an aqueous catalyst phase (Pd(OAc)$_2$ + TPPTS) and the primary products are extracted in situ by a second organic phase (e.g., toluene, methylene chloride, butadiene) which is immiscible with water.

In order to optimize and better understand the regioselectivity and the rate of product formation, the kinetics of this reaction were determined. The kinetic experiments led to a model of the reaction that also explains the different regioselectivities. By ligand variations the regioselectivity of the reaction can be controlled. The ligand with the highest π-acceptor properties (*p*-F-TPPTS) gives the lowest selectivity to the terminal amine **4** whereas the ligand with the highest σ-donor properties (TOM-TPPTS) leads to the terminal amine **4** with a selectivity of 94%.

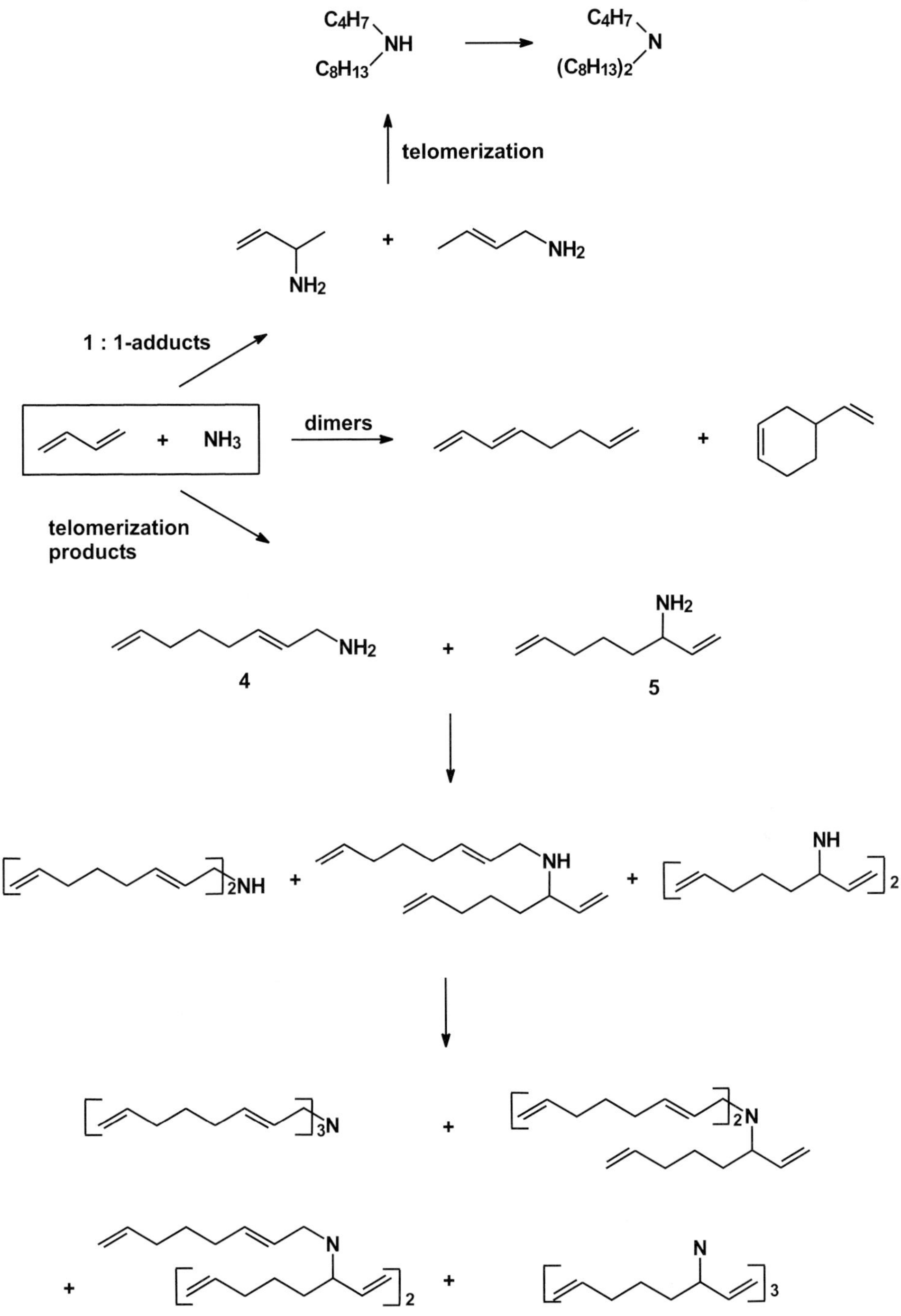

Scheme 1 Possible reactions of butadiene with ammonia.

By lowering the π-acceptor properties of the ligand, the activity of the catalyst system decreases [11].

2.4.4.5.4
Reductive Aminations/Hydroaminomethylations

The reductive amination of carbonyl compounds constitutes a convenient and practical approach to synthesizing primary amines and is generally performed by heterogeneous catalysts [12].

Reductive amination is the last step in a domino reaction called hydroaminomethylation (Scheme 2). One-phase hydroaminomethylations with primary and secondary amines have been described in particular by Eilbracht [13].

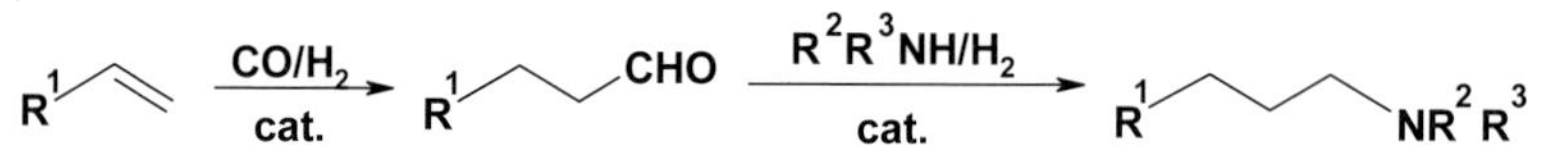

$$R^1 \diagdown \diagup \xrightarrow[\text{cat.}]{CO/H_2} R^1 \diagdown \diagup CHO \xrightarrow[\text{cat.}]{R^2R^3NH/H_2} R^1 \diagdown \diagup \diagdown NR^2R^3$$

Scheme 2 Hydroaminomethylations of olefins.

The hydroaminomethylation of 1-pentene and ammonia, with a Rh/Ir/TPPTS catalyst in an aqueous/organic two-phase system, was developed by Beller et al. [14]. Under standard hydroformylation conditions (130°C, 12 MPa) amines could be isolated in 75% yield. By increasing the ammonia/olefin ratio and by using the extraction effect of the organic solvent as in Ref. [10], the selectivity for primary amines could reach more than 90%. This method could also be used for propene and 1-butene to give butylamine and pentylamine as main products.

Organometallic aqua complexes of iridium such as $[Cp^*Ir(H_2O)_3]^{2+}$ are slightly active catalysts in reductive amination of *n*-butanal when $HCOONH_4$ is utilized as the hydrogen and amine donor [15]. The rate of the reductive amination is dependent on the concentration of $HCOONH_4$ and on pH. The initial TOF is 2 h^{-1}. Recently, an unprecedented ruthenium-catalyzed reductive amination of aldehydes with tertiary amines in an aqueous medium was reported [Eq. (2)] [16]. Treatment of benzaldehyde with tributylamine afforded benzyldibutylamine in 43% yield and dibenzylbutylamine in 7% yield. Several other ruthenium complexes were ineffective for this reaction, and performing the reaction in pure dioxane gave much lower yields.

$$RCHO + NR_3' \xrightarrow[\substack{\text{dioxane/H}_2\text{O1} \\ 180\,°C,\,12\,h}]{[Ru_3(CO)_{12}],\,CO} RCH_2NR_2' + R(CH_2)_2NR' \qquad (2)$$

The synthesis of primary amines in biphasic operation has been developed via selective reductive amination of aromatic and aliphatic carbonyl compounds using aqueous ammonia in the presence of water-soluble transition metal catalysts [17]. The use of $[Rh(cod)Cl]_2$ with TPPTS as catalyst and ammonium acetate in water/ THF afforded benzylamine from benzaldehyde in 86% yield. This method is also feasible for the synthesis of aliphatic primary amines from aliphatic aldehydes in

45–47% yield. Byproducts are secondary amines as well as aldol condensation products, but the formation of alcohol is negligible.

Aqueous ammonia was found to increase the yield of the alcohol but not of the amine in the highly enantioselective hydrogen-transfer reductive amination of acetophenone, as recently described by Kadyrov et al. [18]. All these reactions were performed in methanol/NH_3 with [((R)-tol-binap)RuCl$_2$] as catalyst with a best asymmetric induction of 98% *ee*.

2.4.4.5.5
Allylic Aminations

Allylic substitutions in water were described a few years ago by the groups of Genet and Sinou [19] (see also Section 2.4.5.1), but few results have been obtained so far for allylic amination in water.

A new tetrapodal phosphine ligand, *cis,cis,cis*-1,2,3,4-tetrakis(diphenylphosphino-methyl)cyclopentane (Tedicyp), in association with [PdCl(C$_3$H$_5$)]$_2$, is an extremely efficient catalyst for allylic amination [20]. Addition of diisopropylamine to an allyl acetate [Eq. (3)] is suprisingly higher in water than in THF.

The complex seemed to be very stable in water. A conversion of 98% was observed when a substrate/catalyst ratio of 1.000 was used.

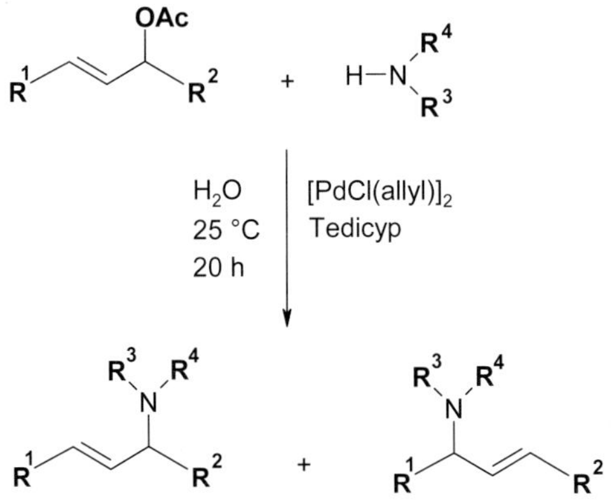

(3)

References

1 (a) A. Ricci, *Modern Amination Methods*, Wiley-VCH, Weinheim **2000**; (b) T. E. Müller, M. Beller, *Chem. Rev.* **1998**, *98*, 675; (c) M. Nobis, B. Driessen-Hölscher, *Angew. Chem., Int. Ed.* **2001**, *40*, 3983.
2 For reviews, see: (a) J. F. Hartwig, *Angew. Chem., Int. Ed.* **1998**, *37*, 2046; (b) D. Prim, J.-M. Campagne, D. Joseph, B. Andrioletti, *Tetrahedron* **2002**, *58*, 2041.
3 G. Wüllner, H. Jänsch, S. Kannenberg, F. Schubert, G. Boche, *Chem. Commun.* **1998**, 1509.
4 X. Huang, K. W. Anderson, D. Zim, L. Jiang, A. Klapars, S. L. Buchwald, *J. Am. Chem. Soc.* **2003**, *125*, 6653.

5 R. KUWANO, M. UTSUNOMIYA, J. F. HARTWIG, *J. Org. Chem.* **2002**, *67*, 6479.

6 S. R. STAUFFER, J. F. HARTWIG, *J. Am. Chem. Soc.* **2003**, *125*, 6977.

7 C. T. BRAIN, J. T. STEER, *J. Org. Chem.* **2003**, *68*, 6814.

8 N. YOSHIMURA, in *Aqueous-Phase Organometallic Catalysis* (Eds.: B. CORNILS, W. A. HERRMANN), Wiley-VCH, Weinheim **1998**, p. 408.

9 (a) T. MITSUYASU, M. HARA, J. TSUJI, *J. Chem. Soc., Chem. Commun.* **1971**, 345; (b) J. TSUJI, M. TAKAHASHI, *J. Mol. Catal.* **1981**, *10*, 107; (c) C. F. HOBBS, D. E. MCMACKINS, MONSANTO Co. US 4.100.194 (1978) and 4.130.590 (1978); (d) Toray Industries (J. TSUJI), JP 75/22014 (1975); (e) R. N. FAKHRETDINOV, G. A. TOLSTIKOV, U. M. DZHEMILEV, *Neftechimija* **1979**, *3*, 468.

10 T. PRINZ, W. KEIM, B. DRIESSEN-HÖLSCHER, *Angew. Chem.* **1996**, *108*, 1835; *Angew. Chem., Int. Ed. Engl.* **1996**, *35*, 1708.

11 T. PRINZ, B. DRIESSEN-HÖLSCHER, *Chem. Eur. J.* **1999**, *5*, 1111.

12 (a) P. N. RYLANDER, *Catalytic Hydrogenation in Organic Synthesis*, Academic Press, New York **1979**, p. 165; (b) P. N. RYLANDER, *Hydrogenation over Platinum Metals*, Academic Press, New York **1967**, p. 292.; (c) A. W. HEINEN, J. A. PETERS, H. VAN BEKKUM, *Eur. J. Org. Chem.* **2000**, 2501.

13 (a) T. RISCHE, L. BÄRFACKER, P. EILBRACHT, *Eur. J. Org. Chem.* **1999**, 653; (b) T. RISCHE, B. KITSOS-RZYCHON, P. EILBRACHT, *Tetrahedron* **1998**, *54*, 2723; (c) C. L. KRANEMANN, P. EILBRACHT, *Synthesis* **1998**, 71; (d) T. RISCHE, P. EILBRACHT, *Synthesis* **1997**, 1331; (e) P. EILBRACHT, L. BÄRFACKER, C. BUSS, C. HOLLMANN, B. E. KITSOS-RZYCHON, C. L. KRANEMANN, T. RISCHE, R. ROGGENBUCK, A. SCHMIDT, *Chem. Rev.***1999**, *99*, 3329.

14 B. ZIMMERMANN, J. HERWIG, M. BELLER, *Angew. Chem., Int. Ed.* **1999**, *38*, 2372.

15 S. OGO, N. MAKIHARA, Y. KANEKO, Y. WATANABE, ORGANOMETALLICS **2001**D, *20*, 4903.

16 C. S. CHO, J. H. PARK, T.-J. KIM, S. C. SHIM, *Bull. Korean Chem. Soc.* **2002**, *23*, 23.

17 T. GROSS, A. M. SEAYAD, M. AHMAD, M. BELLER, *Org. Lett.* **2002**, *4*, 2055.

18 R. KADYROV, TH. H. RIERMEIER, *Angew. Chem.* **2003**, *115*, 5630.

19 J. P. GENET, M. SAVIGNAC, *J. Organomet. Chem.* **1999**, *576*, 305.

20 M. FEUERSTEIN, D. LAURENTI, H. COUCET, M. SANTELLI, *Tetrahedron Lett.* **2001**, *42*, 2313.

2.4.4.6
Alternating Copolymers from Alkenes and Carbon Monoxide

Eite Drent, Johannes A. M. van Broekhoven, and Peter H. M. Budzelaar

2.4.4.6.1
Introduction

The copolymerization of ethylene and carbon monoxide to give alternating copolymers has attracted considerable interest in both academia and industry over recent decades [1, 2]. Attention was focused on aliphatic polyketones such as poly(3-oxotrimethylene) (**1**) because of the low cost and plentiful availability of the simple monomers. The new family of thermoplastic, perfectly alternating olefin/carbon monoxide polymers commercialized by Shell provides a superior balance of performance properties not found in other commercial materials; the an ethylene/propene/CO terpolymer is marketed by Shell under the tradename Carilon®. About the history of polyketones see Refs. [3–11].

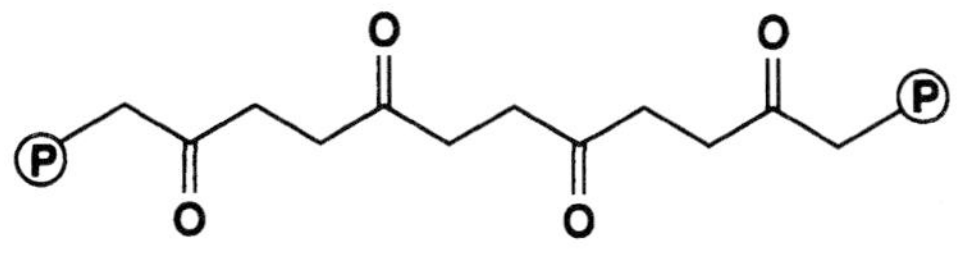

1 polyketone

2.4.4.6.2
Copolymerization of Ethylene and CO

The discovery of efficient catalysts for the copolymerization of alkenes originated from a study of the alkoxycarbonylation of ethylene in methanol (MeOH) to methyl propionate [Eq. (1)].

$$H_2C=CH_2 + CO + MeOH \rightarrow \quad \text{(structure: OMe ester)} \quad (1)$$

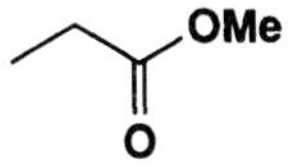

The catalysts were cationic palladium–phosphine systems prepared from palladium acetate, an excess of triphenylphosphine (PPh$_3$), and a Brønsted acid of a weakly or noncoordinating anion (e.g., p-tosylate (OTs$^-$); methanol was used as both the solvent and a reactant. An unexpected change in selectivity was observed upon replacement of the excess of PPh$_3$ by a stoichiometric amount of the *bidentate* 1,3-bis(diphenylphosphino)propane (dppp). Under the same conditions, these modified catalysts led to perfectly alternating ethylene/CO copolymers with essentially 100% selectivity [Eq. (2)] [12–14].

$$n \; H_2C=CH_2 + n \; CO + MeOH \rightarrow \quad \text{(polymer structure: H ... OMe, O, n)} \quad (2)$$

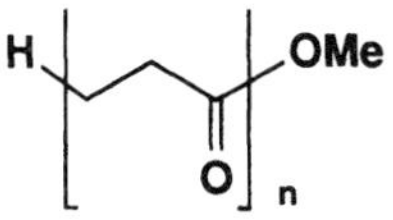

A typical reaction rate would be about 10^4 mol of converted ethylene/mol Pd per hour, to give a polymer with an average molecular weight (M_n) of ~20 000. Under suitable conditions the catalysts are highly stable and total conversions of more than 10^6 mol of ethylene per mol of Pd can be obtained. The product is high-melting (~260 °C) and is insoluble in most organic solvents; it crystallizes and precipitates during copolymerization as a snow-white solid.

Variation of the bidentate ligand results in significant changes in both the reaction rate and the molecular weight of the product. Figure 1 shows the effect of changing the chain length n of the diphosphine, of general formula Ph$_2$P(CH$_2$)$_n$PPh$_2$, on the rate and molecular weight. Many patents deal with more subtle variations of the diphosphine. In addition, several other types of *chelating* ligands (bipyridines [15], bisoxazolines [16], thioethers [17]) can be used. The counteranions also affect reaction rates; highest catalyst activities are obtained with weakly or noncoordinating anions (OTs$^-$, triflate (OTf$^-$), trifluoroacetate (TFA$^-$), BF$_4^-$, ClO$_4^-$, and "organic" anions such as certain tetraaryl borates). Best results are observed in protic solvents, such

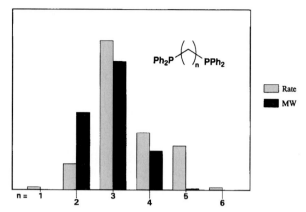

Figure 1 Influence of chain length of ligand $Ph_2P(CH_2)_nPPh_2$ on polymerization rate and molecular weight of polyketone.

as lower alcohols, but the polymerization also proceeds well in some aprotic solvents. The anions can, generally, conveniently be introduced by adding a Brønsted or Lewis acid, as anion source, to palladium acetate [Eq. (3)].

$$L_2Pd(OAc)_2 + 2\ HX \rightarrow L_2PdX_2 + 2\ HOAc \tag{3}$$

Instead of the free acid it is sometimes advantageous to use a metal salt (e.g., of Cu(II) of Ni(II)) to introduce the anions. Preformed complexes of the type L_2PdX_2 [18] and $L_2Pd(R)X$ [19], where L_2 represents a chelating ligand, X a weakly coordinating anion, and R a hydrocarbyl group (e.g., methyl), have also been tested as catalysts. The results are, generally, very similar to those obtained with catalysts prepared in situ.

a) Mechanism of Polymerization

Propagation

The catalytically active species in polyketone formation is thought to be a d^8 square-planar cationic palladium complex $L_2Pd(P)^+$, where L_2 represents the bidentate ligand and Ⓟ is the growing polymer chain. The fourth coordination site at palladium may be filled with an anion, a solvent molecule, a carbonyl group of the chain (vide infra), or a monomer molecule. The two alternating *propagation* steps are migratory insertion of CO in the palladium–alkyl bond [Eq. (4)] [20] and subsequent migratory insertion of ethylene in the resulting palladium–acyl bond [Eq. (5)]. Propagation "errors" (double CO or ethylene insertion) are not observed.

$$\left[L_2Pd\diagdown\diagup\diagdown\diagup Ⓟ \right]^+ +\ CO \longrightarrow \left[L_2Pd\diagdown\diagup\overset{O}{\diagdown}\diagup Ⓟ \right]^+ \tag{4}$$

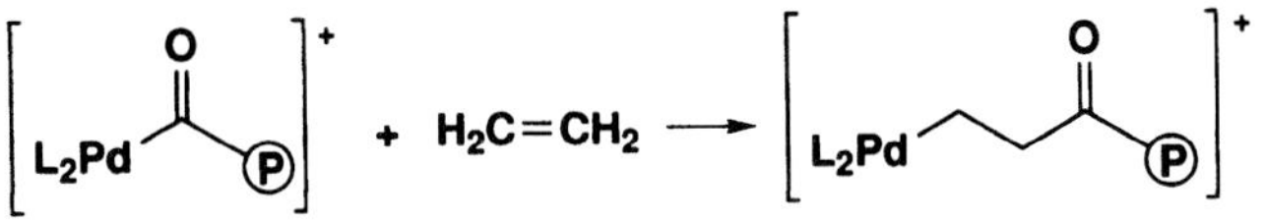

$$\left[L_2Pd\text{-}C(=O)\text{-}(P) \right]^{+} + H_2C{=}CH_2 \longrightarrow \left[L_2Pd\text{-}CH_2CH_2\text{-}C(=O)\text{-}(P) \right]^{+} \qquad (5)$$

Carbon monoxide insertion in a palladium–carbon bond is a fairly common reaction [21]. Under polymerization conditions, CO insertion is thought to be rapid and *reversible*. Olefin insertion in a palladium–carbon bond is a less common reaction, but recent studies involving cationic palladium–diphosphine and –bipyridyl complexes have shown that olefin insertion also, particularly in palladium–acyl bonds, appears to be a facile reaction [22]. Nevertheless, it is likely that olefin insertion is the slowest (rate-determining) and *irreversible* step (vide infra) in polyketone formation.

Initiation and Termination

End-group analysis by ^{13}C-NMR of the ethylene/CO copolymer produced in methanol generally shows the presence of 50% ester (–COOMe) and 50% ketone (–COCH$_2$CH$_3$) groups, in accordance with the average overall structure of the polymer molecule as depicted in Eq. (2). It is not clear a priori which group is the "head" and which is the "tail" of the molecules. Moreover, GC and MS analyses of oligomers produced with certain catalysts [13] show, in addition to the expected keto-ester product (2), the presence of diester (3) and diketone (4) compounds.

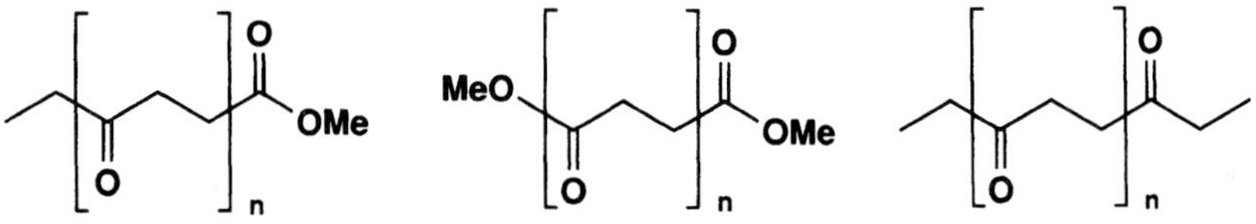

2
keto-ester
$n \geq 0$

3
diester
$n \geq 1$

4
diketone
$n \geq 0$

At low temperatures (below ~85 °C), the majority of the product molecules are keto-esters, with only small but balancing quantities of diesters and diketones. At higher temperatures, the same product molecules are produced in a **2/3/4** ratio close to 2 : 1 : 1. These observations have been explained [13] by assuming two initiation and two termination mechanisms for polyketone formation.

One initiation pathway produces ester end-groups. It starts with a palladium–carbomethoxy species, which can be formed either by CO insertion in a palladium methoxide or by direct attack of methanol on coordinated CO (Scheme 1).

Alternatively, a polymer chain can start by insertion of ethylene in a palladium hydride (vide infra), producing a ketone end-group. Ethylene insertion in palladium hydride and CO insertion in the resulting ethyl complex are both rapid and reversible; it is thought that the second ethylene insertion (in the Pd–acyl) is irreversible and "traps" the acyl complex to start the chain (Scheme 2).

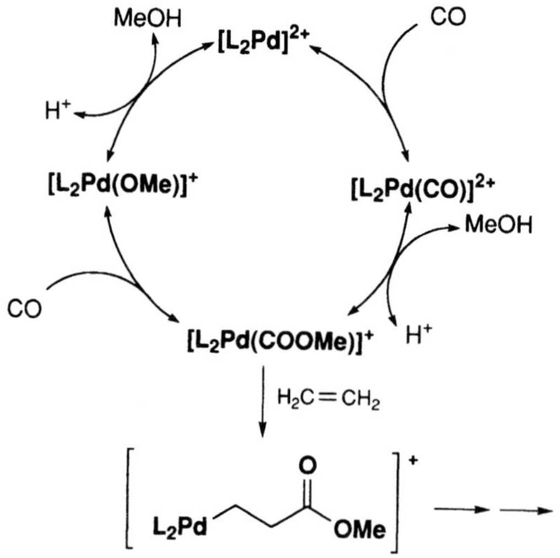
MeOH
$[L_2Pd]^{2+}$
CO
H^+
$[L_2Pd(OMe)]^+$
$[L_2Pd(CO)]^{2+}$
MeOH
CO
$[L_2Pd(COOMe)]^+$
H^+
$H_2C=CH_2$
O
L_2Pd
OMe

Scheme 1

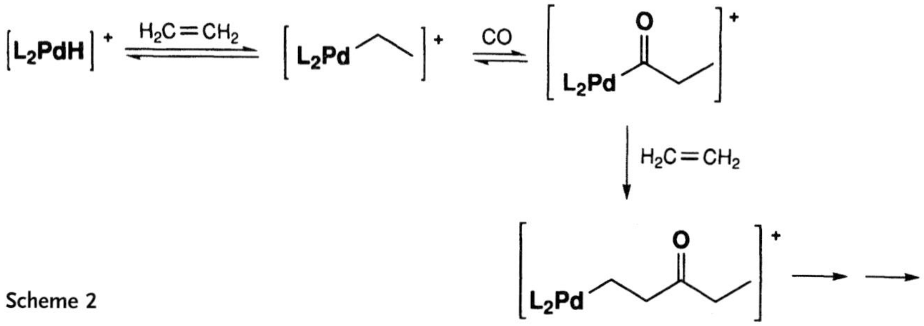
$[L_2PdH]^+$
$H_2C=CH_2$
L_2Pd
CO
O
L_2Pd
$H_2C=CH_2$
O
L_2Pd

Scheme 2

For ethylene/CO copolymerization, two relevant termination mechanisms have been proposed. One mechanism, protonolysis of the palladium–alkyl bond, produces a saturated ketone end-group and a palladium methoxide [Eq. (6)]. The latter can again be converted to a palladium carbomethoxide initiator by CO insertion into the palladium–methoxide bond.

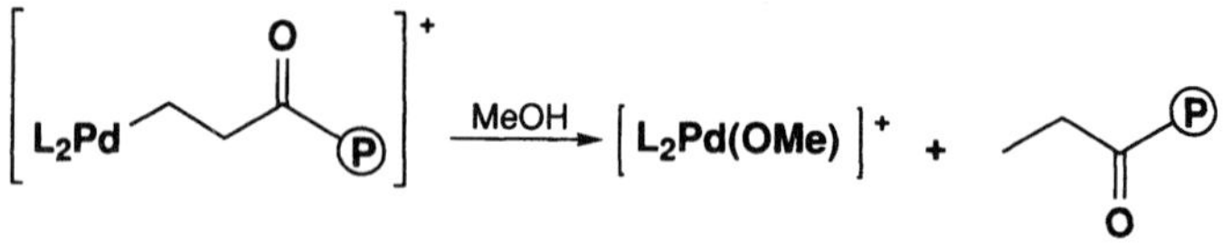
O
L_2Pd
P
MeOH
$[L_2Pd(OMe)]^+$
+
P
O
(6)

A second mechanism, the alcoholysis of the palladium–acyl bond, gives an ester end-group and a palladium hydride species [Eq. (7)], which is again an initiator for the next polymer chain.

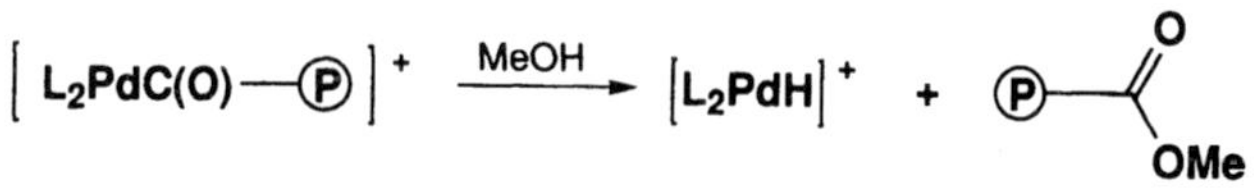
$[L_2PdC(O)—P]^+$
MeOH
$[L_2PdH]^+$
+
P
O
OMe
(7)

Scheme 3 summarizes the formation of the three possible polymeric products of **2**, **3**, and **4** by the two initiation–propagation cycles (*A*) and (*B*). Both cycles produce keto-ester molecules, but the cycles are connected by two "cross"-termination steps to give diester and diketone products.

The formation of substantial amounts of **3** and **4** at the higher temperatures demonstrates that transfer between the cycles is rapid and that both cycles contribute with comparable rates [23].

Copolymers with predominantly ketone end-groups (i.e., **4**) can be produced either by admitting water to the polymerization or by adding some hydrogen [13]. In aprotic solvents diketones can be produced exclusively. This indicates that palladium hydrides, generated via the water-gas shift reaction [Eq. (8)], or by heterolytic hydrogen splitting [Eq. (9)], are indeed efficient initiators and it also shows that protonolysis and/or hydrogenolysis of palladium alky Is can be an efficient termination mechanism.

$$[L_2Pd]^{2+} + H_2O + CO \rightarrow [L_2PdH]^+ + H^+ + CO_2 \tag{8}$$

$$[L_2Pd]^{2+} + H_2 \rightarrow [L_2PdH]^+ + H^+ \tag{9}$$

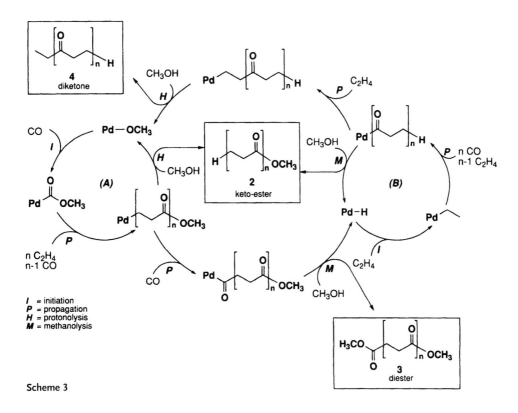

I = initiation
P = propagation
H = protonolysis
M = methanolysis

Scheme 3

b) The Role of Bidentate Ligands

Under conditions of polyketone catalysis, cationic palladium(II) catalysts modified with excess of *monodentate* phosphine and Brønsted acids of weakly coordinating anions selectively give methyl propionate with high rates [Eq. (1)]. Methyl propionate formation can be considered as a combination of polyketone initiation and termination steps without intervening propagation steps. Again, there are two possible catalytic cycles ((A) and (B), Scheme 4). There is no a priori argument to decide which cycle is actually responsible for methyl propionate formation. The absence of the cycle-transfer products diethyl ketone and dimethyl succinate suggests that only one cycle is operative, but it is also possible that both cycles operate in isolation.

The most obvious difference between monodentate and bidentate ligands is that the latter are always *cis* coordinated, whereas the former can also coordinate in a *trans* fashion. If bidentate ligands are used, the starting or growing polymer chain and the "empty" fourth coordination site are always *cis* to each other, which is the most favorable position for insertion reactions. Therefore, olefin insertion in the palladium–acyl is a probable reaction in *bidentate* phosphine complexes. If *monodentate* phosphines are used, both palladium–alkyl and palladium–acyl species prefer a *trans* orientation of the two phosphine ligands, which avoids the unfavorable situation of a Pd–P bond *trans* to a Pd–C bond. At the same time, *cis/trans* isomerization is expected to be rapid because of the presence of excess ligand [24]. It is further expected that both the insertion of ethylene in Pd–H and CO insertion in Pd–alkyl can only occur when the phosphine ligands are *cis* to each other. Immediately after insertion, *cis/trans* isomerization is likely to occur which places the "chain" and the fourth coordination site *trans* and thus opposes further monomer insertions. Therefore, the palladium–acyl can rapidly terminate by alcoholysis of the Pd–acyl bond to give methyl propionate.

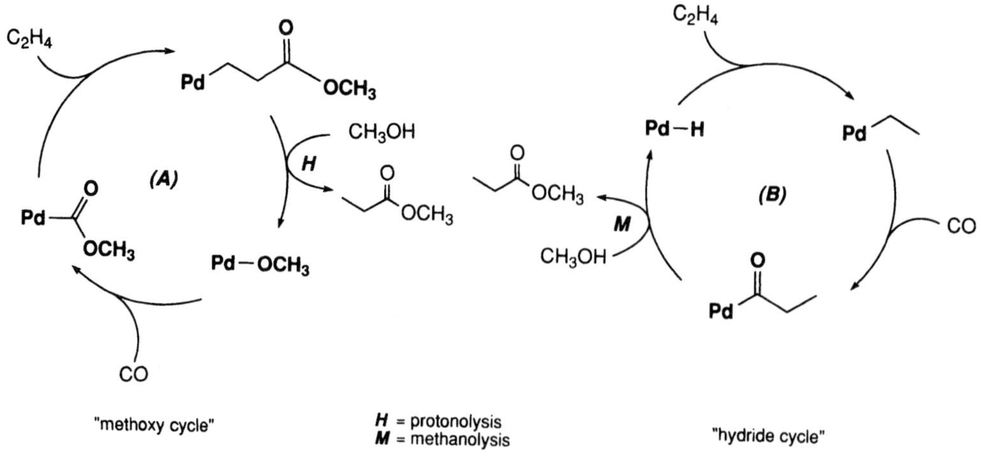

Scheme 4 Two possible mechanisms for methyl propionate formation.

c) The Role of the Anions

Apart from the requirement of *cis*-chelating neutral ligands (L_2), the high-activity catalyst complexes, L_2PdX_2, also require weakly or noncoordinating anions (X^-) [10, 13].

The higher reactivity of catalyst systems formed with such anions is thought to stem at least partly from the easier access of the substrate molecules to the coordination sites at the metal center. Nevertheless, it is suggested that the anionic ligands are actively involved in the catalytic cycle. Their presence in the proximity of the cationic palladium center, forming more or less strongly associated cation–anion pairs, can have a profound effect on the catalysis at that center. A contributing factor may be that less strongly coordinating anions, because of their easier dissociation from the ion-pair, generate a more electrophilic palladium center. The lower electron density on the palladium center may cause a lower binding energy with the comonomers because of less back-donation from metal to ligand. The intermediate palladium species which are involved in the catalytic cycles would therefore be less stable, with the result that, for instance, carbon monoxide, occupying vacant coordination sites at the palladium center, can be displaced by the olefin, and vice versa. At the same time (migratory) transformations between the various intermediate complexes would require lower activation energies and so proceed at a higher rate.

Coordination of the anions to the cationic palladium center may strongly depend on the polarity of the reaction medium. Solvation of the ion-pair by protic solvent molecules, such as methanol, is expected to facilitate cation–anion dissociation and therefore render the ijietal center more electrophilic and more easily accessible for substrate molecules. In relatively apolar solvents, close-contact ion-pairs are generally expected to exist. Anion displacement by substrate molecules may then require the use of noncoordinating anions, such as certain tetraaryl borates [19], with a relatively strong affinity for interaction with the solvent molecules. This will lead to a reduced barrier for displacement of these anions by monomer molecules.

d) Alternation Phenomena

Chain propagation of CO/ethylene copolymerization proceeds by a strictly alternating insertion of CO and olefin monomers in the growing chain. It is safe to assume that *double CO* insertion does not occur for thermodynamic reasons [1c]. However, the complete absence of double ethylene insertions is remarkable because ethylene insertion in a Pd–alkyl species must be exothermic by about 20 kcal mol^{-1} (84 kJ mol^{-1}). The observation of strict alternation is the more surprising since the same palladium catalysts also efficiently dimerize ethylene to butenes [25]. The perfect alternation is maintained even in the presence of very low concentrations of carbon monoxide. When starting a batch polymerization at a high ethylene/CO ratio, error-free copolymer is produced until all the CO is consumed; then the system starts forming butenes (with some catalyst systems at about twice the rate of copolymerization!).

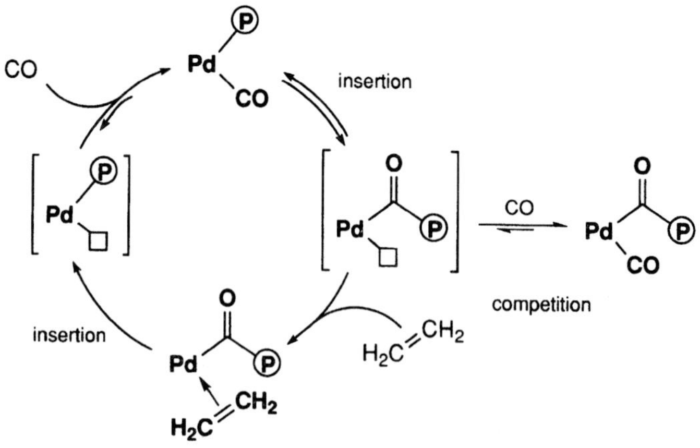

Scheme 5 Competition between CO and ethene coordination in polyketone formation.

One reason for the perfect alternation is probably the stronger coordination of CO to palladium(II), compared with ethylene. Once a palladium–*alkyl* is formed, the stronger CO coordination ensures that the next monomer to insert will usually be a CO molecule (assuming similar insertion barriers). Of course, CO also coordinates more strongly to a palladium–*acyl* but since the CO insertion is thermodynamically unfavorable, there the system will "wait" for an ethylene molecule to displace CO, to coordinate and insert (Scheme 5).

Chain propagation involves alternating *reversible* carbon monoxide insertion in Pd–alkyl species and *irreversible* insertion of the olefin in the resulting Pd–acyl intermediates. The overall exothermicity of the polymerization is caused predominantly by the olefin insertion step. Internal coordination of the chain-end's carbonyl group of the intermediate Pd–alkyl species, together with CO/olefin competition, prevents double olefin insertion, and thermodynamics prevent double CO insertions. The architecture of the copolymer thus assists in its own formation, achieving a perfect chemoselectivity to alternating polyketone.

A more detailed description of mechanistic aspects has been given in a previous article [26].

2.4.4.6.3
Scope of Olefin/CO Copolymerization

One of the unique further features of the catalysts is their ability to catalyze *also* the alternating co- or terpolymerization of higher olefins (both simple aliphatic and heteroatom-functionalized olefins) with carbon monoxide [27–32].

Chemoselectivity of Copolymerization (e.g., perfect alternation) with these olefins is governed by similar factors to those discussed for ethylene/CO Copolymerization, although some differences are noteworthy. Whereas β-H elimination does not take place to a significant extent with ethylene/CO Copolymerization, this termination pathway can play a significant role in Copolymerization with higher olefins. Another difference is that, under certain circumstances and with certain ligands, the polymers can be formed in a polyspiroketal structure (**5**), isomeric with polyketones [33, 34].

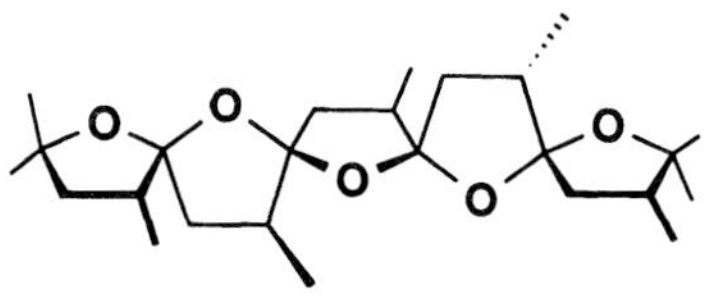

5 polyspiroketal

Apart from chemoselectivity, regio- and stereoselectivity of olefin insertion are also important factors to consider with higher olefin/CO copolymers.

By a suitable choice of ligands (L_2) and anions (X^-) it appears possible to control the regioselectivity of olefin insertion in intermediate Pd–acyl species (1,2- vs. 2,1-insertion) to give polymers with olefin enchainments via CO, varying from regioirregular to completely regioregular (for example, exclusively head-to-tail enchainment). The stereochemistry of olefin insertion can also be controlled to give atactic, isotactic, or syndiotactic olefin/CO copolymers (structures **6** and **7**) [36]. Stereoregularity can be achieved by a chain-end control mechanism [15, 36, 37] or by enantiomorphic site control [16, 28].

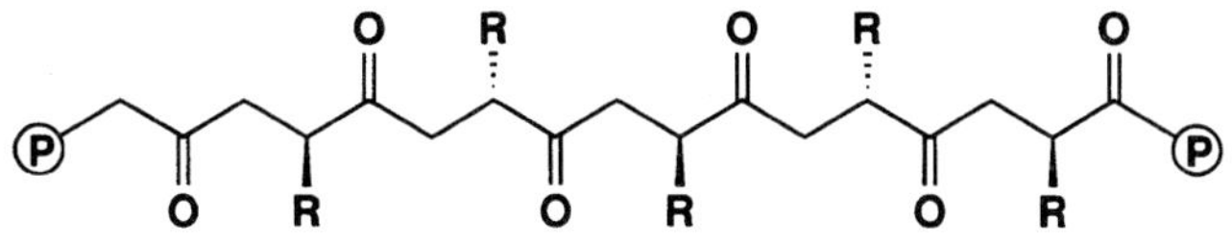

6 isotactic

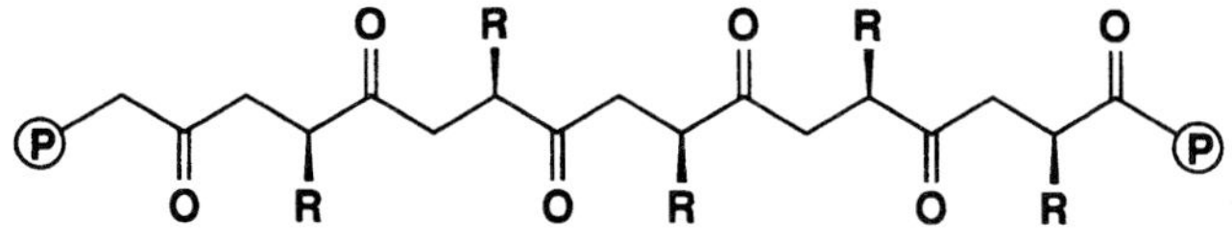

7 syndiotactic

In contrast to polyolefins such as polypropene, polyketones possess true stereogenic centers along the polymer backbone. Therefore, polyketones present a unique opportunity to use simple monomers in combination with chiral, enantiomerically pure palladium catalysts to prepare highly isotactic, optically active polymers (or oligomeric compounds) with main-chain chirality.

2.4.4.6.4
Conclusions

The synthesis of alternating olefin/CO copolymers, discussed here, presents a new chapter in the history of olefin polymerization. Moreover, it constitutes an example of transition metal-catalyzed carbonylation with potentially almost perfect control over selectivity. The cationic palladium(II) complex catalysts derive their ability to activate the nucleophilic substrate molecules from the electrophilic nature of the palladium (d^8) center. The *cis* arrangement of the neutral chelate ligand around the metal center in a square-planar configuration ensures that the polymer's chain-end and incoming monomers will also be in the *cis* configuration required for chain propagation. The electronic and steric properties of both the ligand and polymer chain-end together determine the mode of olefin coordination at the fourth coordination site, which in turn will determine the mode and ease of (higher) olefin insertion during chain propagation. It is likely that the interaction between the polar polymer chain-end and the electrophilic palladium center will play a crucial role not only in achieving the alternating mode of chain propagation (chemo-selectivity), but also in obtaining a high regio- and stereoselectivity for higher olefin insertion in intermediate palladium–acyls.

Although the basic principles of polyketone formation are now reasonably well understood, further studies, both of polymerization characteristics and of the elementary steps underlying polyketone catalysis, will be needed to exploit fully the potential of these selective polymerizations.

References

1 For earlier reviews, see: (a) A. SEN, *Adv. Polym. Sd.* **1986**, *73/74*, 125; (b) A. SEN, *CHEMTECH* **1986**, 48; (c) A. SEN, *Ace. Chem. Res.* **1993**, *26*, 303.

2 (a) C. E. ASH, *J. Mater. Educ.* **1994**, *16*, 1; (b) D. MEDEMA, A. NOORDAM, *Chemisch Magazine* **1995** (March), (in Dutch), 127.

3 DuPont (M. M. BRUBAKER), US 2.459.286 (**1950**); *Chem. Abstr.* **1950**, *44*, 4285.

4 BASF (W. REPPE, A. MAGIN), US 2.577.208 (**1951**); *Chem. Abstr.* **1952**, *46*, 6143.

5 Shell (T. M. SHRYNE, H. V HOLLER), US 3.984.388 (**1976**) *Chem. Abstr.* **1976**, *85*, 178 219.

6 U. KLABUNDE, T. H. TULIP, B. C. ROE, S. D. ITTEL, *J. Organomet. Chem.* **1987**, *334*, 141.

7 BP (B. DRIESSEN, M. J. GREEN, W. KEIM), EP Appl. 407.759 (**1992**); *Chem. Abstr.* **1992**, *116*, 152 623.

8 ICI (A. GOUGH), GB 1.081.304 (**1967**); *Chem. Abstr.* **1967**, *67*, 100 569.

9 (a) Union Oil (D. M. FENTON),US 3.530.109 (**1970**) and 4.076.911 (**1978**); *Chem. Abstr.* **1970**, *73*, 110 466, and **1978**, *88*, 153 263; (b) Shell (K. NOZAKI), US 3.689.460, 3.694.412 (**1972**); *Chem. Abstr.* **1972**, *77*, 152 869, 165 324; and Shell (K. NOZAKI), US 3.835.123 (**1974**); *Chem. Abstr.* **1975**, *83*, 132 273.

10 A. SEN, T.-W. LAI, *J. Am. Chem. Soc.* **1982**, *104*, 3520; A. SEN, T.-W. LAI, *Organometallics* **1984**, *3*, 866.

11 (a) Y. IWASHITA, M. SAKURABA, *Tetrahedron Lett.* **1971**, 2409; (b) G. CONSIGLIO, B. STUDER, F. OLDANI, P. PINO, *J. Mol. Catal.* **1990**, *58*, L9; (c) A. SEN, J. S. BRUMBAUGH, *J. Mol. Catal.* **1992**, *73*, 297.

12 Shell (E. DRENT), EP Appl. 121.965 (**1984**); *Chem. Abstr.* **1985**, *102*, 46 423.

13 E. DRENT, J. A. M. VAN BROEKHOVEN, M. J. DOYLE, *J. Organomet. Chem.* **1991**, *417*, 235.

14 ^{13}C-NMR (9 : 1 HFIPA/C_6D_6, 250 MHz): $ä_{CO}$ 212.9, $ä_{CH3}$ 35.6 (1 : 2). Some small resonances due to end-groups can be identified: –$COCH_2CH_3$, $ä_{CO}$ 217.1, $ä_{CH3}$ 6.5; –$COOCH_3$, $ä_{CO}$ 176.4, $ä_{COH}$, 52.0. The ratio of ester to keto end-groups is close to unity.

15 Shell (E. DRENT), EP Appl. 229.408 (**1986**); *Chem. Abstr.* **1988**, *108*, 6617.

16 (a) M. BROOKHART, M. I. WAGNER, G. G. A. BALAVOINE, *J. Am. Chem. Soc.* **1994**, *116*, 3641; (b) S. BARTOLINI, C. CARFAGNA, A. MUSCO, *Macromol. Rapid Commun.* **1995**, *16*, 9.

17 Shell (J. A. Van Doom, E. Drent), EP Appl. 345.847 (**1989**); *Chem. Abstr.* **1990**, *112*, 199 339.

18 (a) U. Daum, Ph. D. Thesis, ETH Zurich, **1988**; (b) Z. Jiang, G. M. Dahlen, K. Houseknecht, A. Sen, *Macromolecules* **1992**, 25, 2999.

19 M. Brookhart, EC. Rix, J. M. DeSimone, J. C. Barborak, *J. Am. Chem. Soc.* **1992**, *114*, 5894.

20 P. W. N. M. Van Leeuwen, C. F. Roobeek, H. Van der Heijden, *J. Am. Chem. Soc.* **1994**, *116*, 12 117.

21 See, e.g.: P. M. Maitlis, *The Organic Chemistry of Palladium*, Academic Press, London **1971**; P. A. Chaloner, *Handbook of Coordination Catalysis in Organic Chemistry*, Butterworths, London **1986**; A. Yamamoto, *Organotransition Metal Chemistry*, Wiley, New York **1986**.

22 (a) F. Ozawa, T. Hayashi, H. Koide, A. Yamamoto, *J. Chem. Soc., Chem. Commun.* **1991**, 1469; B. A. Markies, J. Boersma, A. L. Spek, G. Van Koten, *Recl. Trav. Chim. Pays-Bas* **1991**, *110*, 133; B. A. Markies, M. H. P. Rietveld, J. Boersma, A. L. Spek, G. Van Koten, *J. Organomet. Chem.* **1992**, *424*, C12; (b) G. P. C. M. Dekker, C. J. Elsevier, K. Vrieze, P. W. N. M. Van Leeuwen, C. F. Roobeek, *J. Organomet. Chem.* **1992**, *430*, 357; (c) J. Brumbaugh, R. R. Whittle, M. Parvez, A. Sen, *Organometallics* **1990**, *9*, 1735.

23 Crossover from (*A*) to (*B*) must have the same rate as termination within (*B*), since the termination rate does not depend on how the chain started. Similarly, crossover from B to (∧4) has the same rate as termination within A. If the ratio of alcoholysis to protolysis is k, the ratio of the products will be **2 : 3 : 4** = $1 + k^2 : k : k$. A ratio of 2 : 1 : 1 implies that both termination steps contribute equally ($k \approx 1$). The absence of **3** and **4** implies that k is either very large or very small, i.e., that only one of the two termination steps contributes.

24 See, e.g.: D. G. Cooper, J. Powell, *Can. J. Chem.* **1973**, *57*, 1634; D. A. Redfield, J. H. Nelson, *Inorg. Chem.* **1973**, *72*, 15.

25 E. Drent, *Pure Appl. Chem.* **1990**, *62*, 661.

26 E. Drent, J. A. M. v. Braekhanen, P. H. M. Budzelaar, in *Applied Homogeneous Catalysis with Organometallic Compounds* (Eds.: B. Cornils, W. A. Herrmann), 2nd ed., p. 333, Wiley-VCH, Weinheim **2002**.

27 (a) Shell (E. Drent, R. L. Wife), EP Appl. 181.014 (**1985**); *Chem. Abstr.* **1985**, *105*, 98 172; (b) Shell (E. Drent), EP Appl. 322.018 (**1988**); *Chem. Abstr.* **1989**, 777, 221 150; (c) Shell (J. A. Van Doom, PK. Wong, O. Sudmeijer), EP Appl. 376.364 (**1989**); *Chem. Abstr.* **1991**, *114*, 24 797; (d) Shell (P. K. Wong), EP Appl. 384.517 (**1989**); *Chem. Abstr.* **1991**, *114*, 103 079; (e) Shell (P. W. N. M. Van Leeuwen, C. F. Roobeek, P. K. Wong), EP Appl. 393.790 (**1990**); *Chem. Abstr.* **1991**, *114*, 103 034.

28 (a) A. Batistini, G. Consiglio, UW. Suter, *Angew. Chem.* **1992**, *104*, 306; *Angew. Chem., Int. Ed. Engl.* **1992**, *31*, 303; (b) M. Barsacchi, A. Batistini, G. Consiglio, UW. Suter, *Macromolecules* **1992**, *25*, 3604; (c) A. Batistini, Ph. D. Thesis, ETH Zurich **1991**; (d) A. Batistini, G. Consiglio, UW. Suter, *Minisymposium New Advances in Polyolefin Polymers*, Div. Polymeric Materials: Science and Eng. Inc., 204th ACS Meeting, Washington DC **1992**, Paper 5b; (e) R. Huter, Diplomarbeit, ETH Zurich **1992**.

29 J. C. W. Chien, A. X. Zhao, F. Xu, *Polym. Bull.* **1992**, *28*, 315.

30 Shell (E. Drent, H. P. M. Tomassen, M. J. Reynhout), EP Appl. 468.594 (**1992**); *Chem. Abstr.* **1992**, *117*, 52 257.

31 Shell (E. Drent), EP Appl. 272.727 (**1988**); *Chem. Abstr.* **1988**, *109*, 191 089.

32 Shell (E. Drent), EP Appl. 463.689 (**1992**); *Chem. Abstr.* **1992**, 776, 129 879.

33 P. K. Wong, J. A. Van Doom, E. Drent, O. Sudmeijer, H. A. Stil, *Ind. Eng. Chem. Res.* **1993**, *32*, 986.

34 A. Batistini, G. Consiglio, *Organometallics* **1992**, *11*, 1766.

35 An isotactic structure is one in which the optically active centers of the repeat units all have the same absolute stereochemistry (G. Natta, P. Pino, P. Corradini, F. Danusso, E. Mantica, G. Mazzanti, G. Moraglio, *J. Am. Chem. Soc.* **1955**, *77*, 1708); in a syndiotactic polymer, neighboring units have opposite stereochemistry. If an isotactic *polyolefin* is drawn in its extended conformation, it will have all its substituents pointing in the same direction. If an isotactic *polyketone* is drawn in its extended conformation, the substituents will alternately point up and down.

36 P. Corradini, A. De Rosa, A. Panunzi, P. Pino, *Chimia* **1990**, *44*, 52.

37 (a) M. Barsacchi, G. Consiglio, L. Medici, G. Petrucci, U. W. Suter, *Angew. Chem.* **1991**, *103*, 992; *Angew. Chem., Int. Ed. Engl.* **1991**, *30*, 989; (b) M. Barsacchi, Ph. D. Thesis, ETH Zurich, **1991**; (c) M. Barsacchi, G. Consiglio, U. W. Suter, *Minisymposium New Advances in Polyolefin Polymers*, Div. Polymeric Materials: Science and Eng. Inc., 204th ACS Meeting, Washington DC **1992**, Paper 32.

2.4.5
Allylic Substitution Reactions

Denis Sinou

Palladium-catalyzed allylic substitution, the so-called Tsuji–Trost reaction, is now a well-used methodology in synthetic organic chemistry, allowing the chemo-, regio-, and stereoselective formation of carbon–carbon bonds as well as carbon–heteroatom bonds under very mild conditions [1–3]. Although it was shown in the late 1980s that palladium(0)-catalyzed reaction of sodium azide with various allyl esters [4] or 1,3-diene monoepoxides [5] occurred in aqueous tetrahydrofuran, leading to allylic azides in quite good yields, the first example of palladium(0)-catalyzed alkylation in a two-phase water/organic solvent system using a palladium complex associated with TPPTS appeared in the literature as early as 1991 [6].

The water-soluble palladium(0) complex was generated in situ generally from $Pd(OAc)_2$ in association with a water-soluble ligand such as TPPTS [6–8], a nitrile being used as the co-solvent. Allylic carbonates such as cinnamyl carbonate reacted with various carbonucleophiles in this two-phase system, giving substitution products in moderate to high yields [Eq. (1)] [6, 9–11]. Since the reaction occurred in neutral medium, only traces of base being generated in the catalytic cycle, the allylic carbonates are stable to hydrolysis. It was noted that the selectivity in the formation of mono- and diallylated compounds was very sensitive to the nature of the carbonucleophile and its pK_a; the acyclic carbonucleophiles such as ethyl acetoacetate, acetylacetone, dimethyl malonate, dicyanomethane, and bis(phenyl-sulfone)methane gave predominantly the monoallylated product, when the cyclic carbonucleophiles such as tetronic acid, dimedone, and barbituric acid gave predominantly the diallylated product.

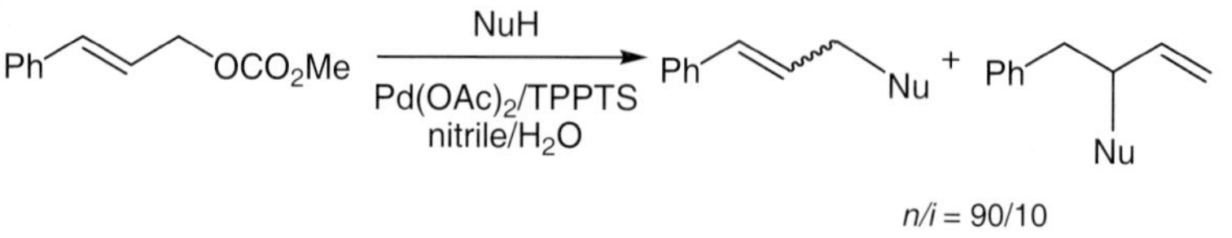

$$Ph\diagup\diagdown\diagup OCO_2Me \xrightarrow[\substack{Pd(OAc)_2/TPPTS \\ nitrile/H_2O}]{NuH} Ph\diagup\diagdown\diagdown Nu \ + \ Ph\diagup\diagdown\diagup \diagup \quad (1)$$

$$n/i = 90/10$$

$$R: CH_3COCH_2CO_2C_2H_5, \ CH_3COCH_2COCH_3, \ NO_2CH_2CO_2C_2H_5$$

A detailed investigation using (*E*)-2-hexenyl methyl carbonate as the π-allyl precursor and ethyl acetoacetate as the nucleophile showed that the regioselectivity

of the reaction is independent of the Pd(OAc)$_2$/TPPTS ratio, the reaction temperature, the water/nitrile ratio, or the nature of the nitrile [9]. However, the catalytic
activity is deeply affected by these parameters, and the highest activity was obtained
for a Pd(OAc)$_2$/TPPTS ratio of 9 : 1 in acetonitrile/water as the solvent mixture.

Alkylation of allylic acetates or vinyl epoxide occurred also under these conditions,
NEt$_3$ or preferably 1,8-diazabicyclo[5.4.0]undec-7-ene (dbu) being used as the base
in the former case.

The alkylation reaction in a two-phase system was also extended to various
heteronucleophiles [11]. Secondary amines (morpholine, benzylmethylamine, etc.)
as well as primary amines (n-butylamine, 2,2-diethylpropargylamine, cyclohexyl-
amine, α-methylbenzylamine, etc.) react for example with (E)-cinnamyl acetate to
give only the monoallylated product in quite good yields [Eq. (2)]. The water-soluble
nucleophiles sodium azide and sodium p-toluene sulfinate react also under these
conditions, giving the corresponding allyl azide and allyl p-toluene sulfone in 92
and 95% yield, respectively.

$$\text{Ph}\diagup\!\!\!\!\diagdown\!\!\!\diagdown\text{OAc} + \text{R}^1\text{R}^2\text{NH} \xrightarrow[\text{CH}_3\text{CN/H}_2\text{O}]{\text{Pd(OAc)}_2/\text{TPPTS}} \text{Ph}\diagup\!\!\!\!\diagdown\!\!\!\diagdown\text{NR}^1\text{R}^2 \quad (2)$$

70-97%

The use of benzonitrile or butyronitrile as the organic solvent allows an easy
recycling of the catalyst, without any decrease in the yields [11].

Allyl chlorides and acetates are reduced to the corresponding alkenes in a two-
phase system heptane/water in the presence of water-soluble palladium complexes
containing ligands such as polyether phosphines, TPPMS, or carboxylic phosphines
in the presence of HCO$_2$Na [Eq. (3)] [12]. The most active catalyst is PdCl$_2$L$_2$ with L
being a polyether phosphine; a mixture of nonenes are obtained in 82% yield,
PdCl$_2$(P-n-Bu$_3$)$_2$ giving lower yields under these conditions.

$$\text{Ph}\diagdown\!\!\!\diagup\text{CH}_2\text{Cl} \xrightarrow[\text{PdL}_2\text{Cl}_2/\text{heptane/H}_2\text{O}]{\text{HCO}_2\text{Na}} \text{Ph-C}_3\text{H}_5 \quad (3)$$

Asymmetric allylic substitution of 1,3-diphenyl-2-propenyl acetate in water or in
an aqueous/organic biphasic medium has been performed in the presence of the
complex obtained from [Pd(η^3-C$_3$H$_5$)Cl]$_2$ and a chiral amphiphilic phosphinite-
oxazoline derived from natural D-glucosamine, enantioselectivity up to 85% ee being
obtained [13]; recycling of the catalyst is possible [Eq. (4)].

$$\quad (4)$$

Ar = C$_6$H$_4$CH$_2$N$^+$Et$_2$MeBF$_4$$^-$ ee = 82%

It has been observed that the rate of palladium-catalyzed allylic alkylation in water is drastically enhanced when the reaction is performed in the presence of surfactant [14]. Enantioselectivity up to 92% is obtained in the reaction of dimethyl malonate with 1,3-diphenyl-2-propenyl acetate when a chiral ligand such as Binap is used in the presence of cetyltrimethylammonium hydrogen sulfate [15, 16].

Water-soluble polymer-bound Pd(0)–phosphine catalyst has also been efficiently used in aqueous or mixed aqueous/organic media, the catalyst being recycled by solvent or thermal preparation methods [17]. Amphiphilic resin-supported palladium–phosphine complexes show high catalytic activity in allylic substitution reactions of various allylic acetates with different nucleophiles in aqueous media [18, 19]. Enantiomeric excess up to 98% is obtained using amphiphilic resin-supported MOP ligand or resin-supported P,N-chelating palladium complexes, the catalyst being recyclable [20, 21]. The catalyst could be recovered by simple filtration and re-used without any loss of activity and enantioselectivity.

A supported aqueous phase system (SAPC; see Section 2.6) has also been developed for allylic substitution. Alkylation of (*E*)-cinnamyl ethyl carbonate by ethyl acetoacetate or morpholine occurs in acetonitrile or benzonitrile using Pd(OAc)$_2$–TPPTS supported on mesoporous or nonporous silica; no leaching of the catalyst has been observed, allowing proper recycling of the catalyst [22–26]. Polyhydroxylated supports such as cellulose and chitosan have also been used successfully in this approach [27–29].

One of the most interesting applications of this palladium-catalyzed alkylation reaction in an aqueous biphasic system was the selective deprotection of amines and alcohols protected by the allyloxycarbonyl group (Alloc). Whereas a variety of nucleophilic species, such as carbonucleophiles, amines, thiols, carboxylates, and hydride donors, have been used in a homogeneous organic medium for intercepting the intermediate π-allyl complexes, Genêt et al. used the aqueous palladium catalyst obtained from Pd(OAc)$_2$ and TPPTS for the catalytic allyl transfer, diethylamine being the allyl scavenger [30–33]. Either nitrile/water or diethyl ether/water is equally suitable for the removal of the alloc moiety from nitrogen or oxygen, the reaction proceeding under very mild conditions with a very high chemoselectivity compared to the usual deprotection techniques. For example, deprotection of alloc-protected primary alcohols, such as (*R*)-citronellol, occurs in a few minutes upon exposure to Pd(OAc)$_2$–TPPTS in CH$_3$CN/H$_2$O, diethylamine being used in a 2–2.5-fold excess [Eq. (5)] [30, 31]. Under these conditions, *t*-butyldiphenyl ether or ester functions are stable. The deprotection of secondary alcohols such as menthol proceeds smoothly.

$$\text{(5)}$$

94%

When the *N*-alloc protecting group of primary amines such as benzylamine is cleaved rapidly under these standard conditions in quantitative yields, the use of a 40-fold excess of diethylamine as the π-allyl scavenger or a five-fold excess of diethylamine in a butyronitrile/water system is necessary for the quantitative deprotection of alloc derivatives of secondary amines.

Deprotection of allyl groups from carboxylic allyl esters is also possible using these conditions [31, 33]. The conditions can be finely adjusted to allow selective differential deprotection of similar protective groups. In a homogeneous CH_3CN/H_2O medium, the facility of cleavage of the allyl group follows the order allyl > cinnamyl > dimethylallyl. However, under biphasic conditions (C_3H_7CN/H_2O), the allyl group of phenylacetic acid allyl ester is still cleaved at room temperature, giving phenylacetic acid in quite good yields, whereas the cinnamyl and the dimethylallyl esters remain intact even after three days of 25 °C. This procedure can also be used for the selective cleavage of allyloxycarbamate in the presence of substituted allyl carboxylate [Eq. (6)]. The allyloxycarbamate is selectively and quantitatively cleaved under homogeneous conditions, in the presence of 1% of palladium complex; treatment of the resulting monodeprotected product with a higher amount of catalyst (5 mol%) gives the free amino acid. Conversely, selective cleavage of an allyl-oxycarbonate could be performed in the presence of an allylcarbamate using successively C_3H_7CN and CH_3CN as the nitrile.

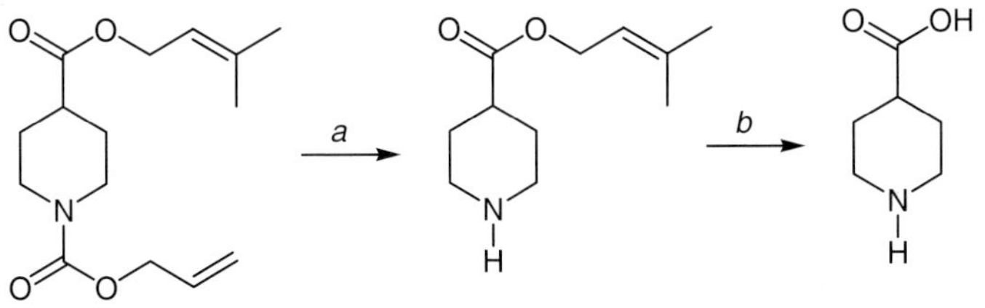

(6)

The use of the water-soluble sodium azide as the allyl scavenger allows the cleavage of allyloxycarbonyl-protected alcohols to occur under essentially neutral conditions [34]. It has been shown that the phosphines $(t\text{-Bu-}p\text{-}C_6H_4)_nP(C_6H_4\text{-}m\text{-}SO_3Na)_{3-n}$ ($n = 1, 2$) are more efficient than TPPTS in the deprotection of long alkyl chain alcohols [35].

The use of water or of a two-phase water/organic solvent system as the reaction medium could also change drastically the selectivity of a given reaction. This dramatic enhancement of selectivity was effectively observed in the allylation of uracils and thiouracils [36, 37]. Whereas the reaction of uracil with (*E*)-cinnamyl acetate in tetrahydrofuran in the presence of $Pd(PPh_3)_4$ gives a complex mixture of mono- and diallylated products, performing the reaction in CH_3CN/H_2O with $Pd(OAc)_2$–TPPTS as the catalyst leads to allylation only at N-1 in quite good yield [Eq. (7)]. In the case of thiouracil, performing the reaction in CH_3CN/H_2O with $Pd(OAc)_2$–TPPTS as the catalyst gave a unique product of monoallylation at sulfur, whereas the use of dioxane as the solvent in the presence of $Pd(PPh_3)_4$ gave a complex mixture of products resulting from allylation at N-1, N-3, and sulfur.

(E) Ph-CH=CH-CH$_2$OAc, cat Pd(OAc)$_2$/TPPTS, dbu, CH$_3$CN/H$_2$O

In conclusion, although the first aim of the use of a water-soluble palladium catalyst in allylic alkylation in a two-phase system was the recycling of the catalyst, this methodology finds quite interesting applications in the deprotection of amines, alcohols, and acids, as well as in the selective alkylation of uracils and thiouracils. More recently the effective use of supported aqueous-phase catalysis, as well as asymmetric alkylation in water in the presence of surfactants or amphiphilic resin-supported phosphines, opens new applications and developments for the future.

References

1 J. Tsuji, *Palladium Reagents and Catalysts: Innovations in Organic Synthesis*, Wiley, Chichester **1995**, p. 290.

2 J. Harrington, in *Comprehensive Organometallic Chemistry* (Eds.: E. W. Abel, F. G. A. Stone, G. Wilkinson), Pergamon Press, Oxford **1995**, Vol. 12, p. 959.

3 E. Negishi, *Handbook of Organopalladium Chemistry for Organic Synthesis*, Wiley, New York **2002**, p. 1669.

4 S. I. Murahashi, Y. Taniguchi, Y. Imada, Y. Tanigawa, *J. Org. Chem.* **1989**, *54*, 3292.

5 A. Tenaglia, B. Waegell, *Tetrahedron Lett.* **1988**, *29*, 4852.

6 M. Safi, D. Sinou, *Tetrahedron Lett.* **1991**, *32*, 2025.

7 C. Amatore, E. Blart, J. P. Genêt, A. Jutand, S. Lemaire-Audoire, M. Savignac, *J. Org. Chem.* **1995**, *60*, 6829.

8 F. Monteil, P. Kalck, *J. Organomet. Chem.* **1994**, *482*, 45.

9 S. Sigismondi, D. Sinou, *J. Mol. Catal.* **1997**, *116*, 289.

10 J. P. Genêt, E. Blart, M. Savignac, *Synlett* **1992**, 715.

11 E. Blart, J. P. Genêt, M. Safi, M. Savignac, D. Sinou, *Tetrahedron* **1994**, *50*, 505.

12 T. Okano, Y. Moriyama, H. Konishi, J. Kiji, *Chem. Lett.* **1986**, 1463.

13 T. Hashizume, K. Yonehara, K. Ohe, S. Uemura, *J. Org. Chem.* **2000**, *65*, 5197.

14 S. Kobayashi, W. W.-L. Lam, K. Manabe, *Tetrahedron Lett.* **2000**, *41*, 6115.

15 C. Rabeyrin, D. Sinou, *Tetrahedron Lett.* **2000**, *41*, 7461.

16 D. Sinou, C. Rabeyrin, C. Nguefack, *Adv. Synth. Catal.* **2003**, *345*, 357.

17 D. E. Bergbreiter, Y.-S. Liu, *Tetrahedron Lett.* **1997**, *38*, 7843.

18 Y. Uozumi, H. Danjo, T. Hayashi, *Tetrahedron Lett.* **1997**, *38*, 3557.

19 H. Danjo, D. Tanaka, T. Hayashi, Y. Uozumi, *Tetrahedron* **1999**, *55*, 14 341.

20 Y. Uozumi, H. Danjo, T. Hayashi, *Tetrahedron Lett.* **1998**, *39*, 8303.

21 Y. Uozumi, K. Shibatomi, *J. Am. Chem. Soc.* **2001**, *123*, 2919.

22 P. Schneider, F. Quignard, A. Choplin, D. Sinou, *New J. Chem.* **1996**, *20*, 545.

23 A. Choplin, S. Dos Santos, F. Quignard, S. Sigismondi, *Catal. Today* **1998**, *42*, 471.

24 S. Dos Santos, Y. Tong, F. Quignard, A. Choplin, D. Sinou, J. P. Dutasta, *Organometallics* **1998**, *17*, 78.

25 S. Dos Santos, F. Quignard, D. Sinou, A. Choplin, *Stud. Surf. Sci. Catal.* **1998**, *118*, 91.

26 S. Dos Santos, F. Quignard, D. Sinou, A. Choplin, *Top. Catal.* **2000**, *13*, 311.

27 F. Quignard, A. Choplin, A. Domard, *Langmuir* **2000**, *16*, 9106.

28 F. Quignard, A. Choplin, *Chem. Commun.* **2001**, 21.

29 P. Buisson, F. Quignard, *Aust. J. Chem.* **2002**, *55*, 73.

30 J. P. Genêt, E. Blart, M. Savignac, S. Lemeune, J. M. Paris, *Tetrahedron Lett.* **1993**, *34*, 4189.

31 S. Lemaire-Audoire, M. Savignac, E. Blart, G. Pourcelot, J. P. Genêt, J. M. Bernard, *Tetrahedron Lett.* **1994**, *35*, 8783.

32 J. P. Genêt, E. Blart, M. Savignac, S. Lemeune, S. Lemaire-Audoire, J. M. Paris, J. M. Bernard, *Tetrahedron* **1994**, *50*, 497.

33 S. Lemaire-Audoire, M. Savignac, E. Blart, J. M. Bernard, J. P. Genêt, *Tetrahedron Lett.* **1997**, *38*, 2955.

34 S. Sigismondi, D. Sinou, *J. Chem. Res. (S)* **1996**, 46.

35 L. Caron, M. Canipelle, S. Tilloy, H. Bricourt, E. Monflier, *Tetrahedron Lett.* **2001**, *42*, 8837.

36 S. Sigismondi, D. Sinou, M. Moreno-Mañas, R. Pleixats, M. Villaroya, *Tetrahedron Lett.* **1994**, *35*, 7085.

37 C. Goux, S. Sigismondi, D. Sinou, M. Moreno-Mañas, R. Pleixats, M. Villaroya, *Tetrahedron* **1996**, *52*, 9521.

2.4.6
Asymmetric Synthesis

Denis Sinou

Since the mid-1980s there have been very important advances in asymmetric organometallic catalysis, enantioselectivities higher than 95% being obtained currently in reactions such as hydrogenation, isomerization, epoxidation, hydroxylation, or allylic substitution [1]. Although water-soluble catalysts have been known and used since the 1970s, it is only recently that this methodology has been extended to the preparation of chiral compounds from prochiral ones.

The enantioselective hydrogenation of some α-amino acid precursors **1** [Eq. (1)] in water or in an aqueous/organic two-phase system has been thoroughly investigated using rhodium or ruthenium complexes associated with chiral water-soluble ligands **3–13**. Some of the most interesting results are summarized in Table 1.

$$\text{(1)}$$

e.g.
a: $R_1 = C_6H_5$, $R_2 = H$, $R_3 = CH_3$; **b**: $R_1 = C_6H_5$, $R_2 = R_3 = CH_3$; **c**: $R_1 = R_3 = C_6H_5$, $R_2 = H$;
d: $R_1 = R_3 = C_6H_5$, $R_2 = CH_3$; **e**: $R_1 = $ 3-MeO-4-AcO-C_6H_3, $R_2 = H$, $R_3 = CH_3$

Table 1 Asymmetric reduction of some α-amino acid precursors **1** at 25 °C in the presence of rhodium complexes containing ionized ligands **3–5** and **7–9**.

Precursor 1	Ligand	Solvent	p_{H_2} [MPa]	Product ee [%] (Config.)	Ref.
1a	(S,S)-3	H$_2$O/AcOEt (1 : 1)	0.1	34 (S)	[3]
1a	(S,S)-4	H$_2$O/AcOEt (1 : 1)	1.5	65 (R)	[3]
1a	(S,S)-5	H$_2$O/AcOEt (1 : 2)	1.0	87 (R)	[3]
1a	(R,R)-7a	H$_2$O	1.4	25 (S)	[6]
1a	(R,R)-7b	H$_2$O	1.4	34 (S)	[6]
1a	(S,S)-8a	H$_2$O	1.4	67 (R)	[6]
1a	(S,S)-8b	H$_2$O	1.4	71 (R)	[6]
1a	(S,S)-9a	H$_2$O	1.4	94 (R)	[6]
1a	(S,S)-9b	H$_2$O	1.4	90 (R)	[6]
1b	(S,S)-3	H$_2$O/AcOEt (1 : 1)	0.1	20 (S)	[3]
1b	(S,S)-4	H$_2$O/AcOEt (1 : 1)	1.5	45 (R)	[3]
1b	(S,S)-5	H$_2$O/AcOEt (1 : 1)	1.0	81 (S)	[3]
1b	(R,R)-7a	H$_2$O	1.4	8 (S)	[3]
1b	(R,R)-7b	H$_2$O/EtOAc/C$_6$H$_6$ (2 : 1 : 1)	1.4	25 (S)	[6]
1b	(S,S)-8a	H$_2$O/EtOAc/C$_6$H$_6$ (2 : 1 : 1)	1.4	45 (R)	[6]
1b	(S,S)-8b	H$_2$O/EtOAc/C$_6$H$_6$ (2 : 1 : 1)	1.4	50 (R)	[6]
1b	(S,S)-9a	H$_2$O/EtOAc/C$_6$H$_6$ (2 : 1 : 1)	1.4	77 (R)	[6]
1b	(S,S)-9b	H$_2$O/EtOAc/C$_6$H$_6$ (2 : 1 : 1)	1.4	74 (R)	[6]
1c	(S,S)-3	H$_2$O/EtOAc (1 : 2)	0.1	13 (S)	[3]
1c	(S,S)-4	H$_2$O/EtOAc (1 : 1)	1.5	44 (R)	[3]
1c	(S,S)-5	H$_2$O/EtOAc (1 : 2)	1.0	86 (R)	[3]
1d	(R,R)-7a	H$_2$O/EtOAc/C$_6$H$_6$ (2 : 1 : 1)	1.4	9 (S)	[6]
1d	(R,R)-7b	H$_2$O/EtOAc/C$_6$H$_6$ (2 : 1 : 1)	1.4	11 (S)	[6]
1d	(S,S)-8a	H$_2$O/EtOAc/C$_6$H$_6$ (2 : 1 : 1)	1.4	54 (R)	[6]
1d	(S,S)-8b	H$_2$O/EtOAc/C$_6$H$_6$ (2 : 1 : 1)	1.4	67 (R)	[6]
1d	(S,S)-9a	H$_2$O/EtOAc/C$_6$H$_6$ (2 : 1 : 1)	1.4	65 (R)	[6]
1d	(S,S)-9b	H$_2$O/EtOAc/C$_6$H$_6$ (2 : 1 : 1)	1.4	58 (R)	[6]
1g	(S,S)-3	H$_2$O/EtOAc (1 : 2)	0.1	37 (S)	[3]
1g	(S,S)-4	H$_2$O/EtOAc (1 : 1)	1.0	58 (R)	[3]
1e	(S,S)-5	H$_2$O/EtOAc (1 : 2)	1.0	88 (R)	[3]
1e	(R,R)-7a	H$_2$O	1.4	42 (S)	[6]
1e	(R,R)-7b	H$_2$O	1.4	67 (S)	[6]
1e	(S,S)-8a	H$_2$O	1.4	76 (R)	[6]
1e	(S,S)-8b	H$_2$O	1.4	79 (R)	[6]
1e	(S,S)-9a	H$_2$O	1.4	93 (R)	[6]
1e	(S,S)-9b	H$_2$O	1.4	88 (R)	[6]

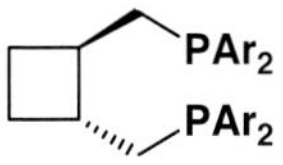
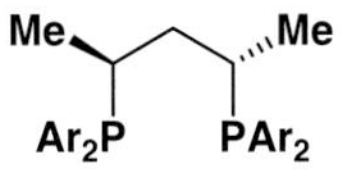
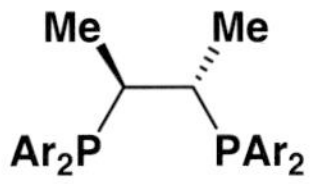
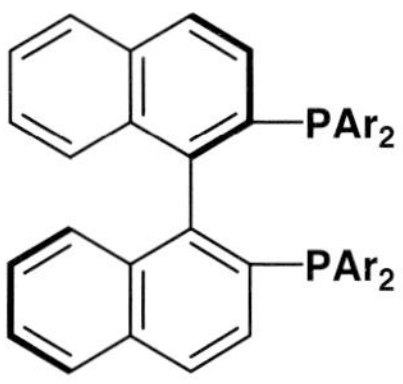

(S,S)-CBD **3** *(S,S)*-BDPP **4** *(S,S)*-CHIRAPHOS **5** *(R)*-BINAP **6**

Ar = 3-NaO_3S-$C_6H_4^-$

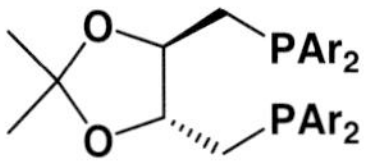
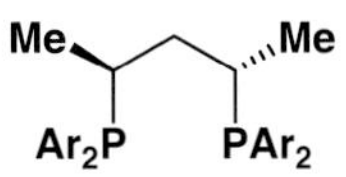
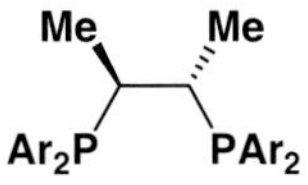

(R,R)-**7** *(S,S)*-**8** *(S,S)*-**9**

a Ar = C_6H_4-4-$NMe_3^+BF_4^-$; **b** Ar = C_6H_4-4-$NHMe_2^+BF_4^-$

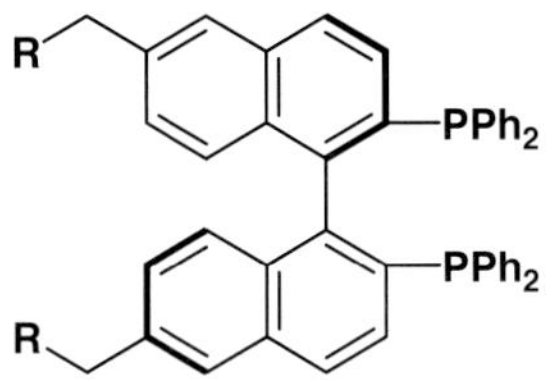
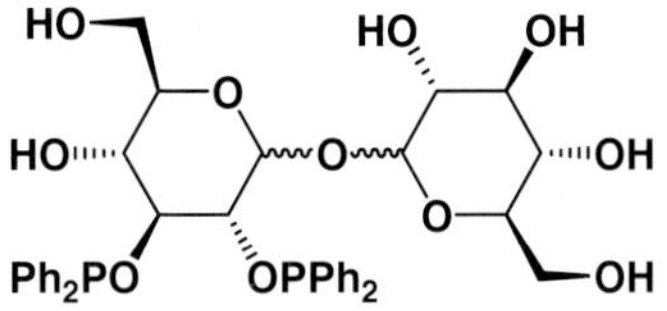

(R)- diamo-BINAP **10** R = $NH_3^+Br^-$

12 α,α
13 β,β

(R)- digm-BINAP **11** R =

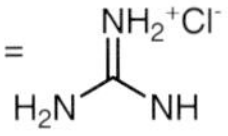

The most investigated chiral ligands were the sulfonated phosphines **3–5** [2, 3] and those possessing a quaternary ammonium function (**7–9**) [4–8]. Rhodium complexes of water-soluble 1,2-diphosphines, such as sulfonated or aminoquaternized ammonium CHIRAPHOS **5** and **9**, reduced unsaturated α-amino acid precursors in water or in a two-phase system with high enantioselectivities (*ee* = 65–96%), close to the values obtained in organic media (Table 1). Conversely, rhodium complexes of water-soluble 1,4-diphosphines **3** and **7**, derived from CBD or DIOP, and 1,3-diphosphines **4** and **8**, derived from BDPP, gave lower enantioselectivities in aqueous phase than the non-water-insoluble analogues in organic solvents (8–34% *ee* for **3** and **7**, and 40–71% *ee* for **4** and **8**, respectively). Later, Hanson and co-workers [9] prepared a surface-active tetrasulfonated chiral diphosphine derived from BDPP which showed similar selectivity but improved reactivity compared to the unmodified BDPP in the reduction of **1b** (with *ee* up to 69%).

Rhodium and ruthenium complexes associated with water-soluble BINAP **6** gave enantioselectivities up to 88% in the reduction of α-acetamidoacrylic acid and its methyl ester [10, 11], quite close to the values obtained in organic solvents [Eq. (2)]; it is to be noted in these cases that the direction of enantioselection was the opposite using the rhodium or the ruthenium catalyst, as in organic solvents.

$$
\underset{\text{NHCOCH}_3}{\overset{\text{CO}_2\text{R}}{\diagup}} \quad \xrightarrow[\text{[catalyst/H}_2\text{O]}]{\text{H}_2 \text{ (0.1 MPa)/25 °C}} \quad \underset{\text{NHCOCH}_3}{\overset{\text{CO}_2\text{R}}{\diagup}} \tag{2}
$$

R = H   cat [Rh]/*(R)*-**6**   ee 70% *(S)*;   cat [Ru]/*(R)*-**6**   ee 68.5% *(R)*
R = CH$_3$ cat [Rh]/*(R)*-**6**   ee 69.5% *(S)*;   cat [Ru]/*(R)*-**6**   ee 82% *(R)*

Carbohydrate-based bisphosphinites **12** and **13** derived from α,α- and β,β-tre-halose associated with rhodium complexes gave enantioselectivities up to 88% in the reduction of **1** in water or in a two-phase H$_2$O/AcOEt system [12–14]; this enantioselectivity was improved to 98% *ee* when the reaction was carried out in H$_2$O/AcOEt/CH$_3$OH.

The lower enantioselectivity observed in water for the reduction of amino acid precursors using water-soluble diphosphines in asssociation with rhodium com-plexes was attributed to solvent effects and to the reaction kinetics in the two solvents [15]. When the reaction was carried out in alcohol/water solvent mixtures, an increase in water content induced a decrease in enantioselectivity. In a systematic study on the influence of various solvents on the enantioselectivity in the reduction of dehydroamino acids, a linear relationship was found between log (%S/%R) and the solvophobicity parameter S$_p$ of various solvents, log (%S/%R) decreasing with increasing S$_p$. However, another possibility could be the presence of a different mechanism than the usual one, occuring via a monohydride species. Effectively, although the biphasic hydrogenation of α-amino acid precursors was shown to be a truly homogeneous process [16], it was also shown that water not only was a solvent, but also had a chemical effect on the reduction [17–19]. Hydrogenation of α-acetamidocinnamic acid methyl ester in AcOEt/D$_2$O in the presence of a rhodium complex associated with a sulfonated ligand such as TPPTS occurred with a 75% regiospecific monodeuteration at the position α to the acetamido and the ester functions, the amount of deuterium incorporation depending on the ligand used [Eq. (3)]. When the reduction was performed under a deuterium atmosphere in the presence of water, hydrogen incorporation occurred at the same position, the overall reaction being a *cis* addition of HD.

$$
\underset{\substack{\text{Ph}\\ \text{D}}}{\overset{\text{H} \quad \text{CO}_2\text{CH}_3}{\diagup}}\underset{\text{NHCOCH}_3}{\overset{\text{H}}{}} \quad \underset{\text{[Rh]/L/D}_2}{\overset{\text{H}_2\text{O/AcOEt}}{\longleftarrow}} \quad \underset{\substack{\text{Ph}\\ \text{1b}}}{\overset{\text{CO}_2\text{CH}_3}{\diagup}}\underset{\text{NHCOCH}_3}{} \quad \underset{\text{[Rh]/L/H}_2}{\overset{\text{D}_2\text{O/AcOEt}}{\longrightarrow}} \quad \underset{\text{Ph}}{\overset{\text{CO}_2\text{CH}_3}{\diagup}}\underset{\text{NHCOCH}_3}{\overset{\text{D}}{}} \tag{3}
$$

It is to be noted that under these conditions the reaction rates are generally lower than in a homogeneous organic phase. However, one of the important points is that these catalytic solutions can be readily recycled, without loss of enantioselectivity, as shown in Table 2. This recycling was performed with little rhodium loss (< 0.1%) in a two-phase system.

Dehydropeptides **14** were also reduced in a two-phase system using [Rh(COD)Cl]$_2$ associated with ligands **3** and **4** [Eq. (4)] [20]; the diastereomeric excess (*de*) of the dipeptide **15** obtained was strongly influenced by the absolute configuration of the

Table 2 Catalyst recycling in the asymmetric hydrogenation of α-amino acid precursors **1** at 25 °C using a rhodium catalyst.

Precursor 1	Ligand	Solvent	p_{H_2} [MPa]	Cycle	Product ee [%] (Config.)	Ref.
1b	(S,S)-3	H$_2$O/AcOEt (1 : 1)	0.1	1	34 (S)	[3]
				2	37 (S)	
1c	(S,S)-3	H$_2$O/AcOEt (1 : 1)	0.1	1	20 (S)	[3]
				2	23 (S)	
1c	(S,S)-5	H$_2$O/AcOEt (1 : 1)	1.0	1	82 (R)	[3]
				2	88 (R)	
				3	87 (R)	
1c	(S,S)-9a	H$_2$O/AcOEt/C$_6$H$_6$ (2 : 1 : 1)	1.4	1	75 (R)	[6]
				2	77 (R)	
				3	77 (R)	
1c	12	H$_2$O/AcOEt (1 : 1)	0.5	1	68 (S)	[11]
				2	66 (S)	
1c	13	H$_2$O/AcOEt (1 : 1)	0.5	1	87 (S)	[11]
				2	85 (S)	

unsaturated substrate. For example, reduction of Ac-Δ-Ph-(S)-Ala-OCH$_3$ using [Rh]-**4** gave a *de* as high as 72% in favor of the (R,S) diastereoisomer, while a *de* of 6% was obtained in the reduction of Ac-Δ-Ph-(R)-Ala-OCH$_3$ in favour of the (R,R) diastereoisomer.

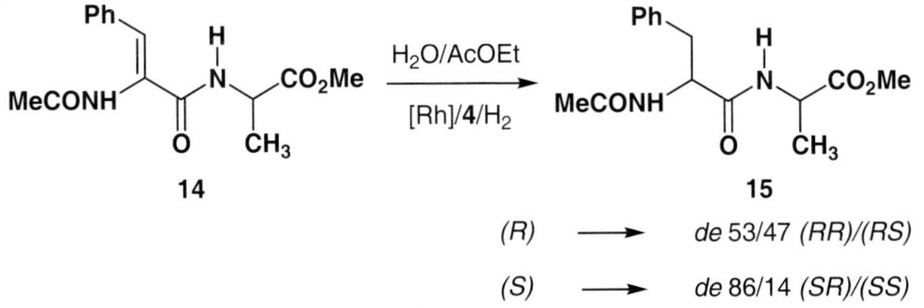

Rhodium complexes of water-soluble polymer ligands **16a,b** (see also Chapter 7) were also quite efficient in the reduction of α-acetamidocinnamic acid and its methyl ester; enantioselectivities up to 89% *ee* being obtained using water and EtOAc/H$_2$O (1 : 1) as the solvents [21].

Reduction of unsaturated acid **17** in a two-phase AcOEt/H$_2$O system using the ruthenium complex associated with ligand **6** allowed the preparation of naproxen (**18**) in 81% *ee*, *ee* values in the range 78–83% being obtained over several recycles of the catalytic solution [Eq. (5)] [22]. The analogous supported aqueous-phase catalyst (SAPC) was also prepared and used in this reduction, an enantioselectivity

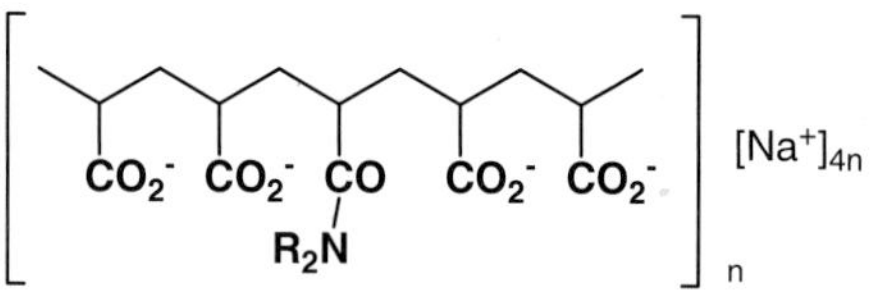

16

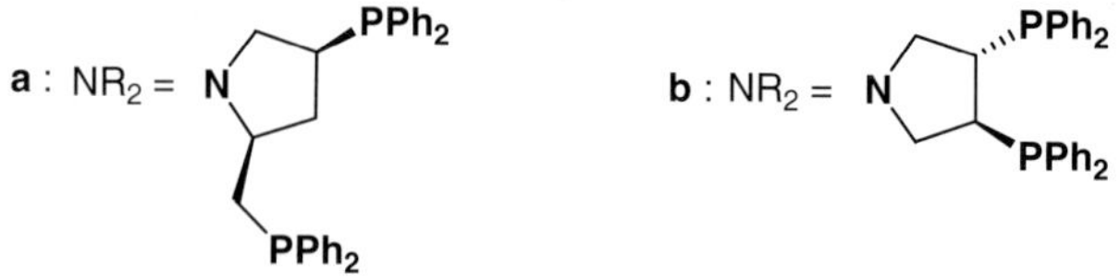

a : NR$_2$ =

b : NR$_2$ =

up to 70% being obtained in ethyl acetate saturated with water; the use of ethylene glycol instead of water as the hydrophilic phase gave *ee* values up to 96%.

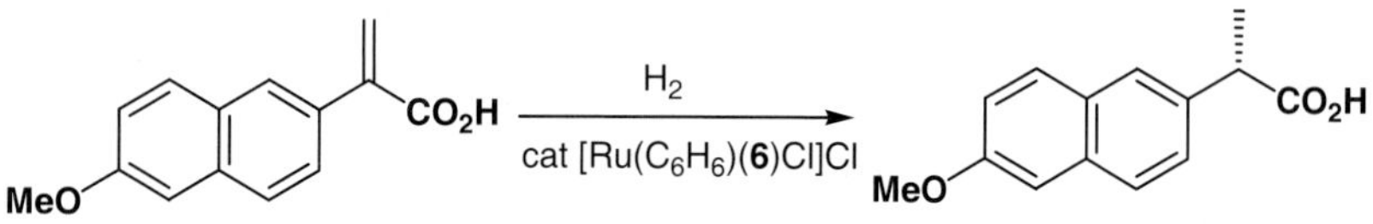

$$\text{(5)}$$

The hydrogenation of various β-keto esters was also performed in water in the presence of ruthenium complexes associated with ligands **10** and **11** [23, 24], affording the corresponding hydroxy ester with enantioselectivities up to 94% [Eq. (6)]. The *ee* values remained unchanged in the first three recycles.

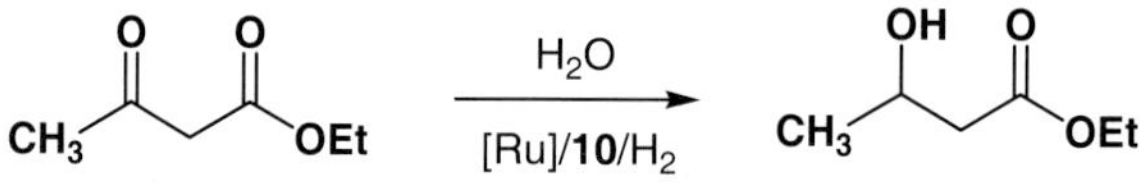

$$\text{(6)}$$

Sinou [25] and de Vries [26, 27] reported the influence of the degree of sulfonation of chiral BDPP on the enantioselectivity in the reduction of prochiral imines. The rhodium complex of monosulfonated BDPP gave *ee* values up to 94% in the reduction of imines **21** in a two-phase AcOEt/H$_2$O system [Eq. (7)], when the tetrasulfonated or disulfonated BDPP gave 34% and 2% *ee* only, respectiveley, in the reduction of the benzylimine of acetophenone.

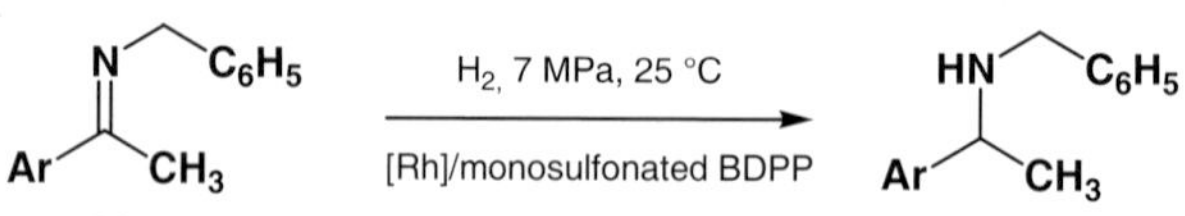

$$\text{(7)}$$

Enantioselectivities up to 43% *ee* were obtained in the hydroxycarboxylation of vinylarenes in the presence of Pd(OAc)$_2$ and tetrasulfonated CBD **3** or BDPP **4**, recycling of the catalyst being possible with no loss in activity and enantioselectivity [28].

Asymmetric palladium-catalyzed alkylation of various allylic acetates occurred in water or in an aqueous/organic biphasic medium, using as ligands phosphines derived from carbohydrates [29], amphiphilic resin-supported MOP and P,N-chelating ligands [30, 31], or BINAP in the presence of surfactants [32]; enantio-selectivities up to 85%, 98%, and 92% have been obtained in the alkylation of 1,3-diphenyl-2-propenyl acetate, respectively.

Rhodium-catalyzed asymmetric addition of aryl borates to cycloalkenones occurred with a high level of enantioselectivity in dioxane (*ee* up to 99%) using BINAP as the ligand when 1 equiv of water was added [33]. The benefit of water was also observed in the asymmetric cyclopropanation of styrene with diazoacetates using chiral bis(hydroxymethyldihydrooxazolyl)pyridineruthenium catalyst (*ee* up to 97%); in this latter case, the utilization of a H_2O/toluene medium allowed recycling of the catalyst [34]. Finally, Kobayashi and co-workers carried out catalyzed asymmetric aldol reactions in aqueous media using a catalyst obtained by mixing $Cu(OTf)_2$ $Pb(OTf)_2$, or $Ln(OTf)_2$, and a chiral bisoxazoline or crown-ether [35–37]; very high chemical yields, as well as diastereo- and enantioselectivities (up to 85% *ee*), have been obtained.

In conclusion, asymmetric catalysis, and particularly asymmetric hydrogenation, occurred in water or in a two-phase system in moderate to high enantioselectivities, allowing a very easy recycling of the catalyst without loss of enantioselectivity. Different techniques have been used in order to solubilize the catalyst and the products in the aqueous phase. It is obvious that in the future this technique will be extended to other asymmetric organometallic-catalyzed reactions.

References

1 (a) *Catalytic Asymmetric Synthesis* (Ed.: I. Ojima), VCH, Weinheim **1993**; (b) H. B. Kagan, in *Comprehensive Organometallic Chemistry* (Ed.: G. Wilkinson), Pergamon, London **1993**, Vol. 8, p. 463; (c) H. Brunner, W. Zettlmeier, *Handbook of Enantioselective Catalysis with Transition Metal Compounds*, VCH, Weinheim **1993**; (d) R. Noyori, *Asymmetric Catalysis in Organic Synthesis*, John Wiley, New York **1994**.

2 F. Alario, Y. Amrani, Y. Colleuille, T. P. Dang, J. Jenck, D. Morel, D. Sinou, *J. Chem. Soc., Chem. Commun.* **1986**, 202.

3 Y. Amrani, L. Lecomte, D. Sinou, J. Bakos, I. Toth, B. Heil, *Organometallics* **1989**, 8, 542.

4 U. Nagel, E. Kingel, *Chem. Ber.* **1986**, 119, 1731.

5 I. Toth, B. E. Hanson, *Tetrahedron: Asymmetry* **1990**, 1, 895.

6 I. Toth, B. E. Hanson, *Tetrahedron: Asymmetry* **1990**, 1, 913.

7 I. Toth, B. E. Hanson, M. E. Davis, *Catal. Lett.* **1990**, 5, 183.

8 I. Toth, B. E. Hanson, M. E. Davis, *J. Organomet. Chem.* **1990**, 396, 363.

9 H. Ding, B. E. Hanson, J. Bakos, *Angew. Chem., Int. Ed. Engl.* **1995**, 34, 1645.

10 K. Wan, M. E. Davis, *J. Chem. Soc., Chem. Commun.* **1993**, 1262.

11 K. T. Wan, M. E. Davis, *Tetrahedron: Asymmetry* **1993**, 4, 2461.

12 S. Shin, T. V. RajanBabu, *Org. Lett.* **1999**, 1, 1229.

13 K. Yonehara, S. Ohe, S. Uemura, *J. Org. Chem.* **1999**, 64, 5593.

14 K. Yonehara, T. Hashizume, K. Mori, K. Ohe, S. Uemura, *J. Org. Chem.* **1999**, 64, 9381.

15 L. Lecomte; D. Sinou, J. Bakos, I. Toth, B. Heil, *J. Organomet. Chem.* **1989**, 370, 277.

16 L. Lecomte, D. Sinou, *J. Mol. Catal.* **1989**, 52, L21.

17 M. Laghmari, D. Sinou, *J. Mol. Catal.* **1991**, 66, L15.

18 J. BAKOS, R. KARAIVANOV, M. LAGHMARI, D. SINOU, *Organometallics* **1994**, *13*, 2951.

19 F. JOÓ, P. CSIBA, A. BENYEI, *J. Chem. Soc., Chem. Commun.* **1993**, 1602.

20 M. LAGHMARI, D. SINOU, A. MASDEU, C. CLAVER, *J. Organomet. Chem.* **1992**, *438*, 213.

21 T. MALMSTRÖM, C. ANDERSON, *J. Chem. Soc., Chem. Commun.* **1996**, 1135; T. MALMSTRÖM, C. ANDERSON, *J. Mol. Catal. A:* **1999**, *139*, 259; T. Malmström, C. Anderson, *J. Mol. Catal. A:* **2000**, *157*, 79.

22 K. T. WAN, M. E. DAVIES, *Nature (London)* **1994**, *370*, 449; K. T. WAN, M. E. DAVIES, *J. Catal.* **1994**, *148*, 1; K. T. WAN, M. E. DAVIES, *J. Catal.* **1995**, *152*, 25.

23 T. LAMOUILLE, C. SALUZZO, R. TER HALLE, F. LE GUYADER, M. LEMAIRE, *Tetrahedron Lett.* **2001**, *42*, 663.

24 P. GUERREIRO, V. RATOVELOMANANA-VIDAL, J.-P. GENÊT, P. DELLIS, *Tetrahedron Lett.* **2001**, *42*, 3423.

25 J. BAKOS, A. OROSZ, B. HEIL, M. LAGHMARI, P. LHOSTE, D. SINOU, *J. Chem. Soc., Chem. Commun.* **1991**, 1684.

26 C. LENSIK, J. G. DE VRIES, *Tetrahedron: Asymmetry* **1992**, *3*, 235.

27 C. LENSIK, E. RIJNBERG, J. G. DE VRIES, *J. Mol. Catal. A:* **1997**, *116*, 199.

28 M. D. MIQUEL-SERRANO, A. AGHMIZ, M. DIEGUEZ, A. M. MASDEU-BULTO, C. CLAVER, D. SINOU, *Tetrahedron: Asymmetry* **1999**, *10*, 4463.

29 T. HASHIZUME, K. YONEHARA, K. OHE, S. UEMURA, *J. Org. Chem.* **2000**, *65*, 5197.

30 Y. UOZUMI, H. DANJO, T. HAYASHI, *Tetrahedron Lett.* **1998**, *39*, 8303.

31 Y. UOZUMI, K. SHIBATOMI, *J. Am. Chem. Soc.* **2001**, *123*, 2919.

32 C. RABEYRIN, C. NGUEFACK, D. SINOU, *Tetrahedron Lett.* **2000**, *41*, 7461.

33 T. HAYASHI, T. SENDA, Y. TAKAYA, M. OGASAWARA, *J. Am. Chem. Soc.* **2001**, *123*, 11 591.

34 S. ISAWA, F. TAKEZAWA, Y. TUCHIYA, H. NISHIYAMA, *Chem. Commun.* **2001**, 59.

35 S. KOBAYASHI, S. NAGAYAMA, T. BUSUJIMA, *Chem. Lett.* **1999**, 71; S. Kobayashi, S. Nagayama, T. Busujima, *Tetrahedron* **1999**, *55*, 8739.

36 S. NAGAYAMA, S. KOBAYASHI, *J. Am. Chem. Soc.* **2000**, *122*, 11 531.

37 S. KOBAYASHI, T. HAMADA, S. NAGAYAMA, K. MANABE, *Org. Lett.* **2001**, *3*, 165.

2.4.7
Biological Conversions

Peter J. Quinn

2.4.7.1
Introduction

Water is the universal solvent in biology. In the order of 80% of the gross weight of living organisms consists of water. The catalysts responsible for mediating the biochemical reactions that create and sustain life depend on an aqueous environment to preserve their stability and catalytic functions. Moreover, their activity is limited to reaction conditions of temperature, pressure, pH, etc., which are compatible with survival of the living organism.

Many biological catalysts, or to give them their correct name, enzymes, often acquire their catalytic functions under these stringent reaction conditions by incorporating transition metals into their catalytic site. The metals are coordinated to ligands, which are constituents of the polypeptide chain, and participate in the formation of transition-state complexes with substrates in the performance of biochemical reactions. While Nature can provide a rich diversity of organometallic catalysts that require an aqueous solvent, it is a challenge to the chemist to duplicate these enzymatic reactions so that they can be exploited and adapted to industrial-scale processes.

2.4.7.2
Biochemical Substrates

Lipids which constitute structural components of living cells and represent a store of metabolic energy contain unsaturated bonds. Chemical modification of these unsaturated fatty acids in living organisms have greatly increased our knowledge of the role of these constituents in biological membranes. Their presence in food, however, presents problems with storage, and catalytic hydrogenation in processing of fats and oils to prevent spoilage and development of off-flavors is a common practice. Such processing, nevertheless, results in the creation of *trans* isomers of fatty acids which are known to present a health risk. Improved practices are actively being sought to meet regulations framed to reduce or eliminate *trans* fatty acids in processed oils.

2.4.7.3
Hydrogenation using Water-Soluble Catalysts

Hydrogenation of unsaturated phospholipids dispersed in aqueous systems using a water-soluble homogeneous catalyst was first reported by Madden and Quinn [1]. The catalyst was a sulfonated derivative of Wilkinson's catalyst which did not appear to affect the structure of bilayers with respect to their permeability barrier properties [2]. The catalyst was found to hydrogenate oil-in-water emulsions and two-phase oil–water systems without the need for organic co-solvents [3]. The reaction rate could be increased significantly by screening the electrostatic charge on the sulfonate groups with inorganic cations added to the aqueous phase. This allowed the catalyst to penetrate into the substrate at the interface; partition of the catalyst from the aqueous into the lipid phase could not be detected.

The reactivity of water-soluble palladium catalyst, $Pd(QS)_2$ (palladium di(sodium) alizarine monosulfonate) has been examined in multilamellar dispersions of unsaturated phospholipids [4]. With substrates of dioleoylphosphatidylcholine there is a transient appearance of *trans* ω9 but no *cis* double bonds were observed when the *trans* ω9 derivative of phosphatidylcholine was used as substrate.

2.4.7.4
Hydrogenation of Biological Membranes

The presence of unsaturated lipids in biological membranes confers a fluid character on the structure and this is integral to its function. Evidence for this has been established by homogeneous catalytic hydrogenation of these lipids in membrane preparations [5, 6] and living cells [7]. Furthermore, the topology of lipids in the membranes of complex organisms or in subcellular membrane preparations has been probed by determining access to hydrogenation catalysts. Water-soluble catalyst complexes, for example, are not readily permeable to membranes, and when added to suspensions of cells or closed vescular structures their action has been shown to be largely restricted to the outer monolayer, at least at short time intervals after commencement of the reaction [8].

An important question as to how cells regulate the fluidity of their membranes by adjusting the proportion of unsaturated lipids has also been addressed by homogeneous hydrogenation methods. This strategy has been used in the unicellular green alga, *Daniella salina*, to define pathways of biosynthesis of unsaturated membrane lipids and the process of redistribution from the site of biosynthesis to the different subcellular membranes [9]. Similar studies have been reported in the regulation of desaturase enzymes in microsomal membranes isolated from yeast [10] and potato tubers [11].

2.4.7.5
Conclusions

The chemical modification of membranes containing unsaturated hydrocarbon substituents has been shown to be a useful tool in the study of the role of these lipids in membrane structure and stability. Homogeneous catalytic hydrogenation of biological membranes in isolated organelles or living cells has developed rapidly over the past few years with the introduction of more active catalytic complexes, especially under conditions of hydrogenation more compatible with living organisms. Advances in targeting catalysts to specific membranes and localizing action to specific membrane sites are likely to be important in future developments.

References

1   T. D. Madden, P. J. Quinn, *Biochem. Soc. Trans.* **1978**, *6*, 1345.
2   T. D. Madden, W. E. Peel, P. J. Quinn, D. Chapman, *J. Biochem. Biophys, Meth.* **1980**, *2*, 1.
3   Y. Drov, J. Manassen, *J. Mol. Catal.* **1977**, *2*, 219.
4   G. M. Brown, B. S. Brunschwig, C. Creutz, J. F. Endicott, N. Sutin, *J. Am. Chem. Soc.* **1979**, *101*, 1298.
5   G. Horváth, M. Droppa, T. Szito, L. A. Mustardy, L. I. Horváth, L. Vigh, *Biochim. Biophys. Acta* **1986**, *849*, 325.
6   L. Vigh, F. Joó, *FEBS Lett.* **1983**, *162*, 423.
7   W. E. Peel, A. E. R. Thompson, *Leuk. Res.* **1983**, *7*, 193.
8   Y. Pak, F. Joó, L. Vigh, A. Katho, G. A. Thompson, *Biochim. Biophys. Acta* **1990**, *1023*, 230.
9   M. Schlame, J. Horváth, Z. Torok, L. Horváth, L. Vigh, *Biochim. Biophys. Acta* **1990**, *1045*, 1.
10 I. Horváth, Z. Torok, L. Vigh, M. Kates, *Biochim Biophys. Acta* **1991**, *1085*, 126.
11 C. Demandre, L. Vigh, A. M. Justin, A. Jolliot, C. Wolf, P. Mazliak, *Plant Sci.* **1986**, *44*, 13.

2.4.8
Other Examples

2.4.8.1
Lanthanides in Aqueous-Phase Catalysis

Shu Kobayashi

2.4.8.1.1
Introduction

Lewis acid-catalyzed reactions have been of great interest in organic synthesis because of their unique reactivities, selectivities, and the mild conditions used [1].Lewis acid-promoted reactions must be carried out under strictly anhydrous conditions despite the general recongnition of the utility of aqueous reactions [2]. The presence of even a small amount of water stops the reaction, because most Lewis acids immediately react with water rather than the substrates and decompose or deactivate; this fact has restricted the use of Lewis acids in organic synthesis.

On the other hand, lanthanide compounds (including scandium and yttrium compounds) were recently found to be stable Lewis acids in water, and many useful aqueous reactions using lanthanide compounds as catalysts have been reported. Lanthanides have larger radii and specific coordination numbers than typical transition metals. They have been expected to act as strong Lewis acids because of their hard character, and to have strong affinity toward carbonyl oxygens [3]. Among these compounds, lanthanide trifluoromethanesulfonates (lanthanide triflates, $Ln(OTf)_3$) were expected to be some of the strongest Lewis acids because of the strongly electron-withdrawing trifluoromethanesulfonyl group. Their hydrolysis was postulated to be slow on the basis of their hydration energies and hydrolysis constants [4]. In fact, while most metal triflates are prepared under strictly anhydrous conditions, lanthanide triflates were reported to be prepared in an aqueous solution [5, 6]. After the finding that lanthanide triflates are stable and act as Lewis acids in water [7], many synthetic reactions using these triflates as catalysts have been developed [8]. This section surveys use of these lanthanides in carbon–carbon bond-forming reactions in aqueous solutions.

2.4.8.1.2
Aldol Reactions

Formaldehyde is a versatile reagent [9]. It has some disadvantages, however, because it must be generated before use from solid polymer paraformaldehyde by way of thermal depolymerization and it self-polymerizes easily [10]. On the other hand, commercial formaldehyde solution, which is an aqueous solution containing 37% formaldehyde and 8–10% methanol, is cheap, easy to handle, and stable even at room temperature [11, 12].

It was found that the hydroxymethylation of silyl enol ethers with commercial formaldehyde solution proceeded smoothly by using lanthanide triflates as Lewis acid catalysts [7, 13]. The reactions were first carried out in commercial formaldehyde solution–THF media. The amount of the catalyst was examined by taking the reaction of the silyl enol ether derived from propiophenone with commercial formaldehyde solution as a model, and the reaction was found to be catalyzed even by 1 mol% ytterbium triflate (Yb(OTf)$_3$) [Eq. (1)].

$$\text{HCHO aq.} \quad + \quad \overset{\text{OSiMe}_3}{\underset{R^1}{\diagup\diagdown}}R^2 \quad \xrightarrow[\substack{\text{H}_2\text{O-THF, rt} \\ 77\text{-}94\%}]{\text{cat. Ln(OTf)}_3} \quad \overset{O}{\underset{R^2}{R^1\diagup\diagdown}}\text{OH} \qquad (1)$$

Ln = Sc, Y, La, Ce, Pr, Nd, Sm, Eu, Gd, Tb, Dy, Ho, Er, Tm, Yb, Lu

Lanthanide triflates are effective for the activation of aldehydes other than formaldehyde [13–15]. The aldol reaction of silyl enol ethers with aldehydes proceeds smoothly to afford the aldol adducts in high yields in the presence of a catalytic amount of scandium triflate (Sc(OTf)$_3$), ytterbium triflate (Yb(OTf)$_3$), gadolinium triflate (Gd(OTf)$_3$), lutetium triflate (Lu(OTf)$_3$), etc. in aqueous media (water–THF). Diastereoselectivities are generally good to moderate. One feature in the present reaction is that water-soluble aldehydes, for instance, acetaldehyde, acrolein, and chloroacetaldehyde, can react with silyl enol ethers to afford the corresponding cross-aldol adducts in high yields. Some of these aldehydes are supplied commercially as water solutions and are appropriate for direct use. Phenylglyoxal monohydrate also works well. It is known that water often interferes with the aldol reactions of metal enolates with aldehydes and that in the cases where such water-soluble aldehydes are employed, some troublesome purifications including dehydration are necessary. Furthermore, salicylaldehyde and 2-pyridinecarboxaldehyde can be successfully employed.

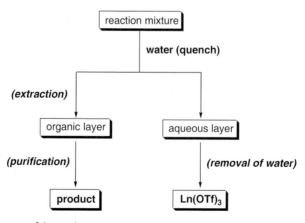

Scheme 1 Recovery of the catalyst.

Lanthanide triflates are more soluble in water than in organic solvents such as dichloromethane. Very interestingly, almost 100% of lanthanide triflate is quite easily recovered from the aqueous layer after the reaction is completed and it can be re-used. The reactions are usually quenched with water and the products are extracted with an organic solvent (for example, dichloromethane). Lanthanide triflates are in the aqueous layer and only removal of water gives the catalyst which can be used in the next reaction (Scheme 1). It is noteworthy that lanthanide triflates are expected to solve some severe environmental problems induced by mineral acid- or Lewis acid-promoted reactions in industrial chemistry [16].

The aldol reactions of silyl enol ethers with aldehydes also proceed smoothly in water–ethanol–toluene [17].

2.4.8.1.3
Mannich-Type Reactions

The Mannich and related reactions provide one of the most fundamental and useful methods for the synthesis of β-amino ketones. Although the classical protocols include some severe side reactions, new modifications using preformed iminium salts and imines have been developed [18].

Mannich-type reactions between aldehydes, amines, and vinyl ethers proceed smoothly using Ln(OTf)$_3$ in aqueous media [19]. The procedure is very simple: in the presence of 10 mol% of Yb(OTf)$_3$, an aldehyde, an amine, and a vinyl ether are combined in a solution of THF–water (9 : 1) at room temperature to afford a β-amino ketone [Eq. (2)]. Commercially available formaldehyde– and chloroacetaldehyde–water solutions are used directly and the corresponding β-amino ketones are obtained in good yields. Phenylglyoxal monohydrate, methyl glyoxylate, an aliphatic aldehyde, and an α,β-unsaturated ketone also work well to give the corresponding β-amino esters in high yields. Other lanthanide triflates can also be used; in the reaction of phenylglyoxalmonohydrate, p-chloroaniline, and 2-methoxypropene, 90% (Sm(OTf)$_3$), 94% (Tm(OTf)$_3$), and 91% (Sc(OTf)$_3$) yields are obtained. In some Mannich reactions with preformed iminium salts and imines, it is known that yields are often low because of the instability of the imines derived from these aldehydes or troublesome treatments are known to be required for their use [20]. The present method provides a useful route for the synthesis of β-amino ketones.

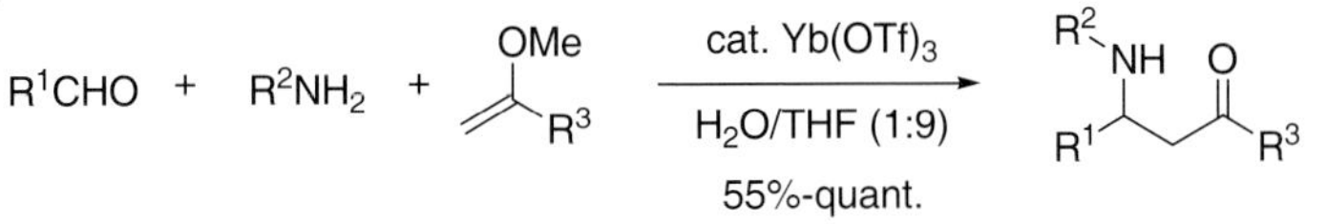

$$R^1CHO + R^2NH_2 + \underset{R^3}{\overset{OMe}{\diagup}} \xrightarrow[\substack{H_2O/THF (1:9) \\ 55\%\text{-quant.}}]{cat.\ Yb(OTf)_3} \underset{R^1}{\overset{R^2\diagdown NH}{\diagdown}}\overset{O}{\diagup}R^3 \qquad (2)$$

A possible mechanism of the present reaction is accompanied by imine formation, and successive addition of a vinyl ether proceeds smoothly in aqueous solution. Use of lanthanide triflates, water-tolerant Lewis acids, is of key importance and essential in this reaction.

2.4.8.1.4
Diels–Alder Reactions

Although many Diels–Alder reactions have been carried out at higher reaction temperatures without catalysts, heat-sensitive compounds in complex and multistep syntheses cannot be employed. While Lewis acid catalysts allow the reactions to proceed at room temperature or below with satisfactory yields in organic solvents, they are often accompanied by diene polymerization and excess amounts of the catalyst are often needed to catalyze carbonyl-containing dienophiles [21].

It was found that the Diels–Alder reaction of naphthoquinone with cyclopentadiene proceeded in the presence of a catalytic amount of a lanthanide triflate in H_2O–THF (1 : 9) at room temperature to give the corresponding adduct in 93% yield (*endo/exo* = 100 : 0) [Eq. (3)] [22].

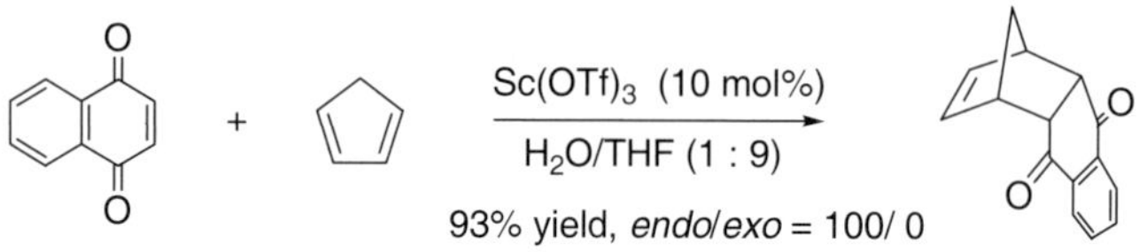

$$\text{(3)}$$

Sc(OTf)$_3$ (10 mol%)
H_2O/THF (1 : 9)
93% yield, *endo/exo* = 100/ 0

2.4.8.1.5
Micellar Systems

Aldol Reactions

Recently, it has been found that scandium triflate (Sc(OTf)$_3$)-catalyzed aldol reactions of silyl enol ethers with aldehydes could be successfully carried out in micellar systems [23]. While the reactions proceeded sluggishly in pure water (without organic solvents), remarkable enhancement of the reactivity was observed in the presence of a small amount of a surfactant [see Eq. (4)].

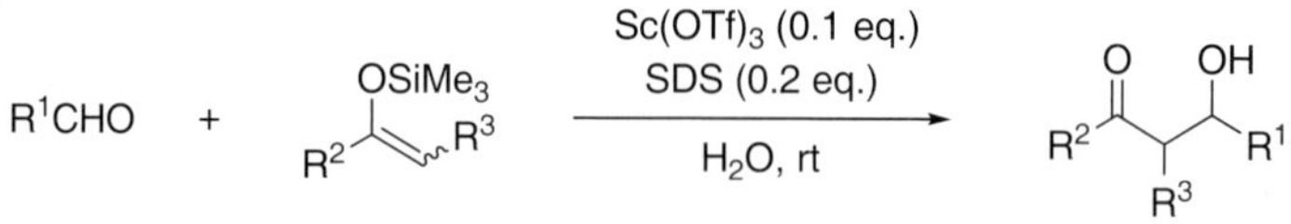

$$R^1CHO \quad + \quad \overset{OSiMe_3}{\underset{R^2 \quad R^3}{\diagup}} \quad \xrightarrow[\substack{H_2O, \text{ rt}}]{\substack{\text{Sc(OTf)}_3 \text{ (0.1 eq.)} \\ \text{SDS (0.2 eq.)}}} \quad \underset{R^3}{\overset{O \quad OH}{R^2 \diagdown R^1}} \quad \text{(4)}$$

Lewis acid catalysis in micellar systems was first found in the model reaction of the silyl enol ether of propiophenone with benzaldehyde. While the reaction proceeded sluggishly in the presence of 0.2 equiv Yb(OTf)$_3$ in water, remarkable enhancement of the reactivity was observed when the reaction was carried out in the presence of 0.2 equiv Yb(OTf)$_3$ in an aqueous solution of sodium dodecyl sulfate (SDS, 0.2 equiv, 35 mM), and the corresponding aldol adduct was obtained in 50% yield. In the absence of the Lewis acid and the surfactant (water-promoted conditions) [11], only 20% yield of the aldol adduct was isolated after 48 h, while a 33% yield of the aldol adduct was obtained after 48 h in the absence of the Lewis acid in an aqueous solution of SDS. The amounts of the surfactant also influenced the reactivity, and the yield was improved when Sc(OTf)$_3$ was used as a Lewis acid catalyst. Judging from the critical micelle concentration, micelles would be formed in these reactions, and it is noteworthy that the Lewis acid-catalyzed reactions proceeded smoothly in micellar systems [24]. Although several organic reactions in micelles were reported, there was no report on Lewis acid catalysis in micelles. Since the amount of the surfactant used in the present case was small, the aldol reaction would *not* proceed *only in micelles*. It was also found that the surfactants influenced the yield, and that Triton X-100 was effective in the aldol reaction (but required a long reaction time), while only a trace amount of the adduct was detected when using cetyltrimethylammonium bromide (CTAB) as a surfactant.

Allylations of Aldehydes

It was also found that the allylation reactions of aldehydes with tetraallyltin proceeded smoothly in micellar systems using $Sc(OTf)_3$ as a catalyst [25]. Utilities of organo-metallic reagents are now well recognized in organic synthesis, and a variety of organometallics have been developed to achieve unique reactivities as well as selectivities [26]. In general, however, most organometallic reagents are hygroscopic, and therefore they are deactivated or decomposed in the presence of even a small amount of water, which sometimes limits their use in organic synthesis. On the other hand, the allylation reaction of 2-deoxyribose (an unprotected sugar) was found to proceed smoothly in water under the influence of 0.1 equiv of $Sc(OTf)_3$ and 0.2 equiv of SDS (sodium dodecyl sulfate) by using tetraallyltin (0.5 equiv) as an allylating reagent [Eq. (5)] [27]. Not only aromatic but also aliphatic and α,β-unsaturated aldehydes reacted with tetraallyltin to afford the corresponding allylated adducts in high yields. Under the present reaction conditions, salicyl-aldehyde and 2-pyridinecarboxaldehyde reacted with tetraallyltin to afford the homoallylic alcohols in good yields. Other unprotected sugars than 2-deoxyribose also reacted directly to give the adducts, which are intermediates for the synthesis of higher sugars [28], in high yields.

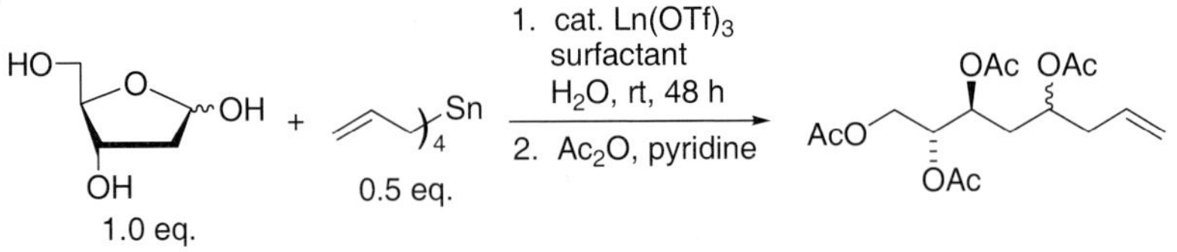

Three-Component Reactions of Aldehydes, Amines, and Allyltributylstannane

The reaction of imines with allyltributylstannane provides a useful route for the synthesis of homoallylic amines [29]. The reaction is generally carried out in the presence of a Lewis acid in organic solvents under strictly anhydrous conditions [30], because most imines, Lewis acids, and the organotin reagents used are hygroscopic and easily decompose in the presence of even a small amount of water [31]. Dispite the utility of aqueous reactions, it was believed that the above reaction would remain difficult to perform in water because of the use of water-sensitive imines, Lewis acids, and organotin reagents. It was found that three-component reactions of aldehydes, amines, and allyltributylstannane proceed smoothly in micellar systems using Sc(OTf)$_3$ as a Lewis acid catalyst [32] [Eq. (6)]. The reaction of benzaldehyde, aniline, and allyltributylstannane was chosen as a model, and several reaction conditions were examined. While the reaction proceeded sluggishly in the presence of Sc(OTf)$_3$ without SDS or in the presence of SDS without Sc(OTf)$_3$, a 77% yield of the desired homoallylic amine was obtained when Sc(OTf)$_3$ and SDS were both present. It was suggested that an imine formed from the aldehyde and the amine rapidly reacted with allyltributylstannane to afford the desired adduct. The effect of the amount of Sc(OTf)$_3$ and SDS was also examined.

$$R^1CHO \ + \ R^2NH_2 \ + \ \diagup\!\!\!\!\diagdown SnBu_3 \ \xrightarrow[\substack{H_2O, \ rt., \ 20 \ h \\ 66\text{-}90\%}]{\substack{Sc(OTf)_3 \ (0.2 \ eq.) \\ SDS \ (0.2 \ eq.)}} \ \overset{NHR^2}{\underset{R^1}{\diagup\!\!\!\!\diagdown}} \diagdown\!\!\!\!= \quad (6)$$

The present three-component reactions of aldeheydes, amines, and allyltributyl-stannane proceeded smoothly in water without using any organic solvents in the presence of a small amount of Sc(OTf)$_3$ and SDS, to afford the corresponding homoallylic amines in high yields. Not only aromatic aldehydes but also aliphatic, unsaturated, and heterocyclic aldehydes worked well [29, 34, 35].

2.4.8.1.6
Asymmetric Catalysis in Aqueous Media

Catalytic asymmetric aldol reactions have emerged as one of the most powerful carbon–carbon bond-forming processes affording synthetically useful, optically active β-hydroxy carbonyl compounds [36]. Among them, chiral Lewis acid-catalyzed reactions of aldehydes with silyl enol ethers are one of the most promising methods. Although several successful examples have been developed since 1990 [37], most of the reactions have to be conducted at low reaction temperatures (e.g., –78 °C) in aprotic anhydrous solvents such as dry dichloromethane, toluene, and propionitrile.

Although lanthanide triflates are the first metal salts which were found to catalyze aldol reactions of aldehydes with silyl enol ethers efficiently in aqueous media, it has been difficult to realize asymmetric versions of lanthanide triflate-catalyzed reactions in such media. Recently, the first example of this type of reaction using chiral bis-pyridino-18-crown-6 (1) has been developed [38]. In the reaction of benz-aldehyde with silyl enol ethers in water–ethanol (1 : 9), the cation size of lanthanide triflates strongly affected the diastereo- and enantioselectivities of the aldol adduct. For the larger cations such as La, Ce, Pr, and Nd, both diastereo- and enantio-selectivities were high, while the smaller cations such as Sc and Yb showed no enantioselection.

A study on the reaction profile of the asymmetric aldol reaction catalyzed by Pr(OTf)$_3$ with **1** revealed that this crown ether-type chiral ligand did not significantly reduce the activity of the metal triflates. This retention of the activity even in the presence of the crown ether containing oxygen and nitrogen atoms is a key to realizing the asymmetric induction in this asymmetric aldol reaction in aqueous media. The X-ray structure of $[Pr(NO_3)_2 \cdot 1]_3[Pr(NO_3)_6]$ shows that the Pr(III) cation is located in the center of the crown ring. The use of a chiral multidentate ligand such as the crown ether is a versatile concept for catalytic asymmetric reactions in aqueous media.

2.4.8.1.7
Conclusions

Lanthanide triflates (Ln(OTf)$_3$) are stable Lewis acids in water and are successfully used in several carbon–carbon bond-forming reactions in aqueous solutions. The reactions proceeded smoothly in the presence of a catalytic amount of the triflate under mild conditions. Moreover, the catalysts can be recovered after the reactions are completed, and can be re-used. Lewis acid catalysis in micellar systems will lead to clean and environmentally friendly processes, and it will become a more important topic in the future. Finally, catalytic asymmetric aldol reactions in aqueous media have been attained using a Ln(OTf)$_3$–chiral crown ether complex as a catalyst.

References

1   (a) *Selectivities in Lewis Acid Promoted Reactions* (Ed.: D. SCHINZER), Kluwer Academic Publishers, Dordrecht **1989**; (b) M. SANTELLI, J.-M. PONS, *Lewis Acids and Selectivity in Organic Synthesis*, CRC, Boca Raton **1995**.
2   (a) C.-J. LI, *Chem. Rev.* **1993**, *93*, 2023; (b) A. LUBINEAU, J. ANGE, Y. QUENEAU, *Synthesis* **1994**, 741.
3   G. A. MOLANDER, *Chem. Rev.* **1992**, *92*, 29.
4   C. F. BAES, JR., R. E. MESMER, *The Hydrolysis of Cations*, John Wiley, New York **1976**, p. 129.
5   K. F. THOM, US 3.615.169 (**1971**); *Chem. Abstr.* **1972**, *76*, 5436a.
6   (a) J. H. FORSBERG, V. T. SPAZIANO, T. M. BALASUBRAMANIAN, G. K. LIU, S. A. KINSLEY, C. A. DUCKWORTH, J. J. POTERUCA, P. S. BROWN, J. L. MILLER, *J. Org. Chem.* **1987**, *52*, 1017; see also: (b) S. COLLINS, Y. HONG, *Tetrahedron Lett.* **1987**, *28*, 4391; (c) M.-C. ALMASIO, F. ARNAUD-NEU, M.-J. SCHWING-WEILL, *Helv. Chim. Acta* **1983**, *66*, 1296; cf. (d) J. M. HARROW-FIELD, D. L. KEPERT, J. M. PATRICK, A. H. WHITE, *Aust. J. Chem.* **1983**, *36*, 483.
7   S. KOBAYASHI, *Chem. Lett.* **1991**, 2187.
8   S. KOBAYASHI, *Synlett* **1994**, 689.
9   (a) B. B. SNIDER, D. J. RODINI, T. C. KIRK, R. CORDOVA, *J. Am. Chem. Soc.* **1982**, *104*, 555; (b) B. B. SNIDER, in *Selectivities in Lewis Acid Promoted Reactions* (Ed.: D. SCHINZER), Kluwer Academic Publishers, London **1989**, p. 147; (c) K. MARUOKA, A. B. CONCEPTION, N. HIRAYAMA, H. YAMAMOTO, *J. Am. Chem. Soc.* **1990**, *112*, 7422; (d) K. MARUOKA, A. B. CONCEPCION, N. MURASE, M. OISHI, H. YAMAMOTO, *J. Am. Chem. Soc.* **1993**, *115*, 3943.
10 Cf. TMSOTf-mediated aldol-type reaction of silyl enol ethers with dialkoxymethanes was also reported: S. MURATA, M. SUZUKI, R. NOYORI, *Tetrahedron Lett.* **1980**, *21*, 2527.
11 Lubineau reported the water-promoted aldol reaction of silyl enol ethers with aldehydes, but the yields and the substrate scope were not yet satisfactory: (a) A. LUBINEAU, *J. Org. Chem.* **1986**, *51*, 2142; (b) A. LUBINEAU, E. MEYER, *Tetrahedron* **1988**, *44*, 6065.
12 (a) T. MUKAIYAMA, K. NARASAKA, T. BANNO, *Chem Lett.* **1973**, 1011; (b) T. MUKAIYAMA, K. BANNO, K. NARASAKA, *J. Am. Chem. Soc.* **1974**, *96*, 7503.

13 S. Kobayashi, I. Hachiya, *J. Org. Chem.* **1994**, *59*, 3590.

14 S. Kobayashi, I, Hachiya, *Tetrahedron Lett.* **1992**, 1625.

15 S. Kobayashi, I. Hachiya, H. Ishitani, M. Araki, *Synlett* **1993**, 472.

16 J. Haggin, *Chem. Eng. News* **1994**, Apr 18, 22.

17 S. Kobayashi, I. Hachiya, Y. Yamanoi, *Bull. Chem. Soc. Jpn.* **1994**, *67*, 2342.

18 E. F. Kleinman, *Comprehensive Organic Synthesis* (Ed.: B. M. Trost), Pergamon Press, Oxford **1991**, Vol. 2, p. 893.

19 S. Kobayashi, H. Ishitani, *J. Chem. Soc., Chem. Commun.* **1995**, 1379.

20 Grieco et al. reported in-situ generation and trapping of immonium salts under Mannich-like conditions: (a) S. D. Larsen, P. A. Grieco, *J. Am. Chem. Soc.* **1985**, *107*, 1768; (b) P. A. Grieco, D. T. Parker, *J. Org. Chem.* **1988**, *53*, 3325, and references cited therein.

21 (a) D. Yates, P. E. Eaton, *J. Am. Chem. Soc.* **1960**, *82*, 4436; (b) T. K. Hollis, N. P. Robinson, B. Bosnich, *J. Am. Chem. Soc.* **1992**, *114*, 5464; (c) W. Carruthers, *Cycloaddition Reactions in Organic Synthesis*, Pergamon Press, Oxford **1990**.

22 S. Kobayashi, I, Hachiya, M. Araki, H. Ishitani, *Tetrahedron Lett.* **1993**, *34*, 3755.

23 S. Kobayashi, T. Wakabayashi, S. Nagayama, H. Oyamada, *Tetrahedron Lett.* **1997**, *38*, 4559.

24 (a) J. H. Fendler, E. J. Fendler, *Catalysis in Micellar and Macromolecular Systems*, Academic Press, London **1975**; (b) *Mixed Surfactant Systems* (Eds.: P. M. Holland, D. N. Rubingh), ACS, Washington, DC **1992**; (c) *Structure and Reactivity in Aqueous Solution* (Eds.: C. J. Cramer, D. G. Truhlar), ACS, Washington, DC **1994**; (d) *Surfactant-Enhanced Subsurface Remediation* (Eds.: D. A. Sabatini, R. C. Knox, J. H. Harwell), ACS, Washington, DC **1994**.

25 S. Kobayashi, T. Wakabayashi, H. Oyamada, *Chem. Lett.* **1997**, 831.

26 For example: *Organometallics in Synthesis* (Ed.: M. Schlosser), John Wiley, Chichester **1994**.

27 As for allylation using tetraallyltin: W. G. Peet, W. Tam, *J. Chem. Soc., Chem. Commun.* **1983**, 853; G. Daude, M. Pereyre, *J. Organomet. Chem.* **1980**, *190*, 43; D. N. Harpp, M. Gingras, *J. Am. Chem. Soc.* **1988**, *110*, 7737; S. Fukuzawa, K. Saito, T. Fujinami, S. Sakai, *J. Chem. Soc., Chem. Commun.* **1990**, 939; A. Yanagisawa, H. Inoue, M. Morodome, H. Yamamoto, *J. Am. Chem. Soc.* **1993**, *115*, 10 356; T. M. Cokley, P. J. Harvey, R. L. Marshall, A. McCluskey, D. J. Young, *J. Org. Chem.* **1997**, *62*, 1961.

28 W. Schmid, G. M. Whitesides, *J. Am. Chem. Soc.* **1991**, *113*, 6674; T. H. Chan, M. B. Isaac, *Pure Appl. Chem.* **1996**, *68*, 919.

29 Y. Yamamoto, N. Asao, *Chem. Rev.* **1992**, *93*, 2207.

30 (a) G. E. Keck, E. J. Enholm, *J. Org. Chem.* **1985**, *50*, 147; (b) Y. Yamamoto, T. Komatsu, K. Maruyama, *J. Org. Chem.* **1985**, *50*, 3115; (c) M. A. Ciufolini, G. O. Spencer, *J. Org. Chem.* **1989**, *54*, 4739; (d) C. Bellucci, P. G. Cozzi, A. Umani-Ronchi, *Tetrahedron Lett.* **1995**, *36*, 7289; (e) M. Yasuda, Y. Sugawa, A. Yamamoto, I. Shibata, A. Baba, *Tetrahedron Lett.* **1996**, *37*, 5951.

31 Cf. (a) P. A. Grieco, A. Bahsas, *J. Org. Chem.* **1987**, *52*, 1378; (b) H. Nakamura, H. Iwama, Y. Yamomoto, *J. Am. Chem. Soc.* **1996**, *118*, 6641.

32 Unpublished so far.

33 In some other substrates, yields were improved a little with increasing amounts of SDS.

34 S. Kobayashi, H. Ishitani, *J. Chem. Soc., Chem. Commun.* **1995**, 1379.

35 S. Kobayashi, S. Nagayama, *J. Org. Chem.* **1997**, *62*, 232.

36 (a) E. M. Carreira, T. Mukaiyama, Aldol reaction, in *Comprehensive Asymmetric Catalysis* (Eds.: E. N. Jacobsen, A. Pfaltz, H. Yamamoto), Springer-Verlag, Berlin **1999**, p. 997; (b) T. D. Machajewski, C.-H. Wong, *Angew. Chem., Int. Ed.* **2000**, *39*, 1352.

37 (a) T. Mukaiyama, S. Kobayashi, H. Uchiro, I. Shiina, *Chem. Lett.* **1990**, 129; (b) S. Kobayashi, Y. Fujishita, T. Mukaiyama, *Chem. Lett.* **1990**, 1455.

38 (a) S. Kobayashi, T. Hamada, S. Nagayama, K. Manabe, *Org. Lett.* **2001**, *3*, 165; (b) T. Hamada, K. Manabe, S. Ishikawa, S. Nagayama, M. Shiro, S. Kobayashi, *J. Am. Chem. Soc.* **2003**, *125*, 2989.

2.4.8.2
Dehalogenation

Mario Bressan

A large number of halogenated organics have been produced commercially in the past few decades and used for a variety of purposes. The quest for environmentally friendly technology in general has developed into a substantial drive to get away from chlorocarbons and halogenated materials altogether, due to the generic deleteriousness associated with them. Dehalogenation treatments of the mess created as a result of past expediency are important issues [1], because of the persistence in groundwater of halogenated compounds: indeed, all of the persistent organic pollutants (POPs) recommended by UN Environment Program to be phased out of production and use are chlorinated organics. Serious problems are still associated with the combustion of chlorinated organics, since they are characterized by a high degree of chemical inertness and thermal stability. Catalytic treatments in situ can be effective remediation systems, although a careful identification of degradative products is necessary.

The ubiquitous presence of methane monooxygenase (MMO) and cytochrome P450 (CyP450) enzymatic systems indicates that they may be of principal importance for the oxidative degradation of halo-organics. In many instances, however, current techniques of bioremediation of contaminated groundwaters involving entire microorganisms are inadequate, because of pollutant concentrations, nutrient limitations, and the lack of membrane permeability for halo-organics (which, coupled with absence of suitable extracellular enzymes, is the accepted basis for a molecule being recalcitrant [2]). MMO is capable of oxidizing a variety of haloalkenes at rates comparable to those of other substrates for the enzyme, the competitive inhibitor tetrachloroethylene being the only substrate not turned over [3]. The rates observed are two orders of magnitude faster than those reported for whole-cell oxidations by nonmethanotrophs and up to ten times faster than comparable oxidations catalyzed by CyP450 containing mixed-function oxidase systems [4].

Compounds with high halogen substitution (i.e., with carbon atoms in high formal oxidation states) are expected to be resistant to the aerobic degradation and more susceptible to the reductive one. Reductive dehalogenation refers specifically to the reaction in which two electrons and a proton act as substrates along with the halogenated compound to yield a reduced product and the corresponding halide [Eq. (1)].

$$RX + 2\,e^- + H^+ \rightarrow RH + X^- \tag{1}$$

CyP450 and other reduced Fe-porphyrins are reported to mediate reductive dehalogenation of haloalkanes and -alkenes [5], and of the aliphatic portion of DDT [6]. Products similar to anaerobic bacteria or CyP450 treatments were obtained by depositing stable ordered film of myoglobin (Mb) and a surfactant on electrodes:

a highly reduced form of Mb was produced, which was successfully used to catalyze reduction of trichloroacetic acid and polyhaloethylenes 87]. Lindane dechlorination is effected by Fe-containing hemin and hematin, by Co-containing protoporphyrin, or by various cobalamins [8]. Vitamin B_{12} was also shown to reductively dechlorinate pentachlorphenol and trichlorophenoxyacetic acid [9]; cobalamins have also been used in the reductive dechlorination of CCl_4, polychlorinated ethanes, polychloro-benzenes, and PCBs [10]. Reductive dehalogenation of chlorinated aliphatic hydrocarbons and Freons was reported to occur with participation of coenzyme F430, a Ni-porphyrinoid present in anaerobic bacteria [11].

Transition metal complexes, in particular metal porphyrins, corrins, and phthalo-cyanines, have been studied as potential remediation catalysts in homogeneous abiotic aqueous systems. Cationic water-soluble Fe-, Co- and Ni-porphyrins with various functional groups in *meso* positions, suitable for immobilization, have been tested as catalysts for reductive dehalogenation of CCl_4 with dithiothreitol: $CHCl_3$, CH_2Cl_2, and CO were found to be breakdown products [12]. Ni(I) octaethyliso-bacteriochlorin has been used as a model of the F430 factor for the reduction and coupling of alkyl halogenides [13]. Photoreductive dehalogenation of aqueous $CHBr_3$ was mediated by CoPcS (PcS = tetrasulfophthalocyanine anion) adsorbed on the positively charged titania surface [14]. Nonspecific biomimetic macrocycles, CoTMPyP (TMPyP = tetrakis(*N*-methyl-4-pyridin)porphyrin cation) and CoPcS, were used as homogeneous and mineral-supported catalysts and were able to reductively dechlorinate CCl_4 in water, even at high concentrations that would inhibit microbial activity [15]. The C–X bond in aliphatic or benzyl halides can be transformed into a C–H bond in a transfer hydrodehalogenation reaction with formate as hydrogen donor and water-soluble Ru catalysts, $RuCl_2$(TPPMS) (TPPMS = *m*-sulfophenyl-diphenylphosphine anion) or $Ru(H_2O)_3(PTA)_3(tos)_2$ (PTA = 1,3,5-triaza-7-phospha-adamantane) [16].

The oldest catalytic oxidative system for dehalogenation is Fenton's reagent: the hydroxyl radical is one of the few chemical species capable of attacking refractory halo-organic compounds, but the scope of the reaction in terms of effective substrate oxidation versus H_2O_2 dismutation is often limited by sensitivity to pH and a narrow H_2O_2/Fe^{2+} ratio [17]. Various water-soluble iron or manganese sulfophenyl-porphyrins catalyzed with exceedingly high activity (up to 20 cycles per second) the oxidative dechlorination of trichlorophenol (TCP) with $KHSO_5$ in aqueous aceto-nitrile [18]. The more easily accessible MnPcS or FePcS catalysts behaved equally in water in the presence of H_2O_2 also, with rates in excess of 0.1 cycles per second: products of dechlorination (up to two chloride ions were released per TCP molecule), of aromatic ring cleavage, and of oxidative coupling have been detected (Scheme 1) [19]. The catalysts, which have also been successfully tested for the dechlorination of chloroanilines [20], maintained their activities when immobilized on cationic resins.

RuO_4 was shown to oxidize PCBs in water [21]. Water-soluble Ru complexes, such as $[Ru(H_2O)_2(dmso)_4]^{2+}$, were effective catalysts for the $KHSO_5$ oxidation of a number of polychlorobenzenes and polychlorophenols, mainly converted into HCl and CO_2 [22]. Replacement of the dmso-"solvated" ruthenium by RuPcS resulted

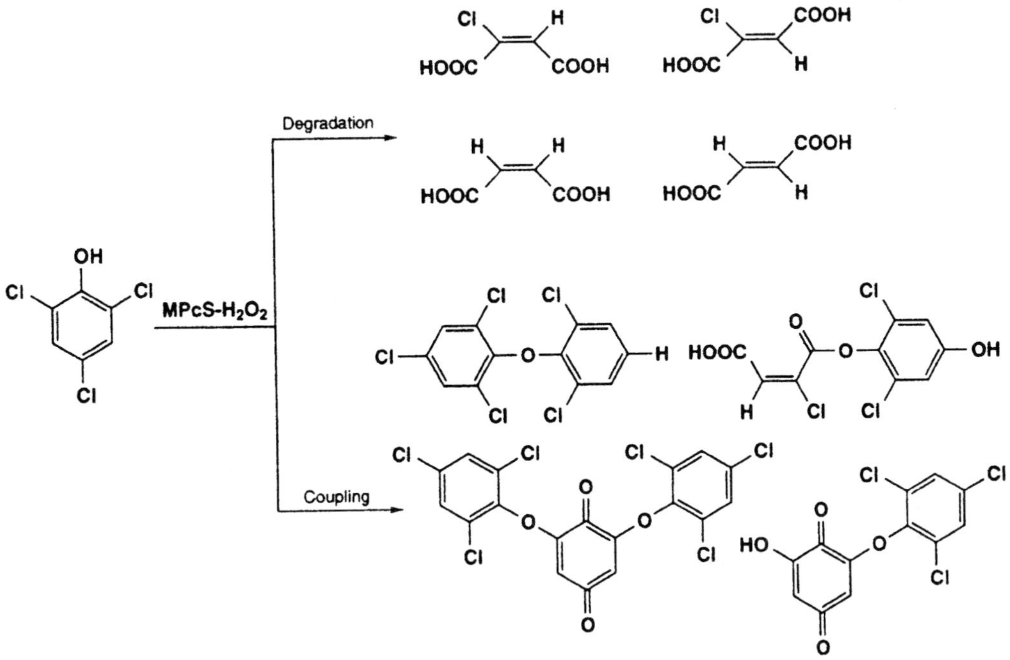

Scheme 1 Oxidation of TCP by MPcS catalysts (M = Mn, Fe) and hydrogen peroxide.

in a definite improvement in the reaction, which could be performed with H_2O_2: oxidation of chlorophenols led to the complete disappearance of the substrates within minutes, with almost quantitative (80%) evolution of inorganic chlorine and massive (50%) formation of carbon dioxide [23]. RuPcS is also able to degrade α-chlorinated olefins to HCl and the appropriate carboxylic acid and/or CO_2, with turnover rates in excess of 1 s^{-1} (Scheme 2) [24].

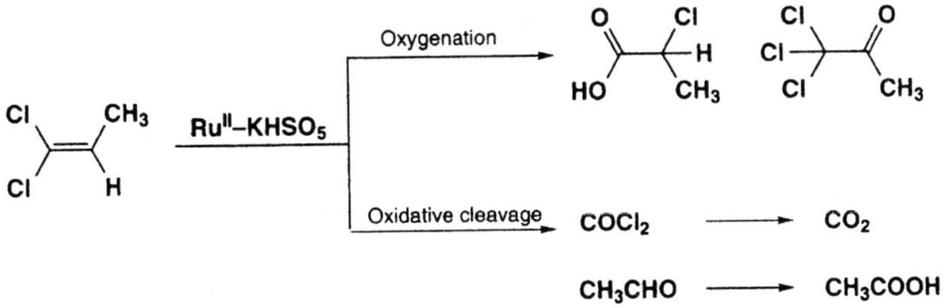

Scheme 2 Oxidation of 1,1′-dichloropropene by Ru(II) catalysts and monopersulfate.

References

1 G. B. Wickramanayake, R. E. Hinchee (Eds.), *Remediation of Chlorinated and Recalcitrant Compounds*, Battelle Press, Columbus, USA **1998**.

2 D. B. Janssen, B. Witholt, in *Metal Ions in Biological Systems* (Eds.: H. Sigel, A. Sigel), Marcel Dekker, New York **1992**, Vol. 28, p. 158; R. B. Winter, H. Zimmermann, *ibid.*, p. 300.

3 B. G. Fox, J. G. Borneman, L. P. Wackett, J. D. Lipscomb, *Biochemistry* **1990**, *29*, 6419.

4 R. E. Miller, F. P. Guengerich, *Biochemistry* **1982**, *21*, 1090.

5 S. Li, L. P. Wackett, *Biochemistry* **1993**, *32*, 9355; C. E. Castro, R. S. Wade, N. O. Belser, *ibid.* **1985**, *24*, 204; R. S. Wade, C. E. Castro, *J. Am. Chem. Soc.* **1973**, *95*, 226.

6 J. A. Zoro, J. M. Hunter, G. Eglinton, *Nature* **1974**, *247*, 235.

7 A.-E. F. Nassar, J. M. Bobbitt, J. D. Stuart, J. F. Rusling, *J. Am. Chem. Soc.* **1995**, *117*, 10 986.

8 T. S. Marks, J. D. Allpress, A. Maule, *Appl. Environ. Microbiol.* **1989**, *55*, 1258.

9 N. Assaf-Anid, L. Nies, T. M. Vogel, *Appl. Environ. Microbiol.* **1992**, *58*, 1057.

10 N. Assaf-Anid, K. F. Hayes, T. M. Vogel, *Environ. Sci. Technol.* **1994**, *28*, 246; C. J. Gantzer, L. P. Wackett, *ibid.* **1991**, *25*, 715.

11 U. E. Krone, R. K. Thauer, H. P. C. Hogenkamp, K. Steinbach, *Biochemistry* **1991**, *30*, 2713; U. E. Krone, K. Laufer, R. K. Thauer, H. P. C. Hogenkamp, *ibid.* **1989**, *28*, 10 061; U. E. Krone, R. K. Thauer, H. P. C. Hogenkamp, *ibid.* **1989**, *28*, 4908.

12 T. A. Lewis, M. J. Morra, J. Habdas, L. Czuchajowski, P. D. Brown, *J. Environ. Qual.* **1995**, *24*, 56.

13 M. C. Helveston, C. E. Castro, *J. Am. Chem. Soc.* **1992**, *112*, 8490.

14 R. Kuhler, G. A. Santo, T. R. Caudill, E. A. Betterton, R. G. Arnold, *Environ. Sci. Technol.* **1993**, *27*, 2104.

15 L. Ukrainczyk, M. Chibwe, T. J. Pinnavaia, S. A. Boyd, *Environ. Sci. Technol.* **1995**, *29*, 439.

16 A. C. Benyei, S. Lehel, F. Joó, *J. Mol. Catal. A:* **1997**, *116*, 349.

17 M. A. Jafar Khan, R. J. Watts, *Water, Air, Soil Pollution* **1996**, *88*, 247; S. W. Leung, R. J. Watts, G. C. Miller, *J. Environ. Qual.* **1992**, *21*, 377; for example: D. L. Sedlak, A. W. Andren, *Environ. Sci. Technol.* **1991**, *25*, 777.

18 G. Labat, J.-L. Seris, B. Meunier, *Angew. Chem., Int. Ed. Engl.* **1990**, *29*, 1471.

19 A. Sorokin, B. Meunier, *J. Chem. Soc. Chem. Commun.* **1994**, 1799; A. Sorokin, J.-L. Seris, B. Meunier, *Science* **1995**, *268*, 1163; A. Sorokin, S. De Suzzoni-Dézard, D. Poullain, J.-P. Noël, B. Meunier, *J. Am. Chem. Soc.* **1996**, *118*, 7410; A. Hadash, A. Sorokin, A. Rabion, B. Meunier, *New J. Chem.* **1998**, *22*, 45.

20 A. Hadash, B. Meunier, *Eur. J. Inorg. Chem.* **1999**, 2319.

21 C. S. Creaser, A. R. Fernandes, D. C. Ayres, *Chem. Ind. (London)* **1988**, 499.

22 A. Morvillo, L. Forti, M. Bressan, *New J. Chem.* **1995**, *19*, 951.

23 M. Bressan, N. d'Alessandro, L. Liberatore, A. Morvillo, *Coord. Chem. Rev.* **1996**, *185–186*, 385; M. Bressan, in *Aqueous Organometallic Chemistry and Catalysis* (Eds.: I. T. Horváth, F. Joó), NATO ASI Series, Kluwer Academic Publishers, Dordrecht **1995**, 173.

24 M. Bressan, L. Forti, A. Morvillo, *J. Chem. Soc. Chem. Commun.* **1994**, 253; M. Bressan, N. Celli, N. d'Alessandro, L. Liberatore, A. Morvillo, L. Tonucci, *J. Organomet. Chem.* **2000**, *593–594*, 416.

2.4.8.3
Various Other Reactions

Fritz E. Kühn, Ana M. Santos, and Wolfgang A. Herrmann

2.4.8.3.1
Isomerizations

Recently, the nickel-catalyzed isomerization of geraniol and prenol has been investigated in homogeneous and two-phase systems. The best results with respect to activity and selectivity have been obtained in homogeneous systems with a bis(cycloocta-1,5-diene) nickel(0)/1,4-bis(diphenylphosphanyl)butane/trifluoroacetic acid combination. Catalyst deactivation occurs in the course of the reaction owing to coordination of the aldehyde group formed to the nickel species or as a result of protonolysis of hydrido- or (π-allyl)nickel complexes [1].

Isomerization of allylic and homoallylic alcohols is also catalyzed by the zwitterionic Rh(I) complex (sulphos)Rh(COD), with sulphos = $[O_3S(C_6H_4)CH_2C-(CH_2PPh_2)_3]^-$, in water/$n$-octane to give the corresponding aldehyde or ketone in high yield and chemoselectivity. A π-allyl metal hydride mechanism was proposed on the basis of various independent experiments in both homogeneous and biphasic systems [2a]. Additionally, a water-soluble rhodium bisphosphine complex was used for the aqueous-phase isomerization of selected allylic alcohols. For the isomerization of cinnamyl alcohol, the catalyst was recycled with slight loss of activity and the optimum phosphine/rhodium ratio was found to be 6 : 1 [2b].

The isomerization of allylic alcohols, being a process of current industrial interest in the geraniol chemistry [3], was furthermore used as a liquid/liquid test reaction for high-throughput screening (HTS) of polyphasic fluid reactions. Nowadays high-throughput synthesis methodologies, such as combinatorial techniques, are applied to the discovery of pharmaceuticals, catalysts, and a multitude of other new materials [4]. The effectiveness of this approach has been demonstrated for restricted libraries in the case of catalysis in a single liquid phase. HTS in one phase is assumed to be unproblematic as long as the reactions are not too fast compared with the micro-mixing rates. However, numerous reactions of interest, e.g., hydrogenation, carbonylation, and hydroformylation, operate in gas/liquid or gas/liquid/liquid systems [5]. Inadequate control of phase and catalyst presentation, a result of nonoptimized agitation, may dramatically affect the estimation of selectivity and reactivity. Therefore a major challenge is to develop special reactors for rapid catalyst screening that would ensure good mass and heat transport in a small volume. The liquid/liquid test reaction, based on the isomerization of allylic alcohols [Eq. (1)] was performed with various Rh-, Ru-, Pd-, and Ni-based catalyst systems in aqueous/n-heptane reaction mixtures with a residence time of 100 s at 80 °C. The best results with respect to conversion have been achieved with $RhCl_3$ and $RuCl_3$/TPPTS catalyst systems [6].

$$R' \diagup\!\!\!\diagdown \overset{OH}{\underset{}{\diagup}} R \xrightarrow[\text{[cat.]}]{\text{water}/n\text{-heptane}} R' \diagup\!\!\!\diagdown \overset{O}{\underset{}{\diagdown}} R \qquad (1)$$

2.4.8.3.2
Aldolizations

As already exemplified in many other cases throughout this book, the main driving force for performing aldolizations in water is the reduction or even exclusion of any harmful organic solvents [7]. Several useful synthetic reactions in an aqueous medium need the presence of an organic co-solvent [8] (see also Section 2.2.3.3). Nevertheless many of these reactions proceed sluggishly in pure water, probably because most of the organic reagents are not completely dissolved. To overcome this problem Lewis acid–surfactant combined catalysts (LASCs) have been developed. In the presence of a catalytic amount of an LASC, organic materials rapidly form a dispersed colloidal system in water and several organic reactions proceed smoothly without the use of organic solvents. In order to solve the problems associated with catalyst recovery, polymer-supported Lewis acid catalysts (**1**) have been developed, which show high activity in water and can be easily recovered and re-used. Among the reactions performed with these polymer-supported catalysts is the aldolization reaction given in Eq. (2), promoted by a scandium-based catalyst (see also Chapter 7). Yields of 98% have been reached by the reaction of benzaldehyde with 1-ethylthio-1-trimethylsiloxy-2-methylpropene at room temperature in water in the presence of 3.2 mol% of catalyst [9a].

$$\text{Ph-CHO} + \text{(OSiMe}_3\text{)(SEt)C=CMe}_2 \xrightarrow[\text{[1]}]{\text{H}_2\text{O/r.t./12 h}} \text{Ph-CH(OH)-CMe}_2\text{-C(O)SEt} \qquad (2)$$

Diphenylboronic acid (Ph$_2$BOH), which is soluble in water, is an effective catalyst for the Mukaiyama aldol reaction in the presence of dodecyl sulfate (SDS) as surfactant. Yields of 93% with *syn/anti* ratios of 94 : 6 have been reached according to Eq. (3). The proposed mechanism of this reaction is shown in Scheme 1 [9b].

$$\text{Ph-CHO} + \text{CH}_2\text{=C(OSiMe}_3\text{)Ph} \xrightarrow[\text{[Ph}_2\text{BOH/SDS]}]{\text{PhCOOH, Solvent}} \text{Ph-CH(OH)-CH}_2\text{-C(O)Ph} \qquad (3)$$

Furthermore, lead(II) and lanthanide(III) complexes were synthesized, which worked well as chiral Lewis acids in aqueous media. Until then, chiral crown ether-based Lewis acids had not been used successfully in catalytic asymmetric reactions. The asymmetric aldol reactions, however, proceed smoothly at −10 to 0 °C in water–alcohol solutions while retaining high levels of diastereo- and enantioselectivity. In most previously established catalytic asymmetric aldol reactions the use of aprotic anhydrous solvents and reaction temperatures of −78 °C were necessary. In the asymmetric aldol reaction using rare earth metal triflates M(OTf)$_3$ and chiral bispyridino-18-crown-6, slight changes in the ionic diameter of the metal cations greatly affect the diastereo- and enantioselectivities of the products. The substituents at the 4-position of the pyridine rings of the crown ether influence the binding ability of the crown ether with the M cations. The binding ability of the crown ether

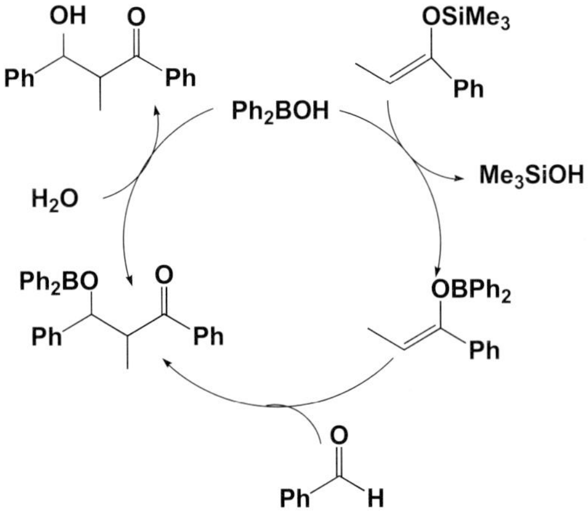

Scheme 1 Mukayama aldol reaction.

with the M cation, however, is important for achieving high selectivities in the asymmetric aldol reactions. Water plays an essential role for the good yield and selectivities. Several aldehyde and silyl enol ethers derived from ketones and thioesters can be applied [10].

Additionally, organocatalytic cross-aldol reactions catalyzed by cyclic secondary amines in aqueous media provide a direct route to a variety of aldols, including carbohydrate derivatives and may warrant consideration as a prebiotic route to sugars [11a].

Dichloroindium hydride, generated by transmetallation between tributyltin hydride and indium trichloride, predominantly reduces unsaturated ketones (enones) with 1,4-selectivity in the presence of aldehydes. Under anhydrous conditions, the successive aldol reaction between the resulting enolates and the remaining aldehydes proceeds with high *anti* selectivity. The stereochemistry, however, is reversed to be *syn*-selective by the use of water and methanol as an additive and solvent, respectively [11b].

2.4.8.3.3
Hydroaminomethylation and Amination

Aliphatic amines are among the most important bulk and fine chemicals in the chemical and pharmaceutical industry [12]. Hydroaminomethylation of olefins to amines presents an atom-economic, efficient and elegant synthetic pathway toward this class of compounds. In hydroaminomethylation a reaction sequence of hydroformylation of an olefin to an aldehyde with subsequent reductive amination proceeds in a domino reaction [Eq. (4)] [13].

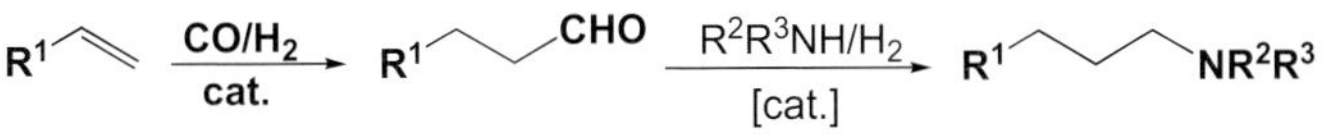

$$R^1 \diagdown \quad \xrightarrow[\text{cat.}]{\text{CO/H}_2} \quad R^1 \diagdown\diagup\text{CHO} \quad \xrightarrow[\text{[cat.]}]{R^2R^3NH/H_2} \quad R^1 \diagdown\diagup\diagdown NR^2R^3 \qquad (4)$$

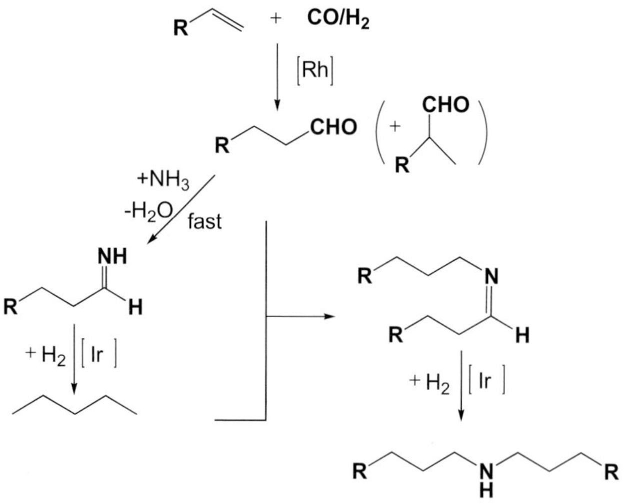

Scheme 2 Selective hydroaminomethylation of alkenes catalyzed by a new Rh/Ir system.

Recently, the highly selective hydroamination of olefins with ammonia to form linear primary and secondary aliphatic amines with a new Rh/Ir catalytic system ([{Rh(cod)Cl}$_2$, [{Ir(cod)Cl}$_2$], aqueous TPPTS solution) has been described (Scheme 2) [14]. The method is of particular importance for the production of industrially relevant, low molecular weight amines.

New methodologies have been reported in recent years for the amination of aromatic halides and triflates according to Eq. (5) with amines to yield aromatic amines of the type ArNRR′ [15]. In contrast to the homogeneous reaction conditions (1–5 mol% Pd(0) catalyst and 1.4 equiv *tert*–butoxide in PhMe at 80–100 °C), by using a two-phase protocol, the separation of products (and unreacted starting material) from the catalyst and subsequent reapplication of the catalyst in further reactions is made facile [16]. The use of the sixfold sulfonated ligand BINAS-6 permits the Pd(0)-catalyzed amination of aromatic halides in water containing single- or two-phase systems [17]. NaOH is used instead of the expensive NaOBu^t as the base. Further advantages are the facile catalyst/product separation and the reusability of the water-soluble Pd(0)/BINAS-6 catalyst. Yields higher than 90% have been reached according to Eq. (6).

$$\text{ArX} + \text{HNRR}^1 \xrightarrow[\text{[Pd}^0\text{L}/\text{NaOBu}^t]}{\text{Toluene, 80--100 °C}} \text{ArNRR}^1 \qquad (5)$$

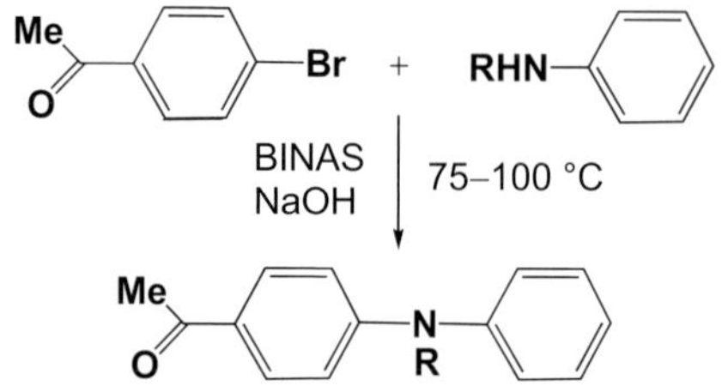

(6)

2.4.8.3.4
Hydrosilylation Reactions

The hydrosilylation of 1-alkenes can be carried out with catalysts of subgroup VIII. Platinum compounds, e.g., the Speier catalyst ($H_2PtCl_6 \cdot H_2O$) and the Karstedt solution, a complex compound of $H_2PtCl_6 \cdot (H_2O)_6$ and vinyl-substituted disiloxanes, are well known and very active catalysts [18]. Several other catalytic systems, e.g., Pt(cod)$_2$, leading to the formation of platinum colloids, have been examined [19]. More recently, hydrosilylation with the Speier catalyst has been examined both under single- and two-phase conditions. The hydrosilylation reaction was thereby optimized for the possibility of technical realization [20].

Different Pt(IV), Pt(II), and Pt(0) catalysts were screened for the hydrosilylation of fatty acid esters [Eq. (7)] containing terminal as well as internal double bonds. The reaction of terminally unsaturated fatty acid esters proceeds smoothly with short reaction times for nearly all the catalysts examined, whereas the Pt(IV) species, and Pt(II) or Pt(0) species with labile ligands, are sufficiently active in the reaction of internally unsaturated compounds. For methyl linoleate, a conjugation of the two internal double bonds before the hydrosilylation was observed. The reaction can be carried out without solvents as well as in solvent systems, allowing catalyst recycling and re-use. In these systems, however, hydrogenation and double bond isomerization are found as side reactions [21].

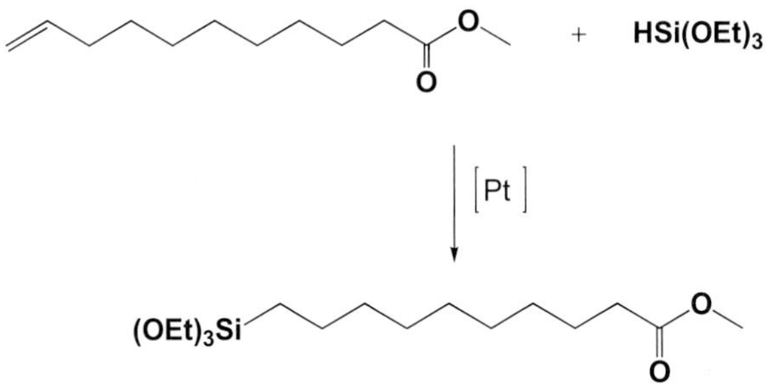

2.4.8.3.5
Thiolysis

Ring opening of 1,2-epoxides with thiol-derived nucleophiles is a well established route to β-hydroxy sulfides that has been applied for the preparation of allylic alcohols, cyclic sulfides, and thioketones, and of important intermediates for the synthesis of natural products and compounds of biological and pharmacological interest [22]. The thiolysis of 1,2-epoxides is usually performed fundamentally in two ways: by using thiolates under basic conditions or thiols in the presence of a variety of activating agents [23]. The Lewis acid catalyzed thiolysis of epoxides in aqueous medium has been investigated recently [24]. The pH dependence of the thiolysis of 1,2-epoxides with thiophenol in water and the influence of a Lewis acid catalyst was examined. InCl$_3$ showed a very high efficiency in catalyzing this process at pH 4. The regioselectivity of the nucleophilic attack is markedly influenced on going from pH 9 to pH 4. A one-pot procedure in water alone, to prepare *trans-2-*(phenylsulfinyl)cyclohexan-1-ol, was reported starting from epoxycyclohexane, via thiolysis and oxidation with TBHP (*tert*-butyl hydroperoxide). An example of the thiolysis of epoxycyclohexane by thiophenol in water is shown in Eq. (8).

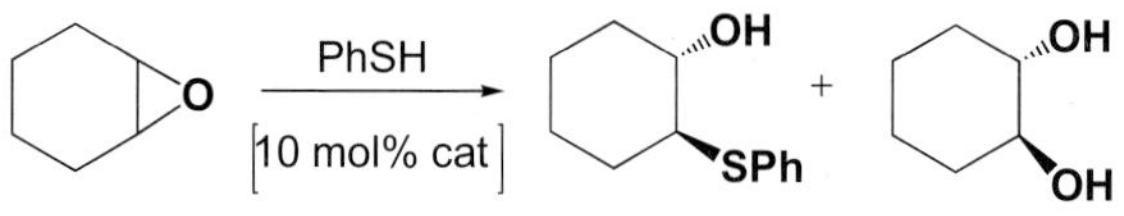

$$\tag{8}$$

2.4.8.3.6
Synthesis of Heterocycles

Easy and efficient access routes to heterocyclic complexes are of significant interest in organic chemistry and its application in synthesis both in the laboratory and industry. During the recent years several novel routes toward such compounds in aqueous media have been described. 2-Substituted 3,4-dihydro-2*H*-1,4-benzoxazines have been prepared in excellent yields and short reaction times through the cyclization of hydroxysulfonamides in water under phase-transfer catalysis conditions (Scheme 3) [25].

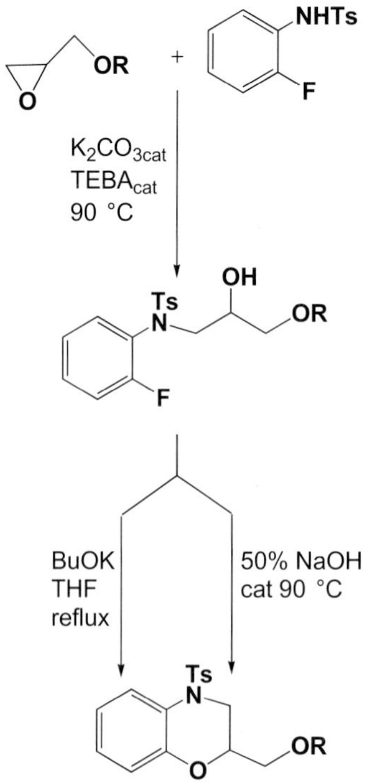

Scheme 3 Cyclization of hydroxysulfonamides in water under phase-transfer conditions.

Reactions of 6-amino-1,3-dimethyluracil with substituted α-ketoalkynes using homogeneous nickel catalysts in aqueous alkaline media afford substituted 2,4-dioxopyrido[2,3-d]pyrimidine derivatives in quantitative yields under very mild conditions. A mechanism has been proposed for the reaction involving the nucleophilic attack by a Ni(0) anion, formed in situ of the triple bond of the substrate [26].

Additionally, the efficient synthesis of substituted phenylalanine-type amino acids using a rhodium-catalyzed, conjugate addition of arylboronic acids has been described. The reactions are run in water and use a low loading (0.5 mol%) of the rhodium catalyst [27].

References

1 H. BRICOUT, E. MONFLIER, J. F. CARPENTIER, A. MONTREUX, *Eur. J. Inorg. Chem.* **1998**, 1739.
2 (a) C. BIANCHINI, A. MELI, W. OBERHAUSER, *New J. Chem.* **2001**, *25*, 11; (b) D. A. KNIGHT, T. L. SCHULL, *Synth. Commun.* **2003**, *33*, 827.
3 S. OTSUKA, T. KAZUHIDE, *Synthesis* **1991**, *9*, 665.
4 (a) B. JANDELEIT, D. J. SCHÄFER, T. S. POWERS, H. W. TURNER, W. H. WEINBERG, *Angew. Chem.* **1999**, *111*, 2648; *Angew. Chem., Int. Ed.* **1999**, *38*, 2494; (b) T. BEIN, *Angew. Chem.* **1999**, *111*, 335; *Angew. Chem., Int. Ed.* **1999**, *38*, 323; (c) D. R. LIU, P. G. SCHULTZ, *Angew. Chem.* **1999**, *111*, 36; *Angew. Chem., Int. Ed.* **1999**, *38*, 36; (d) M. B. FRANCIS, E. N. JACOBSEN, *Angew. Chem.* **1999**, *111*, 987; *Angew. Chem., Int. Ed.* **1999**, *38*, 937; (e) O. LAVASTRE, J. P. MORKEN, *Angew. Chem.* **1999**, *111*, 3357; *Angew. Chem., Int. Ed.* **1999**, *38*, 3163; (f) M. T. REETZ, M. H. BECKER, K. M. KÜHLING, A. HOLZWARTH, *Angew. Chem.* **1998**, *110*, 2792; *Angew. Chem., Int. Ed.* **1998**, *37*, 2647; (g) D. A. ANNIS, O. HELLUIN, E. N. JACOBSEN, *Angew. Chem.* **1998**, *110*, 2010; *Angew. Chem., Int. Ed.* **1998**, *37*, 1907; (h) K. D. SHIMIZU, M. L. SNAPPER, A. H. HOVEYDA, *Chem. Eur. J.* **1998**, *4*, 1885; (i) M. B. FRANCIS, N. S. FINNEY, E. N. JACOBSEN, *J. Am. Chem. Soc.* **1996**, *118*, 8983; (j) K. BURGESS, H. J. LIM, A. M. PORTE, G. A. SULIKOVSKI, *Angew. Chem.* **1996**, *108*, 192; *Angew. Chem., Int. Ed. Engl.* **1996**, *35*, 220.
5 B. CORNILS, W. A. HERRMANN (Eds.), *Applied Homogeneous Catalysis with Organometallic Compounds*, 2nd ed., Weinheim **2002**.
6 C. DE BELLEFON, N. TANCHOUX, S. CARAVIEILHES, P. GRENOUILLET, V. HESSEL, *Angew. Chem.* **2000**, *112*, 3584; *Angew. Chem., Int. Ed.* **2000**, *39*, 3442.
7 See also: (a) P. A. GRIECO (Ed.), *Organic Synthesis in Water*, Blackie A&P, London **1998**; (b) C. J. LI, T. H. CHAN, *Organic Reactions in Aqueous Media*, Wiley, New York **1997**; (c) K. MANABE, S. KOBAYASHI, *Chem. Eur. J.* **2002**, *8*, 4095.
8 S. KOBAYASHI, *Eur. J. Org. Chem.* **1999**, 15.
9 (a) S. NAGAYAMA, S. KOBAYASHI, *Angew. Chem.* **2000**, *112*, 578; *Angew. Chem., Int. Ed.* **2000**, *39*, 567; (b) Y. MORI, K. MANABE, S. KOBAYASHI, *Angew. Chem.* **2001**, *113*, 2900; *Angew. Chem., Int. Ed.* **2001**, *40*, 2816.
10 (a) S. NAGAYAMA, S. KOBAYASHI, *J. Am. Chem. Soc.* **2000**, *122*, 11 531; (b) S. KOBAYASHI, T. HAMADA, S. NAGAYAMA, K. MANABE, *J. Bras. Chem. Soc.* **2001**, *12*, 627; (c) T. HAMADA, K. MANATABE, S. ISHIKAWA, S. NAGAYAMA, M. SHIRO, S. KOBAYASHI, *J. Am. Chem. Soc.* **2003**, *125*, 2989; (d) T. HAMADA, K. MANATABE, S. KOBAYASHI, *J. Synth. Org. Chem.* **2003**, *61*, 445.
11 (a) A. CORDOVA, W. NOTZ, C. F. BARBAS, *Chem. Commun.* **2002**, 3024; (b) K. INOUE, T. ISHIDA, I. SHIBATA, A. BABA, *Adv. Synth. Catal.* **2002**, *344*, 283.
12 G. HEILEN, H. J. MERCKER, D. FRANK, A. RECK, R. JÄCKH, *Ullmanns Encyclopaedia of Industrial Chemistry*, Vol. A2, 5th ed., VCH, Weinheim **1985**.

13 (a) H. Schaffrath, W. Keim, *J. Mol. Catal.* **1999**, *140*, 107; (b) W. Reppe, H. Vetter, *Liebigs Ann. Chem.* **1953**, *582*, 133.

14 B. Zimmermann, J. Herwig, M. Beller, *Angew. Chem.* **1999**, *111*, 2515; *Angew. Chem., Int. Ed.* **1999**, *38*, 2372.

15 (a) J. P. Wolfe, S. Wagaw, S. L. Buchwald, *J. Am. Chem. Soc.* **1996**, *118*, 7215; (b) M. S. Driver, J. F. Hartwig, *J. Am. Chem. Soc.* **1996**, *118*, 7217; (c) M. Beller, T. M. Riermeier, C. P. Reisinger, W. A. Herrmann, *Tetrahedron Lett.* **1997**, *38*, 2073; (d) N. P. Reddy, M. Tanala, *Tetrahedron Lett.* **1997**, *38*, 4807.

16 (a) T. Prinz, W. Keim, B. Driessen-Hölscher, *Angew. Chem.* **1996**, *108*, 1835; *Angew. Chem., Int. Ed. Engl.* **1996**, *35*, 1708; (b) B. Cornils, *Angew. Chem.* **1995**, *197*, 1709; *Angew. Chem., Int. Ed. Engl.* **1995**, *34*, 1575; (c) W. A. Herrmann, C. W. Kohlpaintner, *Angew. Chem.* **1993**, *105*, 1588; *Angew. Chem., Int. Ed. Engl.* **1993**, *32*, 1524.

17 G. Wullner, H. Jansch, S. Kannenberg, F. Schubert, G. Boche, *Chem. Commun.* **1998**, 1509.

18 (a) J. C. Saam, J. L. Speier, *J. Am. Chem. Soc.* **1958**, *80*, 4104; (b) B. D. Karstedt, US 3.775.452 (**1973**).

19 N. L. Lewis, R. J. Uriarte, N. Lewis, *J. Catal.* **1991**, *127*, 67 and references cited therein.

20 A. Behr, N. Toslu, *Chem. Eng. Technol.* **2000**, *23*, 122.

21 A. Behr, F. Naendrup, D. Obst, *Adv. Synth. Catal.* **2002**, *344*, 1142.

22 (a) U. Kesavan, D. Bonnet-Delpon, J.-P. Bégué, *Tetrahedron Lett.* **2000**, *41*, 2895; (b) S. Osaki, E. Matsui, H. Yoshinaga, S. Kitagawa, *Tetrahedron Lett.* **2000**, *41*, 2621; (c) M. Chini, P. Crotti, L. A. Flippin, A. Lee, F. Macchia, *J. Org. Chem.* **1991**, *56*, 7043; (d) P. Meffre, L. Vo Quang, Y. Vo Quang, F. Le Goffic, *Tetrahedron Lett.* **1990**, *21*, 2291; (e) V. Jaeger, W. Huemmer, *Angew. Chem.* **1990**, *102*, 1182; *Angew. Chem., Int. Ed. Engl.* **1990**, *29*, 1171; (f) A. Schwartz, P. B. Madan, E. Mohacsi, J. P. O'Brien, L. J. Todaro, D. L. Coffen, *J. Org. Chem.* **1992**, *57*, 851.

23 (a) M. Romdhani Younes, M. M. Chaabouni, A. Baklouti, *Tetrahedron Lett.* **2001**, *42*, 3167; (b) J. C. Justo De Pomar, A. Soderquist, *Tetrahedron Lett.* **1998**, *29*, 4409; (c) D. Albanese, D. Laudini, M. Penso, *Synthesis* **1994**, 34; (d) P. Raubo, J. Wicha, *Pol. J. Chem.* **1995**, *69*, 78; (e) A. K. Maiti, P. Bhattacharyya, *Tetrahedron* **1994**, *50*, 10 483; (f) J. W. J. Still, L. J. P. Martin, *Synthetic Commun.* **1998**, *28*, 913; (g) M. H. Wu, E. N. Jacobsen, *J. Org. Chem.* **1998**, *63*, 5252; (h) T. Iida, N. Yamamoto, H. Sasai, M. Shibasaki, *J. Am. Chem. Soc.* **1997**, *119*, 4783.

24 F. Fringuelli, F. Pizzo, S. Tortoioli, L. Vaccaro, *Adv. Synth. Catal.* **2002**, *344*, 379.

25 D. Albanese, D. Landini, V. Lupi, M. Penso, *Adv. Synth. Catal.* **2002**, *344*, 299.

26 N. Rosas, P. Sharma, C. Alvarez, A. Cabrera, R. Ramirez, A. Delgado, H. Arzoumanian, *J. Chem. Soc., Perkin Trans.* **2001**, 2341.

27 C. J. Chapman, C. G. Frost, *Adv. Synth. Catal.* **2003**, *345*, 353.

2.5
Commercial Applications

2.5.1
Oxo Synthesis (Hydroformylation)

Boy Cornils

Commercially, the aqueous-phase concept was firstly applied in Ruhrchemie/Rhône-Poulenc's (RCH/RP) process (for the fundamentals, see Section 2.4.1.1). In several units the RCH/RP process has been converting propene to *n*-butyraldehyde and isobutyraldehyde (or butenes to valeraldehydes) since 1984 in the presence of HRh(CO)(TPPTS)$_3$ (with TPPTS = *m*-trisulfonated triphenylphosphine or tris-(sodium-*m*-sulfonatophenyl)phosphine as water-soluble ligand) according to Eq. (1). The output of the units mentioned is approximately 890 000 tpy, which corresponds to roughly 13% of the world's total production.

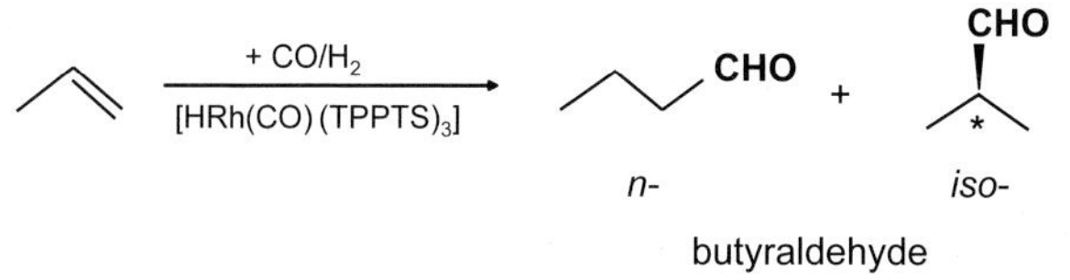

$$\text{(1)}$$

n-   *iso*-

butyraldehyde

This has to be considered against a background of an enormous growth in production capacity of aldehydes that has taken place since 1993. Worlwide oxo capacity in 1993 exceeded 6 MM tpy and increased within five years to more than 9.2 MM tpy [1]. With a share of 75%, C$_4$ products manufactured from propene have by far the pole position. With a share of only 3%, ethylene hydroformylation is of minor importance. Alkenes with medium chain lengths (diisobutene, tripropene, etc.) make up 17% of the total production capacity. Only 5% are used for the hydroformylation of alkenes with > 12 carbon atoms and thus for detergent alcohols. About 45% of the total C$_4$ capacity (or about 52% of the *n*-butyraldehyde capacity) is used for the production of 2-ethylhexanol (2-EH), the major plasticizer alcohol for PVC (dioctyl phthalate, DOP). Approximately 70% of the total hydro-formylation capacity (converting light alkenes such as ethylene, propene, and butenes) is based on the low-pressure oxo processes (LPOs) using phosphine-modified rhodium catalysts. The various processes for the hydroformylation of propene with their catalysts (Rh, Co) are depicted in Table 1.

The process is described in detail in Section 2.4.1.1.3, and the development in Section 2.4.1.1.1. One of the units is depicted on the front cover of this book. Characteristic of RCH/RP's process is how the problem of catalyst deactivation, catalyst recycle, and catalyst removal/work-up has been solved [3].

As has been mentioned earlier, the part of the aqueous catalyst solution which leaves the oxo reactor accompanying (but not dissolved in) the reaction products

Multiphase Homogeneous Catalysis
Edited by Boy Cornils and Wolfgang A. Herrmann et al.
Copyright © 2005 Wiley-VCH Verlag GmbH & Co. KGaA, Weinheim
ISBN: 3-527-30721-4

Table 1 Capacities for C_4 products by various processes (excluding 2-EH).

Process	Central atom of catalyst	Ligand	Capacity [× 1000 t]	[%]
Union Carbide	Rh	TPP[a]	3040	46
BASF	Rh	TPP[a]	900	14
RCH/RP (Celanese)	Rh	TPPTS[b]	890	13
Mitsubishi	Rh	TPP[a]	680	10
Eastman	Rh	TPP[a]	610	9
Neftekhim	Co	–	425	7
Shell	Co	Spec. P[d]	75	1
Unknown[c]			230	–
Total			6850	100

a) Triphenylphosphine.
b) Triphenylphosphine trisulfonate.
c) Not attributable.
d) Special phosphine [2].

passes a phase separator (decanter **2** in Section 2.4.1.1.1, Figure 3) which is a charac-teristic part of the plant shown on the front cover. In this decanter, which ensures spontaneous phase separation, the crude aldehyde formed by hydroformylation according to Eq. (1) is freed from gases and separated into mutually insoluble phases. The catalyst solution, supplemented by an amount of water equivalent to the water content of the crude aldehyde, is recirculated to the reactor. During its active life, the Rh catalyst is mainly situated in the oxo reactor, it is not moved (only part of it flows in a short circuit around the reactor), and no aliquots are withdrawn as in other processes. For this reason, rhodium losses are low – in the range of parts per billion (ppb) – and thus this provides the background for the high economy [4].

Like every technically used and therefore "real" catalyst, the complex HRh(CO)-$(TPPTS)_3$ and the excess ligand TPPTS undergo a degree of decomposition that determines the catalyst's lifetime as measured in years. The catalyst deactivation mechanism is known in detail and depicted in Scheme 1 [5].

The hydroxo complex $(HO)Rh(CO)_2TPPTS$ starts the deactivation cycle as shown in the Scheme 1 [5] (for graphical simplification, Ar_3P is TPPTS and Ar_2P is bis[m-sulfophenyl]phosphine) [6]. The various steps will not be discussed here. Other decomposition products such as the reductively eliminated bis(m-sul-fophenyl)phosphinous acid Ar_2POH (cf. Scheme 1) and the phospine oxides $Ar_2P(OH)(=O)$ and $Ar_2P(=O)(CH[OH]C_3H_7)$ have been identified.

According to Figure 1, the intermittent addition of excess ligands extends the catalyst's lifetime in a sawtooth curve. This addition of ligand compensates for the system-immanent formation of deactivating substances which are brought into the system by the feedstocks. Filters, guard beds, or special precautions to avoid larger sulfur or oxygen inputs and their concentrations are not necessary. Additio-nally, other activity-lowering oxo poisons may be separated with the organic product phase of the decanter and are thus continuously removed at the very point of their formation from the system: any accumulation of activity-decreasing poisons in the

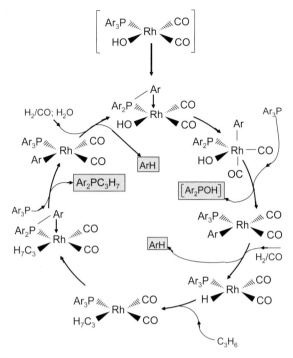

Scheme 1 Deactivation mechanism.

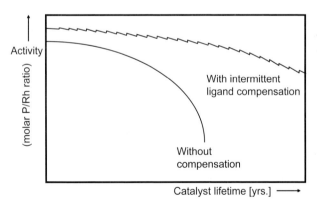

Figure 1 The effect of excess ligand on the catalyst's lifetime.

catalyst solution is prevented. It might be worth mentioning that the Ruhrchemie plant has been supplied over longer periods with syngas manufactured from coal by the TCGP (Texaco coal gasification process) [7].

Catalyst deactivation includes (among other reactions) the formation of inactive Rh species, ligand decomposition, or P–C cleavage by direct oxidative insertion of the rhodium metal for formation of PDSPP (propyl-di[*m*-sulfophenyl]phosphine) acting as strong electron donor reducing the amount of active Rh catalyst.

It turned out to be beneficial to control the P(III)/Rh ratio and the CO partial pressure very carefully.

Catalyst solutions after years of use typically contain 20 mg L^{-1} iron and 0.7 mg L^{-1} of nickel, thus showing no remarkable corrosivity. The Rh content of crude aldehyde is in the ppb range; this corresponds to "real" losses of less than 10^{-9} g kg^{-1} n-butyraldehyde, totalling 5 kg rhodium over a 20-year period and production of five million metric tons of n-butyraldehyde.

The recovery of rhodium from spent catalyst solutions proceeds according to the expertise and the rules of the rare metals producer; the technical know-how includes trade secrets. Because of its relatively low value, TPPTS is not recovered. This may be changed if higher-valued ligands, co-ligands, co-solvents, modifier, surfactant promoters, etc., are necessary for hydroformylation of long-chain alkenes. There are small-scale processes employing aqueous-phase catalysis which use "exotic" ligands such as bi-, tri-, or multidentate phosphines where the situation demands drastic measures such as ligand recovery.

The "real" oxo precatalyst [HRh(CO)(TPPTS)$_3$] is easily made in the oxo reactor by reaction of suitable Rh salts (e.g., rhodium acetate or rhodium 2-ethylhexanoate) with TPPTS without any additional preformation step. After formation of the active species and adjustment of the whole system with water up to the desired P/Rh ratio (ensuring the stability of the catalyst and the desired n/i ratio), the reaction starts.

In total, RCH/RP's process – although neither company still exists: Ruhrchemie went its way through Hoechat AG and Celanese to the present owner Blackstone and the joint venture European Oxo; Rhône-Poulenc became a part of the late Aventis SA and now Sanofi – is still the only oxo version of the fifth-generation hydroformylation processes.

References

1 H.-W. Bohnen, B. Cornils, *Adv. Catal.* 2004, *47*, 1.
2 B. Cornils, W. A. Herrmann, R. Schlögl, C.-H. Wong, *Catalysis from A to Z*, 2nd ed., Wiley-VCH, Weinheim 2003, under "shell oxo process", p. 691.
3 E. Wiebus, B. Cornils, Biphasic Systems: Water–Organic, in B. Tooze, D. J. Cole-Hamilton, *Recovery and Recycle in Homogeneous Catalysis*, Kluwer, Dordrecht 2005.
4 B. Cornils, E. Wiebus, *CHEMTECH* 1995, *25*, 33; B. Cornils, E. Wiebus, *Hydrocarb. Proc.* 1996, March, 63.
5 W. A. Herrmann, C. W. Kohlpaintner, *Angew. Chem., Int. Ed. Engl.* 1993, *32*, 1524.
6 W. A. Herrmann, J. A. Kulpe, W. Konkol, H. Bahrmann, *J. Organomet. Chem.* 1990, *389*, 85; W. A. Herrmann, J. A. Kulpe, J. Kellner, H. Riepl, H. Bahrmann, *Angew. Chem., Int. Ed. Engl.* 1990, *29*, 391.
7 E. Wiebus, B. Cornils, *Chem. Ing. Tech.* 1994, *66*, 916.

2.5.2
Other Reactions

Boy Cornils

Accompanying Ruhrchemie/Rhône-Poulenc's industrial realization of the aqueous biphasic hydroformylation reaction (cf. Sections 2.4.1.1.3 and 2.5.1) and Kuraray's hydrodimerization process (see Section 2.4.4.2), there are some other minor processes for the production of fine chemicals.

Rhodia runs a process for the manufacture of geranylacetone by Rh/TPPTS-catalyzed addition of ethyl acetoacetate to myrcene, a building block for vitamin E (Scheme 1) [1, 2a].

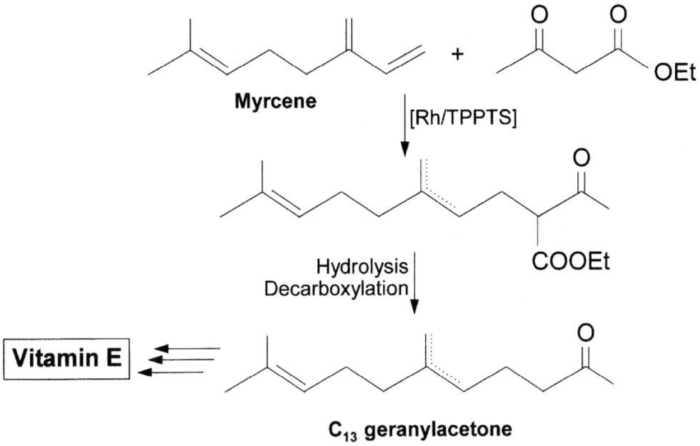

Scheme 1 Rhodia's process for manufacture of geranylacetone.

The process uses TPPTS from Ruhrchemie's production unit. Interestingly, in commercial operation supported aqueous-phase catalysis operation is more effective than the nonsupported catalyst version [2b]. Moreover, TPPTS-modified Ru or Pd catalysts have been proposed for the homogeneous catalyzed hydrogenation step for converting unsaturated into saturated ketones. It is not known how many of these proposals have been realized industrially [3].

Industrial significance has also been acquired by the aqueous biphasic Pd-catalyzed carbonylation of benzyl chloride to phenylacetic acid [Eq. (1)]. Once more, TPPTS is the ligand of choice [4].

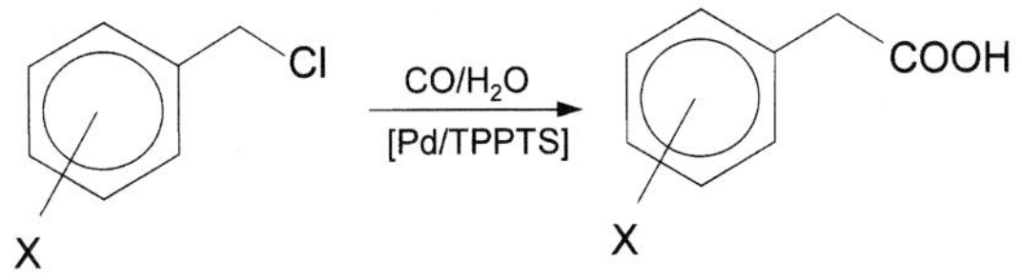

(1)

Mention may be made of the Suzuki coupling of aryl halides and arylboronic acids. In earlier days, TPPMS could be used to convert the brominated or iodinated aromatics; nowadays, cheaper chlorinated aromatics and catalysts based on Pd/TPPTS are used in an aqueous procedure on a commercial scale [Eq. (2)] [5].

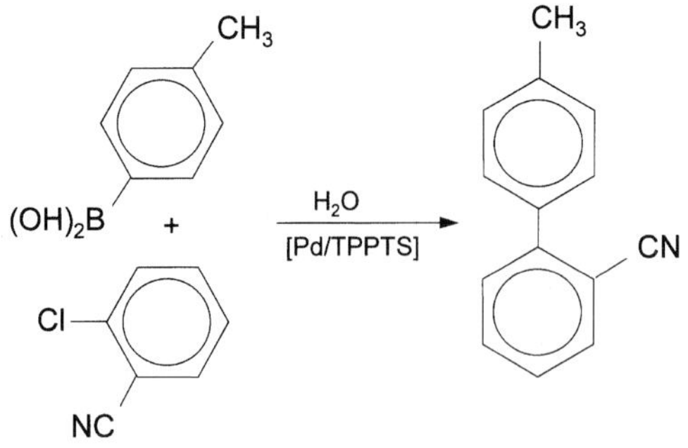

$$(2)$$

References

1 B. CORNILS, W. A. HERRMANN, *Applied Homogeneous Catalysis with Organometallic Compounds*, 2nd ed., Wiley-VCH, Weinheim **2002**, Vol. 2, p. 623.

2 (a) C. MERCIER, P. CHABARDES, *Pure Appl. Chem.* **1994**, *66*, 1509; (b) E. FACHE, C. MERCIER, M. PAGNIER, B. DESPEYROUX, P. PANSTER, *J. Mol. Catal.* **1993**, *79*, 117.

3 (a) K.-C. TIN, N.-B. WONG, R.-X. LI, Y.-Z. LI, X.-J. LI, *J. Mol. Catal. A:* **1999**, *137*, 113; (b) W. BONRATH, T. NETSCHER, *Proceedings of EuropaCat-VI* (August/September 2003, Innsbruck/Austria).

4 C. W. KOHLPAINTNER, M. BELLER, *J. Mol. Catal. A:* **1997**, *116*, 259.

5 Hoechst AG (S. HABER, H. J. KLEINERT), DE 195.27.118 and 195.35.528 (1997).

2.6
Supported Aqueous-Phase Catalysis as the Alternative Method

Henri Delmas, Ulises Jáuregui-Haza, and Anne-Marie Wilhelm

2.6.1
Introduction

As seen in the preceding sections, many attempts have been done to heterogenize the homogeneous catalysts. In general, the immobilized systems never approach the necessary combined activity/selectivity performance levels and tend not to retain the metal complex, causing its leaching to the organic phase. In general, the drawback of all these immobilization methods is that they are not applicable for liquid-phase reactants/products that are miscible with the nonvolatile solvent phase [1]. Only biphasic catalysis, involving water-soluble ligands to maintain the catalytic complex in the aqueous phase, is successfully used in industry [2–4]. But the low solubility of higher olefins in the aqueous catalytic phase represents a serious problem in extending this outstanding technology to a broader variety of substrates (see Section 2.4.1.1) [2, 3, 5]. Several attempts to increase reaction rates with heavy substrates have been investigated: increasing olefin solubility in the aqueous phase using co-solvents, increasing interfacial area and mass transfer via phase tensio-active agents, or trying to locate the catalyst complex at the liquid–liquid interface with amphiphilic ligands or convenient hydrophylic–hydrophobic ligand mixtures. All these techniques would involve significant catalyst leaching, and probably for this reason none of them is known to be used in commercial plants.

The alternative could arise from development of an elegant immobilization method designed specifically to convert liquid-phase reactants: the supported aqueous-phase catalysis (SAPC) [6]. SAPC is a special case of the supported liquid-phase catalysis whose development began according to proposals by Moravec [7] and Rony [8, 9]. SAPC is very promising due to its high capacity in conversion and selectivity, its better stability, and also the very easy recovery of the catalyst from the organic phase. Comprehensive reviews of SAPC are available [1, 10–12].

2.6.2
SAPC: the Concept and Main Applications

SAPC consists in adsorbing on the surface of an hydrophilic solid a thin film of water containing the catalyst precursor. The catalyst should present hydrophilic ligands allowing its dissolution in the water film and thus its anchorage on the support. The hydrophilicity of the ligands and the support creates interaction energies sufficient to maintain the immobilization [1]. The metal atom is oriented toward the organic phase, with the catalytic reaction taking place efficiently at the aqueous–organic interface. The concept of SAPC has been expanded to other hydrophilic liquids such as ethylene glycol and glycerol [13].

Multiphase Homogeneous Catalysis
Edited by Boy Cornils and Wolfgang A. Herrmann et al.
Copyright © 2005 Wiley-VCH Verlag GmbH & Co. KGaA, Weinheim
ISBN: 3-527-30721-4

It has been proven that SAPC can perform a broad spectrum of reactions such as hydroformylation, hydrogenation, and oxidation, for the synthesis of bulk and fine chemicals, pharmaceuticals, and their intermediates. Table 1 summarizes some of the reported applications of SAPC. Although rhodium complexes are the most extensively used in SAPC, the complexes of ruthenium, platinum, palladium, cobalt, molybdenum, and copper have also been employed (Table 1).

Due to interfacial reactions, one of the main advantages of SAPC upon biphasic catalysis is that the solubility of the olefins in the catalytic aqueous-phase does not limit the performance of the supported aqueous phase catalysts (SAPCs); the turnover frequencies (TOFs) are roughly independent of olefin carbon number [17]. This has been shown to be true also for carbon numbers as high as 17 [15].

Table 1 Main applications of SAPC.

Reaction	Metal	Ligand	Substrate	Ref.
Hydroformylation	Rh	TPPTS	Alkenes	[1, 6, 14–33]
		TPPTS	α,β-Unsaturated esters	[34, 35]
		TPPTS	Monoterpenes	[36, 37]
		HexDPPDS	Alkenes	[20]
		Xantphos	Alkenes	[24]
		PAA–PNH	Alkenes	[38]
		Norbos	Alkenes	[39]
		PEI–PNH	Alkenes	[38]
	Co	TPPTS	Alkenes	[40, 41]
		TPrPTS	Alkenes	[20]
	Pt	TPPTS	Alkenes	[42]
	Pt/Sn	TPPTS	Alkenes	[20, 42]
		TBeTS	Alkenes	[20]
		TEtPTS	Alkenes	[20]
		TPrPTS	Alkenes	[20]
		(S,S)-BDPP	Styrene	[20]
Hydrogenation	Rh	(S,S)-BDPP	Alkenes	[20]
		chiraphos	Alkenes	[20]
	Ru	TPPTS	α,β-Unsaturated aldehydes	[43]
		BINAP	Alkenes	[44]
Oxidation	Pd/Cu		Alkenes	[45]
	Mo		Alkenes	[46]
Allylic substitution	Pd	TPPTS	Allyl acetates	[47–49]
Indolization	Rh	TPPTS	Azobenzene	[50]
Asymmetric reactions		BINAP		[51]
Alkoxycarbonylation	Pd	TPPTS	Alkenes	[52]

Concerning the position of the double bond, from the data presented by Arhancet et al. [16] it was clear that the activity of SAPC for the hydroformylation of internal olefins is of approximately the same order of magnitude as for the hydroformylation of terminal olefins. However, the rate of isomerization for the internal olefins is higher than for external ones (e.g., heptene). In the same study, the interesting fact is that the hydroformylation of higher external linear alkenes and dienes gives similar results, although the hydroformylation of dienes takes place sequentially, beginning with the external double bond. A different result was obtained when the role of monoterpenes (linear or monocyclic) in the reactivity of linalool, limonene, and geraniol undergoing hydroformylation by SAPC was investigated [36]. The efficiency of SAPC to achieve hydroformylation of linalool, geraniol and limonene depends strongly on the location of the double bond and of some properties of SAP catalysts, as pore size and hydration ratio. The low reactivity and the lower conversion levels of geraniol and limonene, as compared with linalool, were explained in terms of double bond polarization and of steric hindrance and configuration.

2.6.3
The Supports and the Preparation of the SAP Catalysts

Different inorganic materials have been used as supports in SAPC: glass beads of controlled pore size [6, 14–17, 24, 39–42, 44, 45, 53]; porous [11, 15, 18, 19, 21, 23, 28–32, 35, 36, 38, 39, 43, 48, 49] and nonporous [28, 33, 48] silica nanoparticles; synthetic phosphate [27]; carbon [39], and alumina [15, 39]. It was shown that glass beads, silica, and synthetic phosphate gave the best performance. All these supports have a high specific surface with an average diameter of the pores, in the case of porous supports, between 60 and 345 Å. The use of chitosan as a natural polymeric support of SAP catalysts for the synthesis of fine chemicals has been reported recently [54].

Several methods have been used for preparing the SAP catalysts. According to the preparation procedure, the methods can be classified into two groups: (a) indirect methods, when the support is first impregnated with the hydrosoluble catalytic complex, then dried and rehydrated before use [6, 14, 15, 17, 41, 42, 49]; (b) direct methods, when the support, catalytic complex, and water are mixed at the same time in the reaction system [15, 21, 29]. In general, the best conversions in the hydroformylation of alkenes by SAPC have been obtained by indirect preparation. However, the direct methods being of much easier implementation than the indirect ones, they are most widely used.

2.6.4
Main Factors Influencing the Efficiency of SAPC

Many typical parameters usually investigated in biphasic catalysis, such as temperature, pressure, the excess of water-soluble ligand, and the nature of the reactants and the catalyst, have roughly similar effects on SAPC.

As seen before, a major difference with respect to biphasic catalysis is the low dependence on substrate solubility in the catalytic aqueous phase as the SAPC reaction occurs at the interface. SAPC is strongly dependent on the water content of the solid support. Two types of water content effects have been reported: usually SAPC is efficient over a very restricted hydration range where activity exhibits a clear peak, while only recently a large plateau was observed in a higher hydration range.

Most often a very narrow optimum range of support hydration yielding the best conversion has been found [14–17, 48]. Drastically decreased activity observed at higher or lower water loadings was not due to metal loss through leaching [6, 14–17, 20, 28, 35, 51]. This typical effect of the hydration ratio of the support has been related to three different environments [14]. First, at a very low water content, the catalytic complex is strongly adsorbed to the surface, as evidenced by the lack of mobility seen in the solid-state NMR spectra for the adsorbed complex. The lack of mobility may be responsible for the low catalytic activity observed at low water contents. Furthermore, at low water content the catalysts are very stable. Secondly, at intermediate water contents the catalytic activity reaches a maximum while the complex stability is still enhanced. Thirdly, at high water loading the SAP materials lose both activity and stability and approach the behavior of the unsupported catalyst; this may correspond to a flooded catalyst.

It is clear that this high sensitivity would constitute a serious drawback of the SAPC concept, as crucial measurement and control of the water content would be very difficult during continuous operation or multiple catalyst re-use. Nevertheless, conditions where SAPC is equally efficient over a much wider hydration range have been reported recently [28, 33, 36, 37]. The influence of the hydration rate on the support was studied extensively for the hydroformylation of 1-octene and linalool on several porous silica supports [28, 36, 37]. Data for 1-octene are shown in Figure 1, where the degree of hydration is the ratio of the water content to that corresponding to complete pore filling. Two very different behavior were observed, depending on the support pore characteristics. For S200 and S60, the silica supports with larger

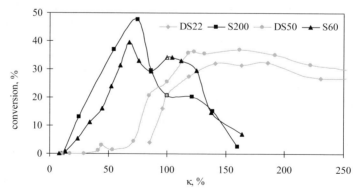

Figure 1 Effect of the degree of hydration (water volume/pore volume) on conversion for various supports [55].

pores, an optimum degree of hydration corresponding to incomplete pore filling was found, while for DS22 and DS50 the large range of approximately constant conversion corresponds to a significant excess of water. The same results were reported for linalool [36, 37]. In both cases, decreased conversion and leaching of rhodium through aqueous droplets were observed when the amount of water was further increased. The same behavior was found during the hydroformylation of 1-hexene with aqueous phosphine–Rh complexes supported on non-porous fumed-silica nanoparticles [33]. A positive effect on the reaction performance was observed for the particle size and surface area of the nanoparticles.

The influence of the hydration ratio and the physico-chemical properties of five supports in the hydroformylation of 1-octene was also studied [28]. The results confirm that the size of the pores and the amounts of water were found to be the determining factors contributing to SAPC efficiency.

Most authors have found the excess of water-soluble ligands to have limited effects on both activity and selectivity for linear/branched aldehydes, whether the optimum hydration ratio zone was narrow or wide. The ratio of linear aldehydes (about 80%) is lower than that observed in biphasic aqueous catalysis.

The main limitation for the use of SAPC is due to the limited stability and reusability of the support–catalytic complex matrix. There are two approaches to solving this problem. The first is the use of more stable ligands. Wan and Davis [44] reported the use of the complex Ru–BINAP in the selective hydrogenation of 2-(6′-methoxy-2′-naphthyl)acrylic acid. After seven cycles the conversion and selectivity remain stable. The xantphos-based complexes of rhodium used in the hydroformylation of olefins have been used successfully as selective and recyclable catalysts [11, 24]. The second option is the addition of basic alkali metal salts such as Na_2CO_3, K_2CO_3, and NaH_2PO_4, which depressed the oxidation of TPPTS [33].

2.6.5
Perspectives and Challenges

Despite promising results, SAPC has not yet been used for large commercial production. The new possibility of efficient SAPC over a wide range of support hydration [27, 28, 33] has technological importance since it significantly improves flexibility by avoiding a strict control of the hydration ratio. Therefore, this modification may open the way to apply SAPC on a commercial scale.

The main technical advantage of SAPC with respect to the usual biphasic catalysis is the much faster reaction rate with poorly water-soluble substrates which may have direct access to the catalyst. Such performances are obtained in relatively mild reaction conditions (temperatures up to 393 K and pressures up to 2 MPa) and with less catalyst loss in the organic phase than in any other boosted aqueous phase catalysis. Nevertheless, one of the most critical points for a SAPC process to be economically successful remains the difficulty in extending widely the reusability of the support–catalytic complex matrix. This difficult catalyst recycling has been attributed to the loss of water from cycle to cycle and to the degradability of the hydrosoluble ligands. Useful data on catalyst leaching and SAPC decreasing efficiency after a large number of re-uses are absent in the open literature. As water content control appears less critical with convenient support, the major improvements should come from more stable water-soluble ligands [11, 24, 51] (e.g., not oxidized like TPPTS) and with even better interaction with the hydrophilic support.

Concerning the fields of application, SAPC has prospects of being used either in the basic chemical industry and petrochemistry, or in fine and pharmaceutical chemistry, in biotechnology, and in the food industry; in these last two it may be associated with enzymatic processes. In basic chemistry and petrochemistry the main target is clearly the hydroformylation of higher olefins. In this case, if catalyst losses are economically acceptable, the best way to improve plant productivity would be to use SAPC in continuous fixed-bed reactors such as the usual trickle-bed reactors. Such operation allows both much higher catalyst support loading, up to 60%, and better reaction performance than slurry reactors, due to plug flow. Based on recently reported data on a TOF of about $500 \, h^{-1}$ and, due to a higher water content, a rhodium concentration up to $20 \, \mu mol \, g^{-1} \, SiO_2$, a fixed-bed reactor could produce aldehydes higher than C_6 in very attractive conditions, at up to $0.5 \, t \, h^{-1} \, m^{-3}$.

Nevertheless major possibilities for using SAPC in the short term can be foreseen in fine and pharmaceutical chemistry, e.g., for the selective hydrogenation of α,β-unsaturated aldehydes such as retinal [43] and to the production of the commercially anti-inflammatory agent naproxen [44].

The hydroformylation of α,β-unsaturated esters by SAPC has also been investigated [34]. Several 2-formylpropanoate esters, which are widely used as intermediates in the synthesis of pharmaceuticals like rifamycin and vitamin E, were obtained using the water-soluble complexes. There are huge prospects for SAPC in the hydroformylation reaction for obtaining molecules having a broad-spectrum therapeutic activity.

The synthesis of many pharmaceutical agents and complex molecules from natural sources is strongly dependent on the availability of intermediates suitable for further structural change. In this context, the hydroformylation by SAPC, especially when catalyzed by rhodium carbonyl complexes, which ensure higher chemo- and regioselectivity with respect to other metal derivatives under comparable reaction conditions, offers a concrete possibility of obtaining a wide variety of molecules endowed with therapeutic activity. As an example, the hydroformylations of styrene and related vinylaromatics to yield 2-arylpropanoic acids (anti-inflammatory and analgesic agents) [54] can be accomplished by SAPC. Recently a new process using a SAP Pd catalyst for the preparation of saturated carboxylic acids by alkoxycarbonylation of olefins was patented [52].

Using the techniques developed to synthesize the organometallic-based SAP catalysts, several enzyme-based SAP catalyst, using porous glass beads and the enzymes polyphenol oxidase and horseradish peroxidase have been studied [53]. These SAP catalysts were active in the reaction of phenol with O_2 or H_2O_2, respectively. Thus, porous enzyme-based SAP catalysts can be synthesized.

2.6.6
Conclusion

Supported aqueous-phase catalysis brings a clear opportunity to overcome the two main limitations of biphasic catalysis for liquid substrates: poor activity and/or significant catalyst losses. SAPC provides relatively high turnover frequencies due to a large interfacial area where the metal is available even for highly hydrophobic substrates. In addition, metal losses are reduced by convenient selection of hydrophilic supports and water-soluble ligands, but the outstanding preservation of Rh in gaseous propene hydroformylation on a water-soluble catalyst is far from being matched again.

As recently reported, the efficiency of SAPC over a much wider hydration range will facilitate its use in industry. Two major commercial applications of SAPC are expected: first, many fine chemicals and pharmaceuticals could be produced in better conditions due to high enantioselectivity; and secondly, hydroformylation of liquid olefins in continuous fixed-bed reactors under mild conditions with relatively high productivity and selectivity is possible.

Nevertheless, the question of costly catalyst losses is not yet perfectly clarified and constitutes the main limitation for the development of SAPC-based processes.

References

1 M. E. Davis, in *Aqueous-Phase Organometallic Catalysis* (Eds.: B. Cornils, W. A. Herrmann), Wiley-VCH, Weinheim **1998**, p. 241.
2 B. Cornils, W. A. Herrmann, in *Aqueous-Phase Organometallic Catalysis* (Eds.: B. Cornils, W. A. Herrmann), Wiley-VCH, Weinheim **1998**, p. 585.
3 B. Cornils, *Org. Process Res. Dev.* **1998**, *2(2)*, 121.
4 E. Kuntz, FR 2.314.910 (**1975**).
5 R. V. Chaudhari, B. M. Bhanage, R. M. Deshpande, H. Delmas, *Nature* **1995**, *373*, 501.
6 J. P. Arhancet, M. E. Davis, S. S. Merola, B. E. Hanson, *Nature* **1989**, *399*, 454.
7 R. Z. Moravec, W. T. Schelling, C. F. Oldershaw, GB 511.556 (**1939**).
8 P. R. Rony, *J. Catal.* **1969**, *14*, 142.
9 P. R. Rony, J. F. Roth, *J. Mol. Catal.* **1975/76**, *1*, 13.
10 M. S. Anson, M. P. Leese, L. Tonks, J. M. J. Williams, *J. Chem. Soc., Dalton Trans.* **1998**, *21*, 3529.
11 P. W. N. M. Van Leeuwen, A. J. Sandee, J. N. H. Reek, P. C. J. Kamer, *J. Mol. Catal. A:* **2002**, *182–183*, 107.
12 U. J. Jáuregui-Haza, A. M. Wilhelm, H. Delmas, *Tecnociencia* **2003**, *5(1)*, 23.
13 Wan, K. T., Davis, M. E. *Nature*, *370*, 449, **1994**.
14 J. P. Arhancet, M. E. Davis, S. S. Merola, B. E. Hanson, *J. Catal.* **1990**, *121*, 327.
15 J. P. Arhancet, M. E. Davis, B. E. Hanson, *J. Catal.* **1991**, *129*, 94.
16 J. P. Arhancet, M. E. Davis, S. S. Merola, B. E. Hanson, *J. Catal.* **1991**, *129*, 100.
17 I. T. Horváth, *Catal. Lett.* **1990**, *6*, 43.
18 Y. Yuan, J. Xu, H. Zhang, K. Tsai, *Catal. Lett.* **1994**, *29*, 387.
19 Y. Yuan, Y. Q. Yang, J. Xu, H. Zhang, K. Tsai, *Chin. Chem. Lett.* **1994**, *5(4)*, 291.
20 I. Toth, I. Guo, B. E. Hanson, *J. Mol. Catal.* **1997**, *116*, 217.
21 Ph. Kalck, M. Miquel, M. Dessoudeix, *Catal. Today* **1998**, *42*, 431.
22 Ph. Kalck, M. Dessoudeix, *Coord. Chem. Rev.* **1999**, *190–192*, 1185.

23 M. Dessoudeix, Ph. D. Thesis ENSCT-INP, Toulouse, France **1999**.

24 A. J. Sandee, V. F. Slagt, J. N. H. Reek, P. C. J. Kamer, P. W. N. M. van Leeuwen, *Chem. Commun.* **1999**, 1633.

25 J. Zhang, W. Cao, H. Shi, J. Duan, *Fenzi Cuihua* **1999**, *13(1)*, 35.

26 H. Liu, H. Chen, Y. Li, P. Cheng, X. Li, *Fenzi Cuihua* **1994**, *8(1)*, 22.

27 M. Dessoudeix, U. J. Jáuregui-Haza, M. Heughebaert, A. M. Wilhelm, H. Delmas, A. Lebugle, Ph. Kalck, *Adv. Synth. Catal.* **2002**, *344(3+4)*, 406.

28 U. J. Jáuregui-Haza, M. Dessoudeix, Ph. Kalck, A. M. Wilhelm, H. Delmas, *Catal. Today* **2001**, *66*, 297.

29 U. J. Jáuregui-Haza, A. M. Wilhelm, H. Delmas, *Latin Am. Appl. Res.* **2002**, *32(1)*, 47.

30 U. J. Jáuregui-Haza, E. Pardillo-Fontdevila, Ph. Kalck, A. M. Wilhelm, H. Delmas, *Catal. Today* **2003**, *79–80*, 409.

31 H. Zhu, Y. Ding, H. Yin, L. Yan, J. Xiong, Y. Lu, H. Luo, L. Lin, *Appl. Catal. A:* **2003**, *245*, 111.

32 Y. Yuan, J. Xu, H. Zhang, Y. Yang, Y. Zhang, K. Tsai, *Cat. Today* **2002**, *74*, 5.

33 Z. Li, Q. Peng, Y. Yuan, *Appl. Catal. A:* **2003**, *239*, 79.

34 G. Fremy, Y. Castanet, R. Grzybek, E. Monflier, A. Mortreux, A. M. Trzeciak, J. J. Ziolkowski, *J. Organomet. Chem.* **1995**, *505*, 11; G. Fremy, E. Monflier, J. F. Carpentier, Y. Castanet, A. Mortreux, *Angew. Chem., Int. Ed. Engl.* **1995**, *34(12–13)*, 1474.

35 G. Fremy, E. Monflier, J. F. Carpentier, Y. Castanet, A. Mortreux, *J. Catal.* **1996**, *162(2)*, 339.

36 U. J. Jáuregui-Haza, R. Nikolova, I. Nikov, A. M. Wilhelm, H. Delmas, *Bulgarian Chem. Comm.* **2002**, *34(1)*, 64.

37 M. Benaissa, U. J. Jáuregui-Haza, I. Nikov, A. M. Wilhelm, H. Delmas, *Catal. Today* **2003**, *79–80*, 419.

38 T. Malmstrom, C. Andersson, J. Hjortkjaer, *J. Mol. Catal. A:* **1999**, *139 (2–3)*, 139.

39 A. Riisager, K. M. Ericksen, J. Hjortkjaer, R. Fehrmann, *J. Mol. Catal. A:* **2003**, *193*, 259.

40 T. Bartik, B. Bartik, I. Guo, B. E. Hanson, *J. Organomet. Chem.* **1994**, *480*, 15.

41 I. Guo, B. E. Hanson, I. Toth, M. E. Davis, *J. Organomet. Chem.* **1991**, *403*, 221.

42 Guo, I., Hanson, B. E., Toth, I., Davis, M. E. *J. Mol. Catal.*, *70*, 363, **1991**.

43 E. Fache, C. Mercier, N. Pagnier, B. Despeyroux, P. Panster, *J. Mol. Catal.* **1993**, *79*, 117.

44 K. T. Wan, M. E. Davis, *J. Catal.* **1994**, *148*, 1.

45 J. P. Arhancet, M. E. Davis, B. E. Hanson, *Catal. Lett.* **1991**, *11*, 129.

46 S. C. Tsang, N. Zhang, L. Fellas, A. M. Steele, *Catal. Today* **2000**, *61*, 29.

47 P. Schneider, F. Quignard, A. Choplin, D. Sinou, *New J. Chem.* **1996**, *20*, 545.

48 S. Dos Santos, Y. Tong, F. Quignard, A. Choplin, D. Sinou, J. P. Dutasta, *Organometallics* **1998**, *17(1)*, 78; S. Dos Santos, F. Quignard, D. Sinou, A. Choplin, *Stud. Surf. Sci. Catal.* **1998**, *118*, 91.

49 A. Choplin, S. Dos Santos, F. Quignard, S. Sigismondi, D. Sinou, *Catal. Today* **1998**, *42*, 471.

50 S. Westernacher, H. Kisch, *Monatsh. Chem.* **1996**, *127(5)*, 469.

51 K. T. Wan, M. E. Davis, US 5.827.794, 5.756.838 (**1998**).

52 J. Seayed, A. M. Seayed, B. R. Sarkar, R. V. Chaudhari, US 6.479.693 (**2002**).

53 M. E. Davis, *CHEMTECH* **1992**, *22*, 498.

54 F. Quignard, A. Choplin, A. Domard, *Langmuir* **2000**, *16(24)*, 9106.

55 U. J. Jáuregui-Haza, Ph. D. Thesis, La Habana, Cuba **2002**.

56 C. Botteghi, M. Marchetti, S. Paganelli, in *Transition Met. Org. Synth.* (Eds.: M. Beller, C. Bolm), Wiley-VCH, Weinheim **1998**, Vol. 1, p. 25.

2.7
The Way Ahead: What Should be Done in the Future?

Boy Cornils and Wolfgang A. Herrmann

As a result of the assessment of all the known techniques regarding multiphase homogeneous catalyses, the aqueous-phase operation is clearly ahead of the other possibilities (cf. Chapter 8). This is mainly due to the attractiveness of the principle as well as to the state-of-the-art attained: the process makes it possible to utilize fully the inherent advantages of homogeneous catalysis. A major aspect of this technology is the tailoring of the organometallic complexes as catalysts (which is nowadays becoming increasingly important for industrial applications) for new reactions, and for new products. The avoidance of selectivity- and yield-reducing operations (such as thermal stress caused by chemical catalyst removal or distillation) to separate products and catalysts makes it possible to use sensitive reactants and/ or to obtain sensitive reaction products from homogeneous catalysis. Figure 2 of Section 2.1 demonstrates the advantages of the aqueous-phase technique over the conventional ones, taking the hydroformylation reaction as the prominent example: the aqueous-phase methodology avoids extensive technical equipment except for the column just behind the reactor – a tremendous cost-saving effect.

It is true that only those technologies that use the separation of products and catalysts with organic/organic or organic/water solvent couples are operative to a larger extent. Beyond that, the aqueous biphasic variety is the only one with a convincing "green" touch – more satisfying than just reading about an "even greener ionic liquid" [2], "toward greener chemistry" [1], the "greening" of reactions [3], "green chemistry gets greener" [4], or "clean solutions for chemical synthesis" [5]. On the other hand there are reasonable doubts about "ionic liquids are not always green" [6] and "you think your process is green – how do you know?" [7a]. And things can hardly get any worse: there are serious publications about the "noninnocent nature" of various ionic liquids and "undesirable transformations" yielding corrosive and/or toxic degradation products [7b]. This all has to be seen a realistic background of, on the one hand, the Ruhrchemie/Rhône-Poulenc process being the only realization of a biphasic process which provides data about its "Life Cycle Assessment" (LCA), "Failure Mode, Effects, and Criticality Analysis" (FMECA), "Fault Tree Analysis" (FA), and "Maintainability Analysis" (MA) [8]. On the other hand it is true that the RCH/RP process has been developed with the aim of better catalyst activity, handling, selectivity, and efficiency, improved process economics, reduction of by-products, etc.; i.e., it takes as a basis catalyst design, process intensification, and the principles of sustainable development – but it does not work under the label "green"!

And it is obvious that only the aqueous biphase technology has a far-ranging claim for wider applicability as far as substrates and the tuning of appropriate catalyst systems (consisting of central atoms *and* ligands) are concerned. Yet it is difficult to believe that the very special Shell approach of an organic/organic separation will be extended to other syntheses.

Multiphase Homogeneous Catalysis
Edited by Boy Cornils and Wolfgang A. Herrmann et al.
Copyright © 2005 Wiley-VCH Verlag GmbH & Co. KGaA, Weinheim
ISBN: 3-527-30721-4

Aqueous, two-phase catalysis is used industrially in a number of processes for the production of bulk-scale chemicals such as *n*-butyraldehyde (in amounts up to 800 000 tpy; in total since the start of the new technology, 5 MM tons) to a series of fine chemicals such as precursor compounds of fragrances or vitamins. Here, the possibility of the spontaneous separation of the aqueous catalyst phase complies in a fortunate way with the usual mode of operation of fine chemicals manufacture: the work in discontinuously operated stirred-tank reactors, the decanting of the reaction mixture from the first run, and the use of the catalyst phase for a subsequent run. Taking a broad view, solution of chemical engineering questions will be a task for the future: from basic problems (where does the reaction between the catalyst, in aqueous solution, and the substrates and reaction products, based in the organic phase, take place?) through to specific questions concerning various aspects of the technology, such as reaction control, alternative concepts of catalyst recycling and work-up, reactor design, use of mixtures of solvents with mixture gaps, combinations with ultrasonic and microwave reactors, and so on. The experiences gained so far with the inclusion of membrane techniques in aqueous-phase processes may give the decisive impetus for their further development which will be necessary for the use of other multiphase operations (e.g., polymer-supported homogeneous catalysts) which will be described in Chapter 7.

Quite another part of work will be the extension of the scope of the reactants. Due to the decrease solubilities of the starting alkenes and products in water with an increasing number of carbon atoms, the bulk-scale technology (of hydroformylation) has so far been restricted to the use of alkenes from propene up to pentenes. The further development of ligands will be one of the fields of activity. For today's exotic ligands such as carbohydrates and dextrins, proteins, porphyrins, paracyclophanes, calix[*n*]arenes, supramolecular compounds (dendrimers, etc.), or templates, once more the use of membrane separation techniques has to be evaluated. Temperature-regulated or thermomorphic variants have to be checked as well as the work on microemulsions or micellar systems. In any case, a look at the bulk-scale operation and its cost situation prohibits most of the proposals discussed. The same is true for advanced or even highly sophisticated ligand developments: it is known that so far none of the so-called "second generation ligands" has paid off its costs. On the whole, on closer inspection, any additional effort increases the costs – it may be an additional process step, an extra or expensive auxiliary (such as co-solvents, co-catalysts, counterions, surfactants, promoter ligands, or other additives), bimetallic or bifunctional catalyst systems, or supplementary amounts of reactant to move or to separate. In the handbook *Aqueous-Phase Organometallic Catalysis* [9], all these lines of development are discussed in detail.

In terms of "green syntheses", the community expects ongoing intensive research work with aqueous biphasic operations. One of the reasons is that water is the only "green solvent" being used industrially. Aspiring users of the others, such as supercritical carbon dioxide or ionic liquids, suffer from lack of experience, and with *all* the other biphasic variants they have to cope with the problem that the demanding catalyst systems need to be tolerated by the second phase – mostly as far as the extremely complicated and costly ligands are concerned.

Additionally, various papers dealing with "effective" (but extremely costly) perfluoro compounds as ligands or with liquids which are hampered by insufficient stability ("which are not always green" – in fact the contrary of a green solvent!) read like fairy tales. They ought to be regarded with scepticism.

The field is in flux. As a consequence of the increases scientific study of aqueous biphasic homogeneous catalysis, an increasing number of commercial applications may be expected in the future. Here one should mention one-pot and tandem reactions which combine the aqueous-phase hydroformylation with subsequent steps such as cyclotrimerization, Michael or Mannich additions, or aldolizations. Another example might be catalyzed reactions in water which use the extremely high enthalpy of the O–H bond. This value (436 kcal mol^{-1}) predestines water as an excellent solvent for carrying out free–radical reactions because in order for a radical to abstract a hydrogen atom from water, the bond enthalpy of the bond formed between a new heteroatom and hydrogen must be greater than that of the O–H bond. Several other processes are the subject of detailed surveys and reviews. Further methodological progress will comprise both the improvement of the technology and extension of the respective reaction to other applications.

The investigation of "hybrid" processes comprising the combination of the aqueous with other biphasic options such as those described in Chapters 3–7 (preferably for hydroformylation reactions) may be included – but will enhance the overall costs. The same is true for other hybrid applications such as SAPC (supported aqueous-phase catalysis) or SOMC (surface organometallic chemistry). The serious drawback of these special variants originates from the notorious "leaching" of the catalyst, concerning the (usually expensive) transition metal as central atom of the catalyst complex, the ligand (which, as a tailor-made compound to meet special demands, may be even more expensive), or both. By following this, no real progress can be expected. Ligands which imply phase-separating properties, such as the "thermoregulating ligands" developed by Jin t al. and as described by Ishihara and Yamamoto in Chapter 4, Section 4.2.6, may be worth developing. PEG has some disadvantages, but the basic idea of bringing in different behavior of the operative phases as part of the properties of the ligand is tempting! Only this (or a similar development) will extend the horizons of the aqueous biphasic technology to higher boiling and less soluble substrates. But once more: in this respect, the whole (commercially inclined) community agrees that additional additives would be too expensive to handle, to remove, and to recycle (not to mention their costs).

From the standpoint of solvents and their environmental friendliness and nontoxicity one can image a hybrid process comprising an "inverted aqueous biphasic catalysis" where the catalyst is immobilized in the organic phase and the polar substrates and products reside in the aqueous phase [10] – but there is no indication for serious industrial research work.

Further methodological progress will comprise both the improvement of the technology and the extension of the respective reactions to other applications. Reaction engineering may be improved by other and new reactor concepts, as demonstrated by the Oxeno researchers [11]. Also, a new understanding of the micro-events is demanded with regard to phase boundaries (or the interfacial processes) and their reciprocal solubility and dissolution of gases, fractional hold-ups of different phases (holdup is the volume fraction occupied by the flowing phase besides the catalyst phase), mass and heat transfer, overall and intrinsic reaction rates, and the coupled influence of mass transfer with chemical reaction. This understanding will justify the preceding research.

References

1 P. Wasserscheid, R. van Hal, A. Bösmann, *Green Chem.* **2002**, *4*, 400.

2 (a) R. T. Baker, W. Tumas, *Science* **1999**, *284*, 1477; (b) S. K. Ritter, *Chem. Eng. News* **2003**, Oct. 13, 66.

3 L. V. R. Bonaga, J. A. Wright, M. E. Krafft, *Chem. Commun.* **2004**, 1746.

4 S. K. Ritter, *Chem. Eng. News* **2002**, May 20, 38.

5 C. A. Eckert, C. Liotta, J. Brown, *Chem. Ind.* **2000**, 7 Feb., 94.

6 R. P. Swatloski, J. D. Holbrey, R. D. Rogers, *Green Chem.* **2003**, *5*, 361.

7 (a) A. D. Curzons, D. J. C. Constable, D. N. Mortimer, V. L. Cunningham, *Green Chem.* **2001**, *3(1)*, 1; (b) J. Dupont, J. Spencer, *Angew. Chem., Int. Ed.* **2004**, *43*, 5296.

8 (a) M. Todinov, *Reliability and Risk Analysis Models*, Wiley, New York **2005**; (b) I. Sutton, *Process Hazard Analysis*, SW Books, Houston/TX **2001**; (c) D. J. Paustenbach (Ed.), *Human and Ecological Risk Assessment*, Wiley, New York **2002**.

9 B. Cornils, W. A. Herrmann (Eds.), *Aqueous-Phase Organometallic Catalysis*, 2nd ed., Wiley-VCH, Weinheim **2004**.

10 M. McCarthy, H. Stemmer, W. Leitner, *Green Chem.* **2002**, *4*, 501.

11 (a) G. Protzmann, K.-D. Wiese, *Erdöl, Erdgas, Kohle* **2001**, *117*, 235; (b) K.-D. Wiese, O. Möller, G. Protzmann, M. Trocha, *Catal. Today* **2003**, *79–80*, 97.

3
Organic–Organic Biphasic Catalysis

Dieter Vogt (Ed.)

Multiphase Homogeneous Catalysis
Edited by Boy Cornils and Wolfgang A. Herrmann et al.
Copyright © 2005 Wiley-VCH Verlag GmbH & Co. KGaA, Weinheim
ISBN: 3-527-30721-4

3.1
Introduction

Dieter Vogt

Many attempts have been made to combine the advantageous features of homogeneous and heterogeneous catalysis. In general, these methods can be combined in two categories as heterogenization and immobilization of molecular catalysts. These involve the deposition of metal complexes on solid carriers or anchoring them onto the outer or inner surface of the support. The heterogenized metal complex catalysts so far have been found unacceptable for large-scale applications because of poor performance or stability. A more successful technique of catalyst immobilization is the separation of the reactants (products) and the catalyst(s) by immiscible liquid phases. This is the concept of biphasic catalysis of which several variants have been developed during recent years (see Chapter 1).

Biphasic systems are built up of (at least) two solvents with limited mutual solubility. Depending on which solvents are chosen, biphasic systems can be divided into six categories:

- In *aqueous–organic* biphasic systems one of the solvents is water and the other is a solvent immiscible with water. The substrate/product mixture may also constitute the organic phase with no added solvent.
- In *organic–organic* biphasic systems both solvents are organic compounds with limited mutual solubility.
- In *fluorous* biphasic systems one of the solvents is a highly fluorinated organic compound and the other the usual organic solvent.
- Using *ionic* liquid biphasic systems means that melts of specially designed, low melting salts are used as one of the liquid phases in combination with an aqueous phase or an immiscible organic solvent.
- Solvents under *supercritical* conditions use the peculiar properties of the super-critical state and its phase behavior.
- Last but not least, catalysis with polymer-bound, water-soluble ligands has been recommended.

All six variations have been introduced in Chapter 1 and the following sections.

In this chapter only the second category will be discussed: organic–organic biphasic systems. However, the distinction between the first two categories is not always clear. In some cases water is added to an organic–organic biphasic system to improve the performance of the system. In that case, depending on the reaction and the amount of water added, the system will be discussed as well. There are also transitions between ionic liquids and water or fluorous solvents, between water and $scCO_2$, or between water-soluble polymers and organic solvents.

Besides the above-mentioned classification into six categories, depending on the solvents chosen, biphasic systems have been classified into up to 12 categories, depending on how and at which moment in the process product(s) and catalyst(s)

Multiphase Homogeneous Catalysis
Edited by Boy Cornils and Wolfgang A. Herrmann et al.
Copyright © 2005 Wiley-VCH Verlag GmbH & Co. KGaA, Weinheim
ISBN: 3-527-30721-4

are being separated [1, 2]. Finally, the effect of temperature on solvent mixtures has to be mentioned. Even the simplest biphasic catalytic system contains at least five components: two solvents, a reactant, a product, and the catalyst. It is therefore not a big surprise that the mutual solubility may vary in a complex way with temperature. Certain biphasic (or multiphasic) systems show a sharp phase transition into a homogeneous phase when the temperature is raised. If the resulting homogeneous liquid phase dissolves both the substrate(s) and the catalyst, then a genuine homogeneous catalytic reaction takes place (without limitation of mass transfer at the liquid–liquid phase boundary). After the reaction is completed, the temperature is lowered and two immiscible liquid phases are obtained again, leading to easy separation of catalyst and product(s). Section 3 of this chapter will be devoted to examples of such systems; another proposal is described in Chapter 2.3.5.

It is noteworthy that the only example of a large-scale industrial process based on organic–organic two-phase catalysis so far, the Shell higher olefin process, SHOP, was at the same time the first example of a two-phase homogeneous catalytic process applied in industry [1, 3]. Then, after the introduction of the successful example of aqueous–organic biphasic catalysis by Kuntz [4, 5] and the development of the Ruhrchemie/Rhône-Poulenc process [6], the organic–organic systems lost interest. It is only very recently that academic groups became interested again in this concept. One reason might be the fact that many more empirical data and much more knowledge are required for a thorough description of such a process. The description of the phase behavior with the corresponding phase diagrams considering all the components of the system is a particular prerequisite. On the other hand, the reaction engineering aspects of liquid–liquid reactions seem to have been well examined [7–12].

3.2
State-of-the-Art and Typical Reactions

3.2.1
Organic–Organic Biphasic Catalysis on a Laboratory Scale

Dieter Vogt

3.2.1.1
Hydrogenation

The hydrogenation of different double bonds has often been performed in biphasic systems. Hydrogenations in the homogeneous phase has often been performed with the Wilkinson [(Ph$_3$P)$_3$RhCl] catalyst. By modification of the ligands a catalyst suitable for biphasic catalysis can be obtained. Da Rosa et al. [13] described a biphasic system containing poly(ethylene oxide) (PEO), *n*-heptane and either dichloromethane (CH$_2$Cl$_2$) or methanol. This liquid biphasic system was tested in the hydrogenation of 1-hexene. As catalyst, Wilkinson's catalyst was used, as well as the cationic rhodium complex [Rh(cod)(dppe)]PF$_6$. The catalyst was used in a ternary mixture of PEO, *n*-heptane, and CH$_2$Cl$_2$. This polymer solution shows phase separation of the UCST (upper critical solution temperature) type. For PEO solutions in CH$_2$Cl$_2$ the phase separation temperature ranges from –80 °C to –40 °C for PEO 3350 concentrations between 1 and 60% (w/w). Therefore, by cooling the reaction system with N$_2$, it was possible to selectively separate the catalyst and the reaction products. Owing to large density differences, this biphasic system was kinetically stable for a longer time, even close to room temperature, allowing easy separation of the liquid phases.

This procedure resulted in an efficient and selective substrate conversion and it was established by spectrophotometry that there was no catalyst leaching to the apolar phase. However, a marked decrease in the catalytic activity was observed after the third cycle. This was probably caused by a continuous loss of free triphenylphosphine ligand present in equilibrium with the rhodium complex, ultimately generating inactive species. To overcome this problem the cationic rhodium complex [Rh(cod)(dppe)]PF$_6$ was tested. In the ternary mixture containing CH$_2$Cl$_2$ this complex showed poor catalytic activity. Using methanol instead clearly increased the activity. This effect of increased activity in methanol is well known for rhodium complexes.

The cationic rhodium complex was used in a homogeneous system of methanol, PEO, and 1-hexene, resulting in a conversion of 54 ± 5% and a selectivity of 90 ± 10% after 2.5 h. Under biphasic conditions, i.e., a system composed of 14 mL methanol, 3.6 g PEO, and 14 mL *n*-heptane, 0.0335 mmol of rhodium complex, 8.01 mmol of 1-hexene, and a hydrogen pressure of 1 MPa, the cationic rhodium complex was tested as well: five runs were conducted with complete conversion of 1-hexene to *n*-hexane and without loss of catalytic activity or selectivity. However, considering

Multiphase Homogeneous Catalysis
Edited by Boy Cornils and Wolfgang A. Herrmann et al.
Copyright © 2005 Wiley-VCH Verlag GmbH & Co. KGaA, Weinheim
ISBN: 3-527-30721-4

the fact that the solvent system has to be cooled to a very low temperature for separation, the energy costs of such a system will prevent a larger-scale application.

Other ligands for the biphasic hydrogenation (and hydroformylation) were investigated by Mieczyńska et al. [14]. The rhodium complexes were formed from the precursor [Rh(acac)(CO)$_2$] and the (water-soluble) phosphine ligand. The phosphine ligands PH and PNS, **1** and **2**, were tested in the hydrogenation reactions of 1-hexene, toluene, *o*-xylene, and cyclohexene. However, only 1-hexene showed considerable conversion by each rhodium complex in a 1-hexene/water mixture. The yield of *n*-hexane for the ligands PNS and PH was determined to be 67% and 42% respectively after 4 h. For the system [Rh(acac)(CO)$_2$/PNS] different co-solvents were added to water in order to improve the yield of hydrogenation. The solvent systems thus investigated are given in Table 1. The resulting biphasic system was composed of 1.5 mL 1-hexene, 1.0 mL Water, and 0.5 mL co-solvent at a rhodium concentration of $5 \cdot 10^{-3}$ M, a hydrogen pressure of 1 MPa, and a temperature of 80 °C. The reaction time varied between 1.5 and 4 h for the different co-solvents. Only for the co-solvents ethanol (83%), diglyme (80%), tetraglyme (73%), and ethylene glycol (75%) was an increase in the yield of *n*-hexane (after 3 h) determined. However, in case of the ethers more isomerization reaction also took place (2-hexene). It was therefore concluded that a mixture of water–ethanol is the best choice for the hydrogenation of 1-hexene.

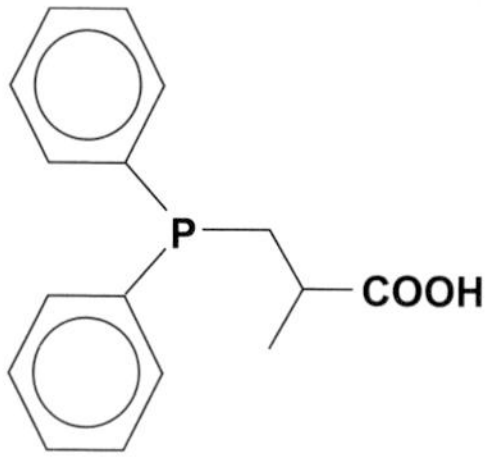

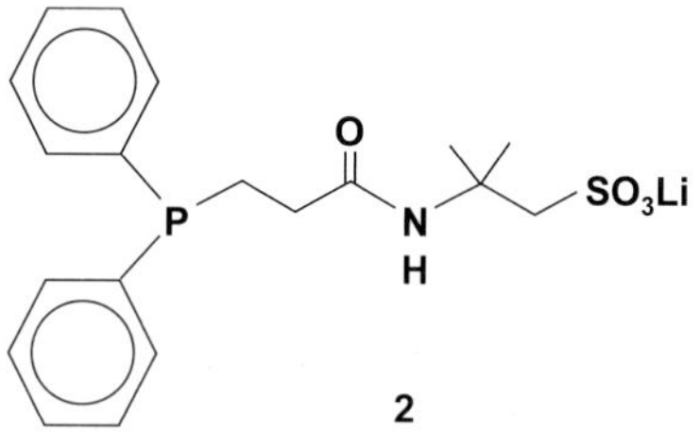

Bianchini et al. [15] reported a sulfonated ligand for use in hydrogenation reactions. They synthesized the complex [Rh(sulphos)(cod)] (where sulphos is structure **3**), which was tested in the biphasic hydrogenation of styrene. Styrene was dissolved in a 1 : 1 (v/v) mixture of methanol and *n*-heptane in the presence of

Table 1 Solvent systems tested according to Mieczyńska et al. [14].

	Solvent system	Ratio [v/v/v]
1	Water–ethanol/1-hexene	2 : 1 : 3
2	Water–diglyme/1-hexene	2 : 1 : 3
3	Water–tetraglyme/1-hexene	2 : 1 : 3
4	Water–ethylene glycol/1-hexene	2 : 1 : 3
5	Water–poly(ethylene glycol)/1-hexene	2 : 1 : 3
6	Water–THF/1-hexene	2 : 1 : 3
7	Water–decamethylene glycol/1-hexene	2 : 1 : 3
8	Water–crown ether/1-hexene	2 : 1 : 3

[Rh(sulphos)(cod)] (1 : 500 catalyst/substrate ratio). The hydrogen pressure was 3 MPa, and the temperature was 65 °C. After 3 h a conversion of more than 90% was obtained and complete disappearance of styrene occurred after a further 2 h of reaction. After cooling the reaction mixture to room temperature, separation of the two phases did not give complete organic product separation, whereas the rhodium catalyst remained in the alcohol phase. Simple addition of water resulted in complete elimination of ethylbenzene (and residual styrene) from the alcohol phase. Remarkable, the catalytic activity of [Rh(sulphos(cod)] in the biphasic system is the same as in pure methanol, indicating that at the reaction temperature a homogeneous phase is formed.

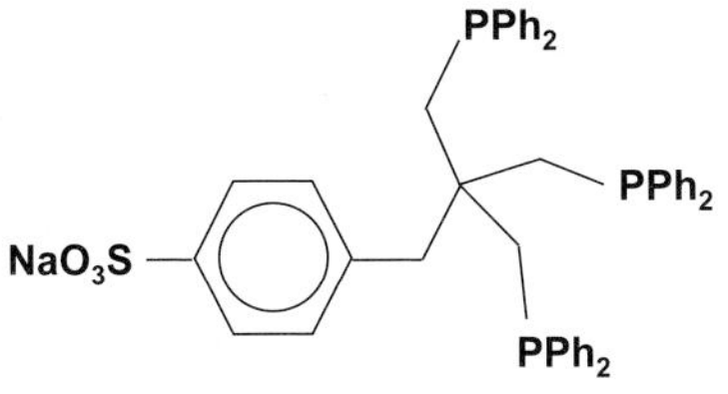

3

These examples demonstrate that research has not passed the proof-of-principle state so far. Future work on biphasic (organic–organic) hydrogenation has to address more realistic substrates. The conditions have to be defined under which biphasic hydrogenation could compete with homogeneous systems.

3.2.1.2
Hydroformylation

There are only a very few reports dealing with the organic–organic biphasic hydroformylation, although aqueous–organic biphasic hydroformylation is harder for water-insoluble (i.e., higher) olefins.

One suitable polar phase for hydroformylation is poly(ethylene glycol) (PEG), in which higher olefins are somewhat more soluble than in water. Ritter et al. [16] described the hydroformylation of 1-hexene catalyzed by two PEG-substituted cobalt clusters: [tris(tricarbonylcobalto)methylidyne]silanetriol and a derivative of it (4 and 5). The reaction was performed in toluene in the case of catalyst 4 or in PEG for catalyst 5. An equal quantity of 1-hexene with respect to PEG was added to the mixture. The autoclave was pressurized with CO/H_2 (1 : 1) to 7 MPa, heated to 120 °C and stirred for 8 h. After the reaction, phase separation took place, which could be promoted by the addition of pentane. Both catalysts resulted in high yields (> 90% based on 1-hexene), but the reaction produced a rather low n/i ratio of 0.75. As usual, the regioselectivity could be enhanced by the addition of phosphines. For each catalyst no formation of alcohols occurred, which makes these catalysts interesting for further investigation, since formation of alcohols is characteristic of cobalt-catalyzed hydroformylation. Although the catalyst could be separated from the product mixture, the recycled catalyst showed a decrease in activity and selectivity.

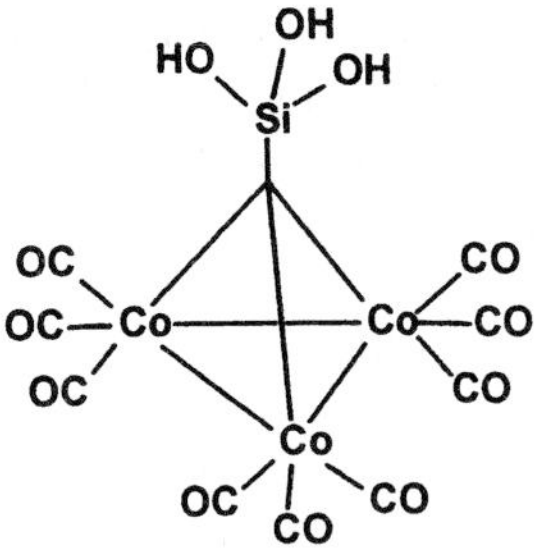

4 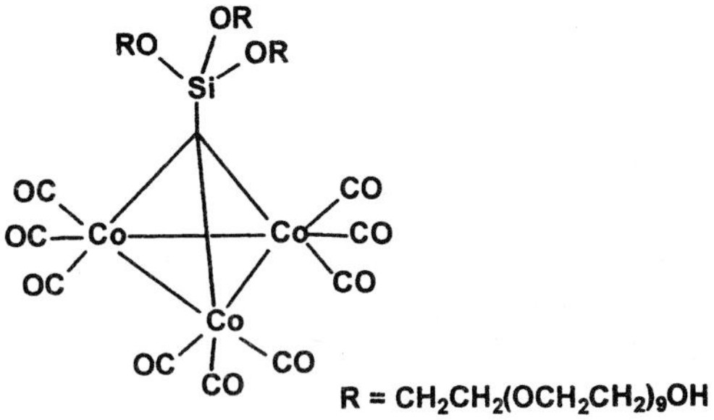   **5**

R = CH$_2$CH$_2$(OCH$_2$CH$_2$)$_9$OH

Bianchini et al. [15] described the use of the rhodium catalyst [Rh(sulphos(cod)] (3) for the hydroformylation of 1-hexene. The catalyst was dissolved in a 1 : 1 (v/v) mixture of methanol–isooctane to which 1-hexene was added (1 : 100 catalyst/ substrate ratio). The CO pressure and the H$_2$ pressure were both 1.5 MPa and the temperature was set to 80 °C. At this temperature the biphasic system had become a homogeneous phase. Cooling the system to room temperature resulted in phase separation. Water was added as a co-solvent, resulting in a 1 : 1 : 1 (v/v/v) mixture of water–methanol/isooctane. This hydroformylation in the absence of water gave aldehydes (heptanal, 2-methylhexanal, and 2-ethylpentanal) and alcohols (1-hepta- nol, 2-methylhexanol, and 2-ethylpentanol) in an overall ratio of only 22 : 78. The total conversion of 1-hexene into aldehydes and alcohols was 76%. If the reaction time was prolonged, eventually only alcohols were formed. The formation of alcohols is rather remarkable, since rhodium catalysts normally produce aldehydes only.

In the presence of water during hydroformylation only heptanal and 2-methyl- hexanal in an overall yield of 54% (based on the aldehyde) were formed. Only traces of the aldehyde remained in the water–methanol phase after cooling to room temperature. The rhodium complex was completely recovered in the water– methanol phase, as was established by ^{31}P NMR spectroscopy and atomic absorption spectrophotometry.

Mieczyńska et al. [14] used the rhodium complexes [HRh(CO)(L)$_3$] of ligands PNa (6) and PNS (2) as catalysts for the hydroformylation of 1-hexene. In the case of the PNa ligand, hydroformylation was performed in a water–ethanol/toluene (7 : 5 : 5, v/v/v) solvent system. The catalyst/substrate ratio was 1 : 800 at a pressure of 2 MPa CO/H$_2$ (1 : 1) and a temperature of 80 °C. The yield of aldehydes after

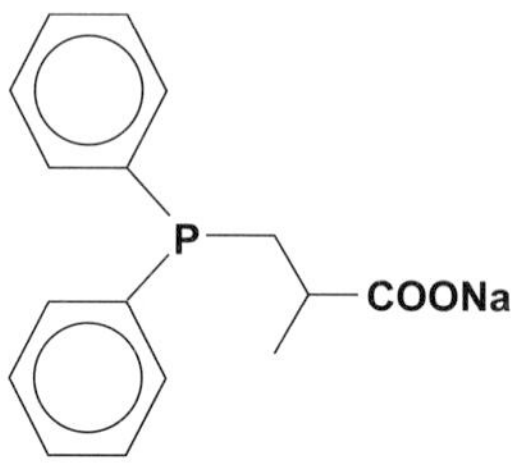

6

2.5 h was 94%, with an n/i ratio of 2.4. In comparison to the homogeneous hydroformylation in toluene the yield increased considerably (from 26% to 94%) whereas the n/i ratio did not change (2.4 in both cases). For the PNS ligand, different water–ethanol/toluene mixtures and water–ethanol/1-hexene mixtures were examined.

For the first solvent mixture it was found that the total yield of aldehydes increased more than proportionally with the amount of ethanol in the polar phase. For a ratio of 2 : 11 : 2 (v/v/v) of water/ethanol/toluene the total yield of aldehydes of 86% was obtained with an n/i ratio of 2.9. For the system water–ethanol/1-hexene similar results were obtained. For a ratio of 11 : 15 : 4 (v/v/v) the total yield of aldehydes was 93% with an n/i ratio of 3.0. Replacement of ethanol by methanol or isopropanol gave almost identical results. However, when pure ethanol was used as the polar phase a total yield of only 77% was obtained. This might be caused by the lower solubility of the catalyst in ethanol, compared with that in water–ethanol mixtures.

The stability of the $[HRh(CO)(PNS)_3]$ catalyst was investigated as well and it was found that during the nine following catalytic cycles the catalytic activity demonstrated by the yield of aldehydes remained practically constant.

The results discussed above show that hydroformylation in organic–organic solvent systems is basically possible. However, much more detailed studies are needed to determine whether it is economically feasible. Aspects such as catalyst deactivation during recycling, reaction time, (undesired) formation of alcohols, and continuous operation should be considered.

3.2.1.3
Hydrogenolysis

Hydroprocessing of fossil fuel feedstocks for the removal of sulfur, nitrogen, and residual metals is a large-scale refinery process. Sulfur in fossil materials is found in various compounds, such as thiols, sulfides, disulfides, thiophenes, benzo-thiophenes, and dibenzothiophenes. The removal of sulfur from these fossil materials is commonly referred to as hydrodesulfurization (HDS), which is important in order to reduce the amount of sulfur introduced into the atmosphere since this contributes to acid rain. Secondly, the poisoning of catalysts in down-stream processing has to be reduced.

HDS catalysts generally consist of (heterogeneous) Mo or W sulfides on alumina supports. However, Bianchini et al. described a two-step procedure for HDS of thiophenes by the hydrogenolysis of thiols, followed by the desulfurization of the thiols by applying their zwitterionic rhodium(I) complex, [Rh(sulphos((cod)] (see previous section) [17]. This complex is soluble in polar solvents, such as methanol and methanol–water mixtures, but not in hydrocarbons. Benzo[*b*]thiophene was chosen as substrate since it is one of the most difficult thiophene derivatives to degrade. Under the mild reaction conditions of the two-step process, the benzene rings of the (di)benzothiophenes were not affected. In the absence of a base, the double bond of benzo[*b*]thiophene was hydrogenated, while in the presence of a base (NaOH) 2-ethylthiophenolate was the major product (Scheme 1).

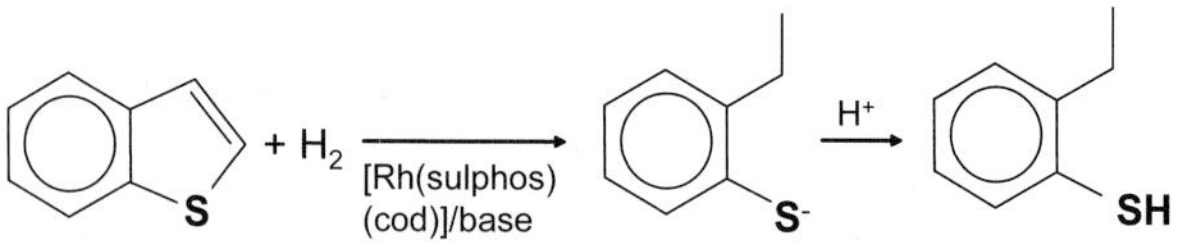

Scheme 1 Hydrogenolysis of benzo[b]thiophene.

Hydrogenolysis of benzo[b]thiophene was performed in 20 mL of solvent mixture (methanol/n-heptane or methanol–water/n-heptane) with 180 mg of NaOH present. The catalyst/substrate ratio was 1 : 100, at an H_2 pressure of 3 MPa, and 160 °C for 5 h. After the reaction, hydrochloric acid was added to obtain 2-ethylthiophenol. Both biphasic systems gave good yields of 2-ethylthiophenol (95% for methanol/n-heptane and 89% for methanol–water/n-heptane). These results were similar to those obtained in the monophasic systems of methanol (93%) and methanol–water (84%). In both biphasic systems, all the 2-ethylthiophenol is found in the polar phase as sodium 2-ethylthiophenolate, leaving the hydrocarbon phase almost completely desulfurized.

The hydrogenolysis to *thiols* can be carried out effectively in a biphasic system, with the catalyst exclusively soluble in the polar phase, thus enabling easy catalyst recycling. However, to introduce this biphasic technique to industrial hydrodesulfurization, much research has to be carried out to design catalysts that are suitable for biphasic catalysis and that contain inexpensive metals (cobalt, ruthenium). Furthermore the catalysts have to tolerate the great thermal and chemical stress of the reaction conditions.

3.2.1.4
Hydrosilylation

One of the most important methods to synthesize organic silicon compounds is the addition of hydrosilanes to double bonds. Production of plasticizers, adhesives, and cosmetic formulation compounds are examples of industrial hydrosilylation products. The most widely used catalyst is hexachloroplatinic acid, H_2PtCl_6 (Speier's catalyst). The method described by Behr et al. [18–20] uses a biphasic liquid–liquid system under mild conditions at ambient pressure and short reaction times.

Hydrosilylation of methyl undec-10-enoate with various silanes has been carried out on a laboratory scale [Eq. (1)]. In a typical experiment 10.0 mmol of methyl undec-10-enoate and 10.0 mmol of the silane were dissolved in 8 mL of cyclohexane,

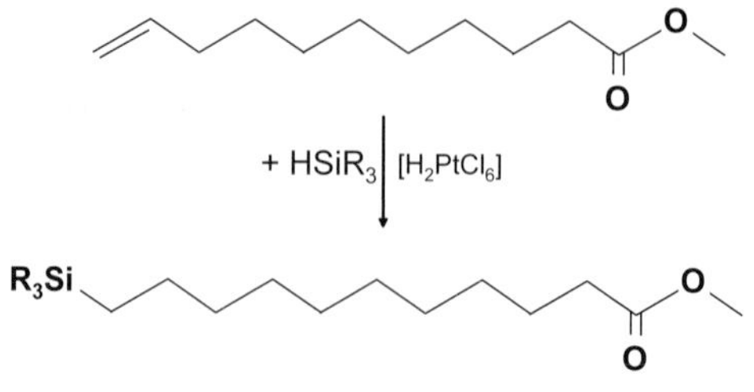

(1)

the apolar solvent. The catalyst H_2PtCl_6 (0.10 mmol) was dissolved in 8 mL of propylene carbonate. The solutions were warmed to 40 °C and combined under intensive stirring. Phase separation was allowed to take place after 2 h. The product was isolated by distillation and analyzed by NMR spectroscopy. The catalyst was recycled without any purification.

Triethoxysilane gave the best results with an excellent selectivity (> 99%) and a good reactivity (77% conversion of fatty acid ester). Alkyl- and arylsilanes gave poor yields, while chlorosilanes attained moderate yields and selectivity. The catalyst activity and selectivity was maintained during five recycle runs. Based on these results thermomorphic solvent systems were developed, which will be described in Section 3.2.3.

In the future, other solvent systems than cyclohexane/propylene carbonate should be investigated to optimize this biphasic hydrosilylation.

3.2.1.5
Oxidation

The direct oxidation of benzene to phenol is of great interest. The selectivity of this reaction is often poor, due to the higher reactivity of phenol towards further oxidation. As a result, over-oxidized byproducts usually occur, such as catechol, hydroquinone, benzoquinone, and tars. Bianchi et al. described a biphasic system using hydrogen peroxide as the oxidant for the selective oxidation of benzene to phenol. $FeSO_4$ was used in the absence of additional ligands, with trifluoroacetic acid as a co-catalyst [21]. The effect of different organic solvents on the selectivity was investigated using water–co-solvent/benzene (45 : 45 : 10, v/v/v) as the reaction medium. The results for different co-solvents are given in Table 2.

Acetonitrile gave the best selectivity, nearly twice as high as for water (21% versus 38% based on H_2O_2) and less over-oxidation to byproducts (2% versus 1% based on H_2O_2). The reason for the increased selectivity is a better solubility of benzene in the resulting polar phase, and phenol is largely extracted into the organic phase. Reducing the contact between phenol, which is soluble in the polar phase, and the catalyst, the biphasic system minimizes over-oxidation.

Table 2 Effect of the solvent system on the oxidation of benzene.

Co-solvent	Fraction of H_2O_2 oxidizing benzene to phenol [%]	Fraction of H_2O_2 oxidizing benzene to double-oxygenated products [%]
H_2O	21	2
MeCN	38	1
EtCN	6	1
tert-BuOH	17	1
Acetone	10	1
Dioxane	7	20
DMF	1	33
n-Octane	3	1

A series of bidentate N⌢N, N⌢O, and O⌢O ligands for iron were tested. 5-Carboxy-2-methylpyrazine-*N*-oxide (**7**) proved to give the most selective catalyst (78% based on H_2O_2).

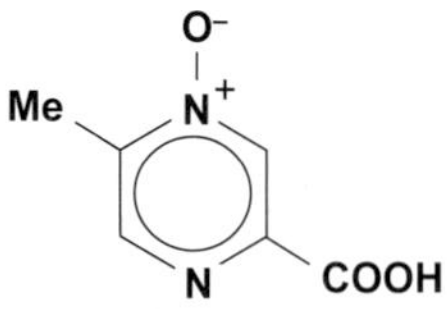

7

The influence of different acid co-catalysts was pronounced. Trifluoroacetic acid gave the best results, increasing the selectivity from 68% in the absence of co-catalyst to 78% based on H_2O_2. Under optimized reaction conditions at a benzene conversion of 8.6%, a selectivity of 97% (based on benzene) and 88% (based on H_2O_2) was obtained. These results are superior to any other iron-based catalyst system reported so far. The recycling of the catalyst in the biphasic system is also easier than in a homogeneous system.

Koek et al. described the biphasic epoxidation of styrene and 1-dodecene using the complex $[Mn^{IV}_2(\mu\text{-}O)_3(L_n)_2]^{2+}$ (**8**). These manganese complexes of cyclo-1,4,7-triazanonane ligands were used with hydrogen peroxide as the oxidant [22]. The catalyst was dissolved in 1.0 mL of dichloromethane, then 1.0 mL of the substrate solution, and 1.0 mL of a 30% aqueous hydrogen peroxide solution were mixed. While stirring, 0.3 mL of methanol was added to allow mixing of the hydrogen peroxide with the dichloromethane. The two phases were mixed and epoxidation was allowed to take place for 16 h. The yields for styrene epoxidation increased from 9% (L_1) to 19% (L_4). 1-Dodecene showed only very little conversion, probably due to low solubility.

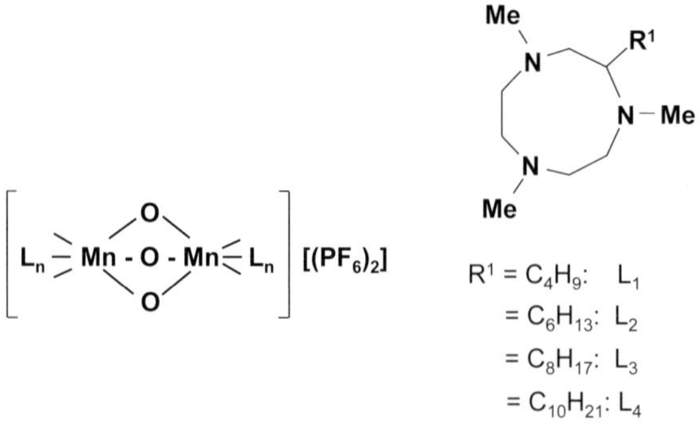

$R^1 = C_4H_9$: L_1
$= C_6H_{13}$: L_2
$= C_8H_{17}$: L_3
$= C_{10}H_{21}$: L_4

8

3.2.1.6
C–C Bond-Forming Reactions

3.2.1.6.1
Oxidative Coupling versus Hydroxylation

Oxidation of arenes by Pd(II), which is typically carried out in acetic acid solution, leads to oxidative coupling and ring or side chain substitution, depicted in Scheme 2.

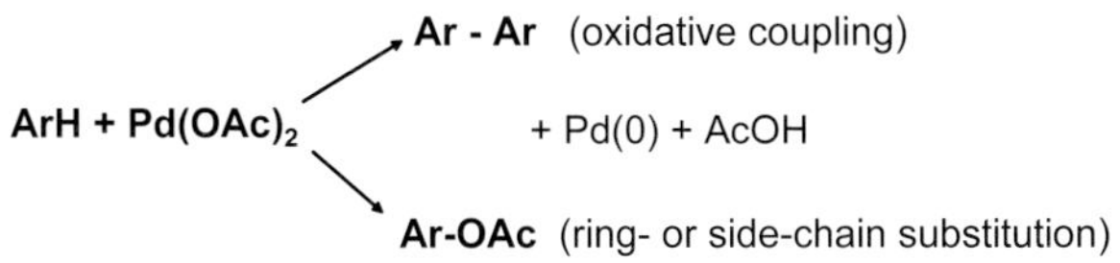

$$\text{ArH + Pd(OAc)}_2 \nearrow \quad \textbf{Ar - Ar} \quad \text{(oxidative coupling)}$$
$$+ \text{Pd(0)} + \text{AcOH}$$
$$\searrow \quad \textbf{Ar-OAc} \quad \text{(ring- or side-chain substitution)}$$

Scheme 2 Oxidative coupling and ring or side chain substitution.

Analogously to the Wacker-type oxidations, the reaction can be made catalytic by reoxidizing the Pd(0) with oxygen. As redox co-catalyst, a hetero-polyacid of formula $H_{3+n}[PMo_{12-n}V_nO_{40}]$ (HPA-n, $n > 1$) can be used. It is known that strong acids favor coupling, while basic additives favor substitution. The oxidative coupling of arenes proceeds by the following sequence [Eqs. (2)–(4)].

$$2 \text{ ArH} + \text{Pd(II)} \rightarrow \text{Ar–Ar} + \text{Pd(0)} + 2 \text{ H}^+ \tag{2}$$

$$\text{Pd(0)} + \text{HPA-}n + 2 \text{ H}^+ \rightarrow \text{Pd(II)} + H_2[\text{HPA-}n] \tag{3}$$

$$H_2[\text{HPA-}n] + \tfrac{1}{2} O_2 \rightarrow \text{HPA-}n + H_2O \tag{4}$$

Burton et al. studied the oxidation of arenes (benzene and toluene) with O_2 catalyzed by Pd(OAc)$_2$ and HPA-n in a biphasic system [23]. The system consisted of an arene phase and a catalyst phase in aqueous acetic acid. Reactions were carried out at 100 °C at an oxygen pressure of 0.5 MPa for about 2 h. The catalyst phase (5.0 mL) consisted of 0.010 M Pd(OAc)$_2$ and 0.050 M HPA-2, dissolved in aqueous acetic acid. About 3 mL benzene or toluene was used as the arene phase. The formation of a Pd-black slurry in the arene phase made phase separation difficult. The reaction products were separated by extraction with water and diethyl ether, the latter containing the products. Addition of water up to 20–30 vol.% increased the conversion of arenes as well as the selectivity to biaryls substantially. Further increases had a negative effect. The explanation for this result can be found on the basis of the mechanism of electrophilic aromatic substitution by Pd(II), which is the rate-determining step. Adding water to acetic acid accelerates the process, but at higher water content the solubility of arenes decreases, causing a lower conversion. Another advantage of the increase in the water fraction is the large increase in the biphenyl/terphenyl ratio, which can be explained by the decreasing solubility of biphenyl. The most important advantage, however, was the increased selectivity for the oxidation of benzene to phenol obtained by increasing the amount of water.

The selectivity to phenol reached 64%, at 6% conversion in a polar phase of AcOH/ H_2O (30 : 70, v/v).

3.2.1.6.2
Heck Reaction

Heck vinylation of aryl halides is one of the most widely used methods for carbon–carbon bond formation. Bhanage et al. described a method for a Heck reaction of iodobenzene in a biphasic system of ethylene glycol/toluene. Several metal–TPPTS complexes were tested [24].

The toluene phase contained the reactants and products and the ethylene glycol phase the metal complex and KOAc. The reaction takes place in the ethylene glycol phase and at the interface [Eq. (5)]. In a typical experiment, 0.1 mmol of catalyst was dissolved in 10 mL of ethylene glycol, mixed with a solution of 10 mmol of iodobenzene and 10 mmol of butyl acrylate in 10 mL of toluene. These two phases were then mixed at 140 °C along with 10 mmol KOAc for several hours. Yields of > 99% at 100% selectivity were obtained for the *trans* compound.

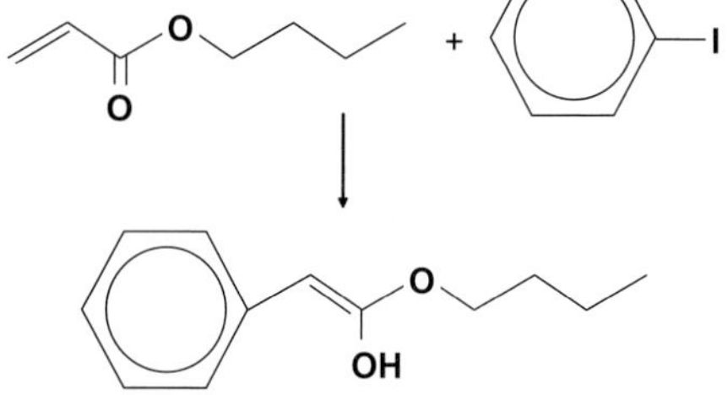

(5)

Another biphasic Heck reaction was described by Beller et al. [25]. The medium consisted of xylene and ethylene glycol. The catalyst was a palladium complex with a carbohydrate-substituted triphenylphosphine (**9** and **10**). Aryl bromide (15 mmol), styrene (22.5 mmol), and NaOAc (16.5 mmol) were suspended in 10 mL of xylene and 10 mL of ethylene glycol. The catalyst precursor (Pd(OAc)$_2$) and ligand (Pd/ligand ratio 1 : 3) were added and the mixture was heated to 130 °C for 20 h. Both ligands A and B showed better results than the TPPTS ligand (cf. Section 2.2.3.2) in the case of activated aryl bromides (for instance, *p*-nitrobromobenzene). However, for deactivated aryl bromides (for instance, 2-bromo-6-methoxynaphthalene) TPPTS proved to generate a more stable and thus more productive catalyst system.

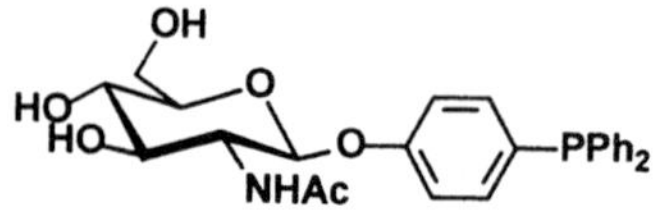

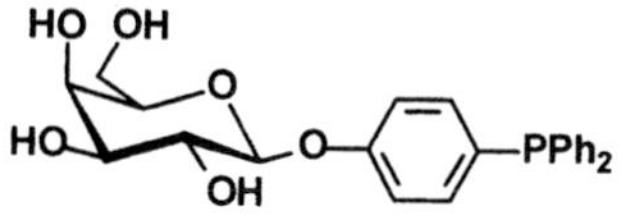

3.2.1.6.3
Suzuki Coupling

Due to the growing importance of asymmetrically substituted biaryl derivates used as drug intermediates, the Suzuki coupling reaction is increasingly important. Mostly, however, large amounts of catalyst are used and the catalyst recycling is often hindered by precipitation of palladium black.

Although the Suzuki reaction often occurs in biphasic mixtures, very few examples are known using catalyst systems soluble in polar phases. Beller et al. used the carbohydrate-substituted triphenylphosphine ligands described in Section 3.2.1.6.2, rendering the catalyst soluble in the polar phase [25]. Two different biphasic mixtures were investigated: ethanol–water/toluene (2 : 1 : 3 and ethanol–water/di-*n*-butyl ether (2 : 1 : 3). Phenylboronic acid (15 mmol), aryl bromide (13.5 mmol), and $N_2CO_3 \cdot 10 H_2O$ (40.5 mmol) were dissolved in 36 mL of the biphasic mixture being investigated and preheated to 60 °C. The catalyst precursor ($Pd(OAc)_2$) and the ligand were added to the mixture and refluxed for 2 h at 78 °C. For the solvent mixture ethanol–water/toluene better conversions were obtained than for ethanol–water/di-*n*-butyl ether. In general yields higher than 50% were obtained, while the TONs ranged between 550 and 9000. In comparison with the TPPTS ligand, the carbohydrate-substituted phosphines performed better or at least equally well.

3.2.1.6.4
Alkylation of Hydroquinone with Isobutene

Timofeeva et al. [26] described the acid-catalyzed alkylation of hydroquinone with isobutene (Scheme 3). Traditionally, the alkylation of hydroquinone with isobutene is carried out in the presence of strong acids such as H_2SO_4, H_3PO_4, sulfonic acid resin, etc. Timofeeva et al. investigated tungsten hetero-polyacids (HPAs) in a biphasic system as alternative catalysts for the alkylation of hydroquinone. The apolar phase consisted of toluene and the polar phase was the liquid HPA dioxane etherate HPA $\cdot$ 39 $C_4H_8O_2 \cdot$ 26 H_2O. The alkylation was performed in a glass reaction vessel equipped with a stirrer and a reflux condenser. The reactor was charged with 20 mL of toluene and 3–9 mL of HPA etherate and heated to 75–95 °C. Hydroquinone (5 g) was added under intensive stirring and isobutene was fed at a rate of 15 mL min^{-1}. After the reaction, the alkylated product (upper phase) was separated and the solvent was distilled to afford a crystalline residue. The residue was dried in vacuum at 80 °C for 2 h. The most efficient catalyst was found to be

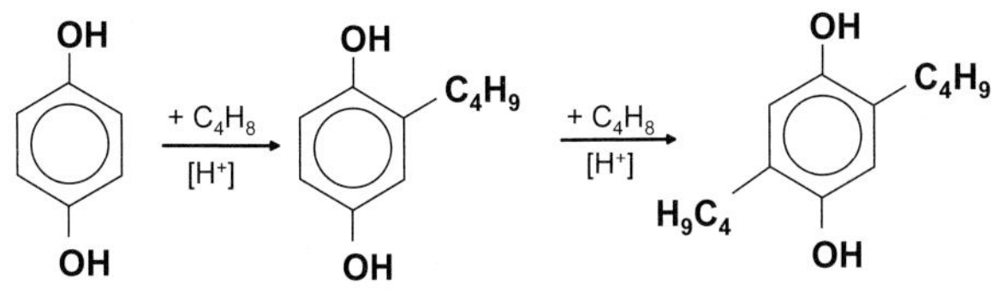

Scheme 3 Alkylation of hydroquinone with isobutene.

the etherate ($PW_{12} \cdot 39\ C_4H_8O_2 \cdot 26\ H_2O$ ($PW_{12} = H_3PW_{12}O_{40}$). Under optimal conditions (85 °C, hydroquinone/isobutene ratio of 0.7) a 71% yield of mono-alkylated hydroquinone was obtained at 90% conversion of hydroquinone.

Raising the temperature resulted in a decreased yield, due to side reactions (double alkylation of hydroquinone and isobutene oligomerization). Since the catalyst was present in the etherate phase and the reactants and products in the toluene phase, the catalyst could be easily recovered and re-used. The recycled catalyst retained the initial activity and selectivity.

In comparison to H_2SO_4 and H_3PO_4, PW_{12} is about twice as efficient based on an equal molar amount of catalyst, and about 100 times more efficient based on an equal number of protons.

3.2.1.7
Isomerization

Bricout et al. [27] described the nickel-catalyzed isomerization of geraniol and prenol in homogeneous and biphasic systems. The catalyst was Ni(dppe)$_2$ with trifluoro-acetic acid as a Brønsted acidic co-catalyst.

The isomerization of geraniol (**11**) into citronellal (**12**) is illustrated in Scheme 4, together with unwanted side reactions that occur.

The isomerization of prenol (**15**) into isovaleraldehyde (**16**) is depicted in Scheme 5, as well as an unwanted side reaction.

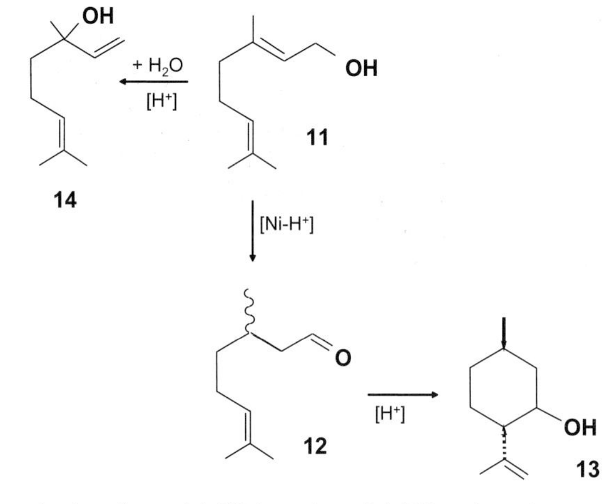

Scheme 4 Isomerization of geraniol (**11**) into citronellal (**12**) and reaction side products (**13** and **14**).

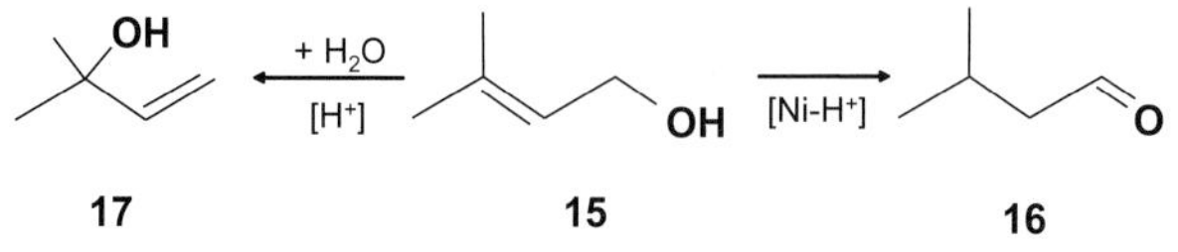

Scheme 5 Isomerization of prenol (**15**) into isovaleraldehyde (**16**) and reaction side product (**17**).

Two biphasic solvent systems were investigated: a water/toluene mixture (2 : 1, v/v) and a water–dimethylformamide/toluene mixture (1 : 1 : 1, v/v/v). The isomerization reaction was carried out at 110 °C for 16 h, with a catalyst/substrate ratio of 1 : 700. For the isomerization of geraniol, in the case of the water/toluene solvent system, a yield of only 3% citronellal was obtained with a selectivity of only 5%. However, side product **14** was obtained with a selectivity of 95%. Adding dimethylformamide to the water/toluene mixture resulted in a yield of 14% citronellal with a selectivity of 74%. Side product **13** was obtained with a selectivity of 11% and side product **14** with 15% selectivity. Reducing the reaction time to 2 h gave a yield of 13% citronellal, but with a higher selectivity of 86%.

Dimethylformamide reduces the polarity of the aqueous phase and consequently suppresses undesirable side reactions. As dimethylformamide is crucial for an effective working catalyst, this system is considered to be organic–organic biphasic and not aqueous–organic biphasic. In the case of prenol isomerization similar yields were obtained for the water–dimethylformamide/toluene solvent system, but with a lower selectivity.

In all biphasic reactions a rapid color change of the polar phase from orange to pale green was observed, indicating irreversible catalyst decomposition. Future research on catalytic isomerization should focus on increased stability of the catalyst.

3.2.1.8
Hydration and Acetoxylation

Acid-catalyzed hydration and acetoxylation of terpenes are important synthetic routes to terpenols and terpene esters that are used in perfumery and flavoring compositions. Such perfume ingredients as dihydromyrcenol (DHM-OH) and dihydromyrcenyl acetate (DHM-OAc) are prepared by the hydration and acetoxylation of dihydromyrcene (DHM) (Scheme 6).

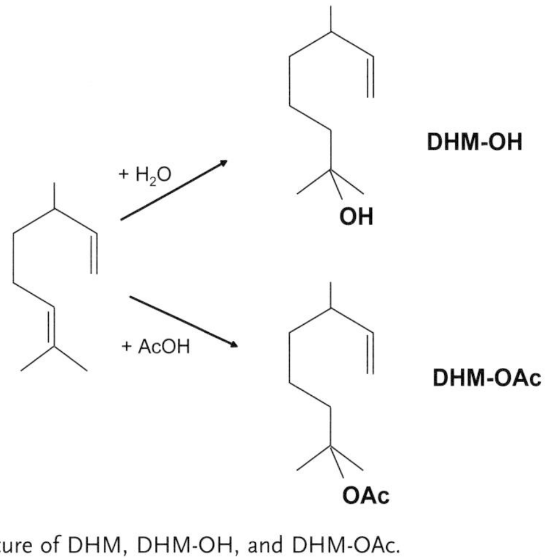

Scheme 6 Manufacture of DHM, DHM-OH, and DHM-OAc.

This process is complicated by acid-catalyzed isomerization and cyclization of DHM. In order to overcome this problem, Kozhevnikov et al. [28] used the biphasic hydration and acetoxylation in the presence of hetero-polyacids (HPAs) of the Keggin series: $H_3PW_{12}O_{40}$ (= PW_{12}) and $H_3PMo_{12}O_{40}$ (PMo_{12}). HPAs are highly soluble in water and polar organic solvents, such as lower alcohols, ketones, carboxylic acids, etc., but only scarcely soluble in hydrocarbons. The biphasic combined acetoxylation and hydration was performed in a DHM/AcOH–water solvent system. DHM (1.5– 4.0 g) was mixed with 8.0 g of a 50–70 wt.% HPA solution in AcOH–water (1 : 1 or 77 : 23, v/v) at 20–30 °C under vigorous stirring for 1–5 h. The upper product phase was separated. A 90% selectivity could be obtained at 21% DHM conversion, when a 77 : 23 (v/v) AcOH–water mixture containing 68 wt.% PW_{12} was used. About 10% of DHM was converted to DHM isomers. The molar ratio DHM-OH/DHM-OAc in the product phase is 80 : 20 – 70 : 30, changing to 96 : 4 – 85 : 5 after work-up.

The PW_{12} catalyst was practically entirely retained in the polar phase, enabling easy and clean catalyst recycling. The catalyst recycling behavior, when tested, showed constant conversion and selectivity during six runs. No decomposition of the PW_{12} took place, as was seen from ^{31}P NMR spectroscopy. PMo_{12} showed a high activity but low selectivity in converting DHM. The quick reduction of PMo_{12} made catalyst separation from the product phase difficult. The Keggin-type HPA, $H_3PW_{12}O_{40}$ (PW_{12}), is thus an excellent catalyst for the hydration and acetoxylation of DHM. The HPA shows a much higher catalytic activity than conventional acid catalysts such as H_2SO_4 and Amberlyst-15 (ion-exchange resin). Because of the high selectivity and efficient recycling of the catalyst without loss of activity, application of the system seems feasible.

3.2.2
Biphasic Catalysis on a Miniplant Scale

Arno Behr

In this section one example of catalysis on a miniplant scale is described. Although the reaction itself is homogeneous, the catalyst separation is done by extraction, making the whole system biphasic. The reaction carried out is the palladium-catalyzed telomerization of 1,3-butadiene and CO_2 to a δ-lactone [Eq. (6)] described by Behr et al. [29, 30]. A possible side product in this reaction is the γ-lactone.

(6)

The reaction is homogeneously catalyzed by a system formed in situ from $Pd(acac)_2$ and tricyclohexylphosphine in acetonitrile. During a batch experiment at 90 °C and 4 MPa, 48% conversion of butadiene could be reached with 95% selectivity to the lactone. The reaction mixture was separated by flash evaporation into a vapor phase consisting only of CO_2, 1,3-butadiene plus acetonitrile and a liquid phase consisting of more than 95% δ-lactone (and side products). The vapor phase, after condensation, was recycled to the reactor, which allowed a closed loop for solvent and nonconverted feedstocks.

The liquid phase was separated by extraction with 1,2,4-butanetriol into a raffinate phase containing the catalyst and an extract phase containing the δ-lactone. The raffinate stream was recycled to the reactor. The extract phase was separated by distillation into a δ-lactone stream and a 1,2,4-butanetriol stream. The bottom stream consisted of 95 wt.% δ-lactone and 5 wt.% 1,2,4-butanetriol. The top stream consisted of 98 wt.% 1,2,4-butanetriol and 2 wt.% δ-lactone (and some catalyst), and was recycled to the extraction unit.

This process was realized in a continuous miniplant for 24 h of operation. The acetonitrile and the unreacted substrates were recycled from the flash evaporator to the reactor. No accumulation of side products was observed; their amount after the continuous experiment was as high as after the batch experiment. Raising the temperature from 100 °C to 160 °C resulted in the isomerization of the δ-lactone to γ-lactones and in the oxidation of the ligand. Therefore it was concluded that the several separation steps should be performed below 100 °C to avoid isomerization and catalyst deactivation. In further investigations, efforts should be made to close the remaining recycle lops.

3.2.3
Thermomorphic Solvent Systems

Arno Behr

The liquid/liquid two-phase technique is limited to processes where the organic reactants are sufficiently dissolved in the polar catalyst phase. If the mass transport between the two phases is scarce, the reaction must be operated in a single homogeneous phase. Behr developed a new concept using "thermomorphic solvent systems" [29–31]. This concept allows a single-phase reaction process combined with an easy catalyst/product separation by the two-phase technique. These new solvent systems consist of a polar solvent **s1** and a nonpolar solvent **s2**, which have to be immiscible. The third, semi-polar, solvent **s3** must be miscible with both **s1** and **s2** and acts as a solubilizer. The miscibility of such ternary solvent mixtures is strongly dependent on the temperature (Figure 1). In the new concept of thermomorphic solvent systems a catalytic reaction is performed at the temperature **T1** in a single-phase homogeneous system, after which a partition into two phases takes place by cooling to temperature **T2**.

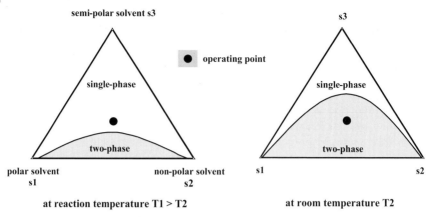

Figure 1 Temperature-dependent miscibility gap of ternary solvent mixtures.

The new recycling concept has been applied to the following model reactions:

- The rhodium-catalyzed co-oligomerization of double unsaturated fatty substances with ethylene was studied in different thermomorphic solvent systems [Eq. (7)] [32]. Using a biphasic system this reaction gives about 40% of products **18** and **19**. Performing the conversion in a system consisting of propene carbonate (**s1**), SFAME (**s2**) (SFAME = sunflower fatty acid methyl ester, which contains 60.5% CML), and dioxane (**s3**), the yield increases to up to 95%. This combination of solvents allows a single-phase process at the reaction temperature and therewith a high catalytic activity without mass-transfer limitation. According to the new concept, product and catalyst are easy to separate. The catalyst can be recycled several times with only small decrease of the product yield.

$$(7)$$

Furthermore temperature-dependent solvent systems containing poly(ethylene glycol) 1000 (PEG 1000) as a polar component were applied to the above-mentioned reaction. High conversion of methyl linoleate and high selectivity of co-oligomers were obtained. By using the novel PEG 1000-based temperature-dependent solvent systems, an easy separation of product and catalyst phase can be achieved. In addition, the effect of ligands on the catalyst recycling was investigated.

• The consecutive isomerization–hydroformylation reaction of *trans*-4-octene yields high conversion and high selectivity of *n*-nonanal in the polar solvent propene carbonate [Eq. (8)] [33].

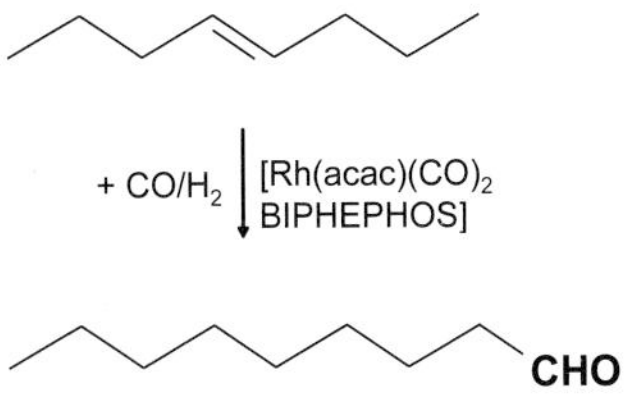

$$+ CO/H_2 \quad \begin{vmatrix} [Rh(acac)(CO)_2 \\ BIPHEPHOS] \end{vmatrix} \tag{8}$$

The catalyst phase can be recovered with the subsequent extraction of the product by the nonpolar solvent dodecane. Thermomorphic solvent systems based on the two solvents with different semipolar components as solubilizer were investigated and used in the hydroformylation mentioned. With this new recycling concept the product phase and the catalyst phase can be separated directly after the conversion without any additional extraction step.

• The hydrosilylation of methyl 10-undecenoate with triethoxysilane catalyzed by anhydrous H_2PtCl_6 was studied in the thermomorphic solvent system propene carbonate (**s1**)/cyclohexane (**s2**)/toluene (**s3**) [Eq. (9)] [18].

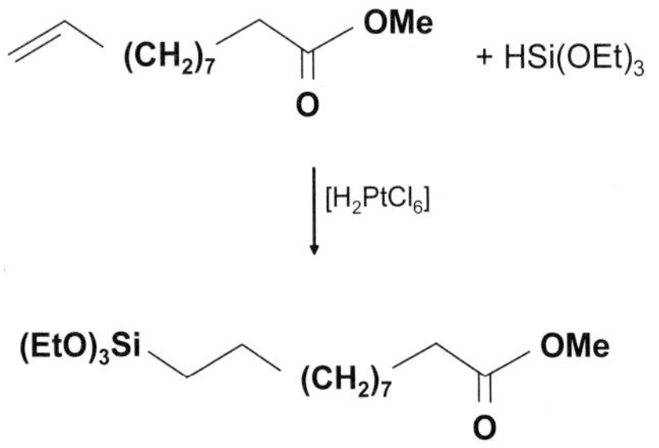

$$\begin{array}{c} (CH_2)_7 \quad OMe \\ O \end{array} + HSi(OEt)_3$$

$$\downarrow [H_2PtCl_6] \tag{9}$$

$$(EtO)_3Si \quad (CH_2)_7 \quad OMe \\ O$$

20

The conversion finishes after 15 s and gives 80% of product **20** at 80 °C under single-phase conditions. Cooling brings about a partition of the ternary solvent mixture. The nonpolar phase (cyclohexane) contains the product, the polar phase (propene carbonate) the catalyst.

3.3
Economical Applications (SHOP Process)

Dieter Vogt

3.3.1
Introduction

The Shell higher-alkene process was undoubtedly the first commercial catalytic process taking benefit from two-phase (but nonaqueous) liquid/liquid technology. In this special case two immiscible organic phases are used to separate the catalyst from the products formed, with the more or less pure products forming the upper phase.

The basis for this nickel-catalyzed oligomerization of ethene goes back to Ziegler and his school at the Max Planck Institute of Mülheim. There the so-called "nickel effect" [35, 36] was found, and Wilke and co-workers learned how to control the selectivity of nickel-catalyzed reactions by use of ligands. Keim introduced $\widehat{PO}$ chelate ligands and on this basis carried out the basic research for the oligomerization process at the Shell research company at Emeryville [37–46]. The whole process was developed in a collaboration between Shell Development, USA, and the Royal Shell Laboratories at Amsterdam in the Netherlands [47–56]. The SHOP is not only a process for ethene oligomerization, but a very efficient and flexible combination of three reactions: oligomerization, isomerization, and metathesis. It was designed to meet the market needs for linear α-alkenes for detergents [57].

The first commercial plant was built at Geismar, LA, USA, in 1977. The development of this plant and that at Stanlow (UK) is summarized in Table 3, together with other oligomerization capacities based on other technology [58].

Table 3 Linear α-alkene capacities via ethene oligomerization.

Technology	Company	Location	Capacity [10^3 tpy]		
			Initial (year)	Expansion (year)	Present total[a]
Ziegler-type	Chevron	Cedar Bayout, TX	125 (1966)	125 (1990)	250
	Ethyl	Pasadena, TX	400 (1971)	55 (1989)	472
	Ethyl	Feluy, Belgium	200 (1992)		200
	Chemopetrol	Czech Republic	120 (1992)		120
	Mitsubishi Kasei Corp.	Kurashiki, Okayama Pref., Japan	50		50
SHOP	Shell	Geismar, LA	200 (1977)	320 (2002) 390 (1989)	910
	Shell	Stanlow, UK	170 (1982)	100 (1989)	270
Zr	Idemitsu Petrochemicals	Ichihara, Chiba Pref., Japan	50 (1989)		50

a) In 1992.

Multiphase Homogeneous Catalysis
Edited by Boy Cornils and Wolfgang A. Herrmann et al.
Copyright © 2005 Wiley-VCH Verlag GmbH & Co. KGaA, Weinheim
ISBN: 3-527-30721-4

The two operational SHOP sites today have a total capacity of nearly 1 million tons of α-alkenes per year. This is about one-half of the total amount made by oligomerization. Today linear α-alkenes are produced mainly by ethene oligomerization because of the high product quality and the good availability of ethene. The wide application and increasing need for short-chain α-alkenes as co-monomers for polymers cause the linear alkene market to continue growing.

3.3.2
Process Description

The oligomerization is carried out in a polar solvent in which the nickel catalyst is dissolved but the nonpolar products, the α-alkenes, are almost insoluble. Preferred solvents are alkanediols, especially 1,4-butanediol (1,4-BD). This use of a biphasic organic liquid/liquid system is one of the key features of the process. The nickel catalyst is prepared in situ from a nickel salt, e.g., $NiCl_2 \cdot 6\,H_2O$, by reduction with sodium borohydride in 1,4-BD in the presence of an alkali hydroxide, ethene, and a chelating P O ligand such as o-diphenylphosphinobenzoic acid (**21**) [45, 54]. Suitable ligands are the general type of diorganophosphino acid derivatives (**22**).

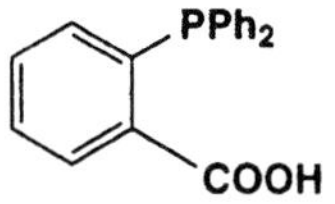

PPh₂

COOH

21

RR'P   COOR"

22

The nickel concentration in the catalyst system is in the range 0.001–0.005 mol% (approx. 10–50 ppm). The oligomerization is carried out in a series of reactors at temperatures of 80–140 °C and pressures of 7–14 MPa. The rate of the reaction is controlled by the rate of catalyst addition [53]. A high partial pressure of ethene is required to obtain good reaction rates and high product linearity [45]. The linear α-alkenes produced are obtained in a Schulz–Flory-type distribution with up to 99% linearity and 96–98% terminal alkenes over the whole range from C_4 to C_{30+} (cf. Table 4) [57].

The shape of the Schulz–Flory distribution and the chain length of the α-alkenes are controlled by the geometric chain-growth factor K, defined as $K = n(C_{n+2})/n(C_n)$ (Figure 2).

Table 4 Comparison of product qualities of technical C_6–C_{18} α-alkenes [59].

Product	Wax-cracking	Quality (wt.% α-alkene)		
		Chevron	Ethyl	SHOP
α-Alkenes	83–89	91–97	63–98	96–98
Branched alkenes	3–12	2–8	2–29	1–3
Paraffins	1–2	1.4	0.1–0.8	0.1
Dienes	3–6	–	–	–
Monoalkenes	92–95	99	> 99	99.9

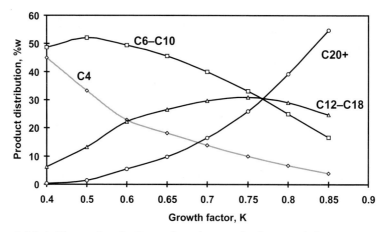

Figure 2 Schulz–Flory product distribution dependence on the chain-growth factor *K*.

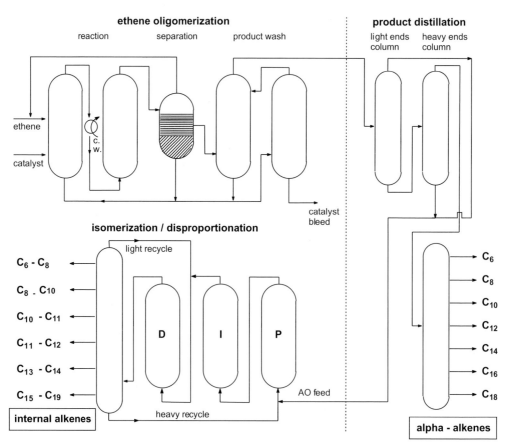

Figure 3 Flow scheme of the Shell higher-alkene process (SHOP).
P = purification; I = isomerization; D = disproportionation; AO = α-alkene; c.w. = cooling water.

For the economy of the whole process it is very important that the K factor can easily be adjusted by varying the catalyst composition. Usually the value is between 0.75 and 0.80. The heat of the reaction is removed by water-cooled heat exchangers between the reactors (Figure 3). In a high-pressure separator the insoluble products and the catalyst solution as well as unreacted ethene are separated.

The catalyst solution is fed back into the oligomerization reactor. Washing of the oligomers by fresh solvent in a second separator removes traces of the catalyst. This improves product quality and the catalyst utilization [60]. Traces of remaining catalyst in the product can lead to the formation of insoluble polyethene during upstream processing, resulting in fouling of process equipment [61].

The formation of insoluble polyethene causes problems also in other parts of the process. During catalyst preparation this can be avoided by adding the preformed, stable nickel complex and the chelate ligand separately to the oligomerization reactor. By this simple change, the catalyst utilization is enhanced markedly, resulting in a significant reduction of nickel salt and borohydride consumption [62–64]. So one major problem is the complete separation of the catalyst. Many attempts have been made to improve this [53]. One approach was using methanol/water solvent mixtures together with sulfonated ligands [65–71]. In the course of this it was shown that the catalyst is not deteriorated by water [72], which might lead to new approaches in the future.

Completely different liquid/liquid two-phase applications for oligomerization of ethene have been reported. Chauvin used ionic liquids as solvents for oligomerization catalysts [73] (cf. Chapter 5). Another approach is the use of perfluorinated solvents together with catalysts bearing perfluorinated ligands (Chapter 4) [74, 75].

Further processing of the product α-alkenes involves separation into the desired product fractions in a series of distillation columns. First the lower C_4–C_{10} α-alkenes are stripped off. In a heavy-ends column the C_{20+} α-alkenes are removed from the desired C_{12}–C_{20} α-alkenes. Finally the middle-range products meeting the market needs are separated into the desired cuts and blends. The very high flexibility of the "SHOP" results from the following steps. The C_4–C_{10} and the C_{20+} fractions are combined to be isomerized to internal linear alkenes and then subjected to a metathesis reaction. Both steps require about 80–140 °C and 0.3–2 MPa. Isomerization is accomplished by a typical isomerization catalyst such as Na/K on Al_2O_3 or a MgO catalyst in the liquid phase [76], where about 90% of the α-alkenes are converted to internal alkenes. Metathesis of the lower and higher internal alkenes gives a mixture of alkenes with odd and even carbon chain lengths. The mixture comprises about 11–15% of the desired C_{11}–C_{14} linear internal alkenes, which are separated by distillation. The undesired fractions can be recycled, feeding the light alkenes directly back to metathesis while the higher-boiling fractions are again subjected to isomerization. Because of the high proportion of short-chain alkenes in the metathesis feed, the double bonds in the end product are shifted toward the chain ends. Altogether the different possibilities of shifting products to the desired chain length and double-bond position makes the SHOP the most elegant and flexible process operating today. It is furthermore one of the larger applications of homogeneous catalysis.

The mechanism of the nickel-chelate complex-catalyzed oligomerization has been investigated in detail by Keim and co-workers [77–86]. Based on these results, the mechanism shown in Scheme 7 was postulated.

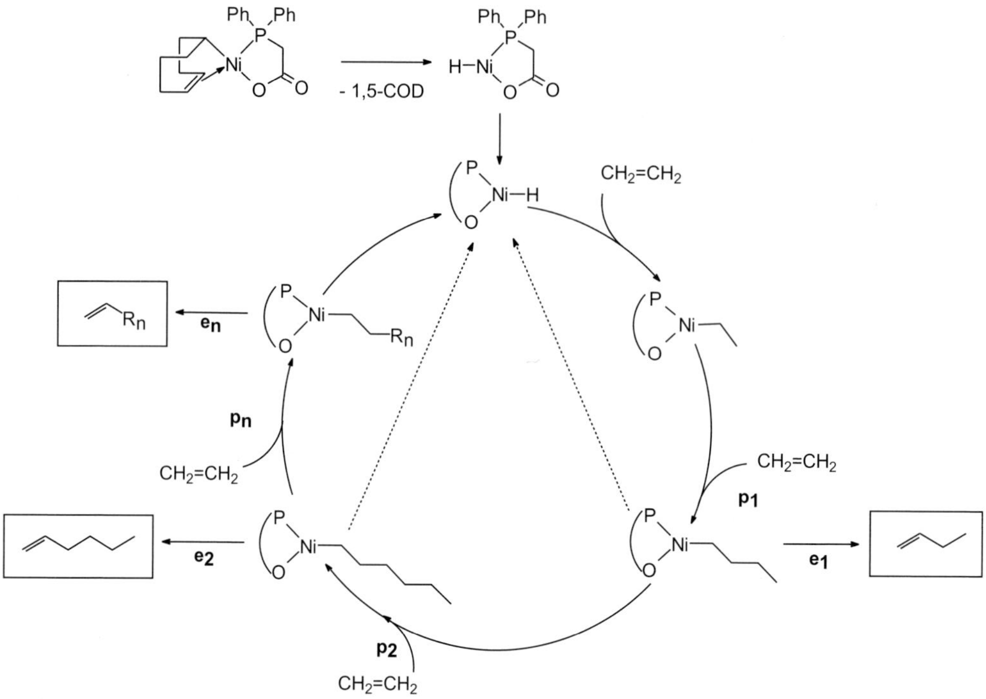

Scheme 7 Postulated mechanism for ethene oligomerization via a P⌒O -stabilized nickel hydride species: p_1, p_2 ... p_n = propagation steps; e_1, e_2 ... e_n = elimination steps.

3.4
Conclusions and Outlook

Dieter Vogt

Biphasic catalysis in general is an attractive technique, combining the advantages of homogeneous and heterogeneous catalysis. Common organic solvents are available in bulk quantities for an acceptable price, giving organic–organic biphasic catalysis an advantage over ionic liquids and fluorous biphasic catalysis.

However, when comparing organic–organic biphasic systems with other categories of biphasic catalysis, there are some difficulties to overcome. There are not that many combinations of immiscible organic solvents for which the catalyst is exclusively soluble in one of the phases. This leads in most cases to application of at least ternary mixtures, which makes a much more thorough study of the phase behavior necessary. However, possibilities given by modern parallel and automated experimentation techniques and their consequent application should help to overcome these hurdles. Academic research on several catalytic transformations has outlined the potential of organic–organic biphasic catalysis. As the substrates and products, as well as the catalyst, will influence the phase compositions and miscibility ranges of such systems, it has to be expected that each catalytic process will require individual optimization for maximum performance.

The success of the SHOP has shown that the difficulties can be overcome, leading to very efficient catalyst recycling.

References

1 A. BEHR, *Chem. Ing. Tech.* **1998**, *70(6)*, 685.
2 B. DRIESSEN-HÖLSCHER, P. WASSERSCHEID, W. KEIM, *Cattech* **1998**, *3(1)*, 47.
3 D. VOGT, in *Aqueous-Phase Organometallic Catalysis* (Eds.: B. CORNILS, W. A. HERRMANN), 2nd ed., Wiley-VCH **2004**, p. 639.
4 Rhône-Poulenc (E. G. Kuntz), FR 2.314.910 (**1976**).
5 E. G. KUNTZ, *CHEMTECH* **1987**, Sept., 570.
6 (a) B. CORNILS, J. FALBE, *4th ISHC*, Leningrad, Sept. **1984**, Proceedings, p. 487; (b) H. BACH, W. GICK, E. WIEBUS, B. CORNILS, *Int. Symp. High-Pressure Chem. Eng.*, Erlangen, Germany, Sept. **1984**, Preprints, p. 129; (c) H. BACH, W. GICK, E. WIEBUS, B. CORNILS, *8th ICC*, Berlin, June **1984**, Preprints, Vol. V, p. 417 (*Chem. Abstr.* **1987**, *106*, 198 051); (d) E. WIEBUS, B. CORNILS, *Chem. Ing. Tech.* **1994**, *66*, 916; H. W. Bohnen, B. Cornils, *Adv. Catal.* **2002**, *47*, 1.
7 G. ASTARITA, *Mass Transfer with Chemical Reaction*, Elsevier, Amsterdam **1967**.
8 L. L. TAVLARIDES, M. STAMATOUDIS, *Adv. Chem. Eng.* **1981**, *11*, 199.
9 L. K. DORAISWAMY, M. M. SHARMA, *Heterogeneous Reactions*, Vol. 2, Wiley, New York **1984**.
10 R. V. CHAUDHARI, A. BHATTACHARYA, B. M. BHANAGE, *Catal. Today* **1995**, *24*, 123.
11 I. HABLOT, J. JENCK, G. CASAMATTA, H. DELMAS, *Chem. Eng. Sci.* **1992**, *47*, 2689.
12 Ruhrchemie AG (B. CORNILS, H. BAHRMANN, W. LIPPS, W. KONKOL), DE 3.511.428 (**1985**).
13 R. G. DA ROSA, L. MARTINELLI, L. H. M. DA SILVA, W. LOH, *Chem. Commun.* **2000**, 33.
14 E. MIECZYŃSKA, A. M. TRZECIAK, R. GRZYBEK, J. J. ZIÓŁKOWSKI, *J. Mol. Catal. A:* **1998**, *132*, 203.

Multiphase Homogeneous Catalysis
Edited by Boy Cornils and Wolfgang A. Herrmann et al.
Copyright © 2005 Wiley-VCH Verlag GmbH & Co. KGaA, Weinheim
ISBN: 3-527-30721-4

15 C. Bianchini, P. Frediani, V. Sernau, *Organometallics* **1995**, *14(12)*, 5458.
16 U. Ritter, N. Winkhofer, H.-G. Schmidt, H. W. Roesky, *Angew. Chem.* **1996**, *108(5)*, 591.
17 C. Bianchini, A. Meli, V. Patinec, V. Sernau, F. Vizza, *J. Am. Chem. Soc.* **1997**, *119(21)*, 4945.
18 A. Behr, F. Naendrup, D. Obst, *Eur. J. Lipid. Sci. Technol.* **2002**, *104*, 161.
19 A. Behr, N. Toslu, *Chem. Eng. Technol.* **2000**, *23(2)*, 122.
20 A. Behr, F. Naendrup, D. Obst, *Adv. Synth. Catal.* **2002**, *344(10)*, 1142.
21 D. Bianchi, R. Bortolo, R. Tassinari, M. Ricci, R. Vignola, *Angew. Chem., Int. Ed.* **2000**, *39(23)*, 4321.
22 J. H. Koek, E. W. M. J. Kohlen, S. W. Russell, L. van der Wolf, P. F. ter Steeg, J. C. Hellemons, *Inorg. Chim. Acta* **1999**, *295*, 189.
23 H. A. Burton, I. V. Kozhevnikov, *J. Mol. Catal. A:* **2002**, *185*, 285.
24 B. M. Bhanage, F.-G. Zhao, M. Shirai, M. Arai, *Tetrahedron Lett.* **1998**, *39*, 9509.
25 M. Beller, J. G. E. Krauter, A. Zapf, S. Bogdanovic, *Catal. Today* **1999**, *48*, 279.
26 M. N. Timofeeva, I. V. Kozhevnikov, *React. Kinet. Catal. Lett.* **1995**, *54(2)*, 413.
27 H. Bricout, E. Monflier, J.-F. Carpentier, A. Mortreux, *Eur. J. Inorg. Chem.* **1998**, 1739.
28 I. V. Kozhevnikov, A. Sinnema, A. J. A. van der Weerdt, H. van Bekkum, *J. Mol. Catal. A:* **1997**, *120*, 63.
29 A. Behr, C. Fängewisch, *Chem. Ing. Tech.* **2001**, *73(7)*, 874.
30 C. Fängewisch, A. Behr, *Chem. Ing. Tech.* **2001**, *73(6)*, 737.
31 C. Fängewisch, A. Behr, *Chem. Eng. Technol.* **2002**, *25*, 143.
32 A. Behr, C. Fängewisch, *J. Mol. Catal. A:* **2003**, *197*, 115.
33 A. Behr, D. Obst, C. Schulte, T. Schosser, *J. Mol. Catal. A:* **2003**, *206*, 179.
34 A. Behr, M. Heite, *Chem. Eng. Technol.* **2000**, *23(11)*, 952.
35 G. Wilke, *Angew. Chem.* **1988**, *100*, 189; *Angew. Chem., Int. Ed. Engl.* **1988**, *27*, 185.
36 G. Wilke, in *Fundamental Research in Homogeneous Catalysis* (Ed.: M. Tsutsui), Plenum, New York **1979**, Vol. 3, p. 1.
37 Shell Dev. Co. (S. R. Baur, H. Chung, P. W. Glockner, W. Keim, H. van Zwet), US 3.635.937 (**1972**).
38 Shell Dev. Co. (S. R. Baur, H. Chung, D. Camel, W. Keim, H. van Zwet), US 3.637.636 (**1972**).
39 Shell Dev. Co. (S. R. Baur, P. W. Glockner, W. Keim, H. van Zwet, H. Chung), US 3.644.563 (**1972**).
40 Shell Dev. Co. (H. van Zwet, S. R. Baur, W. Keim), US 3.644.564 (**1972**).
41 Shell Dev. Co. (P. W. Glockner, W. Keim, R. F. Mason), US 3.647.914 (**1972**).
42 Shell Dev. Co. (S. R. Baur, P. W. Glockner, W. Keim, R. F. Mason), US 3.647.915 (**1972**).
43 Shell Dev. Co. (S. R. Baur, H. Chung, W. Keim, H. van Zwet), US 3.661.803 (**1972**).
44 Shell Dev. Co. (S. R. Baur, H. Chung, C. Arnett, P. W. Glockner, W. Keim), US 3.686.159 (**1972**).
45 E. R. Freitas, C. R. Gum, *Chem. Eng. Prog.* **1979**, *75*, 73.
46 E. L. T. M. Spitzer, *Seifen Öle Fette Wachse* **1981**, *107*, 141.
47 Shell Oil (R. F. Mason), US 3.676.523 (**1972**).
48 Shell Oil (R. F. Mason), US 3.686.351 (**1972**).
49 Shell Oil (A. J. Berger), US 3.726.938 (**1973**).
50 Shell Oil (R. F. Mason), US 3.737.475 (**1973**).
51 Shell (R. F. Mason, G. R. Wicker), DE 2.264.088 (**1973**).
52 Shell Oil (E. F. Lutz), US 3.825.615 (**1974**).
53 Shell Oil (E. F. Lutz), US 4.528.416 (**1985**).
54 Shell (E. F. Lutz, P. A. Gautier), EP 177.999 (**1986**).
55 W. Keim, in *Fundamental Research in Homogeneous Catalysis* (Eds.: M. Graziani, M. Giongo), Plenum, New York **1984**, Vol. 4, p. 131.
56 E. F. Lutz, *J. Chem. Educ.* **1986**, *63*, 202.
57 A. H. Turner, *J. Am. Oil Chem. Soc.* **1983**, *60*, 594.
58 C. S. Read, R. Willhalm, Y. Yoshida, in *Linear Alpha-Olefins, The Chemical Economics Handbook Marketing Research Report*, SRI International, Oct. **1993**, 681.5030 A.

59 A. M. AL-JARALLAH, J. A. ANABTAWI, M. A. B. SIDDIQUE, A. M. AITANI, A. W. AL-SA'DOUN, *Catal. Today* **1992**, *14*, 1.

60 Shell Oil (A. T. KISTER, E. F. LUTZ), US 4.020.121 (**1975**).

61 Shell Oil (C. R. GUM, A. T. KISTER), US 4.229.607 (**1980**).

62 Shell Oil (A. E. O'DONNELL, C. R. GUM), US 4.260.844 (**1981**).

63 Shell Oil (A. E. O'DONNELL, C. R. GUM), US 4.377.499 (**1983**).

64 Shell Oil (E. F. LUTZ), US 4.284.837 (**1981**).

65 Gulf Res. Dev. Co. (D. L. BEACH, J. J. HARRISON), US 4.301.318 (**1981**).

66 Gulf Res. Dev. Co. (D. L. BEACH, J. J. HARRISON), US 4.310.716 (**1982**).

67 Chevron/Gulf Co. (D. L. BEACH, J. J. HARRISON), EP 46.330 (**1982**).

68 Gulf Res. Dev. Co. (D. L. BEACH, J. J. HARRISON), US 4.382.153 (**1983**).

69 Chevron/Gulf Co. (D. L. BEACH, J. J. HARRISON), US 4.529.554 (**1985**).

70 Chevron/Gulf Co. (D. L. BEACH, Y. V. KISSIN), US 4.686.315 (**1987**).

71 Y. V. KISSIN, *J. Polym. Sci., Polym. Chem. Ed.* **1989**, *27*, 147.

72 Chevron Res. Co. (D. L. BEACH, J. J. HARRISON), US 4.711.969 (**1987**).

73 (a) Institut Français du Petrole (Y. CHAUVIN, D. COMMEREUE, I. GUIBARD, A. HIRSCHAUER, H. OLIVIER, L. SAUSSINE), US 5.104.840 (**1992**); (b) Y. CHAUVIN, S. EINLOFT, H. OLIVIER, *Ind. Eng. Chem. Res.* **1995**, *34*, 1149; (c) Y. CHAUVIN, H. OLIVIER, in *Applied Homogeneous Catalysis with Organometallic Compounds* (Eds.: B. CORNILS, W. A. HERRMANN), VCH, Weinheim **1996**, Vol. 1, p. 258.

74 M. Vogt, Ph. D. Thesis, RWTH Aachen, Germany **1991**.

75 I. T. HORVÁTH, J. RÁBAI, *Science* **1994**, *266*, 72.

76 Shell Oil (F. F. FARLEY), US 3.647.906 (**1972**).

77 W. KEIM, F. H. KOWALDT, R. GODDARD, C. KRÜGER, *Angew. Chem.* **1978**, *90*, 493; *Angew. Chem., Int. Ed. Engl.* **1978**, *17*, 466.

78 W. KEIM, A. BEHR, B. KIMBÄCKER, C. KRÜGER, *Angew. Chem.* **1983**, *95*, 505; *Angew. Chem., Int. Ed. Engl.* **1983**, *22*, 503.

79 W. KEIM, A. BEHR, B. GRUBER, B. HOFFMANN, F. H. KOWALDT, U. KÜRSCHNER, B. LIMBÄCKER, F. P. SISTIG, *Organometallics* **1986**, *5*, 2356.

80 W. KEIM, *New J. Chem.* **1987**, *11*, 531.

81 W. KEIM, *J. Mol. Catal.* **1989**, *52*, 19.

82 A. BEHR, W. KEIM, *Arab. J. Sci. Eng.* **1985**, *10*, 377.

83 W. KEIM, *Angew. Chem.* **1990**, *102*, 251; *Angew. Chem., Int. Ed. Engl.* **1990**, *29*, 235.

84 W. KEIM, *Ann. N. Y. Acad. Sci.* **1983**, *415*, 191.

85 U. MÜLLER, W. KEIM, C. KRÜGER, P. BETZ, *Angew. Chem.* **1989**, *101*, 1066; *Angew. Chem., Int. Ed. Engl.* **1989**, *28*, 1011.

52 W. KEIM, R. P. SCHULZ, *J. Mol. Catal.* **1994**, *92*, 21.

4

Fluorous Catalysis

István T. Horváth (Ed.)

Multiphase Homogeneous Catalysis
Edited by Boy Cornils and Wolfgang A. Herrmann et al.
Copyright © 2005 Wiley-VCH Verlag GmbH & Co. KGaA, Weinheim
ISBN: 3-527-30721-4

4.1
Introduction

István T. Horváth

Homogeneous catalysts can be designed at the molecular level to achieve high product selectivity at an economically favorable reaction rate (cf. Chapter 1). The development of facile separation of the homogeneous catalysts from the products is crucial for their industrial applications. Theoretically three types of separation methods can be used for homogeneous catalytic reactions performed in the liquid phase. If the catalyst is a gaseous molecule, it could dissolve in the liquid phase under pressure during reaction and simple depressurization could result in its complete removal from the product. One could imagine a similar system in which a soluble catalyst precursor is activated or turned on by a gaseous molecule under pressure and deactivated or turned off by depressurization after the product formed. The second and most popular approach involves the use of liquid–liquid biphasic systems, in which one of the phases contains the dissolved catalyst and the other the products. Since the formation of a liquid–liquid biphase system is due to the sufficiently different intermolecular forces of two liquids, the selection of a catalyst phase depends primarily on the solvent properties of the product phase at high conversion levels. For example, if the product is apolar the catalyst phase should be polar; and vice versa, if the product is polar the catalyst phase should be apolar. The success of any biphasic system depends on whether the catalyst could be designed to dissolve preferentially in the catalyst phase. Perhaps the most important rule for such design is that the catalyst has to be catalyst phase like, since it has been known for centuries that *similia similibus solvuntur*, or "like dissolves like". The third approach is based on solid–liquid separation. If the product has no or limited solubility in the liquid reaction mixture, it could be continuously removed by precipitation or crystallization. Alternatively, a solid catalyst can be dissolved in the reaction medium at a higher temperature and later separated from the product by lowering the temperature again. It should be noted that the latter approach could also operate in the opposite manner, when a solid catalyst precursor can be dissolved in the reaction medium at a lower temperature and separated from the product by increasing the temperature.

Perfluorinated alkanes, dialkyl ethers, and trialkylamines are unusual because of their nonpolar nature and low intermolecular forces. Their miscibility even with common organic solvents such as toluene, THF, acetone, and alcohols is low at room temperature; thus these materials could form *fluorous biphase systems* [1, 2]. The term *fluorous* was introduced [3, 4], as the analogue to the term *aqueous*, to emphasize the fact that one of the phases of a biphase system is richer in fluorocarbons than the other. Fluorous biphase systems can be used in catalytic chemical transformations by immobilizing catalysts in the fluorous phase. A fluorous catalyst system consists of a fluorous phase containing a preferentially fluorous-soluble catalyst and a second product phase, which may be any organic or inorganic solvent with limited solubility in the fluorous phase (Scheme 1).

Multiphase Homogeneous Catalysis
Edited by Boy Cornils and Wolfgang A. Herrmann et al.
Copyright © 2005 Wiley-VCH Verlag GmbH & Co. KGaA, Weinheim
ISBN: 3-527-30721-4

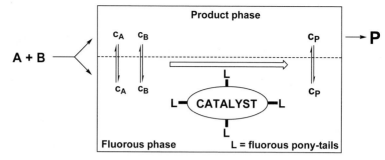

Scheme 1 The fluorous biphase concept for the catalytic conversion of substrates **A** and **B** to product **P**. The attachment of appropriate fluorous ponytails **L** to a homogeneous catalyst ensures that the fluorous catalyst remains in the fluorous phase.

Conventional homogeneous catalysts can be made fluorous-soluble by incorporating fluorocarbon moieties to their structure in appropriate size and number. The most effective fluorocarbon moieties are linear or branched perfluoroalkyl chains with high carbon numbers that may contain other heteroatoms (the "fluorous ponytails"). It should be emphasized that *perfluoroaryl* groups do offer dipole–dipole interactions [5], making perfluoroaryl-containing catalysts soluble in common organic solvents and therefore less compatible with fluorous biphase systems.

The most effective fluorous solvents are perfluorinated alkanes, perfluorinated dialkyl ethers, and perfluorinated trialkylamines. Their remarkable chemical inertness, thermal stability, and nonflammability coupled with their unusual physical properties make them particularly attarctive for catalyst immobilization. Furthermore, these materials are practically nontoxic by oral ingestion, inhalation, or intraperitoneal injection [6]. Although their thermal degradation can produce toxic decomposition products, such decomposition generally begins only at very high temperatures well above the thermal stability limits of most homogeneous catalysts.

A fluorous biphase reaction could proceed either in the fluorous phase or at the interface of the two phases, depending on the solubilities of the substrates in the fluorous phase. When the solubilities of the substrates in the fluorous phase are very low, the chemical reaction may still occur at the interface or appropriate phase-transfer agents may be added to facilitate the reaction. It should be emphasized that a fluorous biphase system might become a one-phase system by increasing the temperature [3]. Thus, a fluorous catalyst could combine the advantages of one-phase catalysis with biphasic product separation by running the reaction at higher temperatures and separating the products at lower temperatures (Scheme 2).

Alternatively, the temperature-dependent solubilities of *solid* fluorous catalysts in liquid substrates or in conventional solvents containing the substrates could eliminate the need for fluorous solvents (Scheme 3) [7, 8].

Because of the well-known electron-withdrawing properties of the fluorine atom, the attachment of fluorous ponytails to conventional catalysts could change

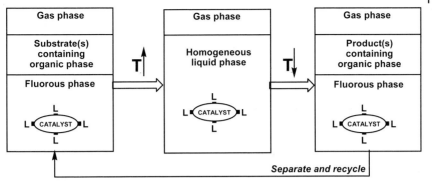

L = Fluorous solubilizing groups

Scheme 2 The temperature-dependent fluorous-liquid/liquid biphase concept.

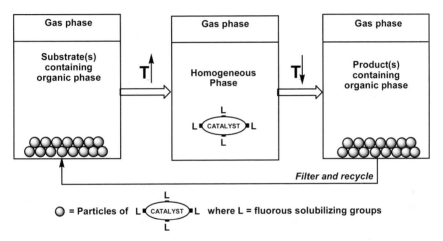

Scheme 3 The temperature-dependent fluorous-solid/liquid biphase concept.

significantly their electronic properties and consequently their reactivity. Insertion of insulating groups before the fluorous ponytail may be necessary to decrease the strong electron-withdrawing effects, an important consideration if catalyst reactivity is desired to approximate to that observed for the unmodified species in hydrocarbon solvents. For example, the first theoretical calculations have shown that the electronic properties of the fluorous phosphines $P[(CH_2)_x(CF_2)_yCF_3]_3$ ($x = 0$, $y = 2$ or 4 and $x = 0$–5, $y = 2$) can be tuned by varying the number of methylene groups between the phosphorus atom and the perfluoroalkyl moiety [9] (Table 4). It was later shown that the electron-withdrawing effect of even five methylene units was observable according to the variation in v_{CO} values of *trans*-[IrCl(CO)L$_2$] complexes [10]. It appears that between eight and ten methylene groups would be needed to insulate effectively the lone pair of the phosphorus atom from the perfluoroalkyl moiety [11].

Table 1 Electronic properties of fluorous phosphines.[a]

Phosphines	P		Protonation energy [eV]	P–H [Å]	∢ HPL [']
	Mulliken population [q]	Lone pair level [eV]			
P[CF$_2$CF$_3$]$_3$	0.83	−11.7	−6.5	1.189	85.9
P[CF$_2$CF$_2$CF$_2$CF$_3$]$_3$	0.83	−11.7	−6.4	1.192	85.4
P[CH$_2$CF$_2$CF$_3$]$_3$	0.62	−10.6	−7.7	1.205	86.3
P[(CH$_2$)$_2$CF$_2$CF$_3$]$_3$	0.48	−9.9	−8.3	1.218	92.3
P[(CH$_2$)$_3$CF$_2$CF$_3$]$_3$	0.40	−9.5	−8.6	1.225	91.8
P[(CH$_2$)$_4$CF$_2$CF$_3$]$_3$	0.38	−9.3	−8.8	1.226	92.0
P[(CH$_2$)$_5$CF$_2$CF$_3$]$_3$	0.36	−9.2	−8.9	1.228	91.8
P[CH$_2$CH$_2$CH$_2$CH$_3$]$_3$	0.33	−8.7	−9.3	1.230	91.7

a) The calculations were performed using the UniChem version of MNDO93 and employed the PM3 parameter set. Full geometry optimizations were performed.

Fluorous catalysts are best suited for converting apolar substrates to products of higher polarity, as the partition coefficients of the substrates and products will be higher and lower, respectively, in the fluorous phase. The net results are no or little solubility limitation on the substrates and easy separation of the products. Furthermore, as the conversion level increases, the amount of polar products increases, further enhancing separation. One of the most important advantages of the fluorous biphase catalyst concept is that many well-established hydrocarbon-soluble catalysts could be converted to fluorous-soluble. In general, fluorous catalysts have similar structures and spectroscopic properties to the parent compounds. The major difference arises from the presence of the fluorous ponytails, which provide a fluorous blanket around the hydrocarbon domain of the catalyst. If the electron-withdrawing effect of the fluorous ponytails on the ligands is not mitigated by insulating groups, the reactivity of the organometallic catalysts could be significantly different. Accordingly, most fluorous analogues of hydrocarbon-soluble catalysts have been prepared by incorporating insulating groups and shown to have comparable catalytic performance with the additional benefit of facile catalyst recycling.

Fluorous catalysis is now a well-established area and provides a complementary approach to aqueous and ionic biphase catalysis [12]. Since each catalytic chemical reaction could have its own *perfectly* designed catalyst (the chemzyme), the possibility to select from biphase systems ranging from fluorous to aqueous systems provides a powerful portfolio for catalyst designers.

References

1 J. H. HILDEBRAND, J. M. PRAUSNITZ, R. L. SCOTT, *Regular and Related Solutions*, Van Nostrand Reinhold, New York **1970**, Chapter 10.

2 C. REICHARDT, *Solvents and Solvent Effects in Organic Chemistry*, 2nd ed., VCH, Weinheim **1990**.

3 I. T. HORVÁTH, J. RÁBAI, *Science* **1994**, *266*, 72.

4 Exxon Research and Engineering Co. (I. T. HORVÁTH, J. RÁBAI), US 5.463.082 (**1995**).

5 R. FILLER, in *Fluorine Containing Molecules* (Eds.: J. F. LIEBMAN, A. GREENBERG, W. R. DOLBIER JR.), VCH, Weinheim **1988**, Chapter 2.

6 J. W. CLAYTON JR., *Fluorine Chem. Rev.* **1967**, *1*, 197.

7 (a) M. WENDE, J. A. GLADYSZ *J. Am. Chem. Soc.* **2001**, *123*, 11 490; (b) M. WENDE, J. A. GLADYSZ, *J. Am. Chem. Soc.* **2003**, *125*, 5861.

8 K. ISHIHARA, S. KONDO, H. YAMAMOTO, *Synlett* **2001**, 1371.

9 I. T. HORVÁTH, G. KISS, R. A. COOK, J. E. BOND, P. A. STEVENS, J. RÁBAI, E. J. MOZELESKI, *J. Am. Chem. Soc.* **1998**, *120*, 3133.

10 L. J. ALVEY, R. MEIER, T. SOÓS, P. BERNATIS, J. A. GLADYSZ, *Eur. J. Inorg. Chem.* **2000**, 1975.

11 H. JIAO, S. LE STANG, T. SOÓS, R. MEIER, K. KOWSKI, P. RADEMACHER, L. JAFARPOUR, J.-B. HAMARD, S. P. NOLAN, J. A. GLADYSZ, *J. Am. Chem. Soc.* **2002**, *124*, 1516.

12 D. BRADLEY, *Science* **2003**, *300*, 2022.

4.2
State-of-the-Art and Typical Reactions

4.2.1
Carbon–Carbon Bond-Forming Reactions

Siegfried Schneider, Carl Christoph Tzschucke, and Willi Bannwarth

Metal-catalyzed cross-coupling reactions are among the most prominent reaction types in synthetic organic chemistry [1]. The search for new chemical entities in medicinal chemistry requires the synthesis of growing numbers of compounds in shorter periods of time. To meet this demand, not only parallel synthesis formats need to be employed, but also the simplification of work-up procedures becomes increasingly important. In catalytic processes this can be accompanied by the recovery and re-use of the catalyst. Several concepts have recently been developed to simplify work-up procedures and to avoid time-consuming purification steps like column chromatography or distillation [2, 3]. They comprise in essence the use of polymer-supported catalysts (see Chapter 7) [4], the use of molten salts (or ionic liquids) as reaction media (see Chapter 5) [5, 6], reactions in scCO$_2$ or water (see Chapters 2 and 6) [7, 8], and the use of fluorous biphasic systems (FBS – see Scheme 1 in Section 4.1) [9–14]. In the FBS concept, fluorous ligands mediate the solubility of the pertinent catalyst in fluorous solvents and the reaction is carried out in a two-phase mixture consisting of a fluorous and an organic solvent. Such two-phase systems often become homogeneous at elevated temperatures (see Scheme 2 in Section 4.1). Lowering the temperature after reaction leads to the re-formation of the two phases. The organic phase contains the product and the fluorous phase the catalyst to be re-used.

Palladium-catalyzed cross-coupling reactions often require relatively large amounts of catalysts which have to be removed from the reaction product. Fluorous palladium complexes offer a solution to this problem, since they are soluble in fluorous solvents and can be readily separated from the organic product by liquid–liquid extractions.

Betzemeier and Knochel performed cross-couplings of aryl iodides with arylzinc bromides in a toluene/1-bromoperfluorooctane mixture in the presence of a Pd catalyst derived from fluorous phosphine **1a** and Pd$_2$(dba)$_3$ (Scheme 1) [15]. The reaction yielded the desired coupling products in excellent yields (87–99%). By using 1.5 mol% of catalyst it was possible to re-use the catalyst up to four times, yielding biphenyl derivative **2**. This cross-coupling reaction could be extended to benzyl- and alkenylzinc bromides forming **3** (76%) and **4** (92%).

The first Heck reaction in fluorous solvents was described by Pozzi et al. with either Pd$_2$(dba)$_3$ or Pd(OAc)$_2$ as palladium source and the fluorous phosphines **1a–1c** as the ligands [16]. Reactions between iodobenzene and methyl acrylate were conducted with 0.5 mol% of catalyst in an acetonitrile/D-100 (mainly *n*-perfluorooctane) mixture. The expected products were formed quantitatively and good

Multiphase Homogeneous Catalysis
Edited by Boy Cornils and Wolfgang A. Herrmann et al.
Copyright © 2005 Wiley-VCH Verlag GmbH & Co. KGaA, Weinheim
ISBN: 3-527-30721-4

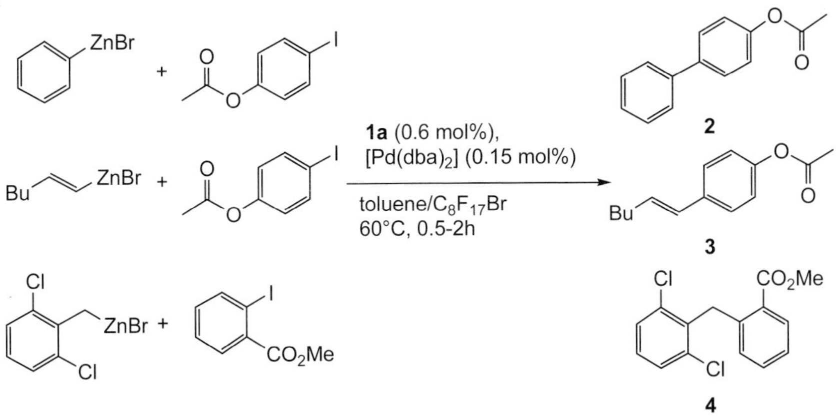

Scheme 1 Cross-coupling of aryl iodides with arylzinc bromides.

selectivity (88–93%) was observed. Recycling of the catalyst was possible, but resulted in lower conversions of the iodobenzene in the second and third runs, respectively.

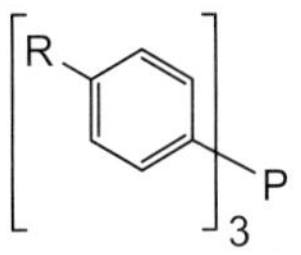

1a: R = C$_6$F$_{13}$
1b: R = OCH$_2$C$_7$F$_{15}$
1c: R = O(CH$_2$)$_2$OCH$_2$CF$_2$[OCF(CF$_3$)CF$_2$]$_p$(OCF$_2$)$_q$OCF$_3$
with p = 3.38 und q = 0.11

Rocaboy and Gladysz prepared a fluorous Schiff base, which was converted to palladacycle **5**. This represents a phosphane-free Pd catalyst which exhibits thermomorphic behaviour [Eq. (1)] [17]. Reactions were performed under homogeneous conditions (DMF, 100–140 °C) without fluorous solvents using (0.68–1.83) · 10^{-6} mol% of palladacycle. After work-up the Heck coupling products were obtained in 49–100% yield. Recycling experiments were done with 0.02 mol% of **5**, using C$_8$F$_{17}$Br as a "carrier" which forms a biphasic mixture with DMF. After phase separation and removal of C$_8$F$_{17}$Br, the catalyst was charged with fresh starting materials and DMF. The results revealed a gradual loss of conversion and yield. The authors assumed that either loss of activity is due to limited stabililty of the catalyst, or that the catalyst is stable but the recycling is not as efficient as anticipated. Reaction rate and transmission electron microscopy indicated the presence of soluble Pd nanoparticles as active catalyst.

Stabilized Pd nanoparticles of compounds featuring perfluorinated chains **6–10** were described by Moreno-Mañas et al. [18, 19]. The Pd nanoparticles were obtained by the reduction of PdCl$_2$ with methanol in the presence of **6–10**, respectively. The presence of such nanoparticles was confirmed by transmission electron microscopy. Due to the stabilization by the perfluorinated ligand, the palladium colloids are soluble in perfluorinated solvents. Pd nanoparticles stabilized by 1,5-bis(4,4'-bis(perfluorooctyl)phenyl)-1,4-pentadien-3-one (**6**) were active in Heck and Suzuki couplings [18].

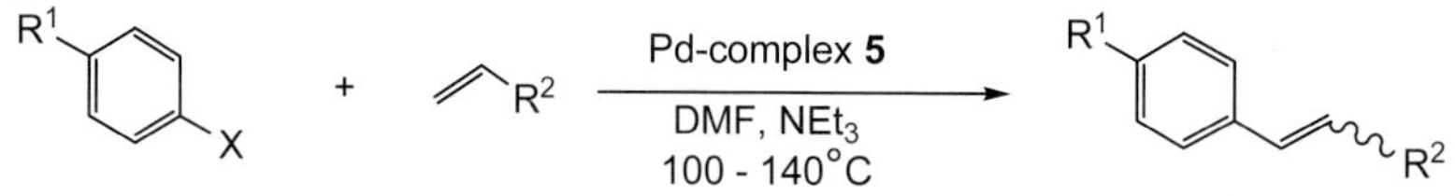

$$R^1 \text{---} X + \text{=} R^2 \xrightarrow[\substack{\text{DMF, NEt}_3 \\ 100\text{ - }140°C}]{\text{Pd-complex 5}} R^1 \text{---} R^2 \qquad (1)$$

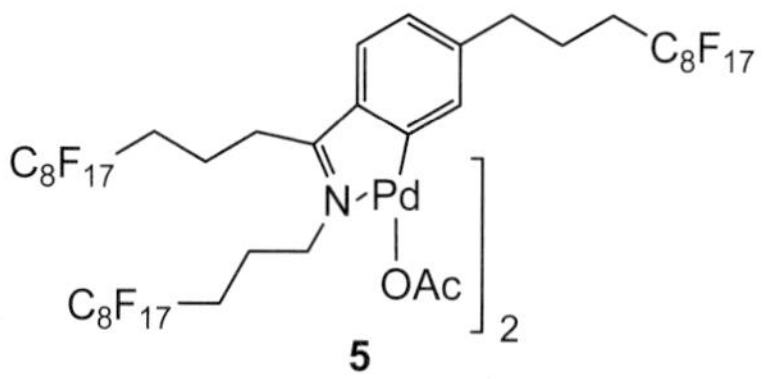

5

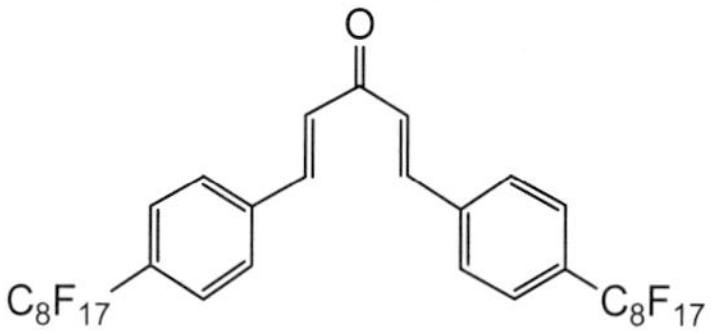

6

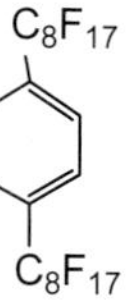

7

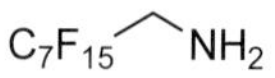

$$C_7F_{15}\text{---}NH_2$$

8

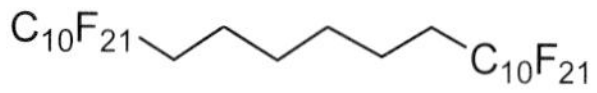

$$C_{10}F_{21}\text{---}C_{10}F_{21}$$

9

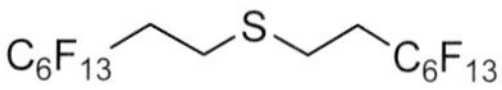

$$C_6F_{13}\text{---}S\text{---}C_6F_{13}$$

10

Schneider and Bannwarth have developed fluorous triphenylphosphine-modified palladium complexes (**11a–11c**) for Stille couplings in DMF/perfluoromethylcyclohexane (1 : 1) using 1 equiv of LiCl as additive at 80 °C for 3–24 h [Eq. (2)] [20].

$$R\text{---}Br + Ar\text{---}Sn(n\text{-}Bu)_3 \xrightarrow[\substack{2.) \text{ KF}}]{\substack{1.) \ 1.5 \ mol\% \ \text{Pd-catalyst (11a-c)}, \\ \text{DMF/CF}_3C_6F_{11}, \text{LiCl}, \\ 80°C, \ 3\text{ - }24h}} R\text{---}Ar + F\text{---}Sn(n\text{-}Bu)_3 \quad (2)$$

12
[39 - 98%]

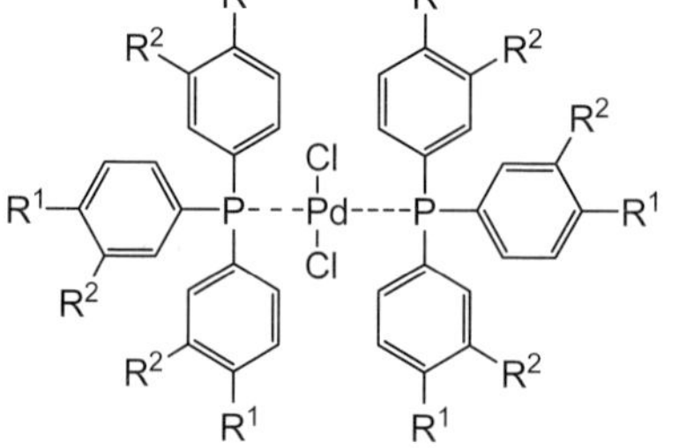

11a: $R^1 = C_8F_{17}$, $R^2 = H$
11b: $R^1 = H$, $R^2 = C_8F_{17}$
11c: $R^1 = C_2H_4C_8F_{17}$, $R^2 = H$
11d: $R^1 = H$, $R^2 = C_2H_4C_8F_{17}$

From the organic phase the C–C coupling products (**12**) were obtained in good yields, while the fluorous phase containing the catalyst was used as such for the next cycle. The catalysts could be used up to six times without significant decrease in yield.

Schneider and Bannwarth applied the same fluorous bis(triphenylphosphane)-palladium dichloride complexes **11a–11d** as catalyst precursors in Suzuki reactions [21] as well. The reactions took place in a H_2O/DMF/PFMCH triphasic mixture with 1.5 mol% of Pd complex. It could be demonstrated that perfluoro-tagged complexes **11a–11d** are highly effective precatalysts for Suzuki couplings under FBS conditions with either electron-rich or electron-deficient bromoarenes and arylboronic acids. The catalysts could be recycled and re-used after phase separation up to six times without a significant decrease in coupling yields. It could be shown in one example that the amount of catalyst could be reduced from 1.5 mol% to 0.1 mol% still resulting in a high yield (> 86%) in the first run but considerable loss of activity in repetitive cycles. Rocaboy and Gladysz prepared perfluoro-tagged dialkyl sulfides $(R^F(CH_2)_n)_2S$ ($n = 2$–16) which are soluble in most fluorous and organic solvents with a $CF_3C_6F_{11}$/toluene partition coefficient of 98.7 : 1.3 for **13** and 96.6 : 3.4 for **14** at 24 °C [Eq. (3)] [22].

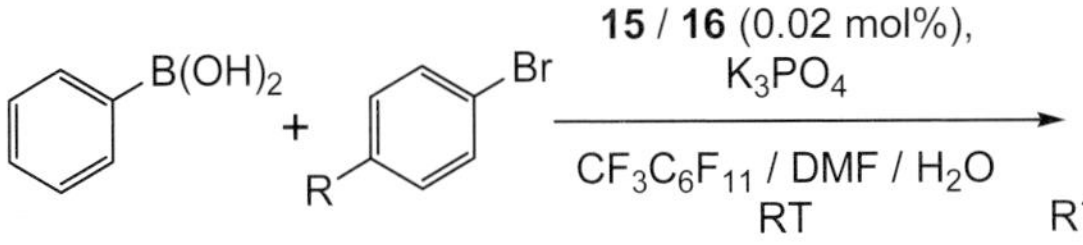

$$ (3) $$

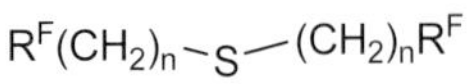

Reaction of **13** or **14** with Na_2PdCl_4 gave Pd complexes **15** and **16** respectively, which are soluble in only a limited range of fluorinated solvents at room temperature. With **15** and **16** as catalyst they were able to achieve turnover numbers of 4500–5000 in Suzuki couplings of aryl bromides and phenylboronic acid in $CF_3C_6F_{11}$/DMF/H_2O in the presence of K_3PO_4. Under fluorous recycling conditions, decreased activities of the catalysts were observed. For this loss of activity the following reasons could be responsible: (a) inefficient recycling of the perfluoro-tagged catalyst; (b) gradual deactivation of the catalyst; (c) slow generation of an active, nonrecyclable heterogeneous catalyst from a homogeneous precursor and recycling of the remaining precursor; (d) generation of a heterogeneous catalyst which is not stable but can be efficiently recycled. The appearance of palladium black in recycling experiments and recent reports on heterogeneous or metallic palladium species lead the authors to favor possibilities (c) and (d) [18, 19, 23, 24].

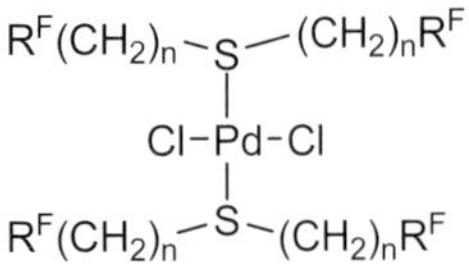

Bannwarth and co-workers have recently developed new protocols for the separation and recycling of perfluoro-tagged catalysts without the need for fluorous solvents [25]. They have employed Pd complexes **11c**, **11d**, and **17**, immobilized by adsorption on fluorous reversed-phase silica gels (FRPSG) **18** and **19**, and demonstrated the application to Suzuki couplings in organic solvents. Coarse-grained silica of 100–300 µm particle size was used with loadings between 0.1 and 100 mg complex per g FRPSG. The coupling of *p*-nitrobromobenzene and phenylboronic acid was carried out using 10 mg Pd complex per g FRPSG with 0.1 mol% palladium. Complete conversions were obtained and recycling was possible by filtration or decantation without a significant decrease of activity. With 0.001 mol% of catalyst, a TON of 131 000 was observed. ICP-MS measurements indicated a leaching of 1.9% of catalyst when adsorbed to FRPSG **18** and 1.6% when adsorbed to FRPSG **19**, respectively. Suzuki couplings with catalyst **11c** on support **18** were performed with different substrates giving high yields for electron-deficient aryl bromides and for aryl iodides. Markert and Bannwarth employed the Pd complexes **11c**, **11d**, and **17** as catalyst precursors for the coupling of bromoarenes with alkynes [Eq. (4)] [26]. The reactions were carried out in a mixture of DMF and perfluoro-dimethylcyclohexane (PFMCH) with 2 mol% of **11c**, **11d**, or **17** and 5 mol% of CuI as co-catalyst in the presence of 2 equiv of Bu₂NH. After the reaction the phases were separated at 0 °C and the fluorous phase containing the catalyst was washed several times with DMF and was re-used as such for the next run. As known for Sonogashira couplings, electron-deficient bromoarenes proved to be good substrates, whereas the coupling of donor-substituted bromoarene resulted in lower yields. Recycling and re-use of the catalyst was possible in most cases but no influence on product yield dependent on the position of the perfluoro tag in the phosphane or the nature of the spacer group was observed.

Perfluoro-tagged Pd complexes **11c**, **11d**, and **17** adsorbed on FRPSG **18** or **19** (2 mol%) were also used for Sonogashira couplings of phenyl acetylene and

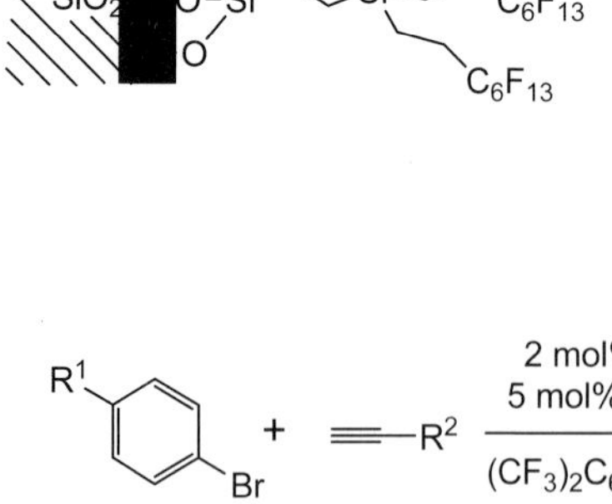

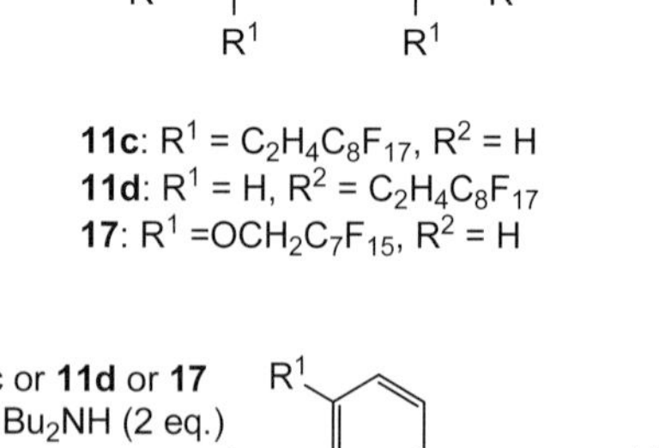

11c: R^1 = C$_2$H$_4$C$_8$F$_{17}$, R^2 = H
11d: R^1 = H, R^2 = C$_2$H$_4$C$_8$F$_{17}$
17: R^1 =OCH$_2$C$_7$F$_{15}$, R^2 = H

R^1 = NO$_2$; COMe; CO$_2$Et, OMe

R^2 = TIPS; Ph; CMe$_2$OH

p-nitrobromobenzene, without the need for fluorous solvents similar to those in the protocol outlined for Suzuki couplings [25]. High yields were obtained for three successive experiments.

Leitner and co-workers described Pd-catalyzed nucleophilic substitutions of allylic substrates with different nucleophiles [27]. They used $Pd_2(dba)_3$ as the palladium source and phosphane **20** as perfluoro-tagged ligand [Eq. (5)]. Reaction between cinnamyl methyl carbonate (**21**) and various nucleophiles (Nu–H) were performed in a THF/C_7F_{14} biphasic mixture. A decrease in conversion was observed only after the ninth run (with 5 mol% Pd complex). By reducing the amount of Pd complex to 1 mol%, five quantitative recyclings were possible. The standard protocol was also applied to the condensation of dimethyl malonate with allyl methyl carbonate, (2-vinyl)butyl carbonate, and cyclohex-2-enyl carbonate. In each case two recyclings were performed without any decrease in conversion.

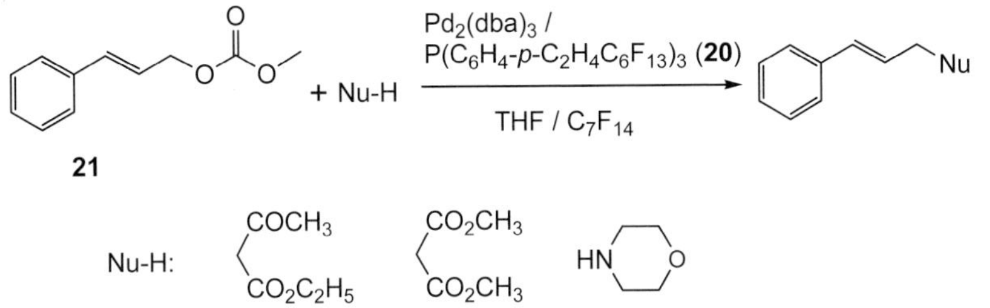

Cyclodimerizations of conjugated enynes **22a–22e** in the presence of perfluoro-tagged Pd catalyst were reported by Saito et al. [Eq. (6)] [28]. Reactions of enynes **22a–22e** were carried out in toluene/hexane/perfluorodecalin with 1 mol% of $Pd_2(dba)_3$ as palladium source and 8 mol% of perfluoro-tagged phosphane **1** giving the desired products (**23a–23e**) in moderate to good yield (43–78%). Recovery of the perfluoro-tagged catalyst was possible up to four times. However, IPC atomic emission analysis of the fluorous phase indicated that the concentration of the palladium species decreased significantly.

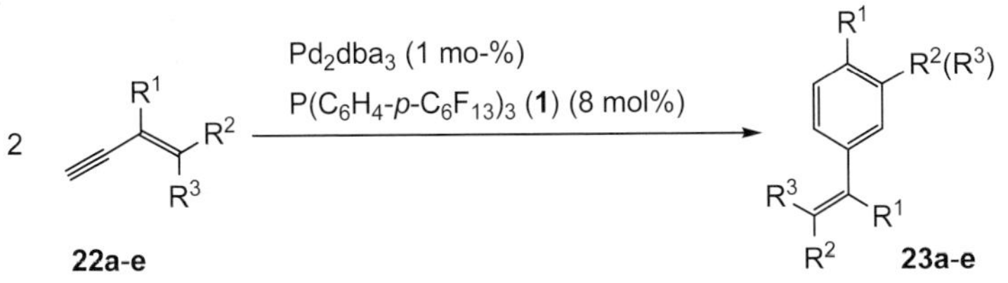

22a: R^1 =H; R^2 = H; R^3 = CO_2Et **23a**: 78% (76, 81, 79, 13)

22b: R^1 = H; R^2 = H; R^3 = CO_2n-C_8H_{17} **23b**: 76% (71, 78, 74, 81, 71, 41)

22c: R^1 = n-C_6H_{13}; R^2 = H; R^3 = CO_2Et **23c**: 74%

22d: R^1 = H; R^2 = CO_2n-C_8H_{17}; R^3 = H   **23d**: 43%

22e: R^1 = n-C_6H_{13}; R^2 = CO_2Et; R^3 = H   **23e**: 53%

Endres and Maas prepared dimeric rhodium(II) carboxylate complexes **24a–24c** from the sodium salts **25a–25c** and rhodium(III) chloride hydrate in ethanol [Eq. (7)] [29, 30].

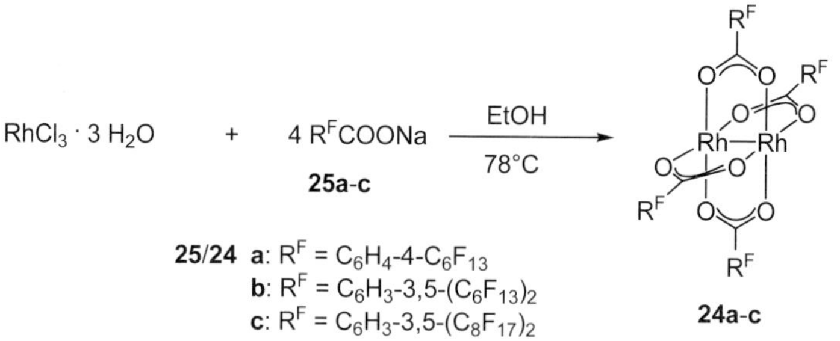
RhCl3 · 3 H2O
+
4 RFCOONa
25a-c
EtOH
78°C
24a-c
25/24 a: RF = C6H4-4-C6F13
b: RF = C6H3-3,5-(C6F13)2
c: RF = C6H3-3,5-(C8F17)2
(7)

Complexes **24a–24c** were applied to cyclopropanations of styrenes with methyl diazoacetate [Eq. (1), Scheme 4.2-12]. Reactions were performed in CH_2Cl_2 with **24a** or in the fluorous/organic hybrid solvent FC-113 ($CCl_2F–CClF_2$) with **24b–24c**. Recovery of the catalyst was achieved by extraction into PFMCH (with **24a**) or by replacement of FC-113 with PFMCH/CH_2Cl_2 (with **24b–24c**) and phase separation, respectively. Yields remained high over four cycles with **24a** or five cycles with **24b–24c**, respectively. Nevertheless, a total loss of 38–56% after five cycles, as shown by gravimetric determination, was observed. The authors attributed this to a partial destruction of the complex. Intramolecular carbenoid C–H insertions of a-diazo-β-keto ester **26** [Eq. (9)] were catalyzed by **24c** with good selectivity for **27a** and good yields [30]. By extraction with PFMCH/CH_2Cl_2, 96% of the catalyst could be recycled.

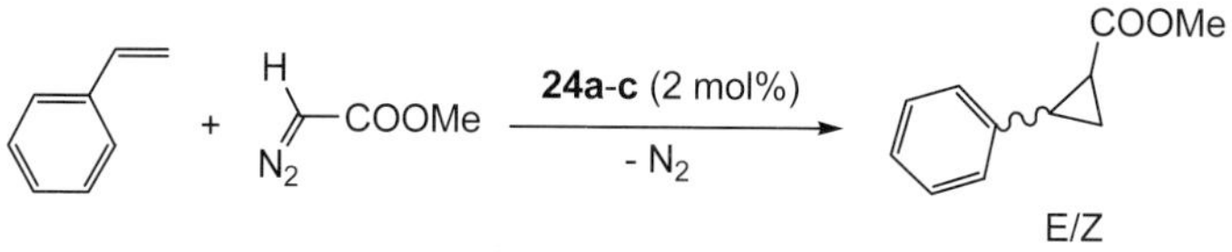
COOMe
24a-c (2 mol%)
- N2
E/Z
(8)

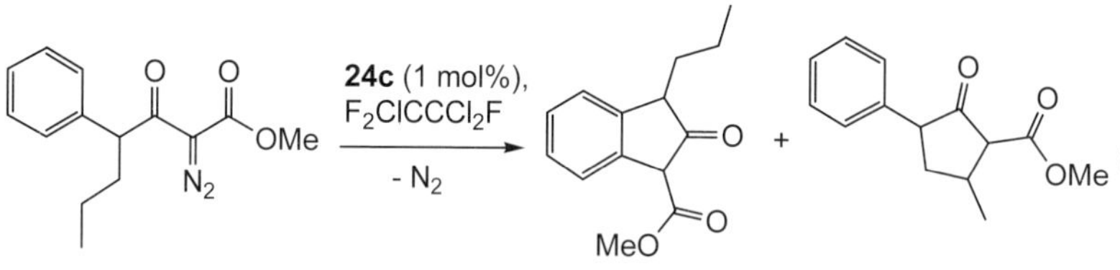
24c (1 mol%),
F2ClCCCl2F
- N2
OMe
MeO
(9)

26 **27a** **27b**

The addition reaction of CCl_4 to methyl methacrylate in DCM was catalyzed by perfluoro-tagged nickel compound **28** as described by van Koten et al. [Eq. (10)] [31]. Unfortunately, **28** did not have an improved affinity for fluorous solvents, which prevented its efficient recycling.

The Friedel–Crafts acylation of arenes with acetic anhydride was efficiently catalyzed by ytterbium tris(perfluoroalkanesulfonyl)methides **29a–29c** [32]. It was demonstrated that catalyst **29c** could be recovered in 96% yield by extraction of the reaction mixture with hot perfluoromethyldecalin and could be re-used in a second run.

$$CH_2\text{=} \quad \overset{O}{\underset{}{\|}}\text{OMe} \; + \; CCl_4 \quad \xrightarrow{\quad 28 \quad} \quad CCl_3\text{-}\overset{Cl}{\underset{}{}}\text{-}\overset{O}{\underset{}{\|}}\text{OMe} \qquad (10)$$

$$Yb[C(SO_2R^{F1})_2(SO_2R^{F2})]_3$$

29a: $R^{F1} = R^{F2} = C_4F_9$
29b: $R^{F1} = R^{F2} = C_6F_{13}$
29c: $R^{F1} = C_6F_{13}$; $R^{F2} = C_8F_{17}$

References

1 (a) F. Diederich, P. J. Stang, *Metal-Catalyzed Cross-Coupling Reactions*, Wiley-VCH, Weinheim **1998**; (b) J. Tsuji, *Palladium Reagents and Catalysts: Innovations in Organic Synthesis*, Wiley, Chichester **1995**.

2 D. P. Curran, *Angew. Chem.* **1998**, *110*, 1230; *Angew. Chem., Int. Ed.* **1998**, *37*, 1174.

3 C. C. Tzschucke, C. Markert, W. Bannwarth, S. Roller, A. Hebel, R. Haag, *Angew. Chem.* **2002**, *114*, 4136; *Angew. Chem., Int. Ed.* **2002**, *41*, 3964.

4 (a) W. Bannwarth, E. Felder, *Combinatorial Chemistry; A Practical Approach*, Wiley-VCH, Weinheim **2000**; (b) S. B. Yang, *Tetrahedron Lett.* **1997**, *38*, 1793.

5 T. Welton, *Chem. Rev.* **1999**, *99*, 2071.

6 P. Wasserscheid, T. Welton, *Ionic Liquids in Synthesis*, Wiley-VCH, Weinheim **2003**.

7 P. G. Jessop, W. Leitner, *Chemical Synthesis Using Supercritical Fluids*, Wiley-VCH, Weinheim **1999**.

8 B. Cornils, W. A. Herrmann (Eds.), *Aqueous-Phase Organometallic Catalysis: Concepts and Applications*, Wiley-VCH, Weinheim **1998**.

9 I. T. Horváth, J. Rábai, *Science* **1994**, *266*, 72.

10 E. de Wolf, G. van Koten, B.-J. Deelman, *Chem. Soc. Rev.* **1999**, *28*, 37.

11 B. Cornils, *Angew. Chem.* **1997**, *109*, 2147; *Angew. Chem., Int. Ed. Engl.* **1997**, *36*, 2057.

12 A. Endres, G. Maas, *Chem. Unserer Zeit* **2000**, *34*, 382.

13 R. H. Fish, *Chem. Eur. J.* **1999**, *5*, 1677.

14 I. T. Horváth, *Acc. Chem. Res.* **1998**, *31*, 641.

15 B. Betzemeier, P. Knochel, *Angew. Chem.* **1997**, *109*, 2736; *Angew. Chem., Int. Ed.* **1997**, *36*, 2623.

16 J. Moineau, G. Pozzi, S. Quici, D. Sinou, *Tetrahadron Lett.* **1999**, *40*, 7683.

17 C. Rocaboy, J. A. Gladysz, *Org. Lett.* **2002**, *4*, 1993.

18 M. Moreno-Mañas, R. Pleixats, S. Villarroya, *Organometallics* **2001**, *20*, 4524.

19 M. Moreno-Mañas, R. Pleixats, S. Villarroya, *J. Chem. Soc., Chem. Commun.* **2002**, 60.

20 S. Schneider, W. Bannwarth, *Angew. Chem.* **2000**, *112*, 4293; *Angew. Chem Int. Ed.* **2000**, *39*, 4142.

21 S. Schneider, W. Bannwarth, *Helv. Chim. Acta* **2001**, *84*, 735.

22 C. Rocaboy, J. A. Gladysz, *Tetrahedron* **2002**, *58*, 4007.

23 M. T. Reetz, E. Westermann, *Angew. Chem.* **2000**, *112*, 170; *Angew. Chem., Int. Ed.* **2000**, *39*, 165.

24 R. M. Crooks, M. Zhao, L. Sun, V. Chechik, L. K. Yeung, *Acc. Chem. Res.* **2001**, *34*, 181.

25 C. C. Tzschucke, C. Markert, H. Glatz, W. Bannwarth, *Angew. Chem.* **2002**, *114*, 4678; *Angew. Chem., Int. Ed.* **2002**, *41*, 4500.

26 C. Markert, W. Bannwarth, *Helv. Chim. Acta* **2002**, *85*, 1877.

27 R. Kling, D. Sinou, G. Pozzi, A. Choplin, F. Quignard, S. Busch, S. Kainz, D. Koch, W. Leitner, *Tetrahedron Lett.* **1998**, *39*, 9439.

28 S. Saito, Y. Chounan, T. Nogami, O. Ohmori, Y. Yamamoto, *Chem. Lett.* **2001**, 444.

29 A. Endres, G. Maas, *Tetrahedron Lett.* **1999**, *40*, 6365.

30 A. Endres, G. Maas, *Tetrahedron* **2002**, *58*, 3999.

31 H. Kleijn, J. T. B. H. Jastrzebski, R. A. Gossage, H. Kooijman, A. L. Spek, G. van Koten, *Tetrahedron* **1998**, *54*, 1145.

32 A. G. M. Barrett, D. C. Braddock, D. Catterick, D. Chadwick, J. P. Henschke, R. M. McKinnell, *Synlett* **2000**, 847.

4.2.2
Hydroformylation and Hydrogenation Catalyzed by Perfluoroalkylated Phosphine–Metal Complexes

Eric G. Hope and Alison M. Stuart

4.2.2.1
Hydroformylation

The hydroformylation of alkenes has been extensively investigated and was exploited commercially for the first time in the 1940s using cobalt-carbonyl catalysts [1]; see Chapter 2. Enhanced hydrogenation of aldehydes to give alcohols coupled with improved selectivity to the linear (*n*-) product has been accomplished by the introduction of phosphine ligands [2] and these technologies are still in use today for the hydroformylation of long-chain alkenes. In the hydroformylation of propene, where the volatile butanal product can be readily distilled directly from the reactor, more selective rhodium-triphenylphosphine catalysts operating under milder reaction conditions have replaced the cobalt-based catalysts in continuous processes operating at 3.5 million tonnes per annum [3]. However, the thermal instability of these rhodium catalysts has precluded their widespread adoption and has led to the evaluation of a number of alternative approaches including fluorous chemistry. In fact, Horváth and Rábai highlighted the potential for fluorous technologies for the hydroformylation of long-chain alkenes in their original disclosure on the fluorous biphase [4].

The relative instability of the rhodium-based catalysts has already been overcome, and the technique exploited commercially, in the Ruhrchemie-Rhône Poulenc hydroformylation of propene, where the sodium salt of a sulfonated triphenyl-phosphine ligand **1** is used to solubilize the rhodium catalyst in the aqueous phase [5]; see Section 2.5.1. However, this approach cannot be extended to the hydro-formylation of longer-chain alkenes because of their very low aqueous solubility, leading to commercially prohibitive low reaction rates. A potential solution to this problem was reported in 1998 in which the sulfonated triphenylphosphine ligand was replaced with triphenylphosphine functionalized with a single, very long, poly(ethylene glycol) ponytail **2** [6] (cf. Section 2.3.5). Here, although the rhodium–phosphine catalyst is exclusively soluble in water at room temperature, on heating the ponytail undergoes a phase transition rendering the catalyst preferentially soluble in the organic phase. Consequently, the catalyst acts on the substrate under classical homogeneous conditions but, on cooling, reversal of the phase transition returns the catalyst to the aqueous phase, allowing the product to be separated by a simple decantation.

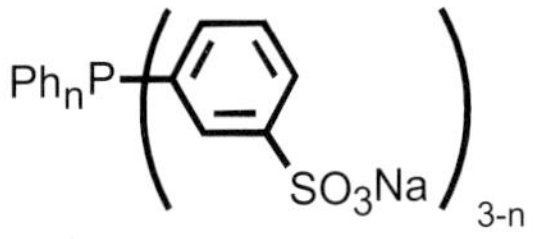

1 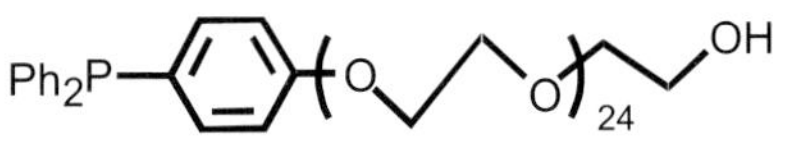   **2**

Ionic liquids, such as 1-butyl-3-methylimidazolium hexafluorophosphate (**3**), are salts that (a) are liquid at room temperature, (b) have extremely low vapor pressures, (c) can be tuned to dissolve organic compounds, and (d) can dissolve ionic catalysts or, in specific examples, the liquid may act as a catalyst as well as the solvent. Consequently, there has been considerable interest in their potential application as alternative solvents for organic synthesis and catalysis (cf. Chapter 5). In initial work on the rhodium-catalyzed hydroformylation of 1-pentene, using the sulfonated phosphine ligand **1** to solubilize the catalyst in the ionic liquid, conversions were disappointingly low [7]. More recently, two groups have reported highly active and regioselective rhodium-catalyzed hydroformylation of 1-octene with Xantphos-type ligands in ionic liquids, {**4**, n/i 21 : 1 and < 0.07% rhodium leaching [8]; **5**, n/i 49 : 1 and < 5 ppb rhodium loss [9]}, where separation of the 1-nonanal product can be achieved again by a simple decantation.

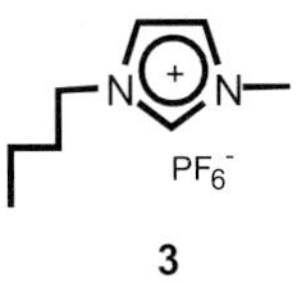

3 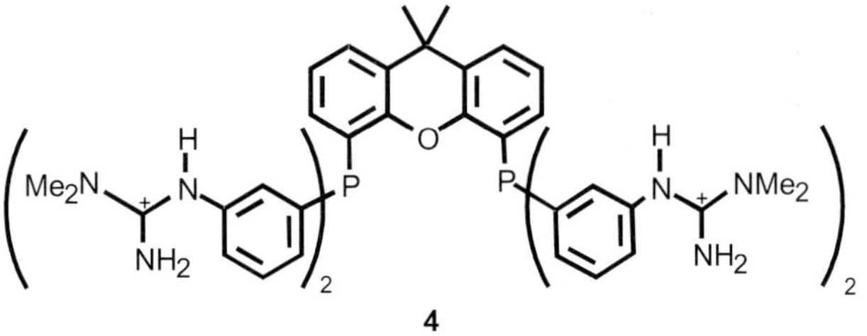   **4**

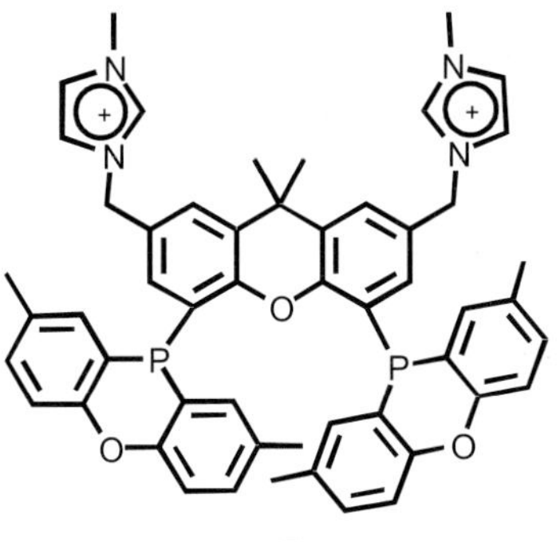

5

Rhodium catalysts for hydroformylation supported on both insoluble (for example, inorganic oxides or polymers) and soluble materials (for example, dendrimers) have been reported. Work on insoluble solid supports has been dogged by loss of activity and high levels of catalyst leaching as a consequence of bonds breaking between the catalyst and support during catalysis (cf. Chapter 7). However, acceptable catalyst leaching (< 100 ppb rhodium) coupled with reasonable activity and excellent regioselectivity (n/i 40 : 1) has been demonstrated for a sol–gel solution incorporating a triethoxysilyl-functionalized Xantphos-type ligand 6 [10]. Dendrimers are large treelike soluble molecules with a globular shape making them suitable for ultrafiltration in which the solvent and reaction product(s) pass through while the dendrimer is retained. Rhodium for hydroformylation can be supported on the "surface" of the dendrimer functionalized with phosphines, and such a system shows enhanced regioselectivity but loss of activity on recycling [11].

$(EtO)_3Si$

N

O

PPh$_2$ PPh$_2$

6

Most metal-containing complexes, particularly rhodium-based hydroformylation catalysts incorporating aryl–phosphine ligands, are virtually insoluble in apolar scCO$_2$. Solubility can be enhanced by the incorporation of the perfluoroalkyl groups characteristic of fluorous chemistry and reaction rates and regioselectivities for the hydroformylation of long-chain alkenes, comparable to those observed under fluorous biphase conditions, have been reported using derivatized rhodium catalysts [12, 13]. In two reports, significantly enhanced rates of reaction have been reported in scCO$_2$ in comparison to those in toluene for the rhodium-catalyzed hydroformylation of acrylic esters using the perfluoroalkylated phosphine 7 [14], or a fluoropolymer ligand [15]. However, although there has been considerable interest in scCO$_2$ as an environmentally friendly solvent for homogeneous catalysis [16, 17], for which removal of the solvent just requires decompression back to the gaseous phase, this does not *per se* overcome the principal issue for homogeneous catalysis outlined above: that is, the separation of product from catalyst (see Chapter 6). In theory, since solubility in supercritical fluids is pressure-dependent, with careful control of the catalyst/substrate/product system it should be feasible to accomplish the desired separation, and this has been achieved with moderate success (< 170 ppb rhodium leaching) in the hydroformylation of 1-octene using the perfluoroalkylated triphenylphosphine 8 [12]. Alternative approaches to the catalyst/product separation problem have combined scCO$_2$ with a supported catalyst (for example 6 [18]) and scCO$_2$ with an ionic liquid [19], both of which are potentially very powerful systems, but in both cases the reaction rates are relatively low.

P—(⟨benzene⟩—C$_6$F$_{13}$)$_3$ P—(⟨benzene⟩—C$_2$H$_4$C$_6$F$_{13}$)$_3$

7   **8**

In the earliest reports of the application of perfluorocarbon solvents and perfluoroalkylated ligands/metal catalysts, Horváth and Rábai outlined the hydroformylation of 1-octene in a toluene/CF$_3$C$_6$F$_{11}$ (PP2) two-phase system at 100 °C under 1 MPa CO/H$_2$ (1 : 1) using a catalyst generated in situ from [Rh(CO)$_2$(acac)] and P(C$_2$H$_4$C$_6$F$_{13}$)$_3$ (**9**) which gave an 85% conversion to aldehydes with an n/i ratio of 2.9 [4, 20]. In the following full paper [21], an in-depth analysis of hydroformylation under fluorous biphase conditions generated a series of important conclusions. Here, the hydroformylation of 1-decene and ethylene were investigated with the same rhodium catalyst (generated in situ) under both batch and semi-continuous conditions at 100 °C and 1.1 MPa CO/H$_2$ (1 : 1) in a 50/50 vol% toluene/PP2 biphase. The long-term stability of this catalyst under these conditions is significantly greater than that for the catalyst based on triphenylphosphine, the regioselectivity is similar, but the catalytic activity is an order of magnitude lower. The reaction, as expected, is first order in both rhodium and alkene and is inhibited by excess phosphine, whereas the regioselectivity increases with phosphine concentration such that the best n/i ratio (7.84) is obtained at a P/Rh ratio of approximately 100 : 1 ([ligand] = 0.3 mol dm^{-3}). The semi-continuous experiments were highly successful with total turnovers of up to 35 000 during nine cycles with only 1.18 ppm (4.2%) loss of rhodium per mole of product(s), which arises from the low solubility of the catalyst in the organic phase.

$$P(C_2H_4C_6F_{13})_3$$

9

It is well established that triarylphosphines give much better regioselectivity than trialkylphosphines in rhodium-catalyzed hydroformylation reactions [3, 22], so rhodium catalysts based upon **7** have been evaluated [23, 24]. Initial screening of the hydroformylation of 1-hexene in a toluene/1,3-(CF$_3$)$_2$C$_6$F$_{10}$ (PP3) two-phase system at 70 °C under 20 MPa CO/H$_2$ (1 : 1) using a catalyst generated in situ from [Rh(CO)$_2$(acac)] and **7** (1 : 3) gave a 98% conversion to aldehydes with an n/i ratio of 3.8. Visual inspection of the solvent system under 2 MPa syngas indicated that the alkene starting materials were miscible with the fluorous solvent at the reaction temperature while the more polar aldehyde products were immiscible, and this work led to an evaluation of the perfluoralkylated rhodium hydroformylation catalyst in the absence of the second organic phase. Here, thorough investigation of the hydroformylation of 1-octene in PP2 between 70 and 90 °C under 2 MPa CO/H$_2$ (1 : 1) using the same rhodium catalyst with metal/ligand ratios of 1 : 3 and 1 : 10 gave 95–98% conversions with n/i ratios of 3.0–6.3. Crucially, rhodium leaching levels, detected at the best regioselectivity (n/i 6.3; conditions rhodium/phosphine 1 : 10; 70 °C), were excellent (80 ppb), indicating that the omission of toluene from the solvent system had enabled the development with excellent retention of rhodium,

Table 1 Rhodium catalyzed hydroformylation of 1-octene.[a]

System	Ligand	Pressure [MPa]	T [°C]	TOF [h⁻¹]	Rate [mol dm⁻³ h⁻¹]	n : i	Rh loss [mg (mg product⁻¹]	Ref.
Homogeneous[b]	PPh$_3$	1.5	95	770	2.0	8.8 : 1	n.a.	[3]
Aqueous biphase[b]	1	5.0	120	400	1.1	19 : 1	< 0.005	[5]
Aqueous biphase	2	5.0	100	182	0.5	n.r.	n.r.	[6]
Ionic liquids	4	3.0	100	50	n.r.	21 : 1	< 0.07% [c]	[8]
Ionic liquids	5	4.6	100	318	1.2	49 : 1	< 0.005	[9]
Supported catalyst	6	5.0	80	287	0.19	40 : 1	< 0.1	[10]
scCO$_2$	8	20.0	65	430	14.2	5.5 : 1	< 0.17	[12]
Supported/scCO$_2$	6	17.0	90	160	n.r.	33 : 1	< 1.2	[18]
Fluorous biphase	9	1.0	100	837	0.1	4.5 : 1	0.12	[21]
Fluorous biphase	7	2.0	70	4400	8.8	6.3 : 1	0.08	[24]

a) n.a. = not applicable; n.r. = not reported.
b) Propene as substrate.
c) Reported as % of catalyst loading.

while both the high reaction rate and good regioselectivity required for commercial application are maintained.

Table 1 summarizes the key catalytic data for these hydroformylation reactions under fluorous biphase conditions alongside data for representative examples from the alternative solvent systems outlined above. In general, the results are comparable; in some cases there are better regioselectivities, in others better reaction rates, in yet others better catalyst retention. In attempts to improve regioselectivity and catalyst retention under fluorous biphase conditions, we and others have been investigating perfluoroalkylated bidentate ligands based upon Xantphos, for example, **10** [25] and BIPHEPHOS [26, 27]. Unfortunately, the introduction of perfluoroalkyl units onto biphenol has, to date, prevented the synthesis of fluorous BIPHEPHOS-type bisphosphite ligands, while **10** with only four perfluoroalkyl groups, although it is active in the rhodium-catalyzed hydroformylation of 1-octene (n/i 16 : 1), is not soluble in fluorous solvents and attempts to increase the number

10

of perfluoroalkyl substituents have not yet been successful. These latest results appear to suggest that the future development of fluorous chemistry in hydroformylation probably rests with simpler, monodentate, ligand systems.

4.2.2.2
Hydrogenation

The hydrogenation of unsaturated organic compounds represents one of the most environmentally benign processes, in that it produces virtually no waste. Since heterogeneous catalysts (for example, Raney nickel, palladium on carbon) are highly effective, homogeneous hydrogenation catalysts will only find application when other factors (for example, substrate incompatibility, enantioselectivity, transfer hydrogenation to avoid the need to use gaseous hydrogen) are important [28]. An industrial example is the rhodium-catalyzed homogeneous enantioselective hydrogenation of dehydroamino acids in the synthesis of L-dopa [29]. In these cases, as for the homogeneous hydroformylation catalysts outlined above, product/catalyst separation is a major issue that has led to the synthesis and evaluation of homogeneous catalysts in a variety of alternative solvents such as water [30–32], ionic liquids [7, 33, 34], and supercritical carbon dioxide [16, 17, 35–39] (see Chapters 2 to 6).

In contrast with the research into hydroformylation under FBS conditions that has been directed toward a commercially important process, publications on hydrogenation under FBS conditions have been focused upon the physical and chemical consequences of using perfluoroalkylated phosphine ligands and fluorous solvents and the ability to recover and recycle the metal catalyst. Horváth et al. [40], using the analogue of Wilkinson's catalyst, [$RhClL_3$], where L = $P(C_2H_4C_6F_{13})_3$ (**9**), studied the hydrogenation of a series of alkenes (2-cyclohexen-1-one, 1-dodecene, cyclododecene, and 4-bromostyrene) in a toluene/PP2 biphase under 0.1 MPa. H_2 at 45 °C affording the hydrogen addition products in 87–98% yields. Although the catalyst activity is significantly poorer than those for conventional homogeneous catalysts, recovery and re-use of the catalyst were illustrated by recharging the fluorous phase with second and third aliquots of substrate and obtaining comparable conversions; however, some catalyst decomposition was also observed. It is well known that alkylphosphines give much less effective analogues of Wilkinson's catalyst than arylphosphines, and perfluoroalkylated arylphosphines have been evaluated by other groups. A soluble fluoropolymer-supported alkyldiphenyl-phosphine is active for the hydrogenation of 1-octene and cyclohexene in a THF/perfluorooctane biphase under 0.2 MPa. H_2 at 25 °C, where re-use seven times shows no loss in activity although rhodium leaching levels have not been measured [41]. Using styrene as a substrate, the catalytic activities of the analogues of Wilkinson's catalyst containing perfluoroalkylated phosphines, for example, **7**, with those of their protio-parents in toluene/hexane : PP3 or fluorobenzene : PP3 biphases under 0.1 MPa. H_2 have been compared at 63.5 or 75 °C respectively, where just the introduction of the fluorous phase had a significant impact upon the rates of reaction, but < 1 ppm rhodium leaching was observed [42]. In line with well-established trends, the incorporation of the electron-withdrawing perfluoroalkyl groups caused a reduction in rate relative to those for the protio-parents; this effect is most pronounced for the trialkylphosphine with the C_2H_4 spacer unit, indicating that it is a poorer electronic insulator than the C_6H_4 group. The most promising results were obtained for a $C_6H_4OCH_2$ spacer group although, even with this unit, complete electronic insulation of the phosphorus atom was not possible. The most effective insulation is reported in a direct comparison of the aryl–silyl spacer ligand (**12**; TOF 870 h^{-1}) with Wilkinson's catalyst (TOF 960 h^{-1}) in the hydrogenation of 1-octene in α,α,α-trifluorotoluene under 0.1 MPa. H_2 at 80 °C [43]. In PP2 (using **12**, 1-octene at 80 °C) the TOF drops to 177 h^{-1}, but on cooling to 0 °C a biphase forms that allows separation of the octane product, and the catalyst phase can be recycled nine times with just 3 ppm (0.12%) rhodium leaching per cycle. The reactivity appears to increase during the subsequent cycles, although this can be ascribed to loss of the perfluorocarbon solvent (ca. 12% per cycle) during phase separation and the nonzero miscibility of PP2 with the product phase. The best catalyst retention (> 99.92%) has been reported following hydrogenation of 1-octyne under 0.1 MPa. H_2 at 40 °C with the cationic [Rh(COD)(**13**)][BF$_4$] in an FC-75/hexane biphase, where the chelating bidentate phosphine ligand **13** has 12 fluorous ponytails [44].

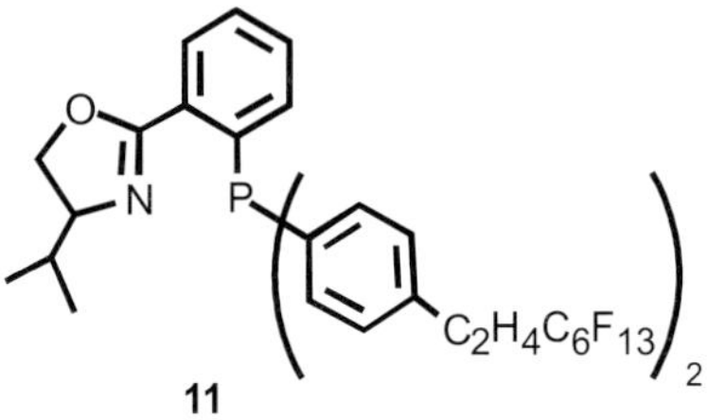

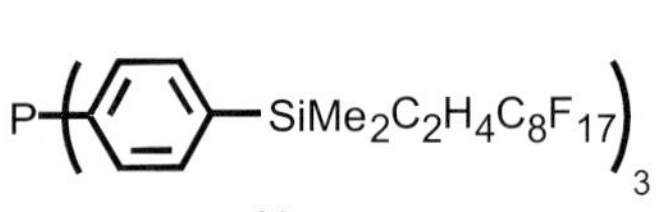

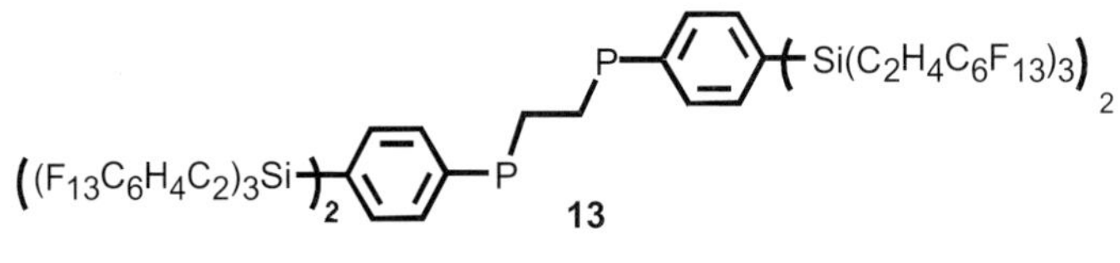

References

1 B. Cornils, in *New Syntheses with Carbon Monoxide* (Ed.: J. Falbe), Springer, Berlin **1980**, Chapter 1.
2 Shell Oil Co (T. H. Johnson), US 4.584.411 (**1985**).
3 C. D. Frohning, C. W. Kohlpaintner, in *Applied Homogeneous Catalysis with Organometallic Compounds* (Eds.: B. Cornils, W. A. Herrmann), VCH, Weinheim **1996**, p. 61.
4 I. T. Horváth, J. Rábai, *Science* **1994**, *260*, 72.
5 C. D. Frohning, C. W. Kohlpaintner, in *Applied Homogeneous Catalysis with Organometallic Compounds* (Eds.: B. Cornils, W. A. Herrmann), VCH, Weinheim **1996**, p. 80.
6 X. Cheng, J. Jiang, X. Liu, Z. Jin, *Catal. Today* **1998**, *44*, 175.
7 Y. Chauvin, L. Mussmann, H. Olivier, *Angew. Chem., Int. Ed. Engl.* **1995**, *34*, 2698.

8 P. Wasserscheid, H. Waffenschmidt, P. Machnitzki, K. W. Kottsieper, O. Stelzer, *Chem. Commun.* **2001**, 451.

9 R. P. J. Bronger, S. M. Silva, P. C. J. Kamer, P. W. N. M. van Leeuwen, *Chem. Commun.* **2002**, 3044.

10 A. J. Sandee, J. N. H. Reek, P. C. J. Kamer, P. W. N. M. van Leeuwen, *J. Am. Chem. Soc.* **2001**, *123*, 8468.

11 L. Ropartz, K. J. Haxton, D. F. Foster, R. E. Morris, A. M. Z. Slavin, D. J. Cole-Hamilton, *J. Chem. Soc, Dalton Trans.* **2002**, 4323.

12 D. Koch, W. Leitner, *J. Am. Chem. Soc.* **1998**, *120*, 13 398.

13 A. M. Banet Osuna, W. Chen, E. G. Hope, R. D. W. Kemmitt, D. R. Paige, A. M. Stuart, J. Xiao, L. Xu, *J. Chem. Soc., Dalton Trans.* **2000**, 4052.

14 Y. Hu, W. Chen, A. M. Banet Osuna, A. M. Stuart, E. G. Hope, J. Xiao, *Chem. Commun.* **2001**, 725.

15 Y. Hu, W. Chen, A. M. Banet Osuna, J. A. Iggo, J. Xiao, *Chem. Commun.* **2002**, 788.

16 P. G. Jessop, I. Ikariya, R. Noyori, *Chem. Rev.* **1999**, *99*, 475.

17 W. Leitner, *Acc. Chem. Res.* **2002**, *35*, 746.

18 N. J. Meehan, A. J. Sandee, J. N. H. Reek, P. C. J. Kamer, P. W. N. M. van Leeuwen, M. Poliakoff, *Chem. Commun.* **2000**, 1497.

19 M. F. Sellin, P. B. Webb, D. J. Cole-Hamilton, *Chem. Commun.* **2001**, 781.

20 I. T. Horváth, J. Rábai, US 5.463.082 (**1995**).

21 I. T. Horváth, G. Kiss, R. A. Cook, J. E. Bond, P. A. Stevens, J. Rábai, E. J. Mozeleski, *J. Am. Chem. Soc.* **1998**, *120*, 3133.

22 J. K. MacDougall, M. C. Simpson, M. J. Green, D. J. Cole-Hamilton, *J. Chem. Soc., Dalton Trans.* **1996**, 1161.

23 D. F. Foster, D. J. Adams, D. Gudmunsen, A. M. Stuart, E. G. Hope, D. J. Cole-Hamilton, *Chem. Commun.* **2002**, 722.

24 D. F. Foster, D. Gudmunsen, D. J. Adams, A. M. Stuart, E. G. Hope, D. J. Cole-Hamilton, G. P. Schwarz, P. Pogorzelec, *Tetrahedron* **2002**, *58*, 3901.

25 D. J. Adams, D. J. Cole-Hamilton, D. A. J. Harding, E. G. Hope, P. Pogorzelec, A. M. Stuart, *Tetrahedron*, **2004**, *60*, 4079.

26 D. Bonafoux, Z. Hua, B. Wang, I. Ojima, *J. Fluorine Chem.* **2001**, *112*, 101.

27 D. J. Adams, D. Gudmunsen, E. G. Hope, A. M. Stuart, *J. Fluorine Chem.* **2003**, *121*, 213.

28 B. K. James, *Homogeneous Hydrogenation*, John Wiley, New York **1973**.

29 W. A. Knowles, *Acc. Chem. Res.* **1983**, *16*, 106.

30 B. Cornils, W. A. Herrmann, in *Applied Homogeneous Catalysis with Organometallic Compounds* (Eds.: B. Cornils, W. A. Herrmann), VCH, Weinheim **1996**, p. 577.

31 W. A. Herrmann, C. W. Kohlpainter, *Angew. Chem., Int. Ed. Engl.* **1993**, *32*, 1524.

32 F. Gassner, W. Leitner, *Chem. Commun.* **1993**, 1465.

33 R. A. Brown, P. Pollet, E. McKoon, C. A. Eckert, C. L. Liotta, P. G. Jessop, *J. Am. Chem. Soc.* **2001**, *121*, 1254.

34 P. J. Dyson, D. J. Ellis, T. Welton, *Can. J. Chem.* **2001**, *79*, 705.

35 P. G. Jessop, T. Ikariya, R. Noyori, *Nature* **1994**, *368*, 231.

36 P. G. Jessop, Y. Hsiao, T. Ikariya, R. Noyori, *J. Am. Chem. Soc.* **1996**, *118*, 344.

37 J. Burk, S. Feng, M. F. Gross, W. Tumas, *J. Am. Chem. Soc.* **1995**, *117*, 8277.

38 J. Xiao, S. C. A. Nefkens, P. G. Jessop, T. Ikariya, R. Noyori, *Tetrahedron Lett.* **1996**, *37*, 2813.

39 S. Kainz, A. Brinkmann, W. Leitner, A. Pfaltz, *J. Am. Chem. Soc.* **1999**, *121*, 6241.

40 D. Rutherford, J. J. J. Juliette, C. Rocaboy, I. T. Horváth, J. A. Gladysz, *Catalysis Today* **1998**, *42*, 381.

41 D. E. Bergbreiter, J. G. Franchina, B. L. Case, *Org. Lett.* **2000**, *2*, 393.

42 E. G. Hope, R. D. W. Kemmitt, D. R. Paige, A. M. Stuart, *J. Fluorine Chem.* **1999**, *99*, 197.

43 B. Richter, A. L. Spek, G. van Koten, B.-J. Deelman, *J. Am. Chem. Soc.* **2000**, *122*, 3945.

44 E. de Wolf, A. L. Spek, B. W. M. Kuipers, A. P. Philipse, J. D Meeldijk, P. H. H. Bomans, P. M. Frederik, B.-J. Deelman, G. van Koten, *Tetrahedron* **2002**, *58*, 3911.

4.2.3
Hydroformylation Catalyzed by Fluorous Triarylphosphite–Metal Complexes

Eric Monflier, André Mortreux, and Yves Castanet

Fluorous biphasic catalysis is a particularly elegant concept for the recycling of homogeneous catalysts as the catalyst and product phases are generally well separated at room temperature and can become homogeneous at higher temperatures [1–3]. Obviously, this behavior allows the combination of the activity and selectivity of homogeneous catalysts with the simplicity of product isolation. The hydroformylation of olefins in a fluorous phase was first reported by Horváth and Rábai in 1994 [1, 4] and discussed in Section 4.2.2.1. Since this pioneering work, the attention of research groups has been focused on the synthesis of new fluorinated ligands showing a better affinity for the fluorous phase and a better *l/b* (*n/i*) ratio. With the aim of keeping the high selectivities and activities observed in classical organic solvents, three classes of ligands have been rapidly developed for hydroformylation under fluorous biphasic conditions: triarylphosphines [5] and triarylphosphites [6–10] bearing one or two perfluoroalkyl group per aromatic ring and more recently, chiral phosphine-phosphite ligands with fluorous ponytails [11].

The behavior of rhodium catalysts generated in situ from fluorous phosphites **1–10** and $Rh(acac)(CO)_2$ has been investigated.

Phosphites **1–6** induce high activity in hydroformylation of terminal alkenes in comparison with the classical ligands PPh$_3$ and P(OPh)$_3$ (Table 1). Nevertheless, considerable differences in activity and selectivity exist according to the nature, the position, and the number of substituents on the aromatic ring of the phosphite. For instance, bulky *ortho*-substituted phosphites lead to catalytic systems that are much more active than those resulting from their *meta* or *para* counterparts. However, these *ortho*-phosphites give lower *l/b* ratios. Indeed, the *l/b* ratios are between 2 and 3, i.e., greatly inferior to those observed with the *meta* and *para* phosphites (ca. **5–8**) and slightly inferior or similar to those obtained with ligands PPh$_3$ or P(OPh)$_3$.

Classically, the reactivity decreases with the size of the substrate and, more unexpectedly, the *l/b* ratio and aldehyde selectivity vary in the same way. Interestingly, internal alkenes are also hydroformylated with significant activity with these catalytic

Table 1 Hydroformylation of higher alkenes under fluorous biphasic conditions with phosphites **1–6**.

Conditions	Substrate	Ligand	TOF[a] [h^{-1}]	l/b	Aldehyde selectivity [%]	Ref.
A	1-hexene	PPh$_3$	900	2.7	93.1	[12]
B	1-hexene	P(OPh)$_3$	1700	2.9	92	[12]
B	1-hexene	1a	4200	0.9	99.2	[12]
B	1-hexene	2a	1300	5.2	96.1	[12]
B	1-hexene	3a	2650	8.4	85.4	[12]
B	2-nonene	3a	880	–	75.1	[12]
C	1-octene	3a	15600	6.3	83.9	[12]
D	1-decene	1b	10000	2.3	71	[6]
D	1-decene	2b	6300	5.8	80	[6]
D	1-decene	3b	3500	5.3	85	[6]
D	1-decene	4	7100	2.9	80	[6]
D	1-decene	5	7900	2.4	46	[6]
D	1-decene	6	6200	2.6	39	[6]
D	1-octene	2b	6900	6.3	87	[6]
E	2-octene	2b	1400	–	75	[6]
E	4-octene	2b	800	–	76	[6]

Conditions:
(A) A solution of [Rh(acac)(CO)$_2$] (10.0 mmol dm^{-3}), ligand (30 mmol dm^{-3}), 1-hexene (8 mmol, 1 cm^3), toluene (4 cm^3) at 70 °C, 2 MPa CO/H$_2$ (1 : 1) for 1 h.
(B) A solution of [Rh(acac)(CO)$_2$] (10.0 mmol dm^{-3}), ligand (30 mmol dm^{-3}), 1-hexene (8 mmol, 1 cm^3) in a mixture of perfluoro-1,3-dimethylcyclohexane (2 cm^3) and toluene (2 cm^3) at 70 °C, 2 MPa CO/H$_2$ (1 : 1) for 1 h.
(C) A solution of [Rh(acac)(CO)$_2$] (2.0 mmol dm^{-3}), ligand (6 mmol dm^{-3}), 1-octene (1 cm^3) in perfluoro-1,3-dimethylcyclohexane (4 cm^3) at 80 °C, 2 MPa CO/H$_2$ (1 : 1) for 1 h.
(D) [Rh(acac)(CO)$_2$] (10 mg, 0.039 mmol), ligand (0.194 mmol), alkene (77.4 mmol), 1H-perfluorooctane (15 cm^3), undecane (internal standard for GC analysis, 1.21 g), 80 °C, 4 MPa CO/H$_2$ (1 : 1).
(E) As conditions D except C$_8$F$_{17}$H (10 cm^3), toluene (10 cm^3).

a) TOF = initial turnover frequency: moles of alkene converted per mole Rh per hour.

systems. Another interesting feature of the process is that better reaction rates and l/b ratios are obtained when the reaction is carried out in the absence of organic solvent.

Phosphites with spacer group (7–9) showed little difference from phosphites 1–6 (see Table 2). Indeed, the *ortho*-substituted phosphites 7 and 9 differ from the *para*-substituted 8 by much higher activities (TOF > 10 000 vs. 3900 h^{-1} respectively), by a lower l/b ratio (2 vs. 3.5), and by a lower aldehyde selectivity. The main difference lies in the fact that the l/b ratio for the *para*-substituted phosphite 8 is much lower than those obtained with phosphites 2 and 3 and close to the one obtained with P(OPh)$_3$.

Experiments made with different fluorous solvents proved that the nature of the fluorous phase has practically no effect on the l/b ratio and on the aldehyde selectivity. On the other hand, this factor greatly influences the activity, since the TOF dropped from 3900 h^{-1} when using 1H-perfluorooctane to 3800, 2500, and 2300 h^{-1} with perfluoromethylcyclohexane (PFMC), perfluoromethyldecalin (PFMD) and per-fluoroperhydrophenanthrene (PFPP) respectively. These observations are explained by the fact that at the reaction temperature (80 °C), PFMD and PFPP are not totally miscible with 1-decene, in contrast with the 1-decene–C$_8$F$_{17}$H and 1-decene–PFMC couple [13].

Other heavy terminal olefins behave similarly to 1-decene (Table 2), giving, for example, with phosphite 8, an l/b ratio of about 3.0 and an aldehyde selectivity of 95%, but the activity drops markedly going from 1-decene to 1-dodecene, presumably again due to their partial solubility with the fluorous solvent C$_8$F$_{17}$H.

Table 2 Hydroformylation of higher alkenes under fluorous biphasic conditions with a fluorous phosphites with spacer group [13].

Phosphite	Substrate	Solvent	Time[a] [min]	TOF[b] [h^{-1}]	l/b	Aldehyde selectivity [%]
7	1-decene	C$_8$F$_{17}$H	15	10000	2.0	85
8	1-decene	C$_8$F$_{17}$H	30	3900	3.5	95
9	1-decene	C$_8$F$_{17}$H	12	11000	2.0	85
8	1-decene	C$_8$F$_{17}$H/Tol.	60	3500	3.0	98
8	1-decene	PFMC	60	3800	3.0	98
8	1-decene	PFMD	90	2500	3.3	95
8	1-decene	PFPP	90	2300	3.6	95
8	1-octene	C$_8$F$_{17}$H/Tol.	60	3600	3.0	95
8	1-dodecene	C$_8$F$_{17}$H/Tol.	60	2600	3.0	94
8	2-decene	C$_8$F$_{17}$H/Tol.	90	1200	–	82
8	4-decene	C$_8$F$_{17}$H/Tol.	150	440	–	77
8	cyclohexene	C$_8$F$_{17}$H/Tol.	–	45	–	100

Conditions:
[Rh(acac(CO)$_2$] (10 mg, 0.039 mmol), phosphite (0.194 mmol), alkene (77.4 mmol), solvent (15 cm^3: pure fluorous solvent or 10 cm^3 C$_8$F$_{17}$H and 10 cm^3 toluene), undecane (1.21 g), 80 °C, 4 MPa CO/H$_2$ (1 : 1).

a) Time required to reach 100% conversion.
b) TOF = initial turnover frequency: moles of alkene converted per mole Rh per hour.

The asymmetric hydroformylation of styrene catalyzed by the (S,R)-**10**/Rh complex was briefly investigated in various fluorous solvents with or without toluene. The nature of the solvent has no significant effect on the catalytic activity of the catalyst. While the regioselectivity was slightly higher than that observed in organic solvent (l/b ratio: 94 : 6 vs. 88 : 12), the enantioselectivity was lower (87 vs. 94%) and an apparent racemization was observed in the course of the reaction [11].

The results obtained with *ortho*-substituted phosphites differ from those of *meta* or *para* counterparts by a remarkably higher initial rate under the same conditions. At the same time, they lead to a more modest l/b ratio and aldehyde selectivity. This difference is probably related to the fact that with the former bulky phosphites, only one phosphite coordinates to the Rh center to give the active species $HRhL_f(CO)_3$ (L_f = fluorous phosphite) [14]. In contrast with the latter phosphites which are less sterically demanding, two phosphites are bonded to rhodium $[HRhL_{f2}(CO)_2]$ as in the classical triphenylphosphine-modified catalyst. Complex $HRhL_f(CO)_3$ is more prone to CO dissociation than $HRhL_{f2}(CO)_2$, resulting in a much higher reaction rate. Due to the large space available with the $HRhL_f(CO)_3$ system in comparison with $HRhL_{f2}(CO)_2$, the reactions giving the branched aldehyde as well the β-H elimination proceed with relative ease, resulting in modest linearity and aldehyde selectivity.

An other important finding is the high l/b ratio observed with *meta*- and *para*-phosphites **2** and **3**. Electron-withdrawing groups attached to the aryl rings of these phosphites afford less basic ligands. Consequently, the electron density on complex $HRhL_{f2}(CO)_2$ decreases, which promotes the olefin insertion on its terminal carbon, giving a linear alkylrhodium intermediate leading to the linear aldehyde. On the other hand, a decrease in the electronic density on the metal also favors the CO dissociation and hence the formation of low-coordination alkylrhodium species, which induce easier β-H elimination and thus olefin isomerization [15].

In the case of phosphites **5** and **6**, the combination of the steric and electron withdrawing effects greatly promotes the β-H elimination, leading to a very low aldehyde selectivity.

Investigations into the recovery and re-use of the catalytic system have been made with phosphites **1b**, **2b**, and **6** on the one hand and with phosphites **7**, **8**, and **9** on the other hand. Figure 1 shows the evolution of the conversion for four reaction cycles with phosphites **1b**, **2b** and **6**. With each phosphite, the activity decreased after each re-use but whereas the decrease was moderate in the case of *ortho*-substituted phosphites **1b** and **6**, in the case of **2b** the activity fell dramatically after the second cycle and practically no activity was observed during the fourth run. Concomitantly, the l/b ratio and the aldehyde selectivity decreased after the first run with phosphite **2b** while they remained practically unchanged with **1b** and **6**.

In the case of phosphites with a spacer group (Figure 2), the activity was maintained or even slightly increased (with phosphite **8**) during the three first cycles. On the other hand, the conversion dropped dramatically during the course of the fourth cycle with *ortho*-substituted phosphite **7**. This decrease in activity that was more important than expected according to the high partition coefficient of the phosphites, and the variation of the l/b ratio observed during the recovery

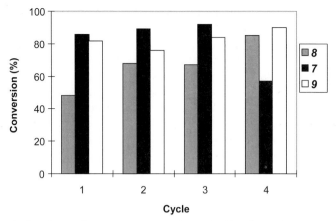

Figure 1 Hydroformylation of styrene: conversion after each of first four cycles with catalyst ligands **1b**, **2b**, and **6**.

experiments, have been interpreted by considering that the phosphites are not stable in the reaction medium. Different modes of decomposition could be envisaged for the fluorous phosphites. The main ones include hydrolysis by water produced by aldehyde condensation, nucleophilic attack on aldehyde, and oxidative cyclization with aldehydes [16, 17]. Aldol condensation of aldehydes could give traces of water that could react with phosphite but no evidence of formation of the expected aldol products was found by GC-MS [12]. To estimate the extent of the other modes of decomposition, the stability of various fluorous phosphites has been studied under hydroformylation reaction conditions. A sample of each phosphite was heated in a mixture of 1-decene/$C_8F_{17}H$ (1 : 1, v/v) or in mixture of 1-decene/undecanal/$C_8F_{17}H$ (1 : 1 : 1, v/v/v). ^{31}P NMR analyses of the fluorous phase after 1 h showed that in the absence of aldehyde, all phosphites remained unchanged. In contrast, in the

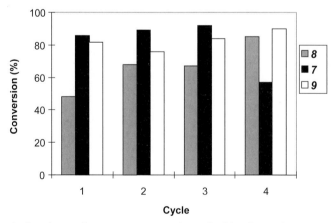

Figure 2 Hydroformylation of styrene: conversion after each of first four cycles with catalyst ligands **7**, **8**, and **9**.

presence of undecanal, a large proportion of the *para-* and *meta*-substituted phosphites **2b** and **8** were converted into oxidation products (30% and 50% of decomposition, respectively) whereas bulky *ortho*-substituted phosphites **1b**, **7**, and **9** appeared more stable (3%, 20%, and 20% of decomposition, respectively) [13]. The decrease in the ligand concentration due to the attack of the phosphites by aldehydes explains the change in *l/b* ratios and the aldehyde selectivity observed with some phosphites. Nevertheless, due to the good stability in particular of phosphites **1b** and **6**, other parameters can be taken into account to explain the decrease in activity observed with these ligands [6, 13].

References

1   I. T. HORVÁTH, J. RABAI, *Science* **1994**, *266*, 72.

2   I. T. HORVÁTH, *Acc. Chem. Res.* **1998**, *31*, 641.

3   A. P. DOBBS, M. R. KIMBERLEY, *J. Fluorine Chem.* **2002**, *118*, 3.

4   I. T. HORVÁTH, G. KISS, R. A. COOK, J. E. BOND, P. A. STEVENS, J. RABAI, E. J. MOZELSKI, *J. Am. Chem. Soc.* **1998**, *120*, 3133.

5   W. CHEN, L. XU, Y. HU, A. M. BANET-OSUNA, J. XIAO, *Tetrahedron* **2002**, *58*, 3889.

6   T. MATHIVET, E. MONFLIER, Y. CASTANET, A. MORTREUX, J. L. COUTURIER, *C. R. Chim.* **2002**, *5*, 417.

7   P. BHATTACHARYYA, D. GUDMUNSEN, E. G. HOPE, R. D. W. KEMMITT, D. R. PAIGE, A. M. STUART, *J. Chem. Soc. Perkin Trans.* **1997**, *1*, 3609.

8   D. J. ADAMS, D. GUDMUNSEN, J. FAWCETT, E. G. HOPE, A. M. STUART, *Tetrahedron* **2002**, *58*, 3827.

9   T. MATHIVET, E. MONFLIER, Y. CASTANET, A. MORTREUX, J. L. COUTURIER, *Tetrahedron Lett.* **1998**, *39*, 9411.

10 T. MATHIVET, E. MONFLIER, Y. CASTANET, A. MORTREUX, J. L. COUTURIER, *Tetrahedron Lett.* **1999**, *40*, 3885.

11 D. BONAFOUX, Z. HUA, B. WANG, I. OJIMA, *J. Fluorine Chem.* **2001**, *112*, 101.

12 D. F. FOSTER, D. GUDMUNSEN, D. J. ADAMS, A. M. STUART, E. G. HOPE, D. J. COLE-HAMILTON, G. P. SCHWARZ, P. POGORZELEC, *Tetrahedron* **2002**, *58*, 3901.

13 T. MATHIVET, E. MONFLIER, Y. CASTANET, A. MORTREUX, J. L. COUTURIER, *Tetrahedron* **2002**, *58*, 3877.

14 A. VAN ROOY, E. N. ORIJ, P. C. J. KAMER, P. W. N. M. VAN LEEUWEN, *Organometallics* **1995**, *14*, 34.

15 P. C. J. KAMER, J. H. N. REEK, P. W. N. M. VAN LEEUWEN, *Rhodium Catalyzed Hydroformylation* (Eds.: P. W. N. M. VAN LEEUWEN, C. CLAVER), Kluwer: Dordrecht **2000**, p. 35.

16 P. W. N. M. VAN LEEUWEN, *Rhodium Catalyzed Hydroformylation* (Eds.: P. W. N. M. VAN LEEUWEN, C. CLAVER), Kluwer: Dordrecht **2000**, p. 233.

17 F. RAMINEZ, J. F. PILOT, C. P. SMITH, S. B. BHATIA, A. S. GULATI, *J. Org. Chem.* **1969**, *34*, 3385.

4.2.4
Fluorous Nitrogen Ligands

4.2.4.1
For Oxidation Reactions

Gianluca Pozzi and Silvio Quici

Selective oxidation reactions catalyzed by transition metal complexes of organic ligands under homogeneous conditions have been the subject of intensive investigations [1] including the titanium-mediated asymmetric epoxidation of allylic alcohols [2]. Nevertheless, two major issues still prevent a widespread application of most of these homogeneous catalytic systems: (a) considerable synthetic efforts are often required for the preparation of suitable organic ligands and (b) the corresponding catalytically active complexes suffer from limited stability in the oxidative environment. Immobilization on organic polymers or inorganic supports represents the most obvious and explored strategy for increasing the lifetime of homogeneous oxidation catalysts and possibly recycling them after reaction completion [3]. In addition, water-soluble organometallic oxidation catalysts have been tested under aqueous–organic biphasic conditions [4, 5], and evidence for some positive effects related to the use of alternative reaction media such as ionic liquids [6], supercritical CO_2, or CO_2-expanded solvents [7] has been presented recently (cf. Chapters 2–6).

Fluorous biphasic (FB) techniques are particularly suited to oxidation reactions, where the substrates are converted to products of greater polarity, in that these are very easily expelled from the fluorous phase [8]. This fact, together with the thermal and chemical inertness of perfluorocarbons and the possible improvement of the catalyst stability due to its confinement in the fluorous phase, made catalytic oxidation reactions one of the first and most appealing targets of fluorous chemistry [9]. In this short review, complementary to the contribution of Vincent et al. (see Section 4.6), attempts at using the FB approach in the case of oxidation reactions catalyzed by metal complexes of bi- and polydentate nitrogen ligands will be highlighted, with emphasis on the epoxidation of alkenes, the aerobic oxidation of alcohols to carbonyl compounds, and the oxidation of organic sulfides to sulfoxides and/or sulfones.

Metal complexes of fluorous tetraarylporphyrins (**1–5**) have been used as catalysts in the epoxidation of alkenes under FB [9] or more traditional conditions [10], depending on their affinity for perfluorocarbons. Free base porphyrins **1–5** were readily metallated with transition metal cations under standard conditions normally employed for their nonfluorous counterparts. In particular, porphyrins **1–4** were metalated with $Mn(OAc)_2 \cdot 4\,H_2O$ in boiling DMF to give their respective Mn(III) complexes **Mn-1–Mn-4** [10], whereas the perfluorocarbon-soluble porphyrin **5** was similarly converted into the cobalt(II) complex **Co-5** by treatment with $Co(OAc)_2 \cdot 4\,H_2O$ [9].

Complexes **Mn-1–Mn-4** (< 60% F) were investigated under aqueous–organic biphasic conditions by using NaOCl or 30% H_2O_2 as oxygen donors [10]. Cyclooctene and 1-dodecene were used as models of reactive and poorly reactive alkenes, respectively, whereas the robust complex {Mn(III)-[5,10,15,20-tetrakis(2,6-dichlorophenyl)porphyrin]} chloride (**Mn-6**) was used as reference catalyst (Table 1). Computational studies, taking into account electronic effects, indicated that the introduction of R_F substituents on the *meso*-phenyl rings should improve the ligand

Table 1 Catalytic epoxidation of alkenes by aqueous NaOCl at pH = 10.[a]

Entry	Catalyst	Substrate	Time [h]	Yield [%]	Selectivity[b] [%]
1	Mn-6	Cyclooctene[c]	3	80	88
2	Mn-1	Cyclooctene[c]	1	8	95
3	Mn-2	Cyclooctene[c]	1	5	95
4	Mn-3	Cyclooctene[c]	3	35	67
5	Mn-4	cyclooctene[c]	3	90	92
6	Mn-6	1-Dodecene	4	33	69
7	Mn-6	1-Decene	3	54	80
8	Mn-6	Hexadec-1-ene	3	18	45
9	Mn-4	1-Dodecene	3	67	96
10	Mn-4	1-Decene	3	65	86
11	Mn-4	Hexadec-1-ene	3	63	70
12	Mn-4	2-Methylundec-1-ene	3	41	41
13	Mn-4	1-Methylcyclohexene	4	15	23
14	Mn-4	Norbornene	4	60	60

a) Reaction conditions: $T = 0\,°C$; solvent = CH_2Cl_2; molar ratio alkene/oxidant/catalyst = 1 : 2 : 0.001.
b) Selectivity = (moles of epoxide)/(moles of substrate converted).
c) Solvent = AcOEt.

stability under oxidizing conditions [11]. The experimental results did not confirm this hypothesis: **Mn-1–Mn-4** underwent extensive decomposition and only poor yields in epoxide were obtained. Apparently, factors not considered in computational studies, such as solvation effects and the steric protection provided by the bulky chlorine atoms in the 2,6-positions of the *meso*-aryl rings, prevailed in determining the catalytic activity. Nevertheless, the introduction of R_F substituents coupled with steric protection had a definite positive effect on the course of the epoxidation of terminal alkenes catalyzed by **Mn-4** which gave consistently higher epoxide yields than the reference catalyst **Mn-6**. This effect was particularly marked in reactions carried out in CH_2Cl_2 as the solvent and using aqueous NaOCl as the oxygen donor [10].

The perfluorocarbon-soluble complex **Co-5** (65% F) proved to be an efficient catalyst for the FB epoxidation of alkenes by molecular oxygen and 2-methylpropanal as reducing agent (Table 2) [9]. Reactions were carried out at room temperature under O_2 at atmospheric pressure, by adding a solution of the catalyst in perfluorohexane to a solution of the substrate in CH_3CN containing an excess of 2-methylpropanal and vigorously stirring the resulting biphasic mixture. The epoxide yields varied from 95% for cyclic substrates to 52% for terminal alkenes and the fluorous phase containing the catalyst could be easily separated and re-used at least three times in the case of the oxidation of cyclooctene. Rather interestingly, the FB approach allowed the use of a much higher substrate/catalyst ratio (1000 : 1) than that reported for the oxidation of (other) alkenes with O_2/aldehyde catalyzed by cobalt complexes of standard tetraarylporphyrins (20 : 1) [12]. Moreover, epoxide yields were higher than those obtained in reactions catalyzed by **Mn-4** under optimized aqueous–organic conditions, except for 1-dodecene (entry 2, Table 2, vs. entry 10, Table 1).

Tetraaza-macrocycle **7** was developed as an alternative to the synthetic demanding porphyrin ligands [13]. At the same time, the triaza-macrocycle **8** bearing three R_F substituents was introduced independently by Fish and co-workers [14]. Metal complexes of these ligands provided new FB catalysts for the oxidative functionalization of hydrocarbons in the presence of *t*-BuOOH and O_2. In particular, oxidation of cyclohexene afforded mixtures of 2-cyclohexen-1-one and 2-cyclohexen-1-ol.

Table 2 FB catalytic epoxidation of alkenes with O_2/2-methylpropanal catalyzed by **Co-5**.[a]

Substrate	Time [h]	Yield [%]	Selectivity[b] [%]
Cyclooctene	3	95	88
1-Dodecene[c]	14	52	87
2-Methylundec-1-ene	5	72	90
1-Methylcyclohexene	4	95	95
Norbornene	5	90	95

a) In CH_3CN/perfluorooctane 1 : 1 v/v. Reaction conditions: $T = 25\ °C$; molar ratio alkene/aldehyde/catalyst = 1 : 2 : 0.001.
b) Selectivity = (moles of epoxide)/(moles of substrate converted).
c) Using 3-methylbutanal instead of 2-methylpropanal.

$R_FCH_2O(CH_2)_2$—N ... N—$(CH_2)_2OCH_2R_F$

$R_FCH_2O(CH_2)_2$—N ... N—$(CH_2)_2OCH_2R_F$

$(CH_2)_3R_F$

$R_F(CH_2)_3$ ... $(CH_2)_3R_F$

7 $R_F = CF_2(OCFCF_2)_q(OCF_2)_pOCF_3$
    $\quad\quad\quad CF_3$
    $\quad\quad \overline{q} = 3.38; \overline{p} = 0.11$

8 $R_F = CF_2(CF_2)_6CF_3$

Perfluoroalkylated bipyridines **9–11** are readily soluble in some organic solvents, for instance CH_2Cl_2, and were tested in the ruthenium-catalyzed epoxidation of *trans*-stilbene with $NaIO_4$ under aqueous–organic biphasic conditions [15], as previously described for 2,2'-bipyridine [16]. In the presence of this ligand, oxidative cleavage of the carbon–carbon double bond strongly affected epoxidation yields at room temperature. The use of fluorous bipyridines **9–11** reduced the incidence of the side reaction and *trans*-stilbene epoxide was obtained in good yields (70–87%) [15]. Since the fluorous affinity of the ruthenium catalysts generated in situ from **9–11** was found to be higher than that of the free ligands, *trans*-stilbene epoxidation was also studied in an aqueous/organic/fluorous triphasic system at 0 °C. Higher epoxide yields (92–96%) were obtained and the fluorous phase could be re-used without addition of $RuCl_3$ for at least three further runs before the epoxide yields decreased significantly.

$CH_2OCH_2C_7F_{15}$

$CH_2OCH_2C_7F_{15}$

$CH_2O(CH_2)_3C_8F_{17}$

$CH_2O(CH_2)_3C_8F_{17}$

$(CH_2)_4C_8F_{17}$

$(CH_2)_4C_8F_{17}$

9   **10**   **11**

Primary and benzylic alcohols were smoothly oxidized to the corresponding aldehydes by O_2 in the presence of catalytic amounts of 2,2,6,6-tetramethylpiperidin-1-oxyl radical (TEMPO, 3.5%) and a Cu(I) complex (2%) generated in situ from CuBr · Me$_2$S and bipyridine **11** [17]. Reactions were carried out at 90 °C in a biphasic $C_8F_{17}Br/C_6H_5Cl$ system affording analytically pure aldehydes in 73–96% isolated yield. The recyclability of the catalyst was demonstrated in the case of the oxidation of 4-nitrobenzyl alcohol to give 4-nitrobenzaldehyde: the fluorous phase was re-used eight times with a slight decrease in aldehyde yield (from 93% to 86%) and no apparent decrease in reaction rate. Secondary alcohols also underwent oxidation to the corresponding ketones under fluorous biphasic conditions [17, 18]. Reaction rates and yields were comparable to those observed with primary alcohols in the case of benzylic, allylic, and cyclic substrates, but in general secondary alcohols were less easily oxidized and higher amounts of TEMPO (up to 10%) were required in order to achieve fast reactions. Sterically hindered secondary alcohols were particularly hard to oxidize and incomplete conversions were observed even using 10% TEMPO. This allowed the selective oxidation of the less sterically hindered isomer in *cis–trans* mixtures of 2-, 3- and 4-substituted cyclohexanols and the easy separation of the unreacted isomer by column chromatography [18].

The first example of FB oxidation of sulfides dates back to 1995: dibenzothiophene and diphenyl sulfide gave the corresponding sulfones in low yields (1.4% and 10%, respectively) upon treatment with O_2 at 100 °C in the presence of a not fully characterized perfluorocarbon-soluble iron–phthalocyanine [19]. Following this earlier report, Co(II)–tetraarylporphyrin **Co-5** and Co(II)–phthalocyanine **Co-12** (cf. Structure) were tested as catalysts for the FB oxidation of methyl phenyl sulfide and *para*-substituted aryl methyl sulfides with O_2 and a sacrificial aldehyde (Table 3) [20].

Co-**12**

Turnover numbers comparable to those obtained in similar FB oxidations catalyzed by nickel complexes of fluorinated 1,3-diketones were observed [21]. Sulfoxides were usually obtained in good yields (50–100%) and selectivities (> 90%) together with variable amounts of sulfones. The latter were the major products both in the oxidation of *p*-nitrophenyl methyl sulfide and *p*-methoxyphenyl methyl sulfide catalyzed by **Co-5**. The absence of any definite relationship between the electronic properties of the *para* substituents and selectivity ruled out the hypothesis of a heterolytic reaction mechanism involving the formation of high-valent oxometal species. Moreover, the addition of a free-radical scavenger was found to inhibit the oxidation process, thus suggesting that acyl and peroxyacyl radicals generated by the action of the cobalt complexes on the sacrificial aldehyde were the true oxidizing agents. As a consequence of the peculiar reaction environment, both **Co-5** and **Co-12** were progressively bleached and the effectiveness of their recycling was limited.

Formation of high-valent oxometal species from (salen)metal complexes and iodosylbenzene (PhIO) and the mechanism of the oxygen transfer from these species to alkyl aryl sulfides have been investigated in detail [22]. Fluorous salen ligands **13** and **14** were synthesized and the corresponding (salen)manganese(III) complexes **Mn-13** and **Mn-14** were evaluated in the oxidation of alkyl aryl sulfides with PhIO under homogeneous and FB conditions, respectively [23].

Table 3 Catalytic oxidation of aryl methyl sulfides with O_2/2,2-dimethylpropanal under FB conditions.[a]

Catalyst	Substrate	Conversion [%]	Selectivity	
			Sulfoxide [%]	Sulfone [%]
Co-5	PhSCH$_3$	82	90	10
Co-5	p-CH$_3$OPhSCH$_3$	100	0	100
Co-5	p-CH$_3$PhSCH$_3$	100	94	6
Co-5	p-ClPhSCH$_3$	67	100	0
Co-5	p-FPhSCH$_3$	100	95	5
Co-5	p-NO$_2$PhSCH$_3$	100	10	90
Co-12	PhSCH$_3$	68	100	0
Co-12	p-CH$_3$OPhSCH$_3$	59	100	0
Co-12	p-CH$_3$PhSCH$_3$	40	100	0
Co-12	p-ClPhSCH$_3$	58	100	0
Co-12	p-FPhSCH$_3$	49	100	0
Co-12	p-NO$_2$PhSCH$_3$	100	15	85

a) In CH$_2$Cl$_2$/perfluorooctane (1 : 1 v/v).
Reaction conditions: $T = 20\,°C$; $t = 4$ h; molar ratio sulfide/aldehyde/catalyst = 1 : 1 : 0.001.

Both complexes were able to catalyze the oxidation of p-substituted methyl phenyl sulfides at a substrate/catalyst molar ratio of 100, with good sulfoxide selectivities (≥ 90%). It should be noted that FB reactions catalyzed by **Mn-14** consistently afforded higher sulfoxide yields than homogeneous reactions catalyzed by **Mn-13** (Table 4). Moreover, three consecutive recyclings of the fluorous layer were performed with no appreciable loss of catalytic activity and selectivity.

Table 4 Catalytic oxidation of aryl methyl sulfides with PhIO catalyzed by achiral (salen)manganese(III) complexes.[a]

Catalyst	Substrate	Conversion [%]	Selectivity	
			Sulfoxide [%]	Sulfone [%]
Mn-13[b]	PhSCH$_3$	70	91	9
Mn-13[b]	p-BrPhSCH$_3$	59	91	9
Mn-13[b]	p-NO$_2$PhSCH$_3$	64	88	12
Mn-14[c]	PhSCH$_3$	95	95	5
Mn-14[c]	p-BrPhSCH$_3$	87	96	4
Mn-14[c]	p-NO$_2$PhSCH$_3$	100	95	5

a) Reaction conditions: $T = 25\,°C$; $t = 5$ h; molar ratio sulfide/oxidant/catalyst = 1 : 1 : 0.01.
b) Homogeneous conditions (CH$_2$Cl$_2$/CH$_3$CN 1 : 1 v/v).
c) FB conditions (CH$_3$CN/perfluorooctane 1 : 1 v/v).

Oxidation of unsaturated compounds with singlet oxygen ($^1O_2$) has been extensively studied because of its considerable synthetic interest [24]. Dye-sensitized photooxidation of triplet oxygen is a practical method for generation of the labile $^1O_2$ reagent on the laboratory scale. However, the presence of the dye and/or of its decomposition derivatives in the final reaction mixture can complicate the purification of the desired organic products. Another point to be considered is the choice of solvent, which must readily dissolve O_2, ensure a relatively long lifetime for the generated $^1O_2$, and be inert to this reagent. Perfluorocarbons fulfill all these requirements and have been used as alternative medium for the photooxidation of alkenes to give hydroperoxides, in the presence of tetraphenylporphyrin (TPPo) dissolved in pyridine as a sensitizer [25]. The biphasic mixture was irradiated while maintaining a constant supply of O_2. After reaction completion the two layers were separated to give an organic phase containing the hydroperoxide (plus TPPo and its decomposition products) and a fluorous phase free from organic compounds.

The issues of sensitizer degradation and purification of the oxidation products were taken into account by DiMagno et al., who used the electron-deficient fluorous porphyrin **15** (5,10,15,20-tetrakis(heptafluoropropyl)porphyrin) as a sensitizer in the photooxidation of cyclohexene and allylic alcohols in CH_3CN/perfluorohexanes [26].

15

In a CCl_4 solution, porphyrin **15** showed increased chemical stability toward $^1O_2$ and hydroperoxides with respect to TPPo. However, physical segregation into the fluorous phase was found to be the most important factor in reducing the incidence of degradation processes. In addition, the FB approach ensured the easy separation of the hydroperoxides from **15** at the end of the reaction. The fluorous layer containing the sensitizer (57–94% of the starting material-depending on reaction conditions) could be re-used without further treatments.

References

1 R. A. SHELDON (Ed.), *Proceedings of the 6th International Symposium on the Activation of Dioxygen and Homogeneous Catalytic Oxidation*, Noordwijkerhout, The Netherlands, April 14–19, 1996, in *J. Mol. Catal. A:* **1997**, *117*, 1.

2 R. A. JOHNSON, K. B. SHARPLESS, in I. OJIMA (Ed.), *Catalytic Asymmetric Synthesis*, 2nd ed., Wiley-VCH, New York **2000**, Chap. 6A, p. 231.

3 D. E. De Vos, B. F. Sels, P. A. Jacobs, *Adv. Cat.* **2001**, *46*, 1.
4 B. Cornils, W. A. Herrmann, in B. Cornils, W. A. Herrmann (Eds.), *Applied Homogeneous Catalysis with Organometallic Complexes*, 2nd ed., Wiley-VCH, Weinheim **2002**, p. 603.
5 R. A. Sheldon, I. W. C. E. Arends, G.-J. ten Brink, A. Dijksman, *Acc. Chem. Res.* **2002**, *35*, 774.
6 J. Dupont, R. F. de Souza, P. A. Z. Suarez, *Chem. Rev.* **2002**, *102*, 3667.
7 G. Musie, M. Wei, B. Subramaniam, D. H. Busch, *Coord. Chem. Rev.* **2001**, *219–221*, 789.
8 I. T. Horváth, J. Rábai, *Science* **1994**, *266*, 72.
9 G. Pozzi, F. Montanari, S. Quici, *Chem. Commun.* **1997**, 69.
10 G. Pozzi, I. Colombani, M. Miglioli, F. Montanari, S. Quici, *Tetrahedron* **1997**, *53*, 6145.
11 A. Ghosh, *J. Am. Chem. Soc.* **1995**, *117*, 4691.
12 A. K. Mandal, V. Khanna, J. Iqbal, *Tetrahedron Lett.* **1996**, *37*, 3769.
13 G. Pozzi, M. Cavazzini, S. Quici, S. Fontana, *Tetrahedron Lett.* **1997**, *43*, 7605.
14 J.-M. Vincent, A. Rabion, V. K. Yachandra, R. H. Fish, *Angew. Chem., Int. Ed. Engl.* **1997**, *36*, 2346.
15 S. Quici, M. Cavazzini, S. Ceragioli, F. Montanari, G. Pozzi, *Tetrahedron Lett.* **1999**, *40*, 3647.
16 G. Balavoine, C. Eskenazi, F. Meunier, H. Rivière, *Tetrahedron Lett.* **1984**, *25*, 3769.
17 B. Betzemeier, M. Cavazzini, S. Quici, P. Knochel, *Tetrahedron Lett.* **2000**, *41*, 4343.
18 G. Ragagnin, B. Betzemeier, S. Quici, P. Knochel, *Tetrahedron* **2002**, *58*, 3985.
19 I. T. Horváth, J. Rábai, US 5.463.082 (**1995**).
20 S. Colonna, N. Gaggero, F. Montanari, G. Pozzi, S. Quici, *Eur. J. Org. Chem.* **2001**, 181.
21 I. Klement, H. Lütjens, P. Knochel, *Angew. Chem., Int. Ed. Engl.* **1997**, *36*, 1454.
22 A. Chellamani, N. I. Alhaji, S. Rajagopal, R. Sevvel, C. Srinivasan, *Tetrahedron* **1995**, *51*, 12 677.
23 M. Cavazzini, G. Pozzi, S. Quici, I. Shepperson, *J. Mol. Catal. A:* **2003**, *204–205*, 433.
24 A. A. Frimer, *Chem. Rev.* **1979**, *79*, 359.
25 R. D. Chambers, G. Sandford, A. Shah, *Synth. Commun.* **1996**, *26*, 1861.
26 S. G. DiMagno, P. H. Dussault, J. A. Schultz, *J. Am. Chem. Soc.* **1996**, *118*, 5312.

4.2.4.2
Synthesis of Fluorous Nitrogen Ligands and Applications, Including Atom-Transfer Radical Reactions

Jean-Marc Vincent, Dominique Lastécouères, María Contel, Mariano Laguna, and Richard H. Fish

Since the seminal paper in 1994 by Horváth and Rábai that introduced the fluorous biphasic catalysis concept (FBC) [1], numerous examples of this methodology for many classical organic reactions have been demonstrated [1, 2]. Unlike the more classical aqueous/hydrocarbon biphasic systems, water-sensitive reactants can also be employed, with the solubility of organic substrates being higher in the perfluorocarbons than in water; therefore, more favorable reaction kinetics could be expected. Moreover, because of the unique thermomorphic properties of the perfluorocarbon solvents, it is possible under FBC reaction conditions, by judiciously choosing the solvent system, to obtain a homogeneous monophasic solution when elevating the temperature of the reaction mixture, while re-formation of the biphasic system occurs at room temperature. In 1996, we initiated an FBC program at the Lawrence Berkeley National Laboratory with the objective of developing new catalytic systems for alkane/alkene oxidation chemistry. Thus, the FBC process appeared as a very attractive procedure for alkane and alkene functionalization reactions, not only for the ability to recycle the catalyst, but also because of the chemical inertness of the fluorocarbon solvents, and their unique characteristic of solubilizing O_2 in large concentrations.

Nitrogen-containing ligands are widely used in coordination chemistry, particularly in the field of oxidation catalysis [3]. Relatively simple synthetic procedures are available to introduce the fluoro ponytails, making these ligands valuable targets for the development of fluorous catalysis with transition metal complexes. Representative examples of fluorous nitrogen ligands that have been prepared and, for most of them, applied to fluorous catalysis chemistry, are **1–10**.

One of the most successful synthetic strategies used to obtain fluorous nitrogen ligands was the direct alkylation of secondary amines (R_2NH) with perfluoroalkyl iodide derivatives, such as $R_{f8}(CH_2)_3I$ ($R_{f8} = C_8F_{17}$) [4, 5]. The three-methylene spacer was necessary, not only to insulate the nitrogen ligating atom from the strong electron-withdrawing effect of the perfluoroalkyl chain, but also to avoid a facile elimination reaction of HI that predominantly occurs during the alkylation step when a two-methylene spacer is used. Ligands 1, 2, and 3 were synthesized in fair to good yields (50, 60, and 40%, respectively) by reaction of the corresponding secondary amines with $R_{f8}(CH_2)_3I$ in DMSO/K_2CO_3 at 90 °C [4, 5]. Interestingly, amongst the three ligands mentioned, the 1,4,7-triazacyclononane (TACN) derivative 1 with a fluorine content of 64.2%, was the only ligand found to be soluble in perfluorocarbons at room temperature. Furthermore, tosylate derivatives have also been shown to be effective alkylating agents. The fluorous peralkylated cyclam 4 was obtained in 60% yield by refluxing $R_fCH_2O(CH_2)_2OTs$ in CH_3CN/Na_2CO_3 for 24 h [6]. The tosylate, $R_{f8}(CH_2)_3OTs$, was recently used in our group in Bordeaux for the preparation of the fluorous, peralkylated polyamino ligands 5 and 6 [7]. The alkylations were conducted by refluxing a CH_3CN/K_2CO_3 solution of the diethylene-triamine or tris(aminoethyl)amine with the tosylate (1.1 equiv/amine group) for 48 h; the ligands 5 and 6 were obtained in 65 and 55% yields, respectively. Interestingly, it has been shown that by using $R_{f8}(CH_2)_3I$, instead of the corresponding tosylate, one can not only increase the yield of 5 from 65 to 79%, but also isolate the ligand as a powder rather than an oil. Moreover, an aromatic nucleophilic substitution on the N-pentafluorophenyl TACN derivatives was used for the preparation of the fluorous TACN 7 [5]. This reaction proceeds efficiently (yield 80%) by reaction of $R_{f8}(CH_2)_3OH$ in a 50% mixture of NaOH/trifluorotoluene at 85 °C in the presence of a phase-transfer agent. Rather surprisingly, this ligand with three fluoro ponytails and a fluorine content of 58.6% was found to be insoluble in perfluorocarbons at room temperature.

Bipyridines are another important class of nitrogen ligands in coordination chemistry and transition metal catalysis. The fluorous bipyridine ligand 8 was conveniently prepared in 40% yield by reacting the dianion, obtained from 4,4′-dimethyl-2,2′-bipyridine, with $R_{f8}(CH_2)_3I$ at low temperature (78 °C) [8a]. Recent studies in Zaragoza/Berkeley provided full synthetic details and spectroscopic characterization of ligand 8 [8b]. The fluorous pyridine ligand 9 was synthesized in 26% yield by reaction of the pyridin-3,5-diacyl chloride with the alcohol $R_{f8}(CH_2)_2OH$, while the pyridine ligand 10 was prepared in 94% yield from pyridine-3-carbaldehyde and $R_{f8}(CH_2)_2OH$ in the presence of CF_3SO_3H [9].

Above, the parameters necessary to solubilize important nitrogen ligands in fluorocarbon media have been established. In a concomitant manner, important metal complexes that are needed as precatalysts for many classical catalytic reactions require special attention with regard to fluorocarbon solubility, simply because of the polar and/or ionic nature of these complexes. Thus, we found in our experience that, in many cases, the counteranion also needed fluoro ponytails to ensure fluorocarbon solubility, even if the metal ion was coordinated to a fluorous-soluble ligand such as 1. Therefore, we [4, 5], and Pozzi et al. [6], have addressed this critical aspect by using fluoro-ponytailed carboxylate ligands as counterions for the metal ions of interest [4, 5]. When fluorous metal carboxylates, such as complexes **11** [Eq. (1)], were reacted with fluorous-soluble ligand **1**, either in situ (Mn^{2+} and Co^{2+} complexes in perfluoroheptane) [4, 5], or via isolation and characterization (Cu^{2+} complex, CH_2Cl_2) [8b], to provide precatalysts **12** for alkene, alkane, and alcohol FBC oxidation, the complexes formed were fully fluorocarbon-soluble [Eq. (1)].

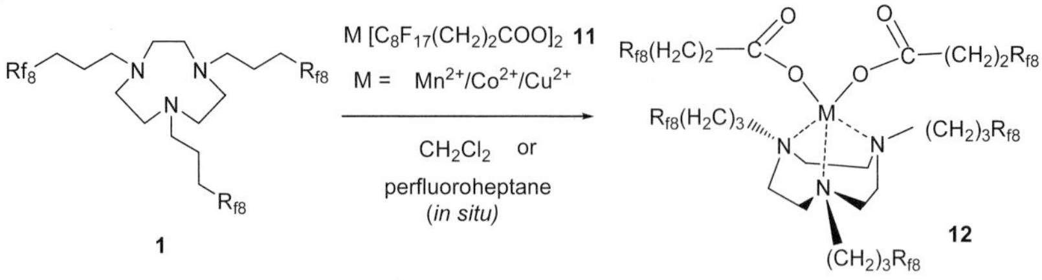

In another example, the reaction of a Cu(I) complex, [CuCl], with ligand R_{f8}-TACN **1**, provided a fully fluorocarbon-soluble complex **13** (fully characterized) [8b], without appended fluoroponytails on the Cu(I) metal ion [Eq. (2)], isolated from trifluoromethylbenzene. This appears to be a general phenomenon with Cu(I) complexes and fluorous ligands, and apparently is predicated on their hydrophobic properties that engender their solubility in hydrophobic solvents, such as fluorocarbons.

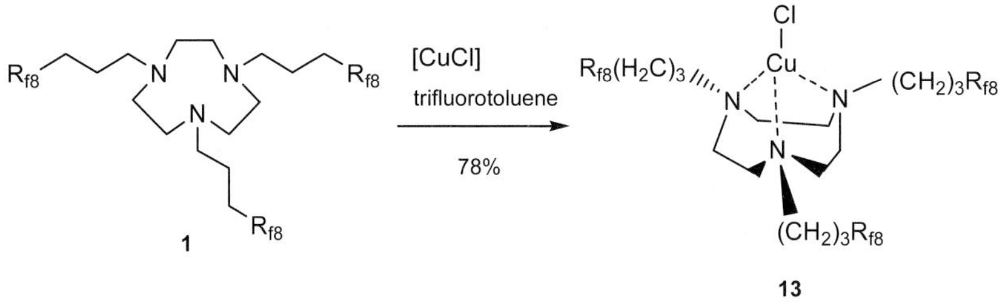

In 1997, we demonstrated unequivocally, possibly for the first time, that indeed this novel FBC approach for separation of the R_f-Mn^{2+} and Co^{2+} precatalysts from the substrates was viable for oxidation of alkanes and alkenes in the presence of the necessary oxidants, TBHP and O_2 gas [4], soon thereafter Pozzi et al. also verified this FBC oxidation chemistry [6]. The R_f-TACN complexes **12** (eq 1) were found to be particularly effective for allylic oxidation of alkenes; the Cu^{2+} analogue [Eq. (1)] and the Cu(I) complex **13** [Eq. (2)] gave comparable results [8b]. We also provided clear evidence that these FBC oxidation reactions occurred via classical autoxidation mechanisms [4, 5]. The limited scope of the substrates studied also showed that allylic oxidation, for example, cyclohexene to cylohexenol and cyclohexenone, was more favorable than alkane functionalization with cyclohexane as the substrate, based on thermodynamic grounds.

The selective oxidation of alcohols to ketones or aldehydes is a very important transformation in organic chemistry. Using the bipyridine ligand **8** (2 mol%), $CuBr \cdot Me_2S$ (2 mol%), and TEMPO (3.5–10 mol%) under O_2 (0.1 MPa) in biphasic perfluorooctyl bromide/chlorobenzene at 90 °C, various primary and secondary alcohols (aromatic and aliphatic) were oxidized to the corresponding aldehydes and ketones in good to excellent yields [10, 11]. The stability of the catalyst was found to be excellent, with no observed decrease in yield and reaction rate during the oxidation of 4-nitrobenzyl alcohol to the corresponding aldehyde, after eight reaction cycles. Further, the fluorous biphase system consisting of $Pd(OAc)_2$ (5 mol%)/10 (20 mol%), in perfluorodecalin/toluene, under O_2 (0.1 MPa) at 80 °C, was another effective catalytic process for the oxidation of primary and secondary alcohols (aliphatic and aromatic) to aldehydes and ketones [9]. Recycling efficiency was also excellent, the yield of isolated acetophenone after five cycles still being 74%, compared to 98% for the first run. The Zaragoza/Berkeley groups have recently studied the mechanism of the FBC oxidation of 4-nitrobenzyl alcohol to 4-nitro-benzaldehyde at 90 °C (single-phase) with precatalyst R_f-TACN-R_f-Cu^{2+} 12 [Eq. (2)], TEMPO, and O_2, by using EPR techniques [8b]. The EPR spectra clearly defined a Cu(II) to Cu(I) redox reaction, and the role of TEMPO and O_2 in this selective conversion to aldehyde from alcohol. Precatalyst R_f-TACN-R_f-Cu^{2+} [Eq. (2)] was solubilized in perfluoroheptane, and then chlorobenzene, TEMPO, and 4-nitro-benzyl alcohol were added to the reaction mixture. The reaction started at 90 °C in the presence of O_2, and after 30 min, followed by cooling to room temperature, an aliquot was removed from the perfluoroheptane phase and immediately frozen at 77 K in an EPR tube. The EPR spectrum shows a narrow central signal at about $g = 2.006$ that can be associated with the TEMPO radical. This was further demonstrated by measuring this radical in perfluoroheptane at 77 K, together with a Cu(II) signal, with $g_{||} = 2.26(1)$ and $g_{\perp} = 2.06(1)$, $A_{||} = 520(5)$ MHz and $A_{\perp} < 50$ MHz. The behavior of the different signals was in accordance with a mechanism proposed by Semmelhack et al. [12]. After 4 h, a 65% yield of aldehyde had been obtained; however, by leaving the reaction for longer periods of time, a ~100% yield of aldehyde was formed, concomitantly with a full recovery of the initial R_f-Cu(II) complex.

Atom-transfer radical addition (ATRA) is a particularly useful radical process for the preparation of lactones and lactams through metal-catalyzed cyclization of unsaturated trichloro esters or amides [13]. One of the most efficient catalysts for both ATRA and atom-transfer-radical polymerization (ATRP) reactions is the copper(I)/pentamethyldiethylenetriamine complex [14]. The fluorous polyamino ligands 5 or 6, associated with copper(I) chloride (1 mol%) and iron powder (10 mol%) catalyzed the intramolecular cyclization of the pent-4-enyl trichloroacetate in almost quantitative yields under FBC conditions [Eq. (3)] [7]. By using a ternary solvent system (perfluoroheptane/trifluorotoluene/1,2-dichloroethane), the reaction was carried out under homogeneous conditions at 80 °C, while phase separation occurred at room temperature enabling facile recycling of the catalyst and recovery of the substrate, with only 1–2% of the copper(I) leaching into the organic phase. Ensuring that no oxygen was introduced during the recycling procedure, the yield of lactone, after the fourth run, was still 91%, making the Cu(I)Cl/5 and Cu(I)Cl/6 complexes highly efficient and recoverable catalysts for atom-transfer radical reactions.

$$\text{Cu(I)Cl / } \mathbf{6} \text{ (1 mol\%)}$$
$$\text{Fe(0) (10 mol\%)}$$
$$\text{20 h, 80 °C}$$
$$\text{98 \%}$$

$$(3)$$

ATRP is a transition metal-mediated living radical polymerization of vinyl monomers that is closely related to ATRA, enabling the synthesis of well-defined and complex macromolecular architectures. The major limitation of ATRP is that the polymer is usually contaminated by the colored transition metal catalyst, therefore requiring purification steps such as column chromatography or precipitation of the polymer. Catalysts grafted onto insoluble supports have been developed to lower the copper content of the final product and for recycling [15]. However, heterogeneous supported catalysts are less efficient than their homogeneous analogues, leading to broader polydispersity (PDi = 1.4–1.5) and lower initiator efficiency. Lower polydispersity and higher recycling efficiency were observed using copper(I) catalysts immobilized on polyethylene-*block*-poly(ethylene glycol), a polymer soluble in toluene above 70 °C but insoluble at room temperature [16]. Using the catalytic system, CuBr/**5**, and ethyl 2-bromoisobutyrate as the polymerization initiator, Haddleton and co-workers have shown that living radical polymerization of methyl methacrylate can be carried out very efficiently under FBC conditions [Eq. (4)] [17]. Interestingly, PMMA was also obtained as a colorless solid after separation of the upper hydrocarbon phase and removal of the volatiles. The catalyst was recycled twice with similar results in terms of kinetics and polydispersity, making this FBC system very attractive for further applications.

$$\text{Cu(I)Br / } \mathbf{5} \text{ (1 mol\%)}$$
$$\text{PFMC, toluene}$$
$$\text{90 °C, 5 h}$$
$$\text{76.6\%, PDi = 1.30}$$

$$(4)$$

References

1 I. T. HORVÁTH, J. RÁBAI, *Science* **1994**, *266*, 72.

2 For leading reviews see: (a) D. P. CURRAN, *Angew. Chem., Int. Ed.* **1998**, *37*, 1174; (b) L. P. BARTHEL-ROSA, J. A. GLADYSZ, *Coord. Chem. Rev.* **1999**, *192*, 587; (c) R. H. FISH, *Chem. Eur. J.* **1999**, *5*, 1677; (d) SPECIAL *Tetrahedron* issue on Fluorous Chemistry, guest editors: J. A. Gladysz, D. P. Curran, *Tetrahedron* **2002**, *58*, 3823.

3 For an excellent review on non-heme, biomimetic iron catalysts for alkane oxidation see M. COSTAS, K. CHEN, L. QUE JR., *Coord. Chem. Rev.* **2000**, 200–202, 517. For representative examples of oxidation catalysts with nitrogen ligands see: (a) J.-B. Vincent, J.-C. Huffman, G. Christou, Q. Li, M. A. Nanny, D. M. Hendrickson, R. H. Fong, R. H. Fish, *J. Am. Chem. Soc.* **1988**, *110*, 6898; (b) D. E. DE VOS, B. F. SELS, M. REYNAERS, Y. V. S. RAO, P. A. JACOBS, *Tetrahedron Lett.* **1998**, *39*, 3221; (d) M. C. WHITE, A. G. DOYLE, E. N. JACOBSEN, *J. Am. Chem. Soc.* **2001**, *123*, 7194.

4 J.-M. Vincent, A. Rabion, V. K. Yachandra, R. H. Fish, *Angew. Chem., Int. Ed. Engl.* **1997**, *36*, 2346.

5 J.-M. Vincent, A. Rabion, V. K. Yachandra, R. H. Fish, *Can. J. Chem.* **2001**, *79*, 888.

6 G. Pozzi, M. Cavazzini, S. Quici, S. Fontana, *Tetrahedron Lett.* **1997**, *38*, 7605.

7 F. De Campo, D. Lastécouères, J.-M. Vincent, J.-B. Verlhac, *J. Org. Chem.* **1999**, *64*, 4969.

8 (a) S. Quici, M. Cavazzini, S. Ceragioli,; F. Montanari, G. Pozzi, *Tetrahedron Lett.* **1999**, *40*, 3647; (b) M. Contel, C. Izuel, M. Laguna, P. R. Villuendas, P. J. Alonso, Fish, R. H. *Chem. Eur. J.* **2003**, *9*, 4168.

9 T. Nishimura, Y. Maeda, N. Kakiuchi, S. Uemura, *J. Chem. Soc., Perkin Trans. I* **2000**, 4301.

10 B. Betzemeier, M. Cavazzini, S. Quici, P. Knochel, *Tetrahedron Lett.* **2000**, *41*, 4343.

11 G. Ragagnin, B. Betzemeier, S. Quici, P. Knochel, *Tetrahedron* **2002**, *58*, 3985.

12 M. F. Semmelhack, C. R. Schimd, D. A. Cortés, C. S. Chou, *J. Am. Chem. Soc.* **1984**, *106*, 3374.

13 F. O. H. Pirrung, H. Hiemstra, W. N. Speckamp, B. Kaptein, H. E. Schoemaker, *Synthesis* **1995**, 458.

14 F. De Campo, D. Lastécouères, J.-B. Verlhac, *Chem. Commun.* **1998**, 2117.

15 (a) G. Kickelbick, H.-J. Paik, K. Matyjaszewski, *Macromolecules* **1999**, *32*, 2941; (b) D. M. Haddleton, D. Kukulj, A. P. Radigue, *Chem. Commun.* **1999**, 99.

16 Y. Shen, S. Zhu, R. Pelton, *Macromolecules* **2001**, *34*, 3182.

17 D. M. Haddleton, S. G. Jackson, S. A. F. Bon, *J. Am. Chem. Soc.* **2000**, *122*, 1542.

4.2.5
Enantioselective Catalysis

4.2.5.1
Under Biphasic Conditions

Denis Sinou

Since the mid-1980s there have been very important advances in asymmetric synthesis via the use of a soluble chiral organometallic catalyst [1]. Although homogeneous organometallic catalysts have many advantages over their heterogeneous counterparts (higher activities and selectivities, mild reaction conditions), one of the major problems is the separation of the products from the soluble catalyst, which is generally a costly and toxic transition metal; this is particularly important for industrial applications. A possible solution to this problem is the heterogenization of the chiral homogeneous catalyst on an inorganic or organic support [2] (see Chapter 7). Another approach is the use of a liquid–liquid two-phase system, the chiral catalyst being immobilized in one phase, the reactants and the products of the reaction being in the other phase. Aqueous–organic systems have been successfully applied [3], and other two-phase systems such as ionic liquids [4] or perfluorohydrocarbons [5] in combination with an organic phase have also been proposed. This section will focus on enantioselective catalysis performed under biphasic conditions, one phase being a fluorous solvent. Catalytic reactions performed in fluorous biphasic systems can effectively show several advantages over classical homogeneous systems or even two-phase systems. One of them is the easy product separation by simple work-up techniques of liquid–liquid extraction, due to the low miscibility of fluorous solvents with common organic solvents, and so the recycling of the catalyst. Moreover, warming the mixture renders the organic and fluorous phases miscible, allowing the reaction to occur under homogeneous conditions [5], and so solving the problems of mass transfer between the two phases. Some examples performed in homogeneous systems, followed by separation of the fluorous catalyst or ligand via extraction with a fluorous solvent, will also be presented, although separation of organic and fluorous compounds by solid-phase extraction with fluorous silica gel is excluded.

The catalytic asymmetric reduction of unsaturated compounds is now a well used methodology in organic synthesis. Enantioselectivities higher than 95% have been obtained using molecular hydrogen from hydrogen donors in the presence of various chiral organometallic complexes [1]. If the asymmetric hydrogenation has been successfully extended to the two-phase water–organic solvent system using water-soluble catalysts [3], the observed enantioselectivities are generally lower than those obtained in the usual homogeneous systems. As phosphorus-based ligands have been extensively used in catalytic hydrogenation, many efforts have been devoted to the synthesis of their fluorous analogues. Klose and Gladysz [6] described the synthesis of the chiral ligand **1** derived from menthol, without any application in catalysis. More recently, a fluorous analogue of BINAP, **2a**, was synthesized by Hope's group [7] and used as a ligand of ruthenium in the asymmetric hydrogenation of dimethyl itaconate; although enantioselectivity up to 95% *ee* was obtained, quite similar to that observed using the original Ru–BINAP complex, recycling of the catalyst was not possible.

$R_f = C_6F_{13}$ or C_8F_{17}

1

a : R = $C_2H_4C_6F_{13}$; **b** : R = $Si(C_2H_4C_6F_{13})_3$

2

Asymmetric transfer hydrogenation of ketones in the presence of soluble transition metal catalysts has been developed [8–10], enantioselectivities up to 99% *ee* being obtained using a ruthenium catalyst bearing mono-*N*-tosylated diphenylethylenediamine as a ligand. Iridium complexes associated with fluorous chiral diimines **3a–3c** or diamines **4a–4b** have also been shown to be effective catalysts in hydrogen-transfer reduction of ketones [11, 12].

Enantioselectivities up to 56% *ee* were obtained using [Ir(COD)Cl]$_2$ associated with fluorous diimines **3a–3c** at 70 °C in the reduction of acetophenone with isopropanol as the hydride source in the presence of Galden D-100 (mainly *n*-perfluorooctane, b.p. 102 °C) as the fluorous solvent. The hydrogen-transfer reduction was extended to other ketones, enantioselectivity of 60% *ee* being obtained

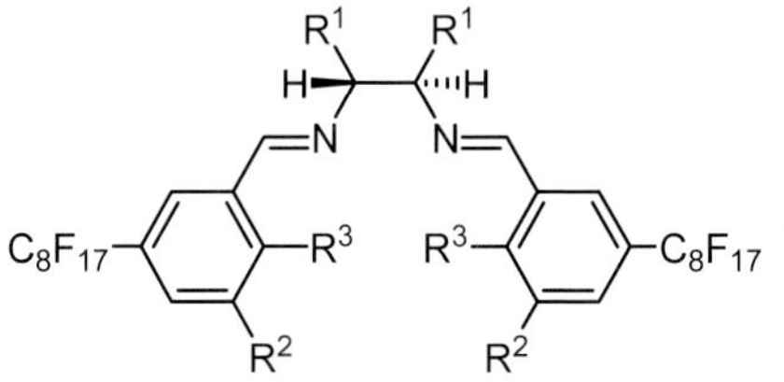

3

a : R^1,R^1 = -(CH$_2$)$_4$- , R^2 = C$_8$F$_{17}$, R^3 = OH
b : R^1 = C$_6$H$_5$, R^2 = *t*-Bu , R^3 = OH
c : R^1,R^1 = -(CH$_2$)$_4$- , R^2 = C$_8$F$_{17}$, R^3 = H

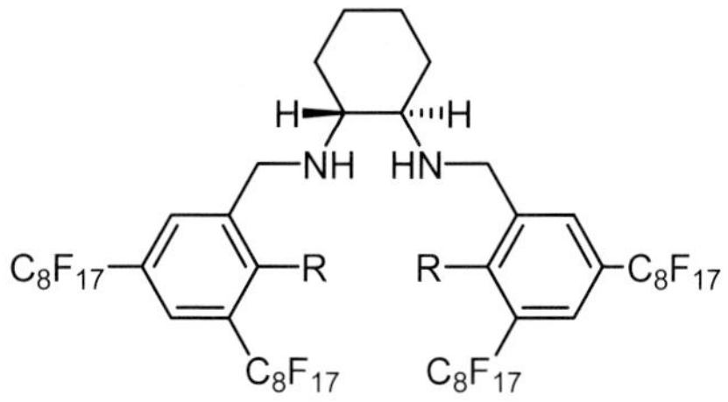

4

a : R = OH ; b : R = H

for ethyl phenyl ketone, for example. However, recycling of the catalyst gave lower activity and enantioselectivity, iridium leaching being very high. In order to circumvent the problem of the recycling of the fluorous catalyst, the chiral fluorous diamines **4a–4b**, obtained by reduction of **3a** and **3c**, were used as ligands of [Ir(COD)Cl]$_2$ in the reduction of acetophenone in the two-phase isopropanol/Galden D-100 system. While ligand **4a** gave low enantioselectivity (23% *ee*) and a very high iridium leaching (51%) in the organic phase, ligand **4b** gave enantioselectivity up to 69, 79, 59, and 58% *ee* for the first, second, third, and fourth cycle, respectively, the iridium leaching being pretty low (less than 4%).

The asymmetric 1,2-addition of diethylzinc to aromatic aldehydes catalyzed by a BINOL–Ti complex occurs with enantioselectivity up to 97% *ee* [13, 14]. Different groups reported the enantioselective carbon–carbon bond formation in a fluorous biphasic system using a titanium–fluorous-BINOL complex. Various chiral fluorous-BINOL ligands **5**, **6**, and **7**, bearing two –Si(C$_2$H$_4$C$_6$F$_{13}$)$_3$ or –Si(C$_2$H$_4$C$_8$F$_{17}$)$_3$ chains [15–17], four C$_4$F$_9$ or C$_8$F$_{13}$ chains [18, 19], or two –C$_2$H$_4$C$_6$F$_{13}$ or –C$_2$H$_4$C$_8$F$_{17}$ chains [20], respectively, have been used in this reaction.

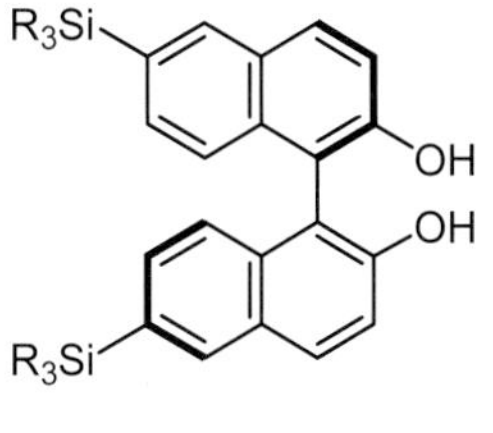

(R)-**5**

a : R = C$_2$H$_4$C$_6$F$_{13}$
b : R = C$_2$H$_4$C$_8$F$_{17}$

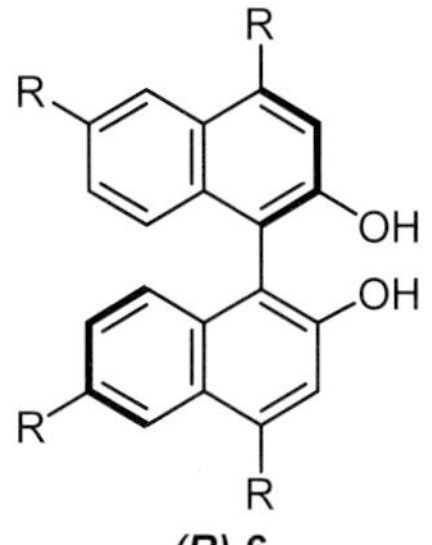

(R)-**6**

a : R = C$_4$F$_9$
b : R = C$_8$F$_{17}$

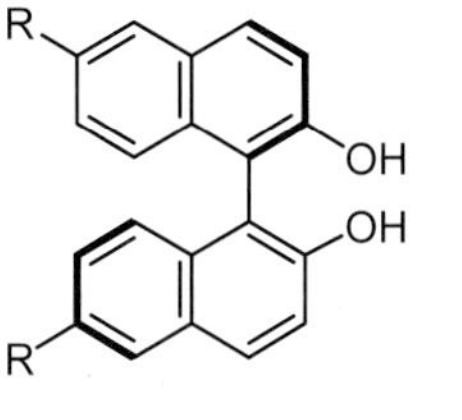

(R)-**7**

a : R = C$_2$H$_4$C$_6$F$_{13}$
b : R = C$_2$H$_4$C$_8$F$_{17}$

Takeuchi and collaborators reported the condensation of benzaldehyde with Et$_2$Zn at 0 °C in the presence of the complex prepared in situ by mixing Ti(O-*i*-Pr)$_4$ and fluorous BINOL **5** [15–17]. When the reaction was performed in a toluene/hexane/

Table 1 Condensation of benzaldehyde [Eq. (1)].

Ligand	Solvent	Yield [%] (cycle)	ee [%] (cycle)	Recovery of ligand [%] (cycle)
5a	Toluene/hexane/FC-72 (3 : 3 : 5)	81 (1), 89 (2), 87 (3), 87 (4), 87 (5)	83 (1), 82 (2), 82 (3), 81 (4), 80 (5)	10 (1), 12 (2), 12 (3), 11 (4), 10 (5)
5a	Toluene/FC-72 (3 : 5)	85 (1), 85 (2), 80 (3)	78 (1), 78 (2), 77 (3)	< 1 (1), < 1 (2), < 1 (3)
5b	Toluene/hexane/FC-72 (1 : 1 : 2)	82 (1), 82 (2), 77 (3)	79 (1), 78 (2), 78 (3)	1 (1), 1 (2), 1 (3)
6a	$C_{11}F_{20}$/hexane (1 : 0.7)	98 (1), 99 (2), 99 (3), 95 (4), 76 (5)	41 (1), 53 (2), 31 (3), 15 (4), 7 (5)	
6b	$C_{11}F_{20}$/hexane (1 : 0 : 7)	69 (1), 80 (2), 79 (3), 76 (4), 80 (5), 79 (6), 80 (7), 79 (8), 79 (9)	54 (1), 57 (2), 58 (3), 55 (4), 60 (5), 58 (6), 57 (7), 56 (8), 55 (9)	

FC-72 system, the enantioselectivity of the obtained alcohol was 80% *ee*, quite similar to that obtained using nonfluorous titanium–BINOL, and remained constant through five consecutive runs, the chemical yield being 80–89%. However, about 10% of the fluorous BINOL was recovered from the organic phase after acidic work-up of the reaction mixture. Since the partial solubilization of the fluorous catalyst in the organic phase was due to the presence of hexane, the use of the fluorous biphasic toluene/FC-72 system gave enantioselectivities up to 78% *ee* (85% yield) and 79% *ee* (82% yield) using ligand **5a** and **5b**, respectively; the enantioselectivity was constant during three consecutive runs, as well as the chemical yields, the leaching of ligand in the organic phase being negligible (less than 1%).

$$C_6H_5\overset{O}{\underset{H}{\bigvee}} + Et_2Zn \xrightarrow[\text{0 °C for 5, or 45°C for 6}]{\text{20 mol% 5 or 6, Ti(O-}i\text{-Pr)}_4} C_6H_5\overset{OH}{\bigvee} \qquad (1)$$

The condensation of Et_2Zn with other aromatic aldehydes in the presence of ligand **5a** or **5b** gave the corresponding alcohols with high enantioselectivity (76–85%) and chemical yields (73–97%). A similar approach was devised by Chan et al. [18, 19], who condensed Et_2Zn with benzaldehyde at 45 °C using the fluorous biphasic system hexane/perfluoromethyldecalin (or $C_{11}F_{20}$) in the presence of the catalyst obtained by mixing Ti(O-*i*-Pr)$_4$ and fluorous BINOL **6** [Eq. (1), Table 1]. When the ligand **6a** was used, the corresponding alcohol was obtained with 98% conversion and 41% *ee*; however, the enantioselectivity of the reaction decreased slowly with the reaction runs, and was lost after six runs. Fortunately, when ligand **6b**, which contains 32 fluorocarbons, was used, the enantiomeric excess of the product (55–60% *ee*) as well as the chemical yields (76–80%) were maintained constant after nine reaction runs. The lower enantioselectivity observed using **6** as the ligand instead of **5** (55–60% *ee* vs. 78–79% *ee*) is probably due to the reaction temperature; in the last case, the critical temperature was 45 °C, lower chemical yield and enantioselectivity being obtained at lower temperature due to the heterogenization of the catalytic system. Similar enantioselectivities as well as chemical yields were obtained in the condensation of Et_2Zn with other aromatic aldehydes in the presence of these ligands **6**, the enantioselectivities remaining constant over three consecutives runs (51–54% *ee* for 4-chlorobenzaldehyde, and 37–40% *ee* for 4-methoxybenzaldehyde).

Chan et al. [19] used ligand **6b** in association with Ti(O-*i*Pr)$_4$ in the condensation of aromatic aldehydes with Et_3Al in the biphasic hexane-perfluoro(methyldecalin) system at 53 °C [Eq. (2)]. Enantioselectivities in the range 76–88% *ee* and chemical yields of 77–82% were obtained during six consecutive runs when fresh titanium was added. When the reaction was extended to the electron-deficient 4-chlorobenz-aldehyde, the yield was the same (59–88% for three runs) and the enantioselectivity a little lower (63–79% *ee* for three runs); for the electron-rich 4-methoxybenzalde-hyde, only 10% of product was obtained with an enantioselectivity of 38%.

$$
\begin{array}{c}
\text{X-}C_6H_4\text{-CHO} + Et_3Al
\xrightarrow[\substack{\text{perfluoromethylcyclohexane/hexane} \\ 53\,°C}]{20\text{ mol\% }\mathbf{6b},\ \text{Ti(O-}i\text{-Pr)}_4}
\text{X-}C_6H_4\text{-CH(OH)CH}_2\text{CH}_3
\end{array} \quad (2)
$$

Zhao and collaborators [20] performed the the condensation of allyltributyltin with benzaldehyde in the presence of the catalyst Ti(O-*i*Pr)$_4$/BINOL **7** in various fluorous biphasic systems [Eq. (3)]. The highest enantioselectivities, up to 90% *ee*, were obtained using the hexane/FC-72 system, the yield being 85%. The ligand was recovered by continuous liquid–liquid extraction and could be re-used in further experiments. The reaction was extended to other aromatic aldehydes; however, only substrates with strong withdrawing groups showed good yields and enantio-selectivities, while aldehydes bearing halides or electron-donating groups gave rather poor yields and enantioselectivities.

$$
C_6H_5\text{CHO} + Bu_3Sn\text{-CH}_2\text{CH=CH}_2
\xrightarrow[\text{solvent, 0 °C}]{10\text{ mol\% Ti(O-}i\text{-Pr)}_4,\ \mathbf{7}}
C_6H_5\text{CH(OH)CH}_2\text{CH=CH}_2 \quad (3)
$$

van Koten et al. [21] synthesized fluorous chiral ethylzinc arene thiolates **8a–8c**. These organometallic complexes are active in the 1,2-addition of diethylzinc to benzaldehyde in hexane, the activity and enantioselectivity being even better than that of the nonfluorous catalyst. Moreover, further experiments showed that they are also active in a two-phase perfluoromethylcyclohexane/hexane medium. The catalyst could be recycled, although a drop in enantioselectivity was observed after two runs: enantioselectivities of up to 92, 92, 76, and 43% *ee* were obtained using ligand **8c** for four consecutive runs [Eq. (4)].

Among the organometallic catalysts used for alkylation and coupling reactions, palladium has a predominant role. Palladium catalysts are used effectively in a

$$C_6H_5\text{-CHO} + Et_2Zn \xrightarrow[\;C_7F_{14}/hexane\ (1:1),\ 15\ h,\ 25\ °C\;]{2.5\ mol\%\ \mathbf{8}} C_6H_5\text{-CHOH-}C_2H_5 \quad (4)$$

$R_f\text{-}C_2H_4Me_2Si\text{—}$ [structure]

8a : $R_f = C_6F_{13}$, R = Me
8b : $R_f = C_{10}F_{21}$, R = Me
8c : $R_f = C_{10}F_{21}$, R = -(CH$_2$)$_4$-

large number of useful transformations in organic chemistry [22]. Surprisingly, there are few examples of applications of chiral palladium complexes in the literature. Nakamura et al. [23] carried out the Heck reaction between 2,3-dihydro-furan and 4-chlorophenyl triflate in the presence of Pd(OAc)$_2$ associated with ligand **2b** in the two-phase benzene/FC-72 system [Eq. (5)]; enantioselectivity up to 93% was obtained, the yield being 39%. Unfortunately, recycling of the catalyst was not possible, due probably to its inactivation by ligand oxidation.

$$[\text{dihydrofuran}] + Cl\text{—}[\text{phenyl}]\text{—OTf} \xrightarrow[\substack{i\text{-Pr}_2NEt/40\ °C \\ C_6H_6/FC\text{-}72}]{Pd(OAc)_2/\mathbf{2b}} [\text{product}] \quad (5)$$

ee 93%

Palladium-catalyzed asymmetric allylic alkylation of 1,3-diphenylprop-2-enyl acetate with carbonucleophiles occurred using fluorous bisoxazolines as the ligands in benzotrifluoride or CH$_2$Cl$_2$ as the solvent with *ee* up to 95% [24]; although recycling of the catalyst was not possible, extraction of the ligand allowed the recycling of the later with the same enantioselectivity.

Pozzi's group has shown that asymmetric epoxidation of prochiral alkenes occurred under fluorous biphasic conditions using various chiral fluorous (salen)-manganese complexes **9** [25, 26]. The chiral (salen)manganese complexes **9a** and **9b**, bearing fluorous alkyl substituents in the 3,3' and 5,5' positions in the ligand, were used in the epoxidation of indene in the two-phase CH$_2$Cl$_2$/D-100 system in the dark at 20 °C under atmospheric pressure of oxygen in the presence of pivalaldehyde, giving the corresponding epoxide with 83 and 77% yield, and 92 and 90% *ee*, respectively. Recycling of the catalyst was possible without loss of the enantioselectivity: *ee* values up to 89 and 92% were obtained, respectively, in a second run. However, very low enantioselectivities were achieved in the epoxidation of other alkenes, such as dihydronaphthalene and benzosuberene, whose structures are very closed to indene, whatever the oxidant used. More recently, chiral fluorous second-generation Mn(salen) complexes **9c** and **9d** were prepared [27, 28]. These complexes took into account the fact that the low *ee* values observed were probably due to the low steric hindrance ensured by the fluorous substituents at the 3,3' and 5,5' positions, as well as their electronic effects. These catalysts were successfully used in the asymmetric epoxidation of dihydronaphthalene system at 100 °C in

9

a : $R^1 \cdot R^1$ = -$(CH_2)_4$-, R^2 = R^3 = C_8F_{17}, X = Cl
b : R^1 = C_6H_5, R^2 = R^3 = C_8F_{17}, X = Cl
c : $R^1 \cdot R^1$ = -$(CH_2)_4$-, R^2 = R^3 = C_6H_2-2,3,4-tri($OC_2H_4C_8F_{17}$)$_3$, X = $OCOC_7F_{15}$
d : $R^1 \cdot R^1$ = -$(CH_2)_4$-, R^2 = C_6H_2-2,3,4-tri($OC_2H_4C_8F_{17}$)$_3$, R^3 = t-Bu, X = $OCOC_7F_{15}$

CH_3CN/perfluorooctane in the presence of PhIO/PNO (pyridine N-oxide) as the oxidant. For example, in the case of **9d**, the highest yield was obtained above 40 °C (76% yield), although the *ee* increased with increasing temperature, the highest value (50% *ee*) being obtained at 100 °C.

The epoxidation reaction using complexes **9c** and **9d** as the catalysts was extended to other alkenes: benzosuberene, 1-methylindene, 1-methylcyclohexene, and triphenylethylene, affording the corresponding epoxides in 68–98% yields, and 50–92% enantioselectivities, very close to the values obtained by Regen and Jan [29] using a Mn(salen) supported on a gel-type resin. The fluorous catalysts could be recycled efficiently, the same activities and enantioselectivities being maintained for three consecutive runs. The lower activity generally observed for the fourth run was mainly due to the oxidative decomposition of the catalyst.

It should be noted that the corresponding Co(salen) complexes have also been used in the hydrolytic kinetic resolution of terminal epoxides, enantioselectivies up to 99% being obtained; however these complexes were never used in a two-phase system [30].

The catalytic enantioselective protonation of a samarium enolate using a C_2-symmetric chiral diol as the catalyst and trityl alcohol as the proton source afforded the corresponding ketone with *ee* up to 93% [31]. The use of (S)-2-bis-[(perfluorohexyl)ethyl-2-methoxy-1-phenylethanol (Rfh$_2$-MPE) and $(C_6F_{13}C_2H_4)_3$OH (or Rf$_3$COH) in a biphasic THF/FC-72 system (3 : 4) at –45 °C gave the ketone in 59% yield and 60% *ee*, although the use of **10a** as the chiral proton source increased the enantioselectivity to 89% *ee* [Eqs. (6)–(8)] [32]. This enantioselective protonation was extended to a samarium enolate derived from cyclohexanone in THF using fluorous alcohol **10b** as the proton source [15]; enantioselectivities of up to 89% have been obtained. The fluoro alcohol was recovered quantitatively by a simple extraction with FC-72 and re-used in five consecutive runs without loss of enantio-selectivity.

Fache et al. used fluorous cinchona derivatives in asymmetric Diels–Alder reactions in $CHCl_3/C_6F_{14}$ (1 : 1); low enantioselectivity (13%) was obtained [33]. Moreover, due to the low fluorine content of the catalyst (45 wt.% F), the reaction probably occurred in the nonfluorous phase.

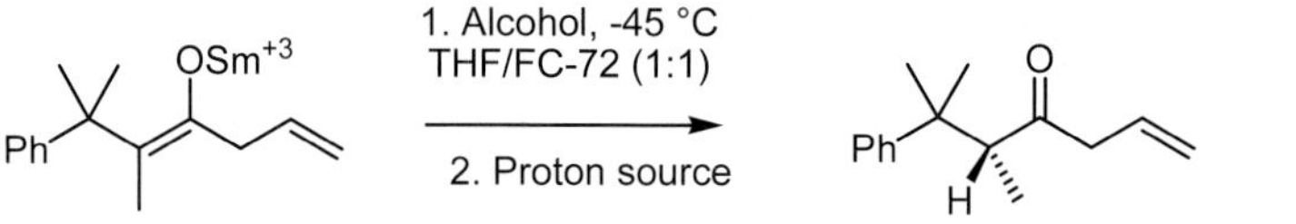

1. Alcohol, -45 °C
THF/FC-72 (1:1)

2. Proton source

(6)

Rfh$_2$-MPE	59 % (60 % *ee*)
10a	55 % (89 % *ee*)

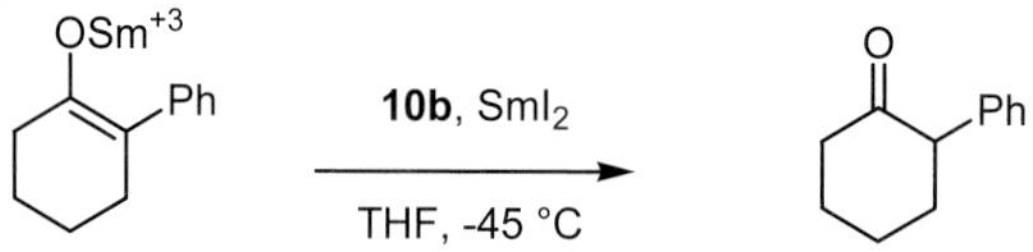

10b, SmI$_2$

THF, -45 °C

(7)

10a : 70 % (87% *ee*)
10b : 73-82 % (81-89% *ee*)

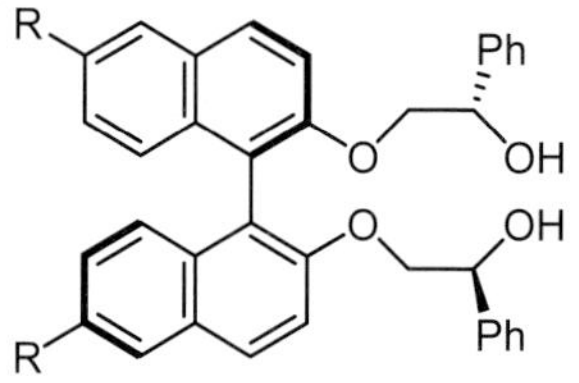

10a : R = H
10b : R = Si(C$_2$H$_4$C$_6$F$_{13}$)$_3$

(8)

References

1   *Comprehensive Asymmetric Catalysis* (Eds.: E. N. JACOBSEN, A. PFALTZ, H. YAMAMOTO), Springer, Berlin **1999**.
2   *Chiral Catalyst Immobilization and Recycling* (Eds.: D. E. DE VOS, I. F. J. VANKELECOM, P. A. JACOBS), Wiley-VCH, Weinheim **2000**.
3   D. SINOU, *Adv. Synth. Catal.* **2002**, *344*, 221.
4   T. WELTON, *Chem. Rev.* **1999**, *99*, 2071; P. Wassersheid, W. Keim, *Angew. Chem., Int. Ed.* **2000**, *39*, 3772.
5   I. T. HORVÁTH, J. RÁBAI, *Science* **1994**, *266*, 72.
6   A. KLOSE, J. A. GLADYSZ, *Tetrahedron: Asymmetry* **1999**, *10*, 2665.
7   D. J. BIRDSALL, E. G. HOPE, A. M. STUART, W. CHEN, Y. HU, J. XIAO, *Tetrahedon Lett.* **2001**, *42*, 8551.
8   R. NOYORI, S. HASHIGUCHI, *Acc. Chem. Res.* **1997**, *30*, 97.
9   R. NOYORI, T. OHKUMA, *Angew. Chem., Int. Ed.* **2001**, *40*, 40.
10 M. J. PALMER, M. WILLS, *Tetrahedron: Asymmetry* **1999**, *10*, 2045.
11 D. MAILLARD, C. NGUEFACK, G. POZZI, S. QUICI, B. VALADÉ, D. SINOU, *Tetrahedron: Asymmetry* **2000**, *11*, 2881.
12 D. MAILLARD, G. POZZI, S. QUICI, D. SINOU, *Tetrahedron* **2002**, *58*, 3971.
13 M. MORI, T. NAKAI, *Tetrahedron Lett.* **1997**, *38*, 6233.
14 F.-Y. ZHANG, C.-W. YIP, R. CAO, A. S. C. CHAN, *Tetrahedron: Asymmetry* **1997**, *8*, 585.
15 Y. NAKAMURA, S. TAKEUCHI, Y. OHGO, D. P. CURRAN, *Tetrahedron* **2000**, *56*, 351.
16 Y. NAKAMURA, S. TAKEUCHI, K. OKUMURA, Y. OHGO, D. P. CURRAN, *Tetrahedron* **2002**, *58*, 3963.
17 Y. NAKAMURA, S. TAKEUCHI, Y. OHGO, *J. Fluorine Chem.* **2002**, *120*, 121.
18 Y. TIAN, K. S. CHAN, *Tetrahedron Lett.* **2000**, *41*, 8813.
19 Y. TIAN, Q. C. YANG, T. C. W. MAK, K. S. CHAN, *Tetrahedron* **2002**, *58*, 3951.
20 Y.-Y YIN, G. ZHAO, Z.-S QIAN, W.-X YIN, *J. Fluorine Chem.* **2003**, *120*, 117.
21 H. KLEIJN, E. RIJNBERG, J. T. B. H. JASTRZEBSKI, G. VAN KOTEN, *Org. Lett.* **1999**, *1*, 853.

22 E. Nigishi (Ed.), *Handbook of Organopalladium Chemistry for Organic Synthesis*, Wiley-Interscience, New York **2002**.

23 Y. Nakamura, S. Takeuchi, S. Zhang, K. Okumura, Y. Ohgo, *Tetrahedron Lett.* **2002**, *43*, 3053.

24 J. Bayardon, D. Sinou, *Tetrahedron Lett.* **2003**, *44*, 1449.

25 G. Pozzi, F. Cinato, F. Montanari, S. Quici, *Chem. Commun.* **1998**, 877.

26 G. Pozzi, M. Cavazzini, F. Cinato, F. Montanari, S. Quici, *Eur. J. Org. Chem.* **1999**, 1947.

27 M. Cavazzini, A. Manfredi, F. Montanari, S. Quici, G. Pozzi, *Chem. Commun.* **2000**, 2171.

28 M. Cavazzini, A. Manfredi, F. Montanari, S. Quici, G. Pozzi, *Eur. J. Org. Chem.* **2001**, 4639.

29 T. S. Reger, K. D. Janda, *J. Am. Chem. Soc.* **2000**, *122*, 6929.

30 M. Cavazzini, S. Quici, G. Pozzi, *Tetrahedron* **2002**, *58*, 3943.

31 Y. Nakamura, Y. Takeuchi, A. Ohira, Y. Ohgo, *Tetrahedron Lett.* **1996**, *37*, 2805.

32 S. Takeuchi, Y. Nakamura, Y. Ohgo, D. P. Curran, *Tetrahedron Lett.* **1998**, *39*, 8691.

33 F. Fache, O. Piva, *Tetrahedron Lett.* **2001**, *42*, 5655.

4.2.5.2
Under Non-biphasic Conditions

Seiji Takeuchi and Yutaka Nakamura

Chiral ligands that contain more than 60 wt.% of fluorine atoms are usually fluorous enough to ensure immobilization of the catalysts in a fluorous phase. However, when the fluorous contents of the ligands are much lower than 60%, the solubilities of the ligands and/or catalysts in fluorous solvents decrease significantly. In such cases, asymmetric reactions cannot be carried out in organic and fluorous biphasic conditions but are conducted in common organic solvents or amphiphilic solvents such as benzotrifluoride (BTF). The products and the ligands and/or catalysts are separated from the products for re-use by fluorous liquid–liquid extraction or solid-phase extraction with a fluorous reverse-phase (FRP) silica gel column [1], depending upon the partition coefficients of the ligands or catalysts. Another option for such ligands is to carry out the reactions in supercritical carbon dioxide (scCO$_2$). The catalysts are recycled successfully by separating the products by the scCO$_2$ extraction method.

Takeuchi, Curran, and co-workers synthesized a fluorous chiral diol, (*R*)-2,2'-bis[(*S*)-2-hydroxy-2-phenylethoxy]-6,6'-bis[tris(1*H*,1*H*,2*H*,2*H*-perfluorooctyl)silyl]-1,1'-binaphthyl ((*R*,*S*)-FDHPEB) (F content = 56%, partition coefficient: benzene/FC-72 = 1 : 32, THF/FC-72 = 19 : 1) and applied it to a SmI$_2$-mediated enantio-selective protonation of 2-methoxy-2-phenylcyclohexanone [2]. The reaction was carried out under the same reaction conditions as those of the original nonfluorous reaction [3]. In the original reaction, the product was separated from the nonfluorous chiral proton source (2 equiv realtive to the substrate) with preparative TLC to give the product in 70% chemical yield and 87% *ee*. In the fluorous version, the product and the fluorous chiral proton source were separated by FC-72 extraction (six times) and more simply by fluorous solid-phase extraction with an FRP silica gel column.

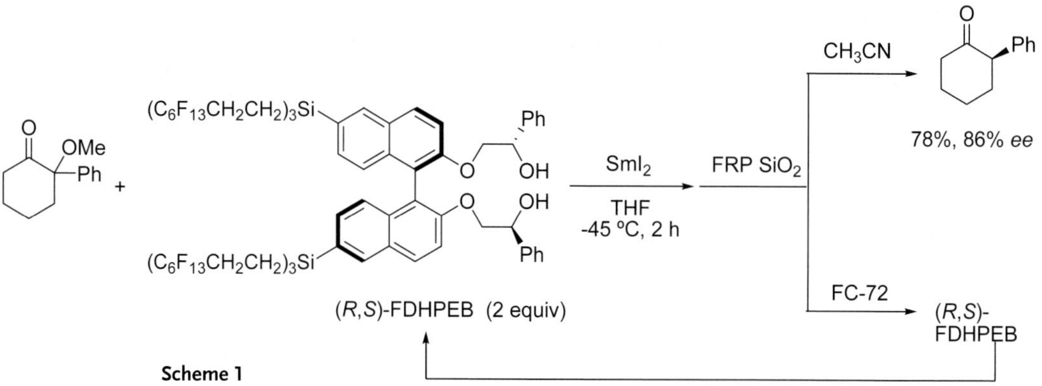

Scheme 1

The recovered (*R,S*)-FDHPEB was used for the next reaction and the reaction was repeated five times. The average chemical yield and enantioselectivity were 78% and 86% *ee*, respectively (Scheme 1). The recovery of (*R,S*)-FDHPEB was quantitative in each run and the recovered (*R,S*)-FDHPEB after the fifth reaction showed the same ^{1}H NMR spectrum as that of the pure compound. When the crude product was analyzed by HPLC with CD and UV detectors, the enantioselectivity was found to reach 95% *ee*. The enantiomeric excess was reduced to 87% *ee* owing to partial racemization during purification of the crude product by a preparative TLC. Quick separation of the product from (*R,S*)-FDHPEB by FRP silica gel revealed these facts.

Nakamura and co-workers reported an enantioselective addition of diethylzinc to benzaldehyde using a fluorous chiral β-amino alcohol, (1*R*,2*S*)-*N*-[4-tris-(1*H*,1*H*,2*H*,2*H*-perfluorooctyl)silyl]benzylephedrine (FBE) (F content = 56%, partition coefficient: CH$_3$CN/FC-72 = 12 : 88, toluene/FC-72 = 41 : 59), as a catalyst. The reaction was carried out in toluene by using Et$_2$Zn in hexane at room temperature [4]. The product was separated from the catalyst by FRP silica gel column and the recovered chiral catalyst was used for the next reaction. The reaction was repeated 10 times and the average chemical yield and enantioselectivity were 88% and 83% *ee*, respectively (Scheme 2). The recovery of the chiral catalyst was almost quantitative in each run and the enantioselectivity and chemical yield did not change significantly throughout the experiments. For an alternative system,

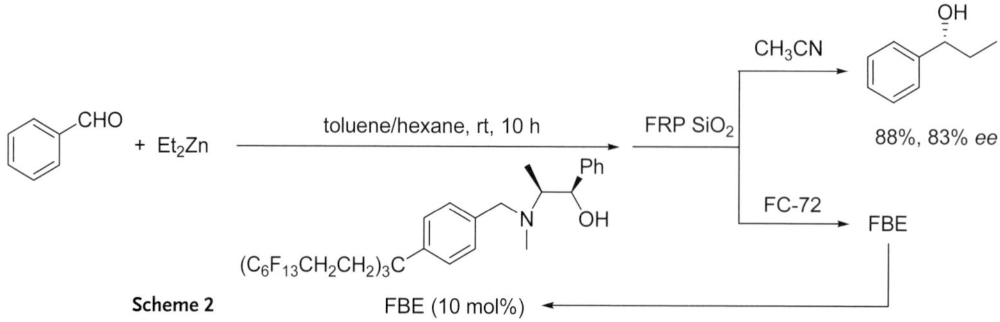

Scheme 2

Soai and co-workers reported that *N*-benzylephedrine, the original source of the fluorous catalyst, and the corresponding polymer-supported compound substituted at the *para* position of benzyl gave the product in 92% *ee* and 89% *ee*, respectively [5]. They recovered another polymer-supported ephedrine catalyst and used it again without any loss in catalytic activity and enantioselectivity (the catalyst being used twice).

In the two reactions described above, the chiral fluorous ligands were recovered quantitatively by FRP silica gel and re-used repeatedly for the reactions. Since no significant drop was observed in chemical yield, enantioselectivity, and recovery throughout the experiments, the reactions can be repeated any number of times until the chiral ligands have been consumed by mechanical losses.

Pozzi et al. used their fluorous chiral salen compounds for the ligands of cobalt(III) complexes in a catalytic hydrolytic kinetic resolution of terminal epoxides [6]. Among them, (1*R*,2*R*)-[*N*,*N*'-bis(3,3'-di-*tert*-butyl-5,5'-diheptadecafluorooctylsalicyliden)-1,2-cyclo-hexanediamine]cobalt(III) (F content = 49%) was most effective for the reaction in the presence of counterion $C_8F_{17}COO^-$. The complex was soluble in neat 1-hexene oxide as well as in common organic solvents such as CH_2Cl_2 and toluene but insoluble in perfluorocarbons at room temperature. Therefore, the reactions were carried out at room temperature without addition of any co-solvent with substoichiometric amount of H_2O under aerobic conditions. In the case of 1-hexene oxide, 1,2-hexanediol and unreacted 1-hexene oxide were isolated by fractional distillation in 47% and 51% chemical yields, respectively, and in enantioselectivities higher than 99% *ee* (in the original nonfluorous reaction, 98% *ee* for the both products at 50% conversion [7]). The nonvolatile residue obtained after the distillation was taken up in toluene and treated with $C_8F_{17}COOH$ in air. The recovered and reactivated catalyst was used for the next reaction and the reaction was repeated four times. Activity of the recovered catalyst was somewhat decreased at the fourth reaction, although the chemical yield and enantioselectivity of the diol were still higher than 46% and 97% *ee*, respectively (Scheme 3). Next, they tried to recycle the catalyst by using fluorous separation methods, liquid–liquid extraction, and solid-phase extraction. *n*-Perfluorooctane, BTF, and CH_3CN were used for the liquid–liquid extraction and the recovered catalyst resulted in 99% *ee* for 1,2-hexanediol although the reaction time was four times longer than the first one. The catalyst

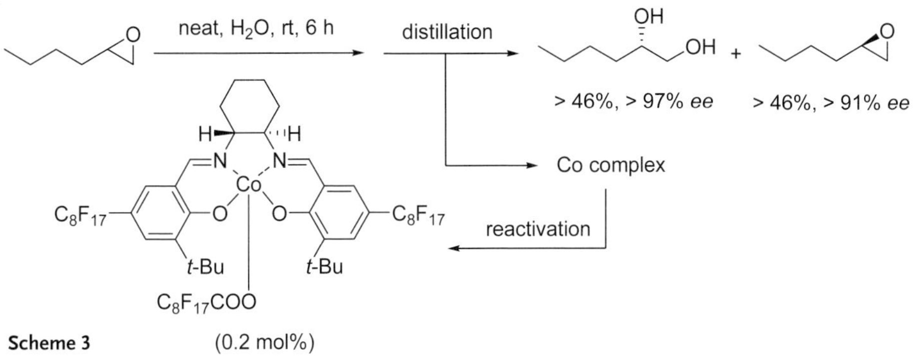

Scheme 3 (0.2 mol%)

recovered by FRP silica gel provided the product in 99% *ee* but the reaction rate was reduced to one-eighth of the first one.

Pozzi, Sinou, and co-workers prepared a fluorous chiral phosphine, (*R*)-2-{bis[4-(1*H*,1*H*-perfluorooctyloxy)phenyl]phosphino}-2′-(1*H*,1*H*-perfluorooctyloxy)-1,1′-binaphthyl (F content = 52%, partition coefficient: *n*-perfluorooctane/toluene = 0.23, *n*-perfluorooctane/CH$_3$OH = 7.42) and used for a chiral ligand of palladium complex in an asymmetric allylic alkylation of 1,3-diphenylprop-2-enyl acetate [8]. The reaction was carried out at room temperature in BTF or toluene and gave the corresponding product in 99% and 88% chemical yields and 81% *ee* and 87% *ee*, respectively after the nonfluorous MOP complex gave the product in 95% yield and 99% *ee* in toluene at 0 °C [9]) [Eq. (1)]. When toluene was used as a solvent, the simple extraction of the reaction mixture with *n*-perfluorooctane (twice) allowed the complete removal of the ligand and of the palladium complex. However, the recovered palladium complex did not have catalytic activity for the reaction.

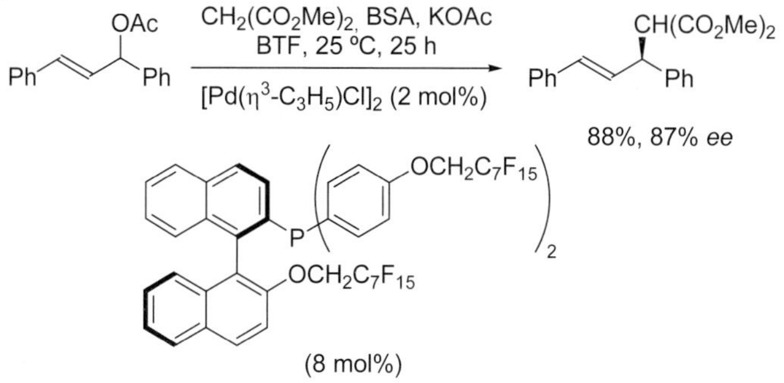

$$\text{(1)}$$

Stuart and co-workers reported the first synthesis of a "light" fluorous BINAP, (*R*)-6,6′-bis(1*H*,1*H*,2*H*,2*H*-perfluorooctyl)-2,2′-bis(diphenylphosphino)-1,1′-binaphthyl (F content = 38%), and its application to a Ru complex catalyzed asymmetric hydrogenation of dimethyl itaconate [Eq. (2)] [10]. The reaction was carried out at ambient temperature under the same reaction conditions as those reported by Noyori [11]. The chemical yield (83%) and enantioselectivity (95.7% *ee*) were similar to those reported (88% and 95.4% *ee*, respectively). However, there was no description of the recovery of the catalyst or ligand.

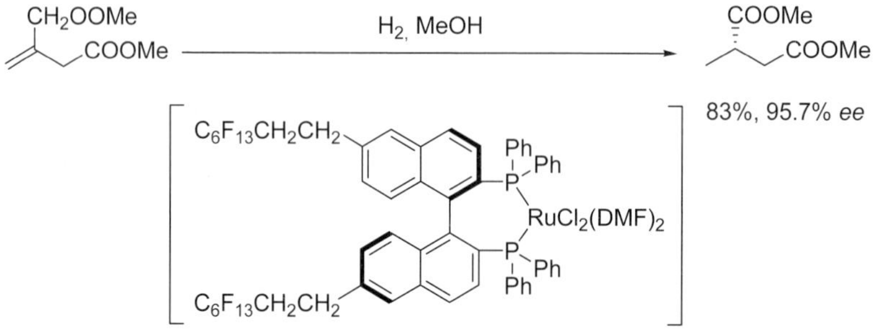

$$\text{(2)}$$

Nakamura and co-workers synthesized a heavily fluorinated chiral BINAP, (*R*)-6,6'- bis[tris(1*H*,1*H*,2*H*,2*H*-perfluorooctyl)silyl]-2,2'-bis(diphenylphosphino)-1,1'-binaphthyl ((*R*)- F_{13}BINAP) (F content = 54%, partition coefficient: benzene/FC-72 = 26 : 74, CH_3CN/FC-72 = 2 : 98) and applied it to an asymmetric Heck reaction [12]. The reaction between 2,3-dihydrofuran and 4-chlorophenyl triflate was carried out under the same conditions as those of original nonfluorous reaction by using F_{13}BINAP in BTF or benzene to provide the corresponding product, 2-(4-chloro-phenyl)-2,3-dihydrofuran, in 59% chemical yield or in 90% *ee* and 92% *ee*, respectively (71% chemical yield and 91% *ee* in the original reaction in benzene [13]) [Eq. (3)]. The reaction rate was about one-third of that in the original reaction. The products and the fluorous chiral ligand were separated by FRP silica gel and about 70% of the chiral ligand was recovered. However, the compound recovered was F_{13}BINAPO and could not be re-used for the next reaction.

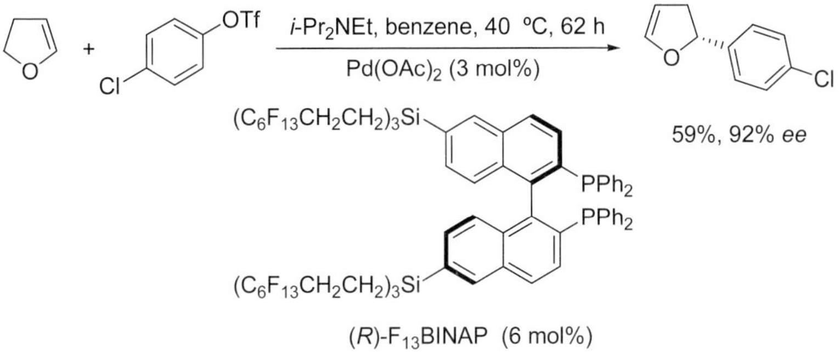

$$(3)$$

59%, 92% *ee*

(*R*)-F_{13}BINAP (6 mol%)

The examples on fluorous chiral phosphine ligands described above indicate that finding a chiral phosphine ligand effective for recycling it by fluorous techniques is still an important challenge.

References

1 D. P. CURRAN, S. HADIDA, M. HE, *J. Org. Chem.* **1997**, *62*, 6714.

2 Y. NAKAMURA, S. TAKEUCHI, Y. OHGO, D. P. CURRAN, *Tetrahedron* **2000**, *56*, 351.

3 Y. NAKAMURA, S. TAKEUCHI, Y. OHGO, M. YAMAOKA, A. YOSHIDA, K. MIKAMI, *Tetrahedron* **1999**, *55*, 4595.

4 Y. NAKAMURA, S. TAKEUCHI, K. OKUMURA, Y. OHGO, *Tetrahedron* **2001**, *57*, 5565.

5 M. WATANABE, K. SOAI, *J. Chem. Soc., Perkin Trans. 1* **1994**, 837.

6 M. CAVAZZINI, S. QUICI, G. POZZI, *Tetrahedron* **2002**, *58*, 3943.

7 E. N. JACOBSEN, *Acc. Chem. Res.* **2000**, *33*, 421.

8 M. CAVAZZINI, G. POZZI, S. QUICI, D. MAILLARD, D. SINOU, *Chem. Commun.* **2001**, 1220.

9 K. FUJII, H. OHNISHI, S. MORIYAMA, K. TANAKA, T. KAWABATA, K. TSUBAKI, *Synlett* **2000**, 351.

10 D. J. BIRDSALL, E. G. HOPE, A. M. STUART, W. CHEN, Y. HU, J. XIAO, *Tetrahedron Lett.* **2001**, *42*, 8551.

11 M. KITAMURA, M. TOKUNAGA, T. OHKUMA, R. NOYORI, *Tetrahedron Lett.* **1991**, *32*, 4163.

12 Y. NAKAMURA, S. TAKEUCHI, S. ZHANG, K. OKUMURA, Y. OHGO, *Tetrahedron Lett.* **2002**, *43*, 3053.

13 F. OZAWA, A. KUBO, T. HAYASHI, *J. Am. Chem. Soc.* **1991**, *113*, 1417.

4.2.6
Liquid/Solid Catalyst-Recycling Method without Fluorous Solvents

Kazuaki Ishihara and Hisashi Yamamoto

Fluorous biphasic catalysis has emerged since the late 1990s as an attractive alternative to traditional catalysis methods [1]. Fluorous techniques take advantage of the temperature-dependent miscibility of organic and perfluorocarbon solvents to provide easier isolation of products and recovery of a fluorinated catalyst. The large-scale use of fluorous solvents, however, has drawbacks: cost, and concern over environmental persistence.

The fluorous biphasic technique involves dissolving a catalyst with long fluorinated alkyl chains in a perfluorocarbon. The reactants are added to an organic solvent that is immiscible with the perfluorocarbon at room temperature, forming a second phase. On heating, the two phases mix and the reaction occurs; on cooling, the fluorinated and organic layers separate. The organic phase can be removed and the product isolated, while the fluorinated catalyst–solvent phase can be re-used.

In 2001, it has been independently reported [2, 3] that the fluorous solvent can be skipped by designing fluorinated catalysts that themselves have a temperature-dependent phase miscibility – that is, solubility – in ordinary organic solvents.

A direct amide condensation catalyst, 3,5-bis(perfluorodecyl)-phenylboronic acid (**1**) has been developed, which can be recovered without using any fluorous solvents [2]. Arylboronic acids bearing electron-withdrawing substituents on the aryl group behave as water-, acid-, and base-tolerant thermally stable Lewis acids and can be easily handled in air. 3,5-Bis(trifluoromethyl)phenylboronic acid (**2**) and 3,4,5-trifluorophenylboronic acid (**3**) are highly effective catalysts for the amide condensation of amines (1 equiv) and carboxylic acids (1 equiv) [4]. To the best of our knowledge, this is the first example of a catalytic and direct amide condensation which does not require excess amounts of substrates. Most of the above homogeneous catalytic reactions require relatively large quantities of arylboronic acid catalysts (1–20 mol%), and trace amounts of the catalysts must be removed from the reaction products. This hampers the application of this methodology to large-scale syntheses. Therefore, we have designed phenylboronic acids **1** and **4** bearing perfluorinated ponytails based on the direct coupling of fluoroalkyl iodides with halobenzenes. Their fluorous boronic acids can be easily recovered by the fluorous biphasic technique [2].

Table 1 Catalytic activities and recovery of arylboronic acid for the direct amide condensation.

ArB(OH)$_2$	Yield of amide [%] [a]	Recovery of ArB(OH)$_2$ [%] [b]
2	59	0
3	60	0
4	39	57
1	47 (95) [c]	> 99
PhB(OH)$_2$	23	0
_[d]	< 2	–

a) Isolated yield.
b) Extraction with perfluoromethylcyclohexane.
c) Yield after heating at azeotropic reflux for 15 h is indicated in parentheses.
d) No catalyst was added.

The catalytic activity of arylboronic acids **1–4** (5 mol%), which promote the model reaction of 4-phenylbutyric acid (1 equiv) with 3,5-dimethylpiperidine (1 equiv) in toluene at azeotropic reflux with removal of water (4 Å molecular sieves in a Soxhlet thimble) for 1 h, and their recoverability by extraction with perfluoromethylcyclohexane, are shown in Eq. (1) and Table 1.

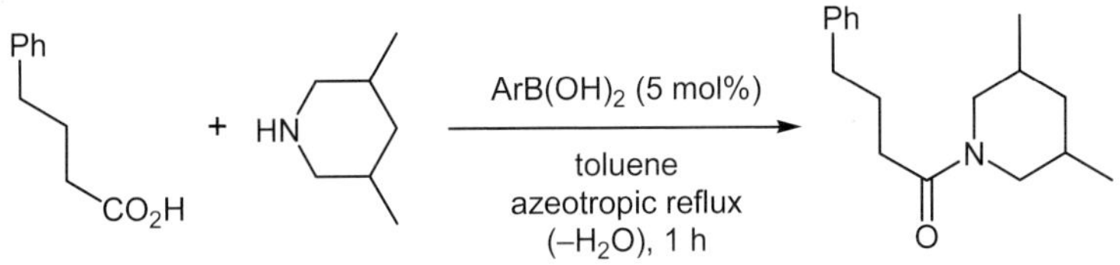

$$\text{(1)}$$

As expected, **1** is more active than **4**, and is recovered in quantitative yield by extraction with perfluoromethylcyclohexane. Although **2** and **3** are more active than **4**, they cannot be recovered by extraction with any fluorous solvents. The amide condensation proceeds cleanly in the presence of 5 mol% of **1**; the desirable amide has been obtained in 95% yield by azeotropic reflux for 15 h. In addition, the corresponding N-benzylamide has been obtained in quantitative yield by heating 4-phenylbutyric acid with benzylamine in the presence of 2 mol% of **1** under azeotropic reflux conditions for 4 h. Based on these results, the re-use of **1** has been examined for the direct amide condensation reaction of cyclohexanecarboxylic acid and benzylamine in a 1 : 1 : 1 mixture of o-xylene, toluene, and perfluorodecalin under azeotropic reflux conditions with removal of water for 12 h [Eq. (2) and Table 2] [5]. After the reaction has been completed, the homogeneous solution is cooled to ambient temperature to be separated in the biphase mode of o-xylene–toluene/perfluorodecalin. The corresponding amide is obtained in quantitative yield from the organic phase. Catalyst **1** can be completely recovered from the fluorous phase and re-used in the recyclable fluorous immobilized phase.

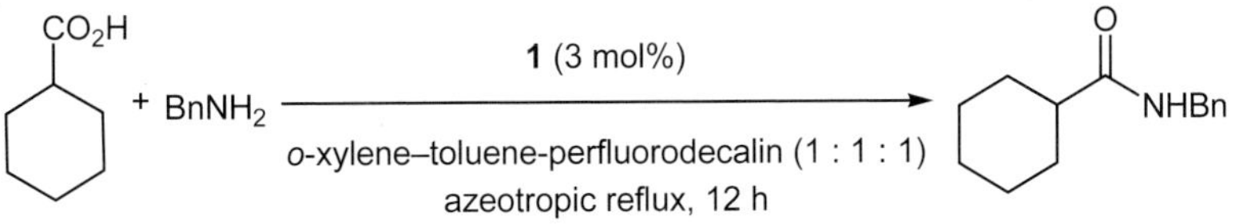

$$\text{(2)}$$

Table 2 Recovery and re-use of **1** in the recyclable fluorous immobilized phase.[a]

Cycle[a]	1	2	3	4	5[c]
Conversion [%][b]	> 99 (99)	> 99	> 99	> 99 (98)	> 99 (99)

a) Reaction conditions: *o*-xylene (2.5 mL), toluene (2.5 mL), and perfluorodecalin (2.5 mL). After the reaction, a solution of the amide in the upper phase was decanted and **1** in the lower phase was recycled successively.
b) Values in parenthesis refer to the isolated yields.
c) Catalyst **1** was recovered in 98% yield from the perfluorodecalin phase.

Table 3 Re-use of catalyst **1** for amide condensation of cyclohexanecarboxylic acid with benzylamine.[a]

Use of 1[b]	1	2	3	4	5	6	7	8	9	10
Conversion [%]	> 99	> 99	> 99	> 99	99	> 99	> 99	> 99	> 99	> 99

a) Reaction conditions: **1** (0.05 mmol), cyclohexanecarboxylic acid (1 mmol), benzylamine (81 mmol), xylene (5 mL). After the reaction, the solution was decanted and the residual catalyst **1** was re-used without isolation (see Scheme 1).
b) Recovered catalyst **1** was used successively (Use 2, 3, 4, ...).

Catalyst **1** is insoluble in toluene and *o*-xylene at room temperature even in the presence of carboxylic acids, amines, and amides. However, the amide condensation catalyzed by **1** proceeds homogeneously under reflux conditions. To demonstrate this advantage of **1** with respect to solubility, we have attempted to re-use **1** (5 mol%) 10 times for the condensation of cyclohexanecarboxylic acid with benzylamine [Eq. (3) and Table 3] [6].

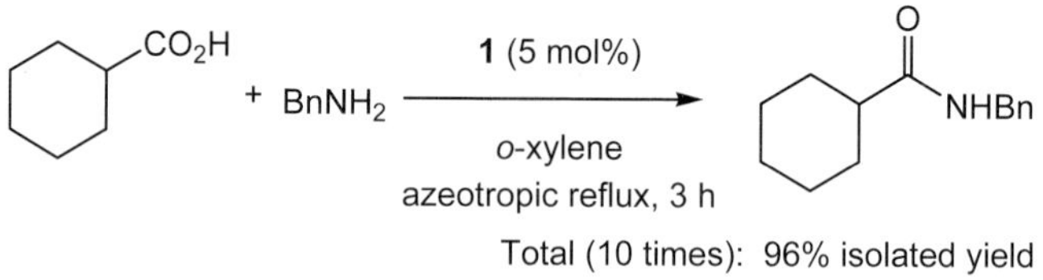

$$\text{Total (10 times): 96% isolated yield}$$

After the reaction mixture has been heated at reflux with removal of water for 3 h, it is allowed to stand at ambient temperature for 1 h to precipitate **1** (Scheme 1). The liquid phase of the resultant mixture is decanted and the residual solid catalyst **1** is re-used without isolation. No loss of activity has been observed for the recovered catalyst, and 26% of **1** remains in the flask in the tenth reaction. This means that 88% of **1** has been retained in each cycle. The total isolated yield of the amide which is obtained in 10 reactions is 96%. Moreover, pure compound **1** can be recovered in 97% yield as a white solid from the above reaction mixture by filtration and washing with toluene [6].

Gladysz's group has also reported the temperature-dependent solubility of the solid phosphine catalyst **5** in octane [3]. Between 20–80 and 20–100 °C, **5** exhibits ca. 60- and 150-fold increases of solubility in octane. Although octane is one of the best organic solvents for dissolving nonpolar fluorous compounds, little **5** can be detected at 0 °C by GC (0.31 mM) or ^{31}P NMR. At 20 °C, millimolar concentration

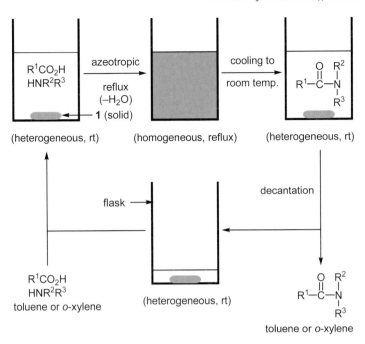

Scheme 1 Recovery of **1** by decantation and its re-use without isolation.

levels are present (1.13 mM, GC; 0.97 mM, NMR). A distinct jump in solubility has been observed near the melting point (19.6 mM, 50 °C), followed by continued increases (63.4 mM, 80 °C; 157 mM, 100 °C). Such a dramatic solubility/temperature dependence suggests an obvious catalyst method. The method has been tested by carrying out a series of additions of alcohols to methyl propiolate (**6**) in octane. Catalyst 5 (10 mol%), benzyl alcohol (2 equiv), and **6** are combined in octane (65 mM in **6**). The sample is kept at 65 °C (8 h) and cooled to –30 °C (arbitrary temperature of a convenient freezer). The precipitated catalyst (in some cases orange-colored) is isolated by decantation. GC analysis of the supernatant indicates an 82% yield of **7**. The recovered catalyst has been used for four further cycles without deterioration in yield, as summarized in Scheme 2.

In a further refinement, Gladysz's group has shown that the above reaction of benzyl alcohol with **6** can be made even greener by not using a solvent at all [3]. Raising the temperature of a mixture of the neat reactants and solid catalyst above the catalyst's melting point of 47 °C yields the addition product. The solid catalyst can be recovered at room temperature and is recyclable with yields consistently above 95%.

We have developed a fluorous super-Brønsted acid catalyst, 4-(1*H*,1*H*-perfluoro-tetradecanoxy)-2,3,5,6-tetrafluorophenylbis(trifluoromethanesulfonyl)methane (**8**), which can be recycled by applying liquid/solid phase separation without fluorous solvents [8] and an organic-solvent-swellable resin-bound super-Brønsted acid, polystyrene-bound tetrafluorophenylbis(trifluoromethanesulfonyl)methane (**9**) [9].

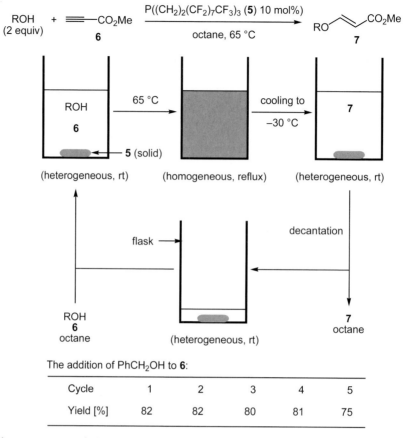

Scheme 2 Recovery of **5** by decantation and its re-use without isolation.

These super-Brønsted acids can be synthesized by using the *para*-substitution reaction of pentafluorophenylbis(trifluoromethanesulfonyl)methane (**10**) with nucleophiles such as sodium alkoxide and alkyllithium as a key step (Scheme 3).

Pentafluorophenylbis(trifluoromethanesulfonyl)methane **10** (47 wt.% F) is soluble in most organic and fluorous solvents. However, it is possible to achieve high fluorous-phase affinity for 4-alkoxy-2,3,5,6-tetrafluorophenylbis(trifluoromethane-sulfonyl)methane by appending "a fluorous ponytail", $OCH_2(CF_2)_nCF_3$ group, to the *para* position of **10** via the nucleophilic *para* substitution reaction. In preliminary experiments, the preparation of 4-hexanoxy- and 4-trifluoroethanoxy-2,3,5,6-tetrafluorophenylbis(trifluoromethanesulfonyl)methane, **11** and **12**, has been examined by reacting a lithium salt of **10** with the corresponding sodium alkoxide in pyridine at room temperature [Eq. (1)] [9]. As expected, **11** and **12** have been obtained in respective yields of 83% and 93%. Fluorous Brønsted acid **13** (59 wt.% F) has been also prepared in 97% yield from a lithium salt of **10** and sodium 1*H*,1*H*-perfluorodecanoxide.

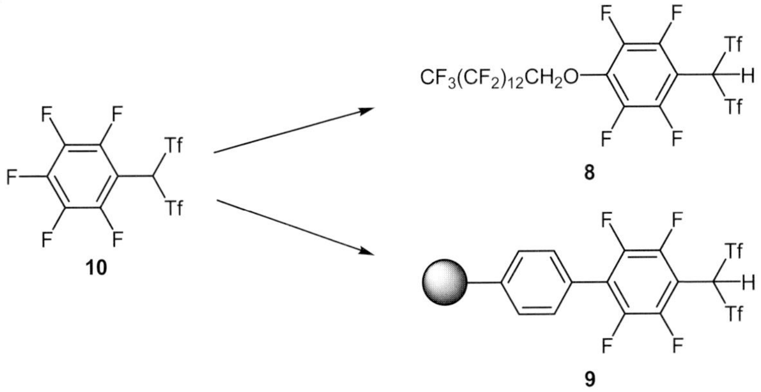

Scheme 3 Preparation of **8** and **9** from **10** via nucleophilic *para*-subsutitution reaction.

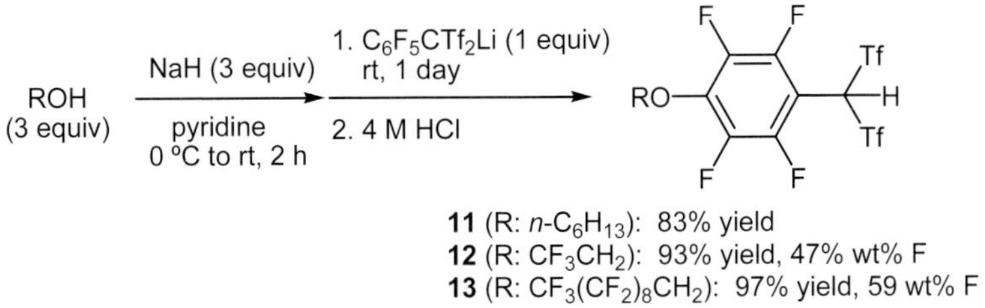

11 (R: *n*-C$_6$H$_{13}$): 83% yield
12 (R: CF$_3$CH$_2$): 93% yield, 47% wt% F
13 (R: CF$_3$(CF$_2$)$_8$CH$_2$): 97% yield, 59 wt% F

Their pK_a values in glacial acetic acid have been measured by the ^{1}H NMR method of Schantl et al. (Table 4) [8, 10]. The Brønsted acidity of **11** is less than that of concentrated H$_2$SO$_4$, while **12** is a superacid like **10**.

To obtain a higher fluorinated Brønsted acid, **3d** (62 wt.% F) has been prepared in 84% yield by heating a lithium salt of **1** and sodium 1*H*,1*H*-perfluorotetradecan-oxide in a 2 : 1 mixed solvent of pyridine and perfluorotributylamine at 70 °C [Eq. (5)]. Perfluorotributylamine has been added to partially dissolve sodium 1*H*,1*H*-perfluorotetradecanoxide.

Table 4 Brønsted acidities of arylbis(trifluoromethanesulfonyl)methane.

	11	*Conc. H$_2$SO$_4$*	12	1
^{1}H NMR (ppm)[a]	6.19	–	6.23	6.21[b]
pK_a in AcOH	11	7.5[b] (7.0)[c]	6.6	1.5[b]

a) ^{1}H NMR chemical shift observed for an acidic proton of ArCHTf$_2$ in CDCl$_3$ is indicated.
b) Reference [8a].
c) Reference [10].

The acetalization of benzaldehyde with 1,3-propanediol has been examined in the presence of 1 mol% of a fluorous super-Brønsted acid, **13** or **8**, at azeotropic reflux in cyclohexane with the removal of water for 3 h [Eq. (6)]. Both solid acids are soluble in cyclohexane under reflux conditions, and promote the reaction well to give the desired acetal in good yields. Post-reaction, **13** has been recovered in 96% yield by precipitation at room temperature. However, **13** can not be recovered in the same manner. Besides this acetalization, **8** is also effective as a fluorous catalyst for the acylation of *l*-menthol with benzoic anhydride [Eq. (7)] and esterification of 3-phenylpropionic acid in methanol [Eq. (8)] [11].

$$\text{PhCHO} + \text{HO}\diagdown\diagup\text{OH} \xrightarrow[\substack{\text{cyclohexane} \\ \text{azeotropic reflux, 3 h}}]{\textbf{13 or 8} \ (1 \ \text{mol\%})} \qquad (6)$$

(1.2 equiv)

The use of **13**: Acetal: 74% yield; recovery of **13**: failed
The use of **8**: Acetal: 86% yield; recovery of **8**: 96%

$$\xrightarrow[\substack{\text{toluene} \\ 70 \,^\circ\text{C, 14 h}}]{\textbf{8} \ (3 \ \text{mol\%})} \qquad (7)$$

+ Bz₂O (1.5 equiv)

> 99% yield; recovery of **8**: 70%

$$\text{Ph}\diagdown\diagup\text{CO}_2\text{H} \xrightarrow[\substack{\text{MeOH} \\ 70\,^\circ\text{C, 7 h}}]{\textbf{8} \ (1 \ \text{mol\%})} \text{Ph}\diagdown\diagup\text{CO}_2\text{Me} \qquad (8)$$

> 99% yield; recovery of **8**: 68%

Fluorous solid catalyst **8** is highly effective for the Mukaiyama aldol reaction [Eq. (9)] and Sakurai–Hosomi allylation reaction [Eq. (10)]. These reactions have been performed at –78 °C and room temperature, respectively, under heterogeneous conditions. Post-reaction, **8** has been recovered in high yield by decanting the liquids at room temperature.

$$\text{PhCHO} + \quad \xrightarrow[\substack{2. \ 1 \ \text{M HCl-THF (1:1)}}]{\substack{1. \ \textbf{8} \ (3 \ \text{mol\%}), \text{ toluene} \\ -78\,^\circ\text{C, 3 h}}} \qquad (9)$$

(1.2 equiv)

82% yield
Recovery of **8**: 92%

$$\diagup\diagdown\text{SiMe}_3 \xrightarrow[\substack{3. \ \text{Stirred at rt, 1 h} \\ 4. \ 1 \ \text{M HCl-THF (1:1)}}]{\substack{1. \ \textbf{8} \ (1 \ \text{mol\%}), \text{CH}_2\text{Cl}_2, \text{ rt, 0.5 h} \\ 2. \ \text{Addition of PhCHO (1 equiv)} \\ \text{at rt over 30 min}}} \qquad (10)$$

(1.5 equiv)

84% yield
Recovery of **8**: 97%

Pentafluorophenylbis(trifluoromethanesulfonyl)methane **10** offers a great advantage over other analogous super-Brønsted acids such as tris(trifluoromethane-sulfonyl)methane, trifluoromethanesulfonimide, and trifluoromethanesulfonic acid from the perspective of synthetic modification. Barrett's [12] and Mikami's groups [13] have independently reported metal tris(perfluoroalkanesulfonyl)methides as fluorous Lewis acids. Similarly, it may be possible to design pentafluorophenyl-bis(perfluoroalkanesulfonyl)methanes. However, it is synthetically more concise and practical to append 1*H*,1*H*-perfluoroalkoxy groups to **10** by a *para*-substitution reaction. In addition, solid acids **8** and **9** are more active catalysts than perfluoresin-sulfonic acids such as Nafion® [8].

Mikami's group has also demonstrated the advantage of the fluorous super-Lewis acids such as lanthanide tris(perfluorooctanesulfonyl)methide and perfluorooctane-sulfonimide complexes with respect to temperature-dependent solubility [13b]. For example, these complexes can be re-used for the Friedel–Crafts acylation reaction without fluorous solvents [Eq. (11)]. After the reaction mixture of anisole has been heated with acetic anhydride in 1,2-dichloroethane in the presence of ytterbium perfluorooctanesulfonimide (10 mol%) at 80 °C for 6 h, the mixture is allowed to stand at −20 °C for 30 min to precipitate the ytterbium complex. The liquid phase is decanted and the residual lanthanide complex is re-used without isolation. No loss of activity is observed for the catalyst recovered. The total isolated yield of the product, which is combined from the three runs, is 78%.

$$\underset{\substack{\text{OMe}\\ \text{1 mmol}}}{\text{(anisole)}} + \underset{\substack{\\ \text{2 mmol}}}{\text{Ac}_2\text{O}} \xrightarrow[\text{ClCH}_2\text{CH}_2\text{Cl, 80 °C, 6 h}]{\text{Yb[N(SO}_2\text{C}_8\text{F}_{17})_2]_3\ (10\ \text{mol}\%)} \underset{\substack{\text{OMe}\\ \text{Total isolated yield: 78\%}}}{\overset{\text{Ac}}{\text{(product)}}} \tag{11}$$

References

1 (a) I. T. Horváth, *Acc. Chem. Res.* **1998**, *31*, 641; (b) M. Cavazzini, F. Montanari, G. Pozzi, S. Quici, *J. Fluorine Chem.* **1999**, *94*, 183; (c) P. Bhattacharyya, B. Croxtall, J. Fawcett, J. Fawcett, D. Gudmunsen, E. G. Hope, R. D. W. Kemmitt, D. R. Paige, D. R. Russell, A. M. Stuart, D. R. W. Wood, *J. Fluorine Chem.* **2000**, *101*, 247.
2 K. Ishihara, S. Kondo, H. Yamamoto, *Synlett* **2001**, 1371.
3 M. Wende, R. Meier, J. A. Gladysz, *J. Am. Chem. Soc.* **2001**, *123*, 11490.
4 K. Ishihara, S. Ohara, H. Yamamoto, *J. Org. Chem.* **1996**, *61*, 4196; (b) K. Ishihara, H. Yamamoto, *Macromolecules* **2000**, *33*, 3511.
5 Perfluorodecalin is not miscible with a non-fluorous solvent, toluene, or *o*-xylene, even under reflux conditions.
6 For recent examples of precipitatable catalysts, see: D. E. Bergbreiter, N. Koshti, J. G. Franchina, J. D. Frels, *Angew. Chem., Int. Ed.* **2000**, *39*, 1040; (b) K. D. Janda, T. S. Reger, *J. Am. Chem. Soc.* **2000**, *122*, 6029; (c) T. Bosanac, J. Yang, C. S. Wilcox, *Angew. Chem., Int. Ed.* **2001**, *40*, 1875.
7 K. Ishihara, A. Hasegawa, H. Yamamoto, *Synlett* **2002**, 1299.

8 (a) K. ISHIHARA, A. HASEGAWA, H. YAMAMOTO, *Angew. Chem., Int. Ed.* **2001**, *40*, 4077; (b) K. ISHIHARA, A. HASEGAWA, H. YAMAMOTO, *Synlett* **2002**, 1296.

9 D. J. BYRON, A. S. MATHARU, R. C. WILSON, *Liquid Crystals* **1995**, *19*, 39. Pyridine is more effective as a solvent in the *para* substitution reaction of a lithium salt of **10** with sodium alkoxides. In contrast, this reaction does not occur smoothly in diethyl ether, which is effective in the *para* substitution reaction with alkyllithiums [8].

10 B. M. RODE, A. ENGELBRECHT, J. Z. SCHANTL, *J. Prakt. Chem. (Leipzig)* **1973**, *253*, 17.

11 In the case of the esterification, the resultant solution is concentrated under reduced pressure, and the crude compounds are diluted in hexane to precipitate **8**. Thus, **8** is recovered by filtration.

12 A. G. M. BARRETT, D. C. BRADDOCK, D. CATTERICK, D. CHADWICK, J. P. HENSCHKE, R. M. McKINNELL, *Synlett* **2000**, 847.

13 (a) K. MIKAMI, Y. MIKAMI, Y. MATSUMOTO, J. NISHIKIDO, F. YAMAMOTO, H. NAKAJIMA, *Tetrahedron Lett.* **2001**, *42*, 289; (b) K. MIKAMI, Y. MIKAMI, H. MATSUZAWA, Y. MATSUMOTO, J. NISHIKIDO, F. YAMAMOTO, H. NAKAJIMA, *Tetrahedron* **2002**, *58*, 4015.

4.3
Concluding Remarks

István T. Horváth

As has been demonstrated in Chapter 4, fluorous catalysts are well suited for converting apolar substrates to products of higher polarity, as the partition coefficients of the substrates and products will be higher and lower, respectively, in the fluorous phase. The net results are little or no solubility limitation on the substrates and easy separation of the products. Furthermore, as the conversion level increases, the amount of polar products increases, further enhancing separation. One of the most important advantages of the fluorous biphase catalyst concept is that many well-established hydrocarbon-soluble catalysts could be converted to fluorous-soluble. In general, fluorous catalysts have similar structures and spectroscopic properties as the parent compounds.

The major difference arise from the presence of the fluorous ponytails, which provide a fluorous blanket around the hydrocarbon domain of the catalyst. If the electron-withdrawing effect of the fluorous ponytails on the ligands is not mitigated by insulating groups, the reactivity of the organometallic catalyst could be significantly different. Accordingly, most fluorous analogues of hydrocarbon-soluble catalysts have been prepared by incorporating insulating groups and have been shown to have comparable catalytic performance with the additional benefit of facile catalyst recycling.

Fluorous catalysis is now a well-established area and provides a complementary approach to aqueous- or ionic-biphase catalysis and the other possibilities of multiphase homogeneous catalysis (not to mention combinations of the different processes). Since each catalytic chemical reaction could have its own perfectly designed catalyst (e.g., a chemzyme), the possibility of selecting from biphase systems ranging from fluorous to aqueous systems provides a powerful portfolio for catalyst designers.

Multiphase Homogeneous Catalysis
Edited by Boy Cornils and Wolfgang A. Herrmann et al.
Copyright © 2005 Wiley-VCH Verlag GmbH & Co. KGaA, Weinheim
ISBN: 3-527-30721-4